INSIDE MAPINFO PROFESSIONAL®

The Friendly User Guide to MapInfo Professional

THIRD EDITION

Larry Daniel
Paula Loree
Angela Whitener

ONWORD PRESS
THOMSON LEARNING

Australia Canada Mexico Singapore United Kingdom United States

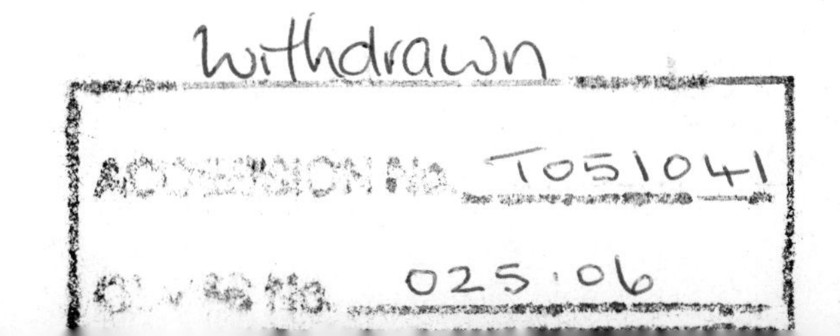

INSIDE MapInfo Professional

LARRY DANIEL, PAULA LOREE, AND ANGELA WHITENER

Publisher:
Alar Elken

Executive Editor:
Sandy Clark

Acquisitions Editor:
James Gish

Managing Editor:
Carol Leyba

Development Editor:
Daril Bentley

Editorial Assistant:
Jaimie Wetzel

Executive Marketing Manager:
Maura Theriault

Executive Production Manager:
Mary Ellen Black

Production Manager:
Larry Main

Manufacturing Coordinator:
Betsy Hough

Technology Project Manager:
David Porush

Cover Design:
Cammi Noah

SAN 694-0269

Printed in Canada
10 9 8 7 6 5 4 3 2

For more information, contact
OnWord Press
An imprint of Thomson Learning
Box 15-015
Albany, New York 12212-15015

Or find us on the World Wide Web at
http://www.onwordpress.com

Library of Congress Cataloging-in-Publication Data
Daniel, Larry, 1961–
Inside MapInfo Professional / Larry Daniel, Paula Loree, and Angela Whitener.—3rd ed.
p. com
ISBN 0-7668-3472-7
1. Geographic information systems—Data processing. 2. MapInfo. I. Loree, Paul, 1967– II. Whitener, Angela, 1961– III. Title.

G70.212.D36 2002
910'.285—dc21

2001021002

NOTICE TO THE READER

Publisher does not warrant or guarantee any of the products described herein or perform any independent analysis in connection with any of the product information contained herein. Publisher does not assume, and expressly disclaims, any obligation to obtain and include information other than that provided to it by the manufacturer.

The reader is expressly warned to consider and adopt all safety precautions that might be indicated by the activities herein and to avoid all potential hazards. By following the instructions contained herein, the reader willingly assumes all risks in connection with such instructions.

The publisher makes no representation or warranties of any kind, including but not limited to, the warranties of fitness for particular purpose or merchantability, nor are any such representations implied with respect to the material set forth herein, and the publisher takes no responsibility with respect to such material. The publisher shall not be liable for any special, consequential, or exemplary damages resulting, in whole or part, from the readers' use of, or reliance upon, this material.

About the Authors

Larry Daniel is president and chief executive officer of Conclusive Strategies in Austin, Texas, a direct marketing and software development firm specializing in decision support. Larry has been involved with commercial GIS applications for 15 years, and has served as vice president of the Castillo Company in Phoenix, Arizona, and director of GIS at MPSI in Tulsa, Oklahoma. He is frequently both a speaker at symposia and other events and contributor to professional journals, including *Business Geographics, GIS World, GeoInfo Systems, Earth Observation Magazine, Computer-Aided Engineering,* and *Design Management.*

Paula Loree is Director of Research, Franchise Finance Corporation of America (FFCA) in Scottsdale, Arizona, which specializes in the convenience store and automotive aftermarket industries. In addition, she manages the company's extensive GIS. She has worked with several Fortune 500 companies to help incorporate MapInfo Professional and other analytical tools into their decision-making processes. Prior to her work at FFCA, Paula was a consultant in the GIS/decision support arena, Director of Research for the Castillo Company, and a market analyst for Northern Automotive Corporation (now CSK Auto).

Angela Whitener is vice president of software development for IntelleVue, a MapInfo strategic partner in Tulsa, Oklahoma. Prior to IntelleVue, she was a senior systems consultant for MPSI, Inc., in Tulsa, and software engineer for E-Systems in Dallas, Texas. She has also coauthored three additional MapInfo titles by OnWord Press: *MapBasic Developer's Guide* (with Breck Ryker), *Minding Your Business with MapInfo* (with Jeff Davis), and *Mapping with Microsoft Office* (with Bill Creath).

Acknowledgments

Thanks to Paula Loree and Angela Whitener, who agreed to participate in the original book project; their effort, energy, and attitude continue to be remarkable. I especially appreciate the continuing commitment by Thomson Learning/OnWord Press to publishing valuable materials for GIS professionals.

Thanks to my business partner, Fred Barber; staff, such as Brad Harvey, Amy Sabalesky, and Karen Ontiveros; as well as recent clients, such as Gary Fong at American Isuzu, Steve DesMarais at Household Finance, and Phil Hutchinson at McCoys, who have understood geographic business advantages and worked together to create innovative business solutions.

Most of all, a most special thank-you to my family, Mellie, Melanie, Lianna, Dwight, and Meredith. My "world" revolves around them, and I have felt blessed to have it so.

—Larry Daniel

Thank you to Larry Daniel, without whose perseverance and friendship the third edition of this book would not exist. A special thank-you to our publisher, Thomson Learning/Delmar Publishers, and especially Jim Gish and Jaimie Wetzel, for making this project thoroughly a joy. Thank you to MapInfo and FFCA for giving me the tools to work on this project, and to FFCA's president Chris Volk for his encouragement and faith in me. Most of all, thank you to my husband Mark and our son Remington for their faith, support, and love.

—Paula Loree

A special thanks to the staff at Thomson Learning, who have shown understanding to professionals juggling hectic schedules and book author responsibilities.

I am also grateful to my family, who have offered support and encouragement throughout each of my book writing projects. And thank you, Keith, my loving husband, who continues to encourage me to explore my potential.

—Angela Whitener

CONTENTS

INTRODUCTION

CONGRATULATIONS ON SELECTING MAPINFO PROFESSIONAL and for choosing the third edition of *Inside MapInfo Professional* to help you explore its use. MapInfo Corporation is the market leader in desktop mapping software. This book will help validate your organization's good judgment in choosing MapInfo Professional.

Inside MapInfo Professional is not simply a rehash of the MapInfo manual and online help system by writers who are not familiar with the software. This book contains the knowledge that comes from time spent on the front lines of public and private sector organizations in helping individuals, departments, corporations, and government agencies learn how to effectively use geographic information systems (GIS) and desktop mapping software. Our work ranges from helping businesspeople and government agency professionals who are just getting started with GIS to teaching developers how to customize GIS software to fit their unique environments. We have included tips and techniques we have learned over time–many of which are either not in the manuals or are difficult to find.

Considerations for the Third Edition

Since issuing the first edition of *Inside MapInfo Professional*, many events have changed the nature of the MapInfo product, and the desktop mapping world in general. With the release of MapInfo Professional 6.0, the product became more polished and more complete than ever before. Moreover, with the continued growth of digital mapping, users are ever thinking of more ways to use the product.

Perhaps the most powerful agent of change has been the Internet. Today, users can surf around web sites and sample a wide variety of software applications. Mapping has been prominent on the Net. Users who previously saw maps as little more than the tattered sheets in vehicle glove compartments can now access specialized maps that show them the nearest ATM or the directions to promotional events with little more effort than a few keystrokes. Maps on the Internet have quickly generated a greater *society-wide* appreciation for how prevalent mapping can be in our everyday lives.

MapInfo Corporation has responded to the increase in general interest by expanding its line of products to accommodate new users on new platforms. Recognizing that new users are fearful of technical complexities, MapInfo has worked hard to deliver wizards for operations such as reporting, charting, and querying from data files. Recognizing that users will no longer be seeking data solely in a traditional workstation PC environment, MapInfo has pioneered leading-edge products on many different platforms for bringing mapping to the Internet.

The seasoned user may also be aware that the mapping industry itself has undergone a significant amount of change. Most significantly, there has been a major consolidation of data providers, and thus a reduced set of established vendors from whom one can obtain data. To counter the reduced availability and competitiveness of data options, MapInfo and various business partners have reinvigorated their data efforts. Many data considerations are chronicled in the revised and expanded appendix, Data Fundamentals and Sources.

MapInfo Corporation remains a very solid, viable, one-stop option for desktop mapping and analysis support. Customers now use the products in a way that surpasses previous desktop mapping usage. Amid the winds of change, MapInfo has stood committed to maintaining its leadership role in the desktop mapping world. Change will undoubtedly continue for those using geotechnologies in the future, and MapInfo Corporation looks like a safe bet for pushing the envelope even further in terms of enhancing ease of use and providing tools for new and better applications.

Book Organization

Inside MapInfo Professional is organized in three parts: Preliminaries (Part I), Tutorial (Part II), and Case Studies (Part III). Chapter 1 (Part I) presents background information on the software and sets the stage for moving into the tutorial section of the book. Part II, Tutorial, contains eight chapters describing MapInfo Professional's basic functionality, and three chapters dedicated to advanced functionality and a customization primer.

The chapters of Part II include series of tutorials and summary exercises that allow you to try first-hand the concepts, techniques, and functionality discussed. Part III, Case Studies, presents the use of MapInfo in retail, telecommunication, outdoor advertising, home appraisal, and railway applications. The appendix, Data Fundamentals and Sources, presents a brief description of data considerations, as well as remarks about other software in the MapInfo product family. The following sections provide a chapter-by-chapter description of parts I through III.

Part I: Preliminaries

Chapter 1 reviews recent milestones in desktop mapping and geographic information systems (GIS), and in integrating graphic and tabular data.

Part II: Tutorial

MapInfo Professional is a program of great capability with a wealth of features. The program can be straightforward and easy to learn if approached logically. The tutorial chapters are organized in functional sections to help you learn MapInfo quickly and efficiently. The component "learning segments" of these sections are presented as hands-on sequentially numbered tutorials. Many chapters in the tutorial section of the book end with an exercise that brings the content of a given chapter together conceptually as you perform a summary exercise, typically involving the techniques and functionality acquired through the series of tutorials leading to the exercise.

Chapter 2, The Whirlwind Tour, contains a sample project that shows you a MapInfo project from beginning to end. Data for the sample project and exercises in subsequent chapters are included in the *samples* directory on the companion CD-ROM.

Chapters 3 through 9 focus on MapInfo's basic tools and functions. Topics include managing the MapInfo environment, queries and browsing, map display options, creating thematic maps, creating graphs, editing attribute and graphical data, and preparing hardcopy output.

Chapters 10 through 12 focus on advanced topics and programming considerations. Chapter 10 addresses tools such as hot links, geocoding, redistricting, buffering, real-time map updates, statistics, and Crystal Reports, a report writing tool incorporated into MapInfo Professional in version 5.0. In Chapter 11, the focus is on metadata, ODBC (open database connectivity), OLE embedding, generating seamless map layers, and using raster images. Chapter 12 is dedicated to customizing MapInfo Professional with MapBasic, MapInfo's programming language and development environment.

Part III: Case Studies

Part III, Case Studies, presents five organizations currently using MapInfo Professional in five different industries. The first case study describes two new products developed by First American Flood Data Services, Inc., based on MapInfo: PropertyView, a nationwide property and lifestyle information system, and the California Property Disclosure data center. Next in the lineup is Eller Media Company, a Phoenix, Arizona, based company specializing in outdoor advertising. Eller Media uses MapInfo to select billboard locations.

The third case study features applications based on MapInfo developed by two-time MapInfo reseller of the year, Integration Technologies (Newport Beach, California), for Grubb and Ellis, one of the nation's largest publicly traded commercial real estate services firms. Number four presents the concept of automated housing appraisals as implemented by MarketMap of South Africa. The final case study describes data and application solutions developed by DeskMap Systems, an Austin, Texas, based MapInfo reseller that has focused on the rail industry.

Appendix

The book's revised, updated, and expanded appendix, Data Fundamentals and Sources, contains descriptions of MapInfo data and software products, as well as key management considerations. Data is often the greatest aspect of investing into mapping systems, and we believe that the new text describing purchasing considerations will help you do it in a more cost-effective and error-free manner. Numerous tables organized by area of specialty serve as indexes to a host of data/information sources by Internet address.

Typographical Conventions

TIP: *Tips on functionality usage, shortcuts, and other information aimed at saving you time and toil appear like this.*

NOTE: *Information on features and tasks that is not immediately obvious or intuitive appears in notes.*

WARNING: *Warnings appear like this and inform you of actions that may lead to consequences that are unacceptable, such as unintentionally and irrecoverably deleting data.*

The names of MapInfo functionality interface items (such as menus, windows, menu items, tools, toolbars, icons, and dialog box options) are capitalized according to the conventions of the software. Command sequences (clicking on a function immediately followed by another function selection) are separated by a pipe (|). Examples follow:

- Using the Frame tool, draw three frames for a map, browser, and graph. Click on None to indicate the window linked to each frame.
- Select Query | SQL Select on the Main menu bar.

User input, and names for files, directories, tables, fields, column names, and so on are italicized. Examples follow:

- Set the Graph dialog box to graph the *Pop_1980* and *Pop_1990* columns of the table.

- Open a Browser window of the *States* table.

General function and keyboard keys appear enclosed in angle brackets.

<Shift>

<S>

Key sequences, or instructions to hold a key down while clicking the mouse or pressing another key, are linked with a plus sign.

<Ctrl>+click

MapBasic code lines are shown in a monospaced typeface, an example of which follows.

```
Set Window FrontWindow( ) ScrollBars On Autoscroll On
Set Map
 CoordSys Earth Projection 9, 62, "m", -96, 23, 20,
  60, 0, 0
 Center (133351.1535.1852579.532)
```

Companion CD-ROM

The companion CD-ROM is organized into three levels, or directories. A brief description of the files in each directory follows.

- *samples:* Data files used in examples and exercises throughout Part II, Tutorial, as well as software utilities provided in Chapter 12.
- *midata:* MapInfoDATA product files. See description in the "MapInfoDATA Products" section that follows.
- *target:* OnTarget Mapping product files. MapInfo acquired OnTarget Mapping in 1999 and has since integrated these data into its product line. See description in the "OnTarget Mapping Products" section that follows.

NOTE: *Copying the data files in the samples directory to your hard drive is recommended before you begin to work with the tutorials and exercises in the Tutorial chapters. Copy the files to a working directory such as c:\mapinfo\exercise* or c:\mapinfo\data*. Working with the files on your hard drive instead of the companion CD-ROM will speed up operations, and permit you to modify or edit the data files as you proceed through tutorials and exercises.*

Data files in the *midata* and *target* directories are provided by respective data vendors. In most cases, you will likely wish to view these data files straight from their respective companion CD-ROM directories. If, however, you want to use or modify files in these directories, copy them to your hard disk.

Workspace and Program Files

Several workspace and MapBasic program files included on the companion CD-ROM are referenced in the tutorials and exercises. In the instances where the workspace (*.wor*) and program files contain disk drive and path specifications, you may have to insert the appropriate specifications for your system. To insert these changes, you can load the *.wor* or program file into any text editor or word processor and save it as an ASCII text file.

For instance, assume that the drive specification in the file is D:, but your CD-ROM drive is E:. In this event, you will have to change D: to E: inside the file. Another example would be that a file contains a directory path that does not match the one you wish to use on your system. Again, you will have to change the path specifications inside the file to match yours. Nearly all files in the *samples\programs* subdirectory on the companion CD-ROM contain disk drive and path specifications. (See the last page of the book for copyright information pertaining to data files.)

MapInfoDATA Products

The companion CD-ROM includes several files of demographic and cartographic data that can be used with MapInfo Professional in order to provide more data files for experimentation after you finish working through the tutorials and exercises in the book. The data sets, supplied by the MapInfoDATA unit of MapInfo Corporation, follow.

- 1990 census tract boundaries for San Mateo County, California
- 1990 demographic data for census tracts in San Mateo County, California
- StreetInfo 4.0 files for San Mateo County, California

1990 Census Tract Boundaries

This data set (located in the *census* subdirectory of *midata*) was developed by MapInfo using the U.S. government's TIGER 1994 boundary files. Although the data used to develop these boundaries consist largely of public-use data, a substantial amount of work was undertaken by MapInfo to ensure the accuracy of the files and the compatibility of the data in MapInfo Professional.

The boundary files can be accessed using the File | Open Table command in MapInfo Professional. The boundaries will be displayed automatically and can be utilized to perform numerous geographic analyses. (See the last page of the book for copyright information pertaining to data files.)

1990 Demographic Data

Numerous 1990 demographic variables are included in the *census* subdirectory of *midata* on the companion CD-ROM. For a list of the field names and descriptions of the data in each field, see the table that follows.

Provided for each census tract in San Mateo County, the demographic data were derived from the 1990 U.S. Census by The Polk Company, a MapInfo Data Partner with over 125 years of data collection experience. Although much of the source data are in the public record, Polk invests a substantial amount of energy in aggregating the data to a usable format and ensuring compatibility with MapInfo software and data products. These demographic data represent a subset of the demographic data available through MapInfoDATA. Additional demographic data include a wide variety of current-year estimates and five-year projections of the characteristics of the population of geographic areas.

To utilize the data, use the File | Open Table command in MapInfo Professional. Follow the suggestions in the tutorials to join the demographic data with boundary data to create thematic maps. (See the last page of the book for copyright information pertaining to data files.)

Fields of Companion CD-ROM *midata* Directory *census* Subdirectory

Item	Field Name	Format	Content
1	ID	{A15}	Polygon ID (concatenated FIPS code)
2	ycoord	{F9.4}	Latitude
3	xcoord	{F9.4}	Longitude
4	medage90	{F4.1}	1990 Median age total pop
5	MEDAGEA90	{F4.1}	1990 Median age total adult pop
6	medageF90	{F4.1}	1990 Median age female pop
7	medagem90	{F4.1}	1990 Median age male pop
8	wmedage90	{F4.1}	1990 Median age White pop
9	wmedagea90	{F4.1}	1990 Median age adult White pop
10	BMEDAGE90	{F4.1}	1990 Median age Black pop
11	BMEDAGEA90	{F4.1}	1990 Median age adult Black pop
12	imedage90	{F4.1}	1990 Median age American Indian/ Eskimo/Aleut pop
13	imedagea90	{F4.1}	1990 Median age adult American Indian/Eskimo/Aleut pop
14	AMEDAGE90	{F4.1}	1990 Median age Asian/Pacific Islander pop
15	amedagea90	{F4.1}	1990 Median age adult Asian/ Pacific Islander pop

16	omedage90	{F4.1}	1990 Median age other pop
17	omedagea90	{F4.1}	1990 Median age adult other pop
18	wmpop90	{I9}	1990 Total white male pop
19	WFPOP90	{I9}	1990 Total white female pop
20	bmpop90	{I9}	1990 Total black male pop
21	BFPOP90	{I9}	1990 Total black female pop
22	impop90	{I9}	1990 Total Indian male pop
23	ifpop90	{I9}	1990 Total Indian female pop
24	AMPOP90	{I9}	1990 Total Asian/Pacific Islander male pop
25	afpop90	{I9}	1990 Total Asian/Pacific Islander female pop
26	ompop90	{I9}	1990 Total other male pop
27	OFPOP90	{I9}	1990 Total other female pop
28	P0TO490	{I9}	1990 Total pop age 0-4
29	P5TO990	{I9}	1990 Total pop age 5-9
30	P10TO1490	{I9}	1990 Total pop age 10-14
31	P15TO1990	{I9}	1990 Total pop age 15-19
32	P20TO2490	{I9}	1990 Total pop age 20-24
33	P25TO2990	{I9}	1990 Total pop age 25-29
34	P30TO3490	{I9}	1990 Total pop age 30-34
35	P35TO3990	{I9}	1990 Total pop age 35-39
36	P40TO4490	{I9}	1990 Total pop age 40-44
37	P45TO4990	{I9}	1990 Total pop age 45-49
38	P50TO5490	{I9}	1990 Total pop age 50-54
39	P55TO5990	{I9}	1990 Total pop age 55-59
40	P60TO6490	{I9}	1990 Total pop age 60-64
41	P65TO6990	{I9}	1990 Total pop age 65-69
42	P70TO7490	{I9}	1990 Total pop age 70-74
43	P75TO7990	{I9}	1990 Total pop age 75-79
44	P80TO8490	{I9}	1990 Total pop age 80-84
45	P85OVR90	{I9}	1990 Total pop age 85+
46	P0TO590	{I9}	1990 Total pop age 0-5
47	P6TO1390	{I9}	1990 Total pop age 6-13
48	P14TO1790	{I9}	1990 Total pop age 14-17
49	P18TO2090	{I9}	1990 Total pop age 18-20

50	P21TO2490	{I9}	1990 Total pop age 21-24
51	P25TO3490	{I9}	1990 Total pop age 25-34
52	P35TO4490	{I9}	1990 Total pop age 35-44
53	P45TO5490	{I9}	1990 Total pop age 45-54
54	P55TO6490	{I9}	1990 Total pop age 55-64
55	P65TO7490	{I9}	1990 Total pop age 65-74
56	P75TO8490	{I9}	1990 Total pop age 75-84
57	P6TO1190	{I9}	1990 Total pop age 6-11
58	P12TO1790	{I9}	1990 Total pop age 12-17
59	XP0TO590	{F6.2}	1990 % Total pop age 0-5
60	XP6TO1390	{F6.2}	1990 % Total pop age 6-13
61	XP14TO1790	{F6.2}	1990 % Total pop age 14-17
62	XP18TO2090	{F6.2}	1990 % Total pop age 18-20
63	XP21TO2490	{F6.2}	1990 % Total pop age 21-24
64	XP25TO3490	{F6.2}	1990 % Total pop age 25-34
65	XP35TO4490	{F6.2}	1990 % Total pop age 35-44
66	XP45TO5490	{F6.2}	1990 % Total pop age 45-54
67	XP55TO6490	{F6.2}	1990 % Total pop age 55-64
68	XP65TO7490	{F6.2}	1990 % Total pop age 65-74
69	XP75TO8490	{F6.2}	1990 % Total pop age 75-84
70	XP85OVR90	{F6.2}	1990 % Total pop age 85+

StreetInfo 4.0

The street data included in StreetInfo 4.0 (in the *street* subdirectory of *midata* on the companion CD-ROM) are derived from the 1995 TIGER files released by the U.S. government. Click on the *setup.exe* file to install the data files. For a demo, click on *demo.exe* or *demo32.exe.*

Product specialists at MapInfo Corporation reengineered the data to make it more accurate, as well as more accessible. In addition, several ease-of-use features have been incorporated into the data. Most notable is the distribution of data into a commonsense configuration of layers that can be browsed and opened individually, in sets, or all at once within MapInfo Professional. StreetInfo includes the 20 layers of data described in the following table.

The StreetInfo data files on the companion CD-ROM pertain to San Mateo County, California. In addition to the diverse cartographic layers,

StreetInfo contains a layer of address ranges that can be used within MapInfo Professional to geocode files; that is, append latitude and longitude coordinates to existing address data. (See the last page of the book for copyright information pertaining to data files.)

Addressed/Segmented Streets	City Boundaries
Highways	Landmarks
Highway shields	Geocoding boundaries
Railroads	Water boundaries
Municipal points	Rivers
Cultural point locations	Voting district boundaries
Natural point locations	Native American lands
Minor civil divisions (mcd)	Elevation contours
Sub-mcd boundaries	School district boundaries (up to 3 layers)

On Target Mapping Products

The *target* directory on the companion CD-ROM contains sample files originally provided by On Target Mapping, a provider of data and software solutions that was based in Monroeville, PA, and since acquired by MapInfo Corporation. Located in *target/clec* are samples of the company's CLECInfo product for San Mateo County, California. A competitive local exchange carrier (CLEC) is a long distance or independent telecom service provider who, due to deregulation, now competes in offering local service. CLECInfo is a map database that includes switch locations.

Samples of On Target Mapping's ExchangeInfo product for San Mateo County, California, are found in *target/exchange.* ExchangeInfo helps businesses and public agencies identify and define coverage areas based on the telephone numbers of clients or constituents. The product provides the wire center boundary and numbering plan. ExchangeInfo is useful in many applications by numerous businesses, including telemarketing, emergency services, ATM deployment, credit unions, and retail operations.

Claritas Data

Claritas Inc. is recognized as a leading supplier of high-quality, geodemographic data. Data used throughout the tutorials and located on the companion CD-ROM has been graciously provided by Claritas and is intended for the sole purpose of learning and instruction associated with this book. Any other use of the data requires prior permission from Claritas. For more information on Claritas and its products, see Case Study 2 and the appendix, and visit the Claritas website at *http://www.claritas.com.*

PART I
PRELIMINARIES

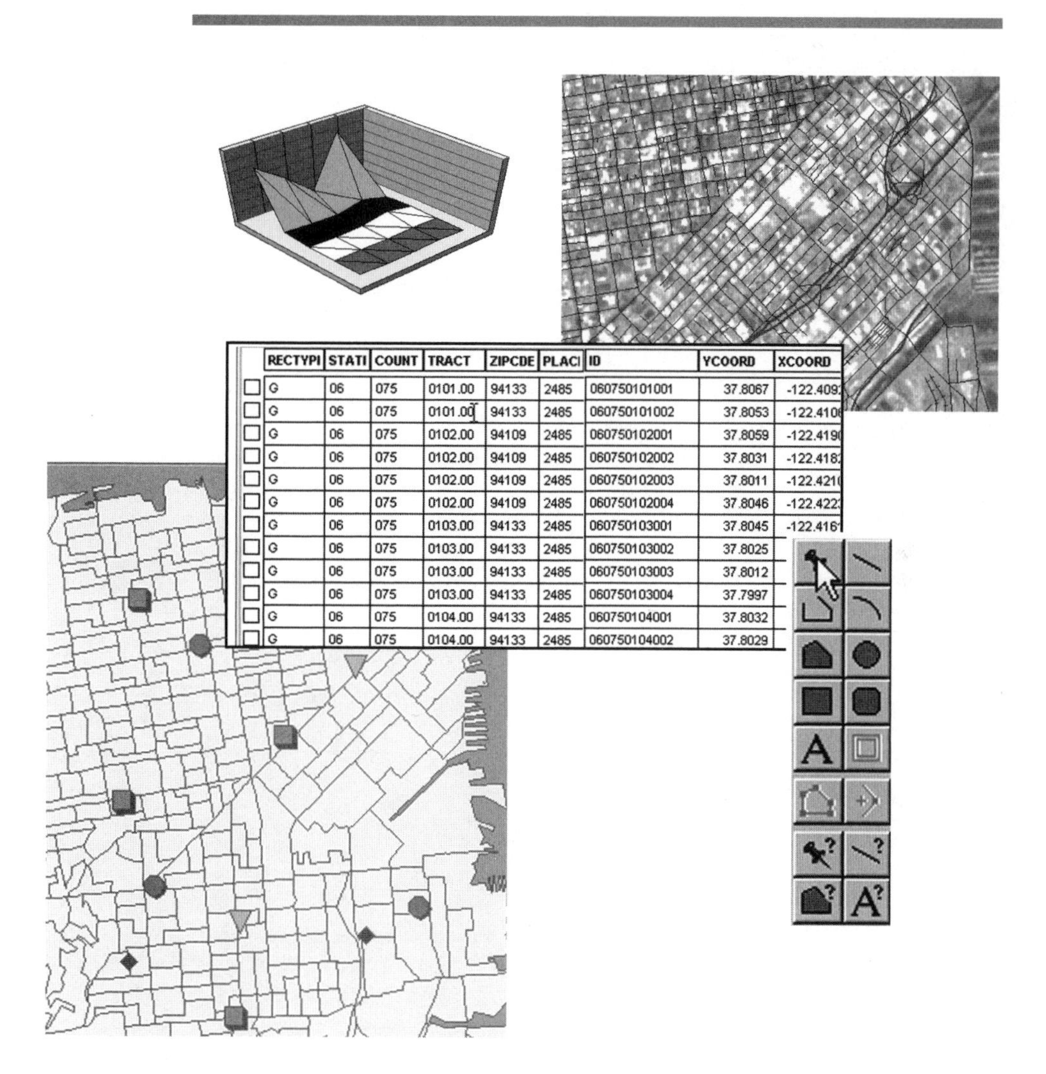

RECTYPI	STATI	COUNT	TRACT	ZIPCDE	PLACI	ID	YCOORD	XCOORD
G	06	075	0101.00	94133	2485	060750101001	37.8067	-122.409
G	06	075	0101.00	94133	2485	060750101002	37.8053	-122.410
G	06	075	0102.00	94109	2485	060750102001	37.8059	-122.419
G	06	075	0102.00	94109	2485	060750102002	37.8031	-122.418
G	06	075	0102.00	94109	2485	060750102003	37.8011	-122.421
G	06	075	0102.00	94109	2485	060750102004	37.8046	-122.422
G	06	075	0103.00	94133	2485	060750103001	37.8045	-122.416
G	06	075	0103.00	94133	2485	060750103002	37.8025	
G	06	075	0103.00	94133	2485	060750103003	37.8012	
G	06	075	0103.00	94133	2485	060750103004	37.7997	
G	06	075	0104.00	94133	2485	060750104001	37.8032	
G	06	075	0104.00	94133	2485	060750104002	37.8029	

CHAPTER 1

INTRODUCING MAPINFO PROFESSIONAL AND DESKTOP MAPPING

IN TODAY'S FAST-PACED ENVIRONMENT, desktop mapping and geographic information systems (GIS) are increasingly making the spatial dimensions of the world accessible, and bringing relationships between the spatial dimension and other variables into focus. Since the initial version of this book, MapInfo Corporation has continued to assert itself as the leader in the desktop mapping industry.

MapInfo Professional's powerful, easy-to-use functionality has brought desktop mapping within reach of tens of thousands of professionals in public and private organizations. In this third edition, we hope to demonstrate how desktop mapping can be used to change your work and leisure activities, as well as stimulate and recharge your ideas for doing so.

Computerization of the Spatial Dimension

Curiously, MapInfo has achieved its success in a period in which the U.S. public is seemingly *less* knowledgeable about geographic issues. In contrast to the explorers of yesteryear, who carried both map and compass, U.S. society has strayed from spatial literacy. Although a resurgence of geography programs in secondary curriculums has been underway for several years, and geography as a discipline has become better appreciated through commercial programs such as "Where in the world is Carmen San Diego?," the fact is that the U.S. public is still surprisingly ignorant about geography.

Ironically, geography bears increasing relevance to many contemporary private and public institutions. The real estate industry, for example, with "location, location, location" as its motto, accounts for over a quarter of all wealth in the United States. State and local governments, which often employ at least 10% of the labor force in respective capitals, focus on problems dominated by spatial issues. In the current era of diminishing resources and environmental concerns, geography is increasingly a "bottom line" factor in decision-making in both the private and public sectors.

Historically, preparation of maps (the means of communicating geographic knowledge) has been very tedious and time consuming. The time and effort involved in preparing maps has been one of the major factors explaining why most people are uninspired by geography. Manual cartography traditionally involved days or weeks of painstaking effort to produce a single map. In addition, maps represented data in a rather inflexible format because they could not be modified to present new data or to present old data in new ways.

However, the nature of creating maps has changed. Just as computer pioneers began in the 1960s and 1970s to automate the way we access and use financial, administrative, and other data sources, so too did they begin to examine how computers might be applied to representing and automating cartographic data. Systems emerged to depict graphic data in raster, and then digital, format. Early breakthroughs defined new graphic data structures for rapidly retrieving and displaying maps. Work on GIS (specifically, integrating graphics and tabular data into a single system) resulted in what may be called the forerunner elements of today's mapping systems.

Once tabular and graphic technologies were integrated, a variety of new mapping capabilities began to emerge. Thematic maps depicting tabular data such as income and population counts or customer densities could be generated automatically.

Software engineers began to examine how address records and digital street files might be combined to locate households and businesses. Software systems could then scan databases to determine which residences, customers, stores, or facilities were located within a given proximity to a particular address. Users could draw 1- or 2-mile "buffer zones" and view other geographic records pertaining to locales found therein. Users could then determine which customers would be affected if sales territories were adjusted, or examine the effects of legislative redistricting. Diverse spatial concepts were transformed and the foundation was laid for a new type of innovative and efficient spatial processing.

Enter MapInfo

In the 1970s and 1980s, the advance of mapping and GIS was hampered by limited cost-effective computer processing power and a lack of pertinent geographic computer-readable files. It was not until the late 1980s, with the major increases in power and declining cost of PC hardware, that the average home or business computer user would have systems powerful enough for working with maps.

The release of the 1990 Census data and TIGER (Topologically Integrated Geographically Enhanced Reference) files by the U.S. Census Bureau set the potential of desktop mapping in motion. Under the leadership of companies such as MapInfo Corporation (who dedicated itself to developing the software interface and improving access to government data files), mapping technology has matured, and off-the-shelf data files are now abundant.

MapInfo has been particularly effective in bringing the technology to leading-edge applications. In 1992, MapInfo introduced the first major GIS Windows software. In 1994, MapInfo formed an alliance with Microsoft Corporation that, among other things, resulted in the incorporation of mapping functionality in the popular spreadsheet software, Microsoft Excel. In more recent years, MapInfo has aggressively implemented "wizards" that increase ease of use, and has developed strategies for support of mapping on the World Wide Web.

Since its inception, MapInfo Corporation has maintained a commitment to retaining an "open" environment. MapInfo data are not only compatible with versions of the program across platforms but are exportable from a wide range of other sources. MapInfo Corporation has created MapBasic, a flexible programming language that enables all types of users to develop powerful programs and to link MapInfo with other Windows applications. More recently, MapInfo's development of MapX and MapXtreme continue the pattern of providing easy access to mapping and data to the typical computer user.

Partly because of MapInfo's commitments in these areas, data providers and utility developers have grown along with MapInfo. At present, MapInfo's third-party product catalog includes over 300 products.

Desktop mapping and GIS technologies have radically reduced the amount of time and effort required to generate maps, and MapInfo Corporation has become a flagship leading the way. MapInfo's efforts have taken the technology from the specialist's domain to the general public, making spatial analysis more accessible than ever before.

Desktop GIS Basics

The sections that follow discuss the integration of graphic and tabular data, the nature of graphic data used with MapInfo, raster imagery, and the nature of tabular data used with MapInfo. An example of data integration is also provided.

Integrating Graphic and Tabular Data

To understand how graphic and tabular data are integrated is to understand the power of GIS. In MapInfo, graphic data (*vector* and *raster* images) permit images to appear on a map. Tabular data sets consist of numeric, statistical, and textual information stored with the graphic data or used in conjunction with the graphic data to enable the user to apply GIS analytical capabilities to applications. Examples of graphic (vector) data and tabular data appear in the following illustrations.

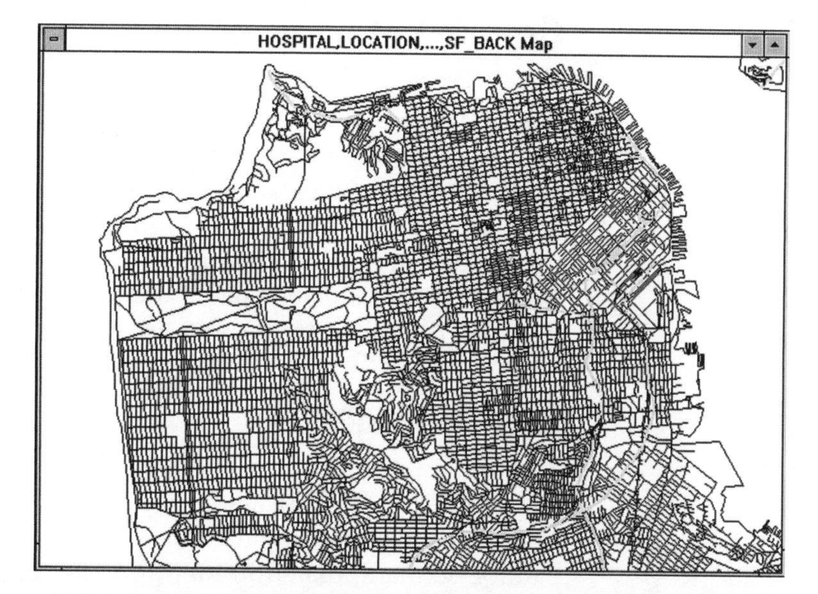

Example of graphic data.

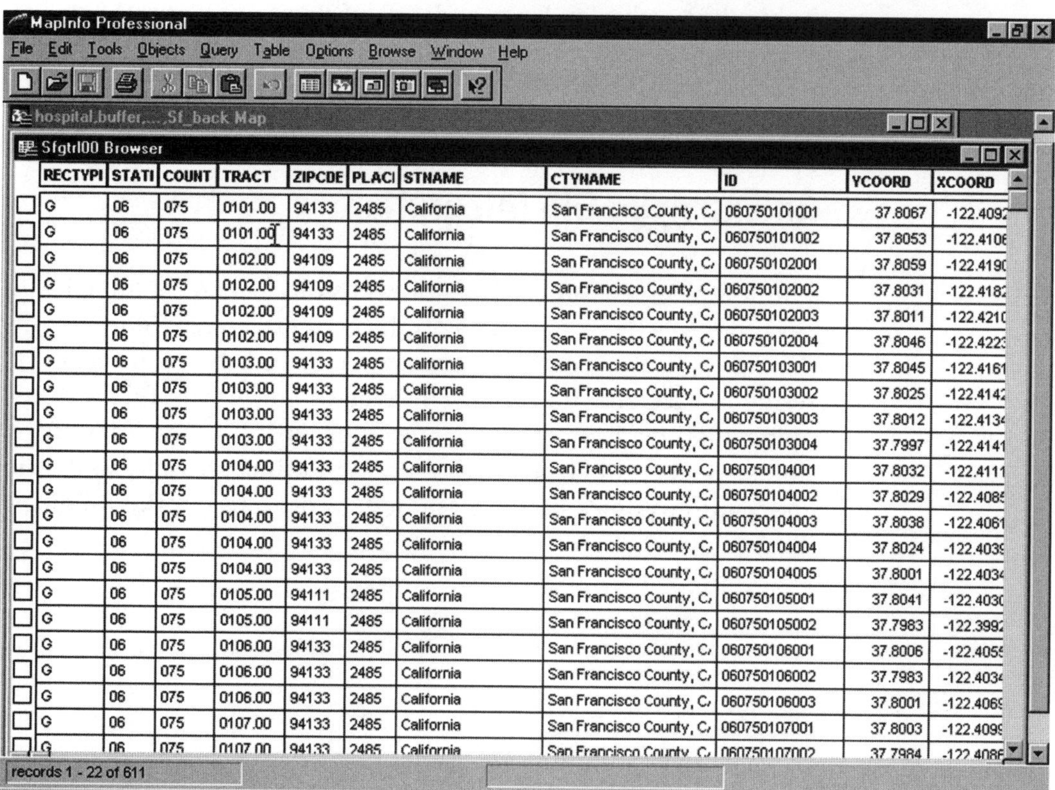

RECTYPI	STATI	COUNT	TRACT	ZIPCDE	PLACI	STNAME	CTYNAME	ID	YCOORD	XCOORD
G	06	075	0101.00	94133	2485	California	San Francisco County, C	060750101001	37.8067	-122.4092
G	06	075	0101.00	94133	2485	California	San Francisco County, C	060750101002	37.8053	-122.4106
G	06	075	0102.00	94109	2485	California	San Francisco County, C	060750102001	37.8059	-122.4190
G	06	075	0102.00	94109	2485	California	San Francisco County, C	060750102002	37.8031	-122.4182
G	06	075	0102.00	94109	2485	California	San Francisco County, C	060750102003	37.8011	-122.4210
G	06	075	0102.00	94109	2485	California	San Francisco County, C	060750102004	37.8046	-122.4223
G	06	075	0103.00	94133	2485	California	San Francisco County, C	060750103001	37.8045	-122.4161
G	06	075	0103.00	94133	2485	California	San Francisco County, C	060750103002	37.8025	-122.4142
G	06	075	0103.00	94133	2485	California	San Francisco County, C	060750103003	37.8012	-122.4134
G	06	075	0103.00	94133	2485	California	San Francisco County, C	060750103004	37.7997	-122.4141
G	06	075	0104.00	94133	2485	California	San Francisco County, C	060750104001	37.8032	-122.4111
G	06	075	0104.00	94133	2485	California	San Francisco County, C	060750104002	37.8029	-122.4085
G	06	075	0104.00	94133	2485	California	San Francisco County, C	060750104003	37.8038	-122.4061
G	06	075	0104.00	94133	2485	California	San Francisco County, C	060750104004	37.8024	-122.4039
G	06	075	0104.00	94133	2485	California	San Francisco County, C	060750104005	37.8001	-122.4034
G	06	075	0105.00	94111	2485	California	San Francisco County, C	060750105001	37.8041	-122.4030
G	06	075	0105.00	94111	2485	California	San Francisco County, C	060750105002	37.7983	-122.3992
G	06	075	0106.00	94133	2485	California	San Francisco County, C	060750106001	37.8006	-122.4055
G	06	075	0106.00	94133	2485	California	San Francisco County, C	060750106002	37.7983	-122.4034
G	06	075	0106.00	94133	2485	California	San Francisco County, C	060750106003	37.8001	-122.4069
G	06	075	0107.00	94133	2485	California	San Francisco County, C	060750107001	37.8003	-122.4099
G	06	075	0107.00	94133	2485	California	San Francisco County, C	060750107002	37.7984	-122.4086

records 1 - 22 of 611

Example of tabular data.

Graphic Data

Vector images consist of coordinate information. The coordinates incorporated in the images permit the MapInfo user to query and process vector image data. Examples of vector data include census block groups and street line segments. You can perform spatial queries because *all components of this type of data are linked to a physical geographical location.* Spatial queries constitute the mechanism by which disparate data can be linked.

Raster Images

Raster images are basically picture files, such as an aerial photograph that has been transformed into a digital file. Raster data cannot be manipulated in MapInfo, although a raster image can be used as a display layer and linked to vector data. When you zoom in to a raster image to the finest level of detail, you see a series of boxes within a grid, with each cell assigned a grayscale or color value. Raster images are often used as map backgrounds.

An Example: GIS Integration

The following example shows how customers can be linked with zip codes strictly through geography. A query is created to identify the zip code areas containing customers, and the customers who reside in each zip code area. This type of query is useful for developing mailing lists, creating customer profiles, and identifying similar pockets of customers to target in other areas. First, the zip code and customer files are opened, as depicted in the following illustration.

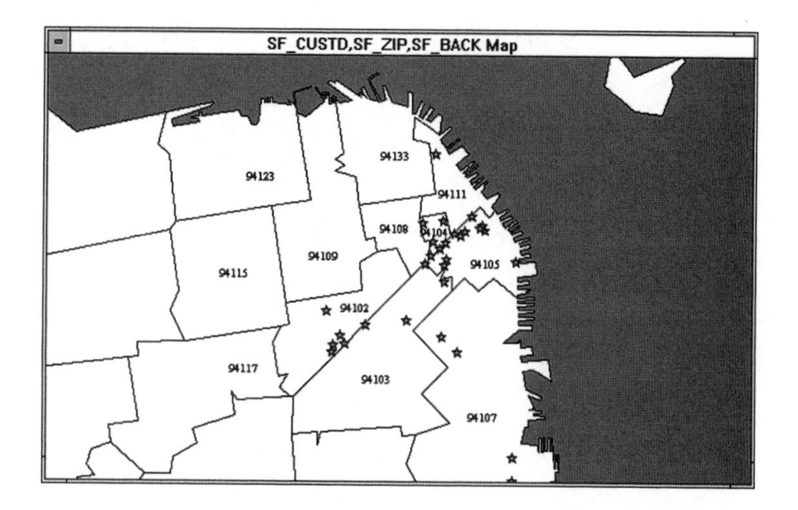

Zip code and customer file map.

Next, you would set up a spatial query in MapInfo Professional to identify the zip code areas that contain customers. The results of the query are shown in the following illustration. The highlighted zip codes on the map contain customers. The browser window at the right shows customer names and the zip code areas in which they reside.

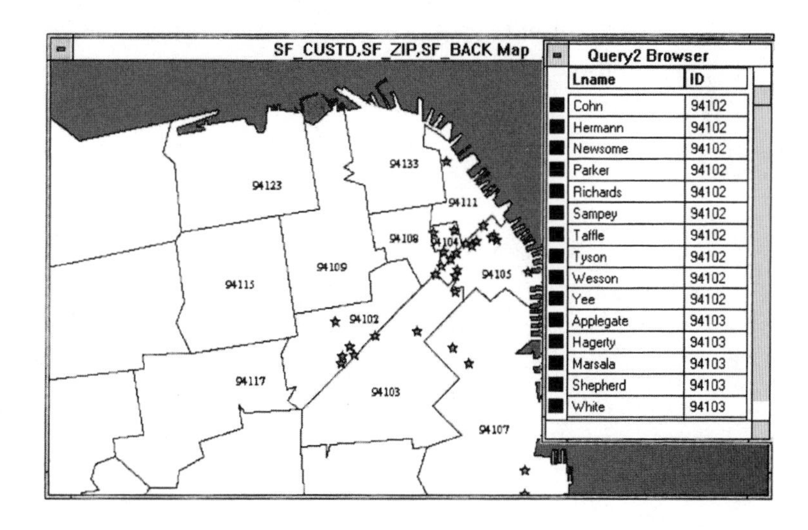

Map and Browser showing customer locations, names, and zip codes.

In addition to spatial queries, graphic information allows you to visualize data. For instance, you can see the median income range for a particular census tract, last year's sales by store location, and so on. Visualization is a powerful tool because it allows you to quickly get your point across with a single map, as opposed to forcing the user to sort through pages and pages of data.

Tabular Data

The use of the "median income" data in the previous section is an example of how "tabular data" links into MapInfo to provide the content for analysis. Geographic data without tabular data is simply distance relationships; geographic data used in tandem with tabular data in MapInfo equates to *knowledge.*

Tabular data may reside in the same "table" (MapInfo's data structure nomenclature) as the graphic data, or it can be linked either by reference to the graphic data's unique ID or by spatial queries (if the data has been geocoded). Examples of tabular data include median income, traffic counts, population, sales figures, and just about any other quantifiable information you can imagine.

INSIDE MapInfo shows you how the geographic world is linked to tabular data. The book endeavors to demonstrate the operations of the software and reveal to you the power of the resulting analysis.

Summary

This book conveys how the power of GIS, and MapInfo Professional in particular, can be used to expand your spatial horizons. Maps no longer take weeks to generate. Data collection no longer need take weeks or months. Graphics and tabular data do not have to be managed separately. In a world in which a picture is worth a thousand words, desktop mapping can be employed to help businesses, government, and nonprofit organizations execute penetrating analyses quickly, easily, and effectively.

Part II
Tutorial

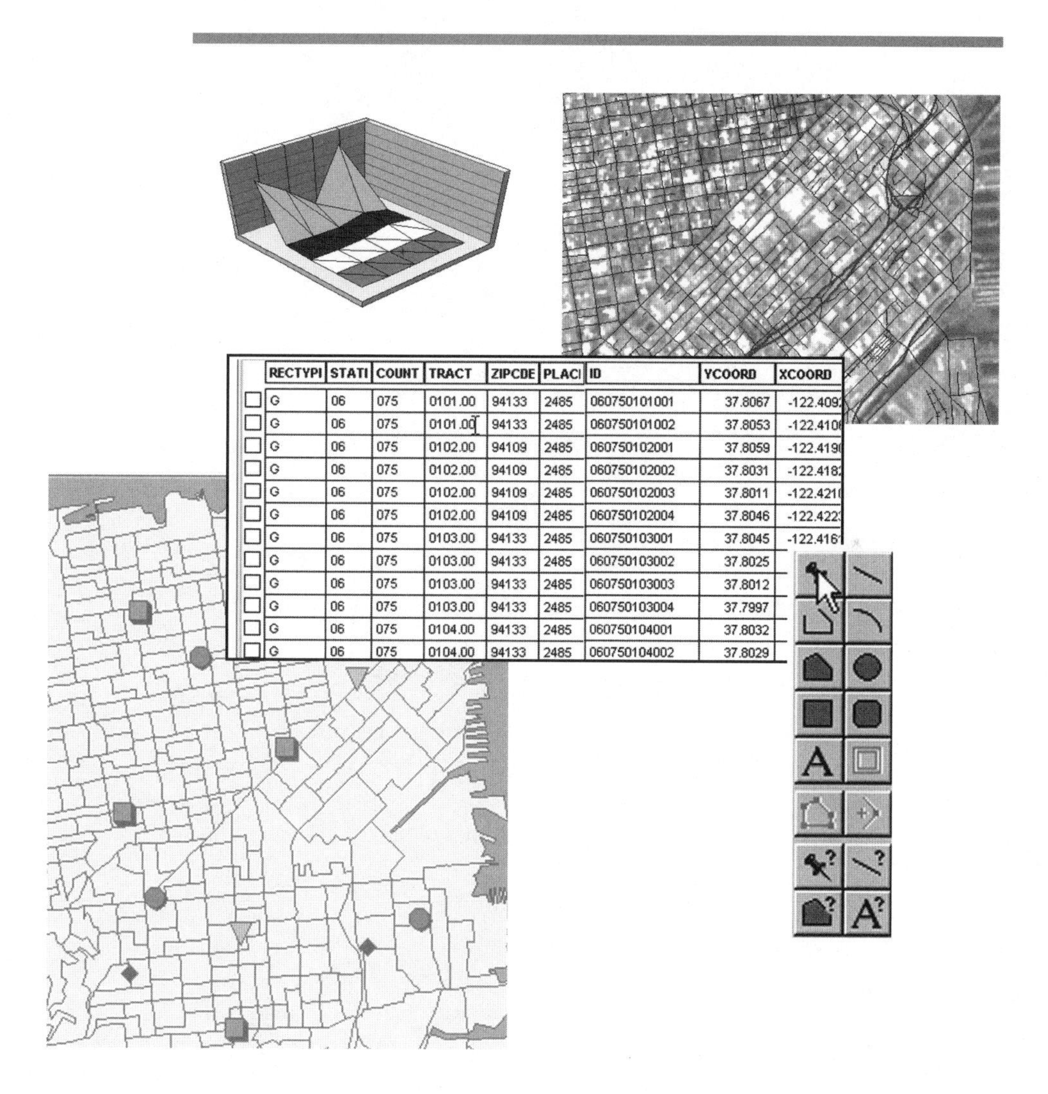

RECTYPI	STATI	COUNT	TRACT	ZIPCDE	PLACI	ID	YCOORD	XCOORD
G	06	075	0101.00	94133	2485	060750101001	37.8067	-122.409
G	06	075	0101.00	94133	2485	060750101002	37.8053	-122.410
G	06	075	0102.00	94109	2485	060750102001	37.8059	-122.419
G	06	075	0102.00	94109	2485	060750102002	37.8031	-122.418
G	06	075	0102.00	94109	2485	060750102003	37.8011	-122.421
G	06	075	0102.00	94109	2485	060750102004	37.8046	-122.422
G	06	075	0103.00	94133	2485	060750103001	37.8045	-122.416
G	06	075	0103.00	94133	2485	060750103002	37.8025	
G	06	075	0103.00	94133	2485	060750103003	37.8012	
G	06	075	0103.00	94133	2485	060750103004	37.7997	
G	06	075	0104.00	94133	2485	060750104001	37.8032	
G	06	075	0104.00	94133	2485	060750104002	37.8029	

CHAPTER 2 WHIRLWIND TOUR: A FIELD TRIP WITH MAPINFO PROFESSIONAL

AS MENTIONED EARLIER, GIS TECHNOLOGY INTEGRATES graphic and tabular data and makes possible the simultaneous analysis of disparate data. To demonstrate the power of MapInfo Professional, a typical business site selection problem and GIS solution are presented in this chapter. The following series of tutorials permits you to participate in a real-world MapInfo analysis from beginning to end.

Do not worry if the pace seems a bit hectic. The rest of the overall tutorial (chapters 3 through 12) examines in detail all MapInfo functions covered in this sample project.

Locating Satellite Pediatric and Geriatric Clinics in San Francisco

Assume that San Francisco county administrators want to reduce patient loads on existing hospitals. Because children and senior citizens visit the hospitals more than other age groups, planners decide that specialty clinics for these two age groups would be the most useful measure toward relieving overburdened hospitals. The target population for geriatric clinics is age 55 and over, whereas the target population for pediatric clinics is under age 18. For purposes of this exercise, you will assume that no pediatric or geriatric clinics currently exist in San Francisco County.

Although this satellite clinic location problem is hypothetical, it readily lends itself to a GIS-based solution. The primary objective is to establish satellite clinics in areas where large numbers of prospective clients reside.

Data Overview

You will use data table files provided on the companion CD-ROM for the following series of tutorials. The files are listed in table 2-1, which follows.

Table 2-1: Companion CD-ROM Files Used for Chapter 2 Tutorials

File Name	Description
Sf_group.tab	Block group boundaries; the geographical unit used in the analysis.
Sf_strts.tab	Street file in graphic format to display on the San Francisco map. This file will be used near the end of the analysis, when you are ready to locate a site.
Sfgtrl00.tab	Demographics to be used in conjunction with the block group boundaries to identify potential clinic sites.
Sf_landm.tab	Hospital information to be used for excluding certain areas from consideration.
Sf_bay.tab	Adds a blue background for the bay shown on the map.
Sf_back.tab	Background color for the base map.

Some of these table files are graphical and others are tabular. When you open a tabular file, you see data organized in tabular form. MapInfo Professional refers to such tables as *browsers.* When you open a graphical table, you will see graphical data displayed on a map as points, lines, or regions (areas, or polygons).

Procedural Outline

The following is an outline of the procedures you will pursue to identify optimum sites for the pediatric and geriatric clinics.

- Define the target population.
- Buffer existing hospitals to locate unsaturated areas.
- Visually scan areas for potential sites.
- Perform radial queries.
- Perform statistical analysis.

Identifying Potential Pediatric Clinic Sites

The first step in identifying potential sites is to define the target population. Pediatric clinics should be located in areas containing a large population of children (people under age 18). In contrast, the geriatric clinics should be located in areas containing a large population of people age 55 and over. Next, ideal target locations for both types of clinics are densely populated areas, such as downtown San Francisco. Finally, the clinics should be located in areas of stable or increasing population growth. The analysis will be based on the 1998 estimates in the demographic data file.

NOTE: *It is recommended that you create a new subdirectory under the main MapInfo directory for storage of the files created during tutorials and exercises in this book. Use the File Manager to create the new subdirectory. For the subdirectory name, you might use* Tutorials *or* Exercises.

To begin this series of tutorials, perform tutorial 2-1, which follows.

▼ *TUTORIAL 2-1: OPENING FILES TO BEGIN A PROJECT*

1 Click on File | Open Table from the main menu bar, or click on the Open Folder icon on the Standard toolbar. (The default location for the Standard button pad is below the menu bar. The toolbar location can be changed.)

2 The Open Table dialog box, shown in the following illustration, displays drives, directories, and file names. Verify that your CD-ROM drive is active, and select the *samples* directory on the companion CD-ROM.

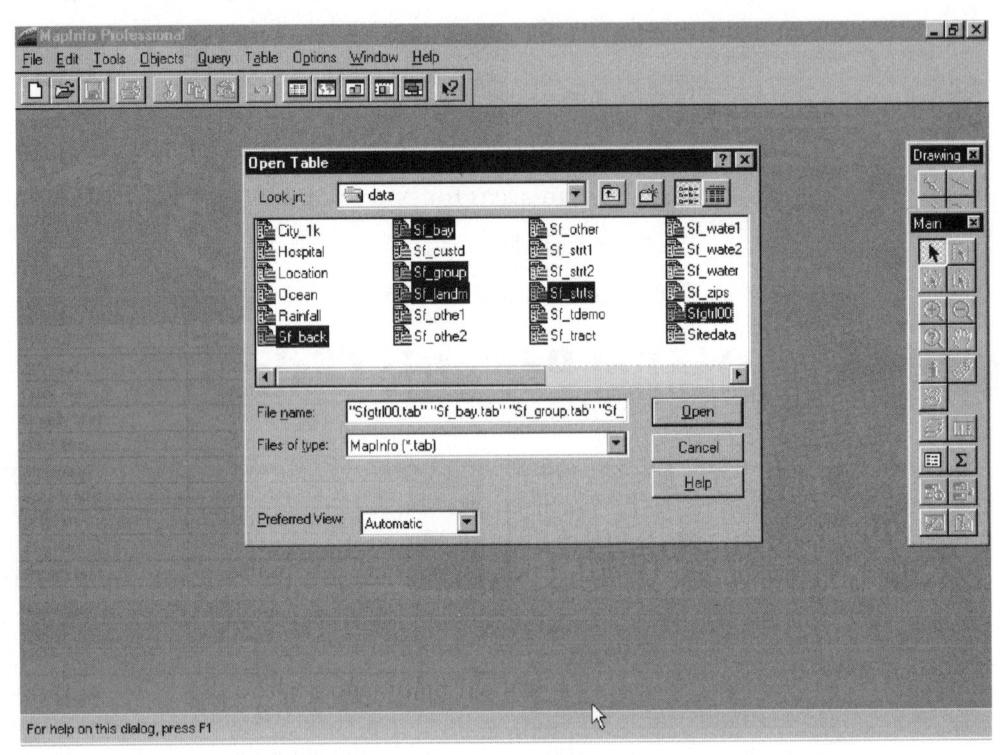

Open Table dialog box.

3 To open the files you need, click on File | Open Table for each of the following file names: *Sf_group, Sf_strts, Sf_landm, Sf_bay,* and *Sf_back.*

TIP 1: *Similar to most other Windows applications, in MapInfo Professional you can simultaneously open multiple files by one of two methods. If the files you want to open are sequential, hold down the <Shift> key and click on the first and last files. MapInfo will open the first and last files and all files between. If the files you want to open are not sequential, hold down the <Ctrl> key and click on the desired files.*

TIP 2: *Selecting items in MapInfo Professional is consistent with selecting items in Windows. To select an object, simply point the mouse cursor at the object and press or click the left mouse button.*

After opening the five files, the map on your screen should resemble that shown in the following illustration.

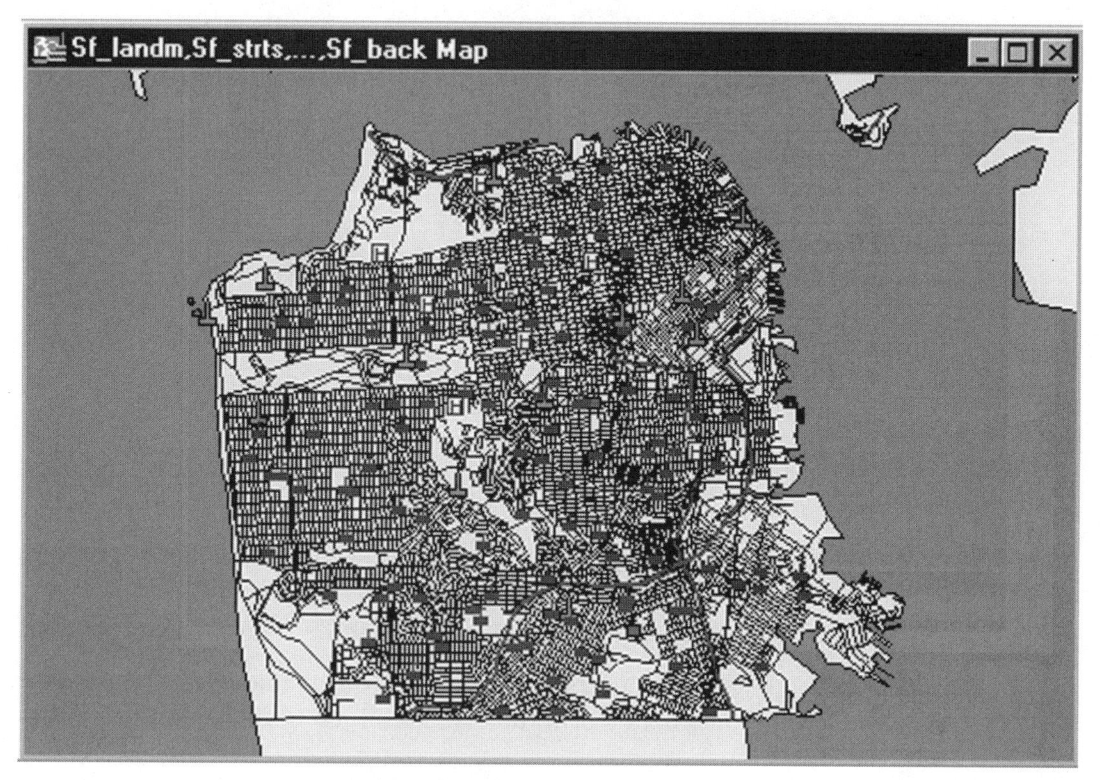

Map resulting from opening five files.

If the map on your screen does not resemble this map, you need to *reorder* the map layers; that is, change the drawing order of the map layers. MapInfo places each mappable (graphical) table in a layer, and assigns each layer a place in the map drawing order. The layer at the top of the list will be drawn last. For example, if the landmark layer (*Sf_landm*) is below the block group layer (*Sf_group*) in the layer listing, the landmarks will not be visible. Reorder the layers by performing tutorial 2-2, which follows.

▼ TUTORIAL 2-2: CHANGING THE DRAWING ORDER OF MAP LAYERS

1. Select Map | Layer Control from the main menu bar. (You can also access the Layer Control dialog box by clicking on the right mouse button while in the Map window and selecting the Layer Control option.) The Layer Control dialog box, shown in the illustration at right, will appear.

Layer Control dialog box.

2. Reorder the files as shown in the previous illustration by selecting a layer, and then using the Up or Down buttons under the Reorder option to change the layer's position.
3. When the layers are reordered, click on the OK button, and the screen will redraw.

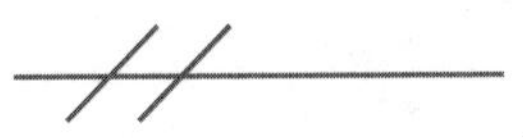

If only a portion of the map is visible, or the map is not focused on San Francisco, perform tutorial 2-3, which follows.

▼ TUTORIAL 2-3: ESTABLISHING THE LAYER VIEW

1. Select Map | View Entire Layer from the main menu bar.
2. Select the *Sf_strts* layer from the View Entire Layer dialog, shown in the illustration at right. The screen will redraw.

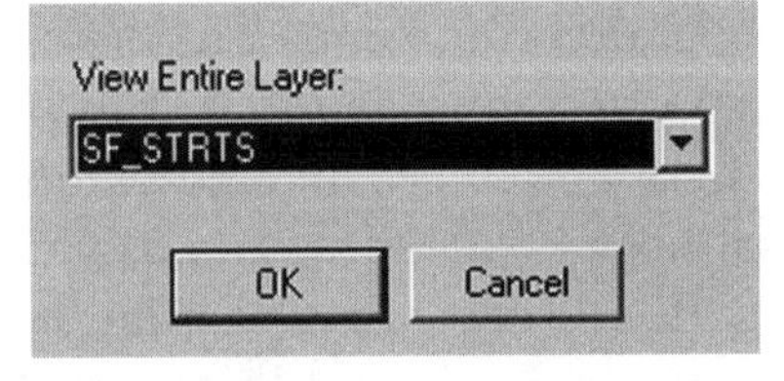

View Entire Layer dialog box.

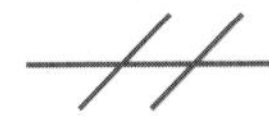

Hospital locations are part of the landmark table (*Sf_landm*) opened previously. A hospital is depicted on the map by a blue H enclosed in a white box. The next task is to isolate the hospitals from the other landmarks to simplify the analysis. To isolate the hospitals, you need to issue a query based on the *Sf_landm* file. Tutorial 2-4, which follows, takes you through this process.

▼ TUTORIAL 2-4: QUERYING TO ISOLATE LAYER ELEMENTS

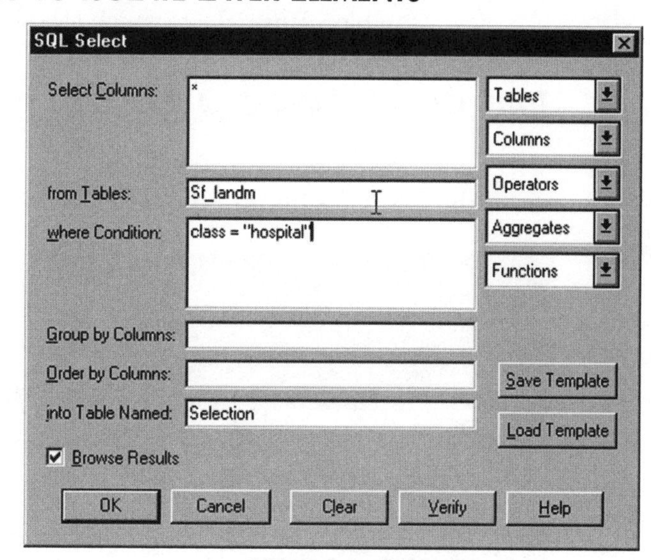

SQL Select dialog box.

1. From the main menu bar, select Query | SQL Select. This will open the SQL Select dialog box, shown in the illustration at right. Select the *Sf_landm* table from the Tables list on the right side of the dialog box. The name of the table should appear in the "from Tables" box.
2. Click inside the box next to the "where Condition" box. Next, select *class* from the Columns list. The *class* column is inserted in the "where Condition" box.
3. Position the cursor after the word *class* and key in an equals sign and the word *hospital* enclosed in double quotes (= *"hospital"*).
4. Click on the Verify button at the bottom of the dialog box to ensure that the syntax of the query statement is correct.
5. Verify that the Browse Results option (lower left) is activated. If a check mark appears to the left of Browse Results, the option is activated.
6. Click on OK to issue the query statement. MapInfo Professional will process the query statement, highlight all hospitals on the map, and open a Browser window containing a list of the hospitals. The results of the query will be placed in a temporary table named *Selection.*

TIP: *You can access query results at any time during the current session by noting the query number at the top of the resulting Browser window, selecting Window | New Browser Window, and selecting the appropriate query number from the list.*

At this juncture, you will save the query results as a new file. Although MapInfo can work with and save temporary tables (i.e., query

results), for the purpose of this exercise you will work with saved tables. Tutorial 2-5, which follows, takes you through this process.

▼ TUTORIAL 2-5: SAVING QUERY RESULTS AS A NEW FILE

1 Select File | Save Copy As from the main menu bar. The following illustration shows the query results and the Save Copy As dialog box.

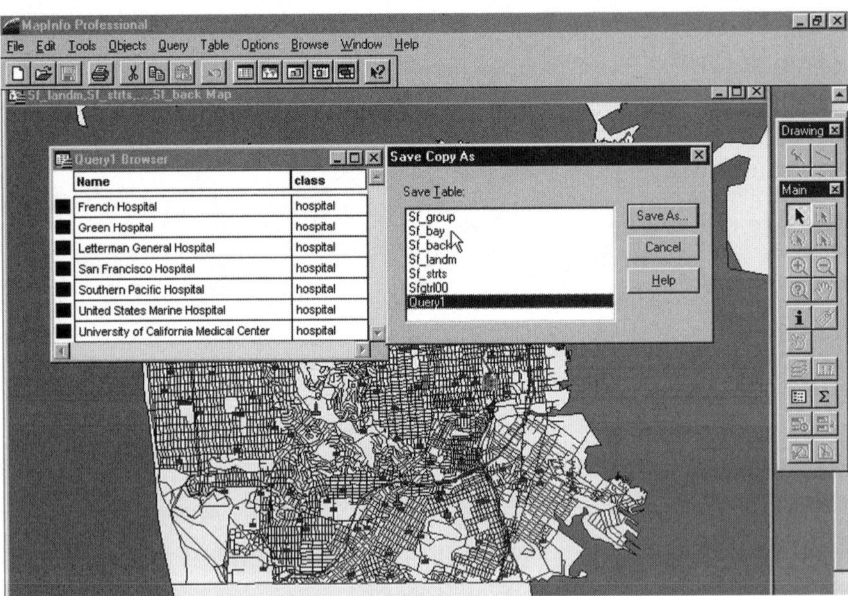

The Query1 Browser list of hospitals and the Save Copy As dialog box.

2 Select Query1 (the name of the query just created) from the Save Table list box.

3 Click on the Save As button to access the Save Copy of Table As dialog box, shown in the illustration at right.

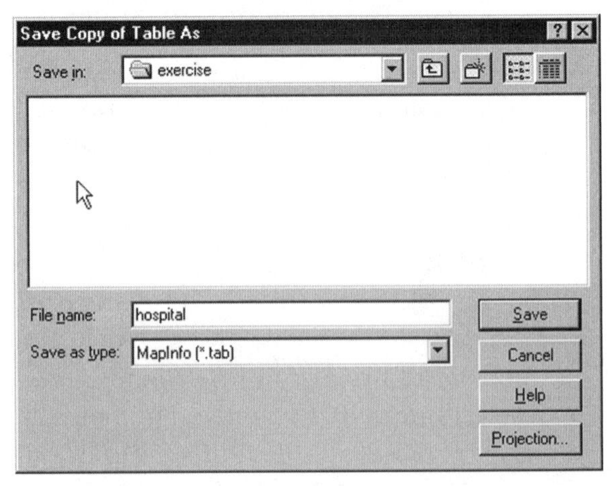

Save Copy of Table As dialog box.

4 Now you need to name the file containing the highlighted hospital landmarks and store it in the Exercises (or Tutorials) directory you created earlier on your hard

disk. Under File Name, type in *hospital* as the name of the new table you are creating.

5 Click on OK to save the file.

6 To add the new *hospital.tab* table to the map, select File | Open Table from the main menu bar, and then select *hospital.*

7 To remove the *Sf_landm* table from the map and eliminate the overlap with the *hospital* table, select File | Close Table, and then *Sf_landm.* Click on the OK button and the screen will redraw.

At this point, you may need to reorder the map layers. Continue with the following steps.

8 Select Map | Layer Control from the main menu bar to access the Layer Control dialog box, shown in the following illustration.

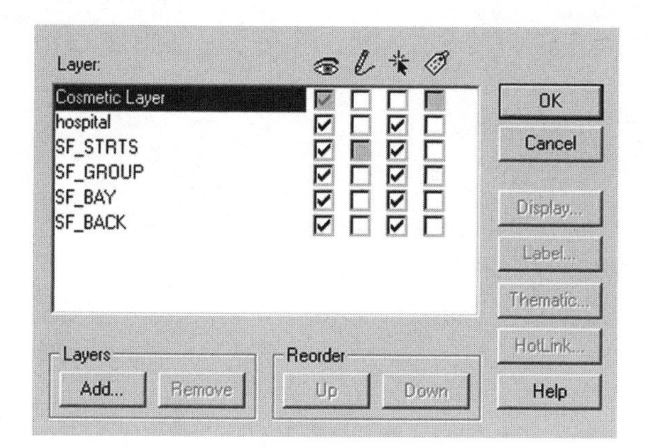

Layer Control dialog box.

9 Reorder the files as shown in the previous illustration by selecting Up or Down under the Reorder option.

10 When the layers are in order, click on OK and the screen will redraw.

The map on your screen should now resemble that shown in the following illustration.

Map showing reordered layers.

Buffer Hospitals

All files required to begin the analysis are in place. Now you will identify the competition. In the absence of pediatric clinics in San Francisco, the only competitors are existing hospitals. Because you do not wish to locate a new clinic within a half-mile of an existing hospital, you will place a half-mile buffer around the hospitals. To place the buffers, you need to return to the Layer Control dialog box to make the cosmetic layer editable. Tutorial 2-6, which follows, takes you through the process of placing buffers.

▼ *Tutorial 2-6: Placing Buffers*

1 Select Map | Layer Control.

2 Highlight the cosmetic layer at the top of the layer list in the dialog box.

3 Place a check mark in the box under the pencil icon to make the cosmetic layer editable. This action will automatically place a check mark in the Selectable box (under the pointer icon) and disable the Selectable attribute while the layer is editable.

4 Click on the OK button and return to the main menu.

At this point, you will select all hospitals and then create buffers around these objects on the map. Continue with the following steps.

5 Select Query | SQL Select from the main menu bar. The previous query you created should appear.

6 Change the name of the table to *hospital* (from *Sf_landm*), deselect the Browse results (remove the check mark), and click on the OK button (see the following illustration for help, which shows the SQL Select dialog box). All hospitals on the map should be highlighted.

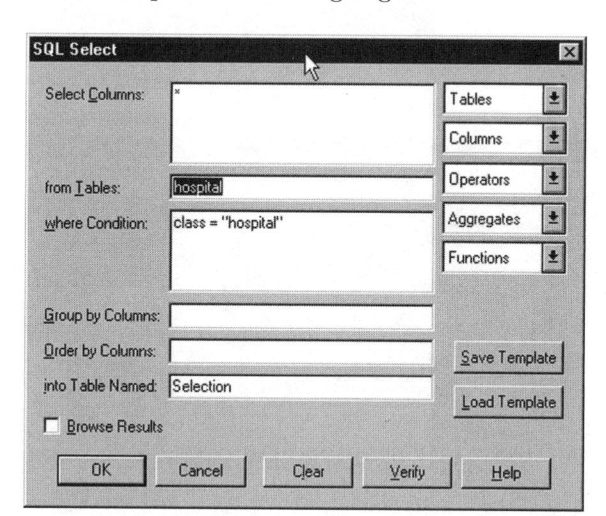

SQL Select dialog box.

7 Now you are ready to create the buffers. Select Objects | Buffers from the main menu bar to access the Buffer Objects dialog box.

NOTE: *If the Buffer option is not enabled, you need to verify that the cosmetic layer is editable. Select Map | Layer Control and proceed, as explained previously.*

8 Check the map to ensure that all hospitals are still highlighted (selected) by clicking on the map window's title bar. If all hospitals are not selected, repeat steps 5 through 7. Click anywhere on the Buffer Objects dialog box, shown in the following illustration, to continue.

9 Duplicate the settings in the Buffer Objects dialog box, shown at right. Make sure you select .5 as the radius Value (for a half-mile radius), and the "One buffer for each object" option. If you select "One buffer of all objects," MapInfo will merge the buffers where they overlap.

10 Click on OK. The map on your screen should redraw to resemble that shown in the following illustration. The buffers are highlighted with hatch patterns.

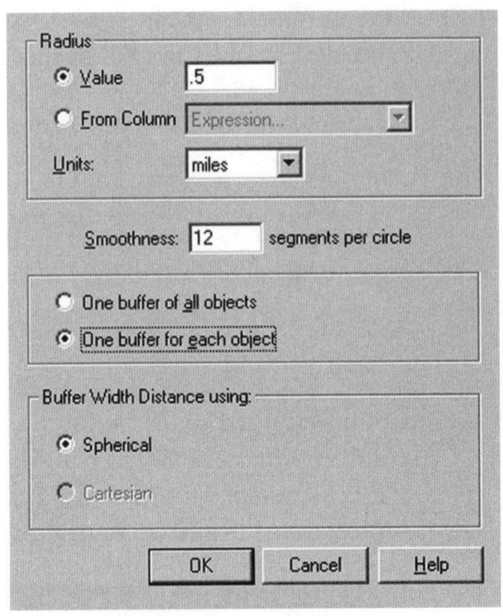

Buffer Objects dialog box.

Map of San Francisco with highlighted half-mile buffers.

With the buffers in place, you can quickly and easily view the areas where new clinics cannot be located. In order to use these buffers throughout the analysis, you will save them as a new file. Tutorial 2-7, which follows, takes you through this process.

▼ TUTORIAL 2-7: SAVING BUFFERS AS A NEW FILE

1 From the menu bar, select Map | Save Cosmetic Objects.

2 Save the trade area buffers to a new table named *buffer*. To do this, select <New> under Transfer Cosmetic Objects to bring up the Save As dialog box. This file will automatically be added to the current map.

3 You will probably need to reorder the layers again so that the hospitals will appear on top of the buffers. (Select Map | Layer Control to ensure that the hospital table appears above the buffer table in the layer listing.)

4 To improve redraw performance, turn off visibility of the street layer by clicking on the visibility box (under the eye icon) and removing the check mark, as shown in the following illustration.

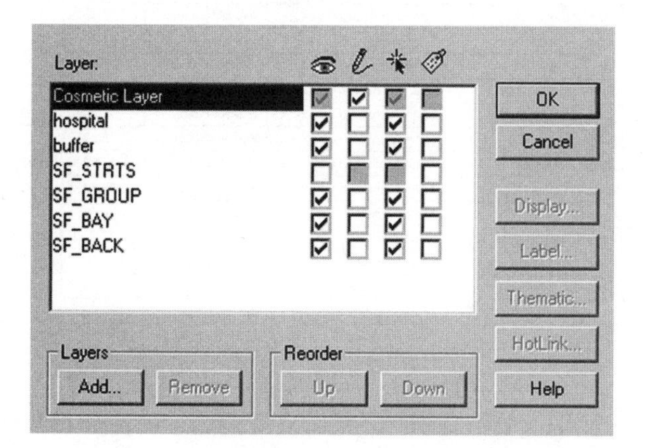

Layer Control dialog box showing current layer order and visibility turned off for the Sf_strts *layer.*

You are now ready to identify potential sites for the clinics; that is, areas with a high concentration of children. Continue with the following steps.

5 Select File | Open Table from the main menu bar.

6 Select the *Sfgtrl00.tab* file. Because this file contains tabular data, the data appear in a Browser window, rather than a graphic window.

7 Scroll through the Browser to examine the data available in the *Sfgtrl00* file. Population age group information appears in columns to the far right. The

block group name is called ID in the table. The ID column will permit you to link the tabular data in this file with the boundaries file (*Sf_group*) opened previously.

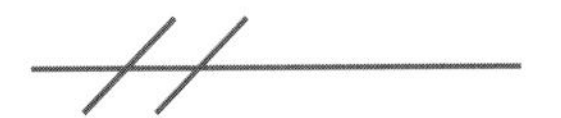

To close the Browser and return to the map, click on the X in the upper right corner of the Browser window, shown in the following illustration.

MapInfo Professional

File Edit Tools Objects Query Table Options Browse Window Help

hospital,buffer,...,Sf_back Map

Sfgtrl00 Browser

RECTYPI	STATI	COUNT	TRACT	ZIPCDE	PLACI	STNAME	CTYNAME	ID	YCOORD	XCOORD
G	06	075	0101.00	94133	2485	California	San Francisco County, C	060750101001	37.8067	-122.4092
G	06	075	0101.00	94133	2485	California	San Francisco County, C	060750101002	37.8053	-122.4106
G	06	075	0102.00	94109	2485	California	San Francisco County, C	060750102001	37.8059	-122.4190
G	06	075	0102.00	94109	2485	California	San Francisco County, C	060750102002	37.8031	-122.4182
G	06	075	0102.00	94109	2485	California	San Francisco County, C	060750102003	37.8011	-122.4210
G	06	075	0102.00	94109	2485	California	San Francisco County, C	060750102004	37.8046	-122.4223
G	06	075	0103.00	94133	2485	California	San Francisco County, C	060750103001	37.8045	-122.4161
G	06	075	0103.00	94133	2485	California	San Francisco County, C	060750103002	37.8025	-122.4142
G	06	075	0103.00	94133	2485	California	San Francisco County, C	060750103003	37.8012	-122.4134
G	06	075	0103.00	94133	2485	California	San Francisco County, C	060750103004	37.7997	-122.4141
G	06	075	0104.00	94133	2485	California	San Francisco County, C	060750104001	37.8032	-122.4111
G	06	075	0104.00	94133	2485	California	San Francisco County, C	060750104002	37.8029	-122.4085
G	06	075	0104.00	94133	2485	California	San Francisco County, C	060750104003	37.8038	-122.4061
G	06	075	0104.00	94133	2485	California	San Francisco County, C	060750104004	37.8024	-122.4039
G	06	075	0104.00	94133	2485	California	San Francisco County, C	060750104005	37.8001	-122.4034
G	06	075	0105.00	94111	2485	California	San Francisco County, C	060750105001	37.8041	-122.4030
G	06	075	0105.00	94111	2485	California	San Francisco County, C	060750105002	37.7983	-122.3992
G	06	075	0106.00	94133	2485	California	San Francisco County, C	060750106001	37.8006	-122.4055
G	06	075	0106.00	94133	2485	California	San Francisco County, C	060750106002	37.7983	-122.4034
G	06	075	0106.00	94133	2485	California	San Francisco County, C	060750106003	37.8001	-122.4069
G	06	075	0107.00	94133	2485	California	San Francisco County, C	060750107001	37.8003	-122.4099
G	06	075	0107.00	94133	2485	California	San Francisco County, C	060750107002	37.7984	-122.4086

records 1 - 22 of 611

Content of Sfgtrl00 *file seen in Browser window.*

To identify potential sites, you will create a thematic map depicting the concentration of the target group: residents under age 18. The *Sfgtrl00* file does not contain a column titled "below 18." An expression for selecting this age group must be created when defining the thematic map. To create the thematic map, perform tutorial 2-8, which follows.

▼ TUTORIAL 2-8: CREATING A THEMATIC MAP

1 Select Map | Create Thematic Map from the main menu bar. The Step 1 of 3 dialog box appears, shown in the following illustration.

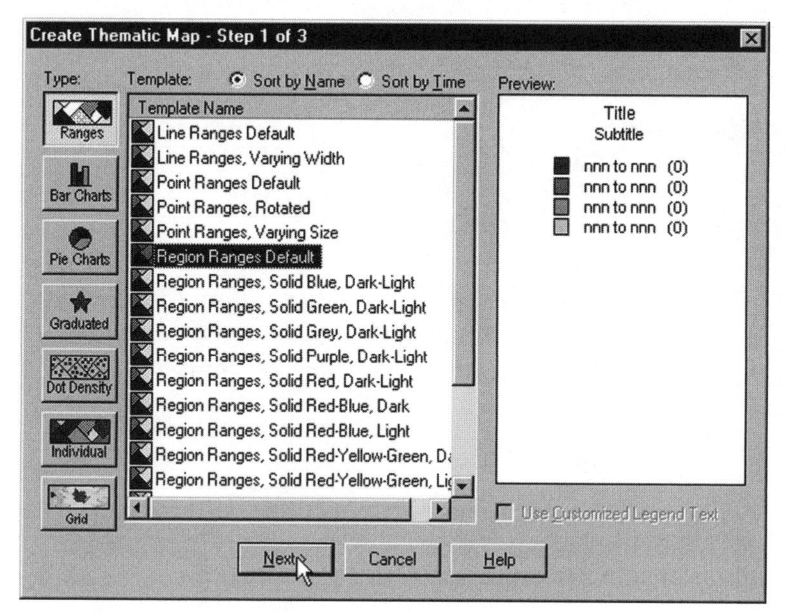

First dialog box in the Create Thematic Map series.

2 Select Ranges as the type of thematic map to create, and Region Ranges Default as the template.

3 Upon selecting a type of thematic map, click on the Next button, and another dialog box appears. In this dialog box, shown in the illustration at right, you will select the block group (*Sf_group*) table as the table to be shaded for the thematic map.

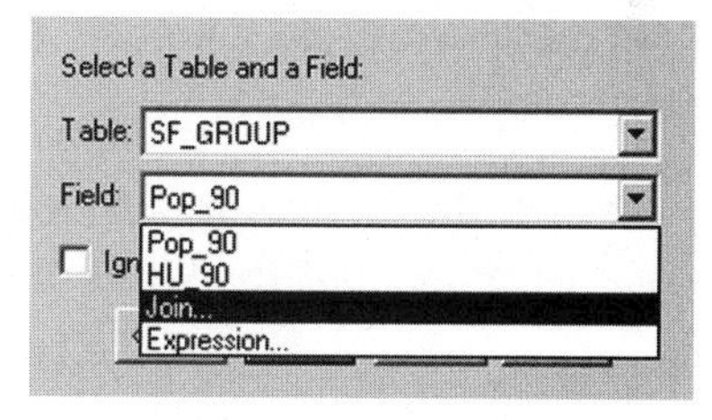

Second dialog box in the Create Thematic Map series.

4 Scroll through the Field list, and select Join. (The Join selection is near the bottom of the list.)

5 Selecting Join activates a dialog box that prompts you for the name of the table to join to *Sf_group*. Select the *Sfgtrl00* table, and then click on the Join button.

6 Duplicate the selections in the Specify Join area, as shown in the following illustration. In other words, join the tables through the BlockGroup field in the *Sf_group* table with the ID field in the *Sfgtrl00* table, and then click on OK.

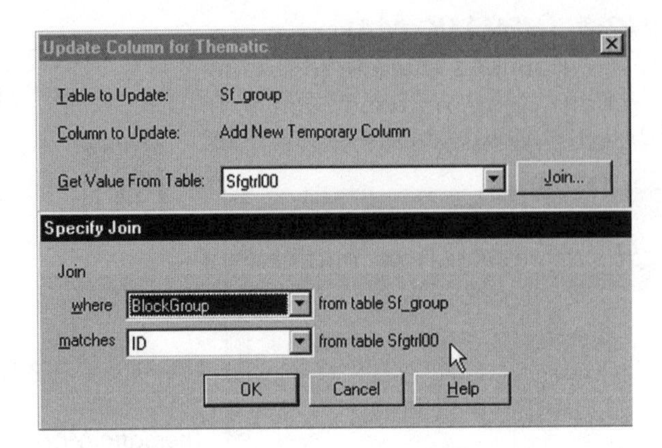

Update Column for Thematic dialog box showing join fields for tables.

7 From the Of option (below Calculate) in the Update Column for Thematic dialog, select Expression. At this point, the Expression dialog box appears. To create the expression for the population under 18, sum the following fields in the *Sfgtrl00* table.

- TPA98_0_4
- TPA98_5_9
- TPA98_10_14
- TPA98_15_17

TPA represents "Total Population Aged," and 98 represents 1998 population estimates. Next, 0_4, 5_9, 10_14, and 15_17 represent age groups. To create the "under age 18" population group, simply add the values in the four age range fields.

8 In the Expression dialog, key in the expression shown in the following illustration, or use the Column and Operator buttons to create the expression.

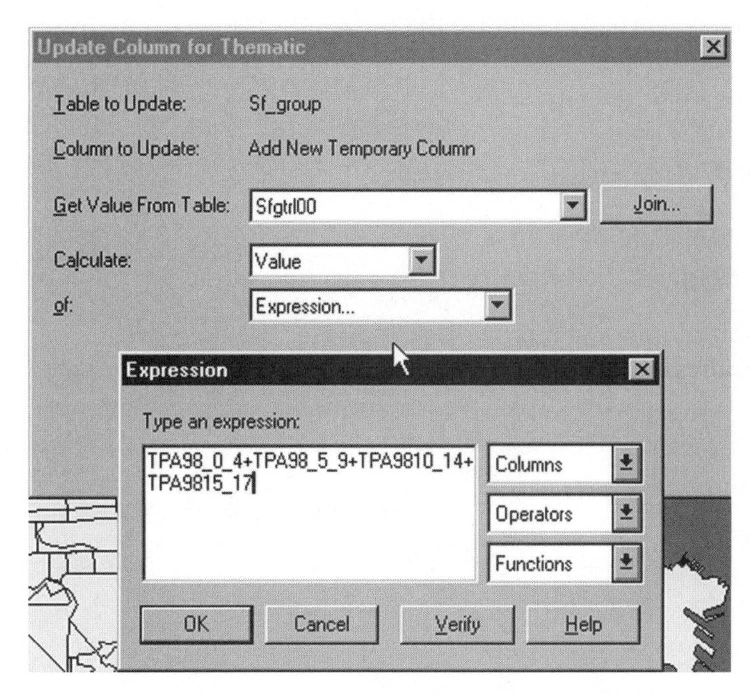

Expression dialog box showing the expression for adding the four age groups.

9 Click on the OK button to return to the Step 2 of 3 dialog box, and then click on the Next button to continue. At this point, the Step 3 of 3 dialog box appears (shown in the following illustration), which will allow you to customize the thematic map prior to implementation.

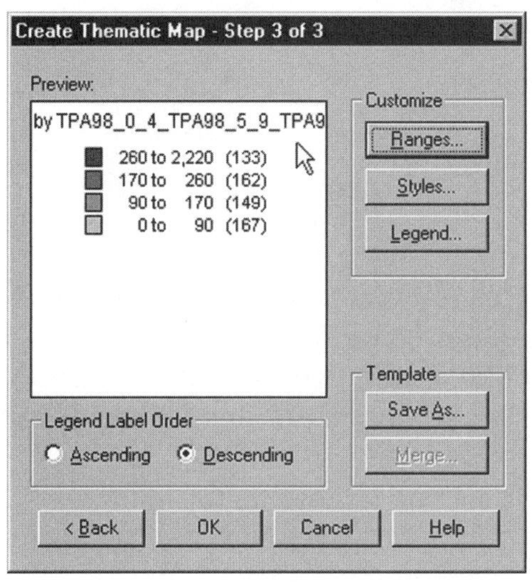

Third dialog box (Step 3 of 3) in the Create Thematic Map series used for presenting data selected in expression.

First, you will customize the ranges for the proportions of the population age 18 and under. Tutorial 2-9, which follows, takes you through this process.

▼ Tutorial 2-9: Establishing Ranges

1. In the Step 3 of 3 dialog box, click on the Ranges button.
2. In the Customize Ranges dialog box, shown in the illustration at right, select Custom as the range Method.
3. Change # of Ranges to 5. The Recalc button appears at the bottom of the dialog box. Select Recalc and the five ranges will appear.
4. Define the ranges, as shown at right, by entering the maximum and minimum values for each range. Use the following ranges: 1-250, 250-500, 500-750, 750-1000, and 1000-2220. These ranges were selected after viewing the initial ranges automatically defined by MapInfo Professional.

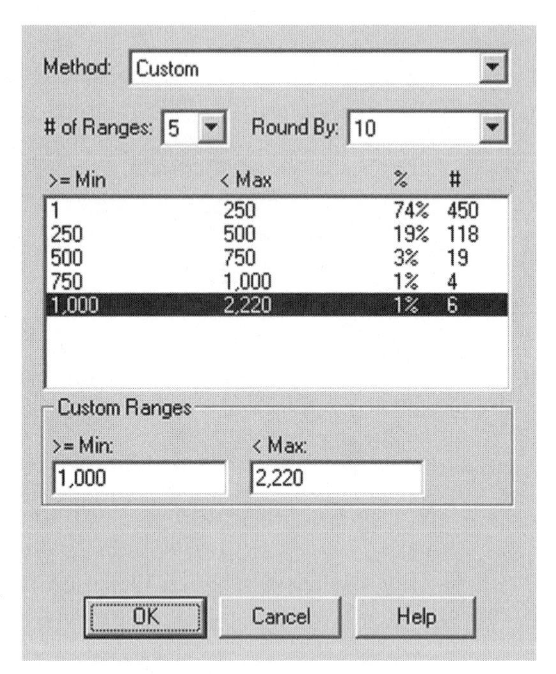

Customize Ranges dialog box.

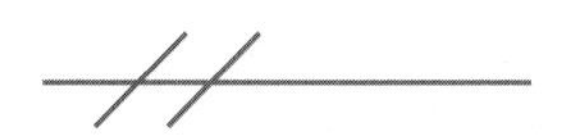

After entering these ranges, select the ReCalc button to calculate the percentage and count of Block Groups that fall within each range. Select OK to return to the Create Thematic Map - Step 3 of 3 dialog box. At this point, you can alter the appearance of the ranges on the map by selecting Styles (in this dialog box). The Customize Range Styles dialog box is shown in the following illustration.

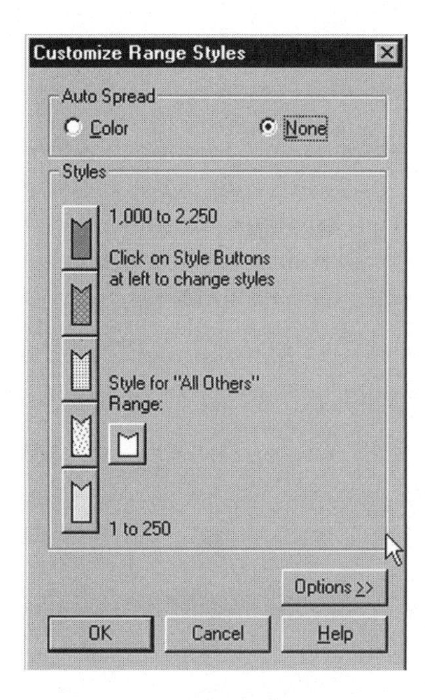

Customize Range Styles dialog box.

If you select the Color option under Auto Spread, and then click on the top region styles color box, range colors are changed from the color scheme of the chosen template to shades of the selected color. At this point, you can select a different color or pattern for each range via the Color box. Select red to depict "hot spots" (areas of high concentration for the target market), orange for the second highest range, yellow for the third range, green for the fourth, and blue for "cold spots" (areas of least potential). Click on OK to return to the Step 3 of 3 dialog box.

You can also customize a range legend to improve the map's appearance by selecting Legend from the Step 3 of 3 dialog box. To customize the legend as shown in the following illustration, perform tutorial 2-10, which follows.

▼ TUTORIAL 2-10: CUSTOMIZING A LEGEND

1 Click on the Legend button.

2 Change the title to *Population Under 18.*

3 Change the title font (click on the Aa box) to Times New Roman, 10 point, and italic.

4 Change the subtitle to *in San Francisco.*

5 Change the subtitle font (Aa box) to Times New Roman, 8 point, and italic.

6 Change the ranges to reflect the fact that the values in each range are greater than or equal to (>=) the minimum value and less than (<) the maximum value. If the range is currently labeled 1-250, the actual inclusive range values are 1-249. Change all values, as shown in the following illustration, so that each range is transparent to your audience.

7 Click on the "All Others" range under Range Labels and then uncheck the Show this Range option so that this range will not appear in the Legend window.

8 Change the font for the Range Labels to Times New Roman, 8 point.

9 Select New Legend Window under the Into Window option. This will bring up the Customize Legend dialog box, shown in the following illustration.

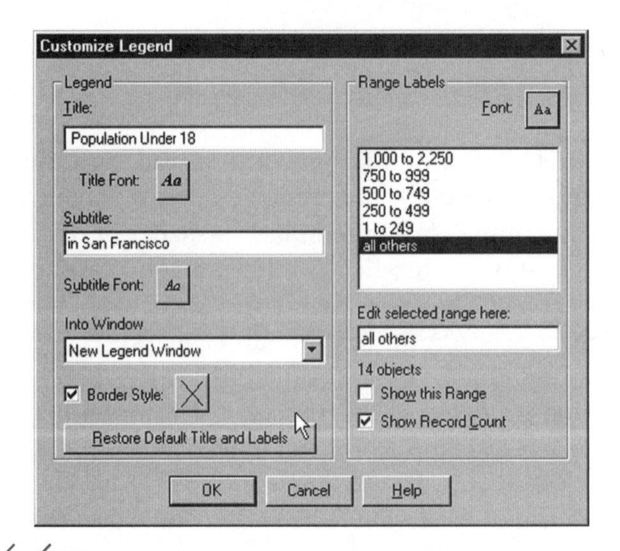

Customize Legend dialog box containing suggested legend and range labels.

The following illustration shows the map depicting different concentrations of the population under age 18. To locate potential sites for a pediatric clinic, you can visually scan the map for hot spots, or high concentrations of the target population. The dark areas (red in a color map) show high concentrations of the under-18 population.

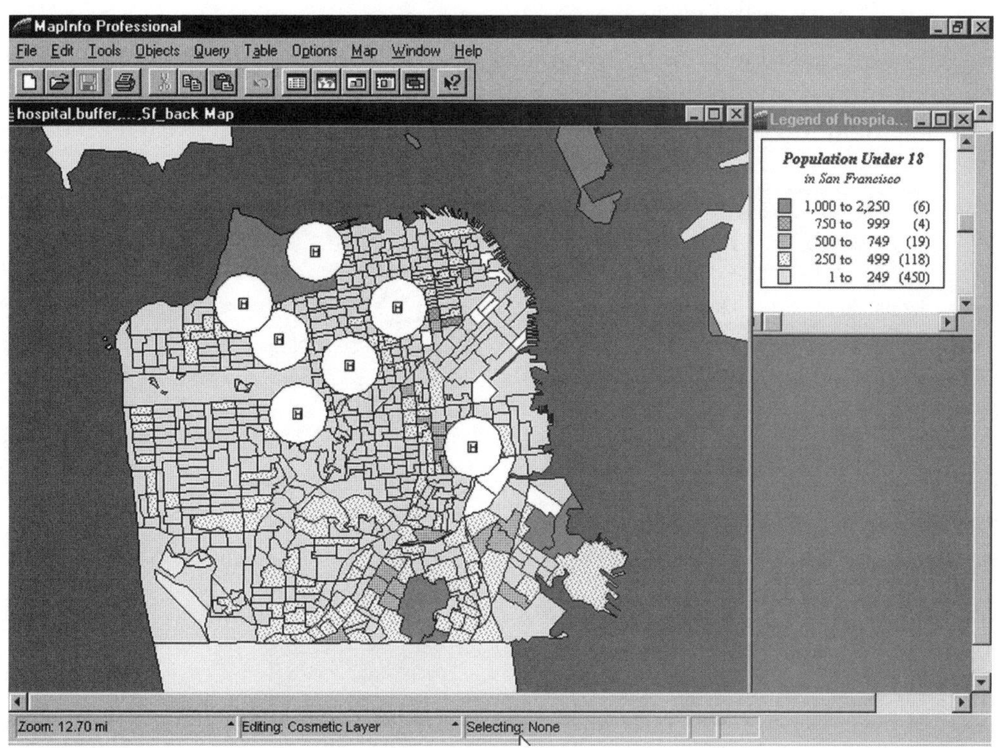

Map of San Francisco showing relative concentrations of population under age 18 by block group.

To locate potential sites for pediatric clinics on the new map, you will create a new table. Tutorial 2-11, which follows, takes you through this process.

▼ TUTORIAL 2-11: CREATING A TABLE FOR LOCATING SITES

1 Select File | New Table from the main menu bar, or click on the New Table icon (blank sheet of paper) on the Standard toolbar.

2 Select Add to Current Mapper in the New Table dialog box, shown in the illustration at right, and then select the Create button to start defining the table.

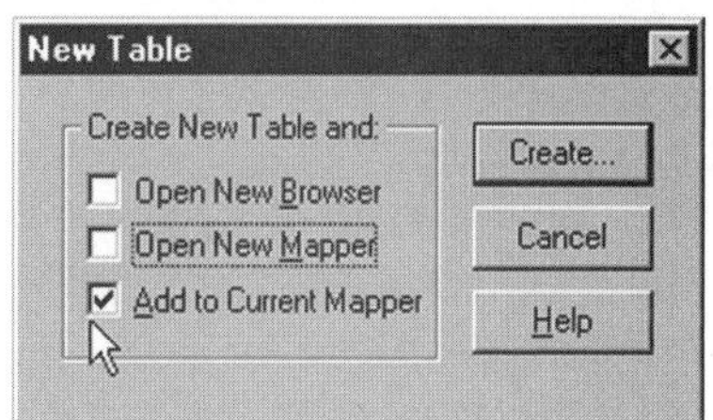

New Table dialog box.

3 Define the table in the New Table Structure dialog box by using the following columns: *id_no*, *type*, and *potential_no*. (See the following illustration.) Type in *id_no* under Name to

define the first field in the new table. Be sure to use the underscore character (_) instead of a space between *id* and *no* because MapInfo Professional does not allow spaces in field names. Select Character as Type, and set Width to 10. The field will now allow up to 10 characters to be entered for each record.

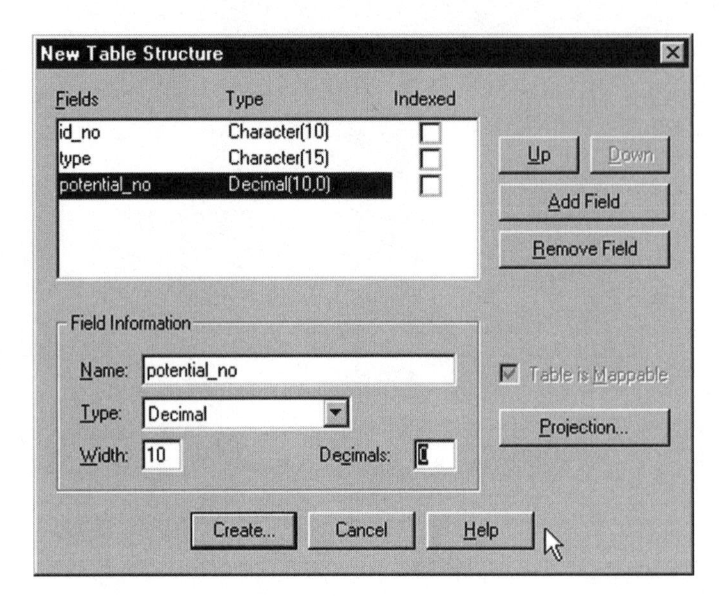

New Table Structure dialog box.

4 Click on Add Field to define the next field. Under Name, key in *type*. Select Character as the data Type for this field, and set Width to 15.

5 Click on Add Field to define the final field. Name this field *potential_no*. Set the Type for *potential_no* to Decimal, Width to 10, and Decimals to 0.

6 Click on Create after all fields are defined. The table structure should update to resemble that shown in the previous illustration.

7 MapInfo will prompt you for a table name. Name the file *sites* in the Create New Table dialog box. Be sure to save this table to the *Exercises* (or *Tutorials*, if you have named it that) directory on your hard drive.

The *id_no* column in the *sites* table allows you to assign a unique number to each site for tracking purposes. Next, the *type* column allows you to use this table for researching the area for sites of other types of clinics (e.g., geriatric clinics). Thus, you could place both pediatric and geriatric clinic sites in a single file. Finally, the *potential_no* column allows you to track potential clients within a one-mile radius of a particular site. The one-mile radius is the estimated trade area for each clinic. Alternatively, you could select block groups individually in a rectangular form or customized polygonal shape if you wanted to invest more time in precisely defining the trade area.

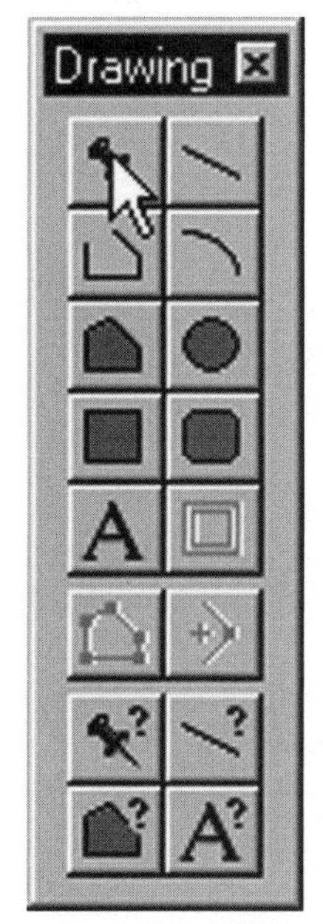

Drawing toolbar.

Once the *sites* table has been defined and saved, check Layer Control to verify that the table has been added to the map and is editable (i.e., that there is a check mark under the pencil icon for the *sites* layer). The *sites* table should have been automatically added to the map and made editable. Next, click on the Symbol Style tool (pushpin with question mark) from the Drawing toolbar. If the Drawing toolbar (shown at left) is not on your screen, select Options | Toolbars on the main menu bar, and click on Drawing.

The Symbol Style dialog box, shown in the following illustration, prompts you to select a symbol, color, size, background, and effects for the *sites* table. Select yellow squares from the MapInfo 3.0 Compatible symbol set (which will disable background and effects options) to depict potential pediatric clinic sites.

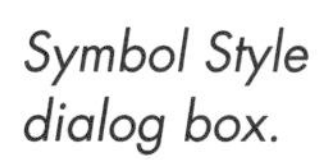
Symbol Style dialog box.

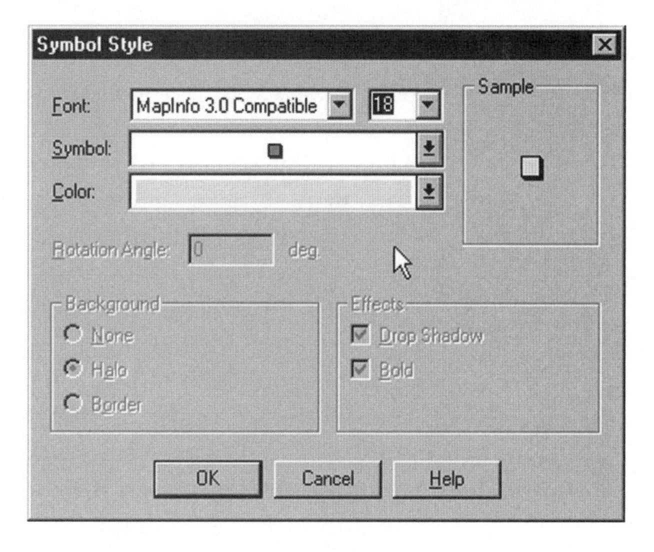

Next, click on the Symbol tool (the pushpin) on the Drawing toolbar and select a likely hot spot from the map. A visual scan has helped you identify three potential sites for initial investigation.

The sites were selected by clicking the mouse on the map. Create the three potential sites on your screen, as shown in the following illustration, by clicking on the map in each location.

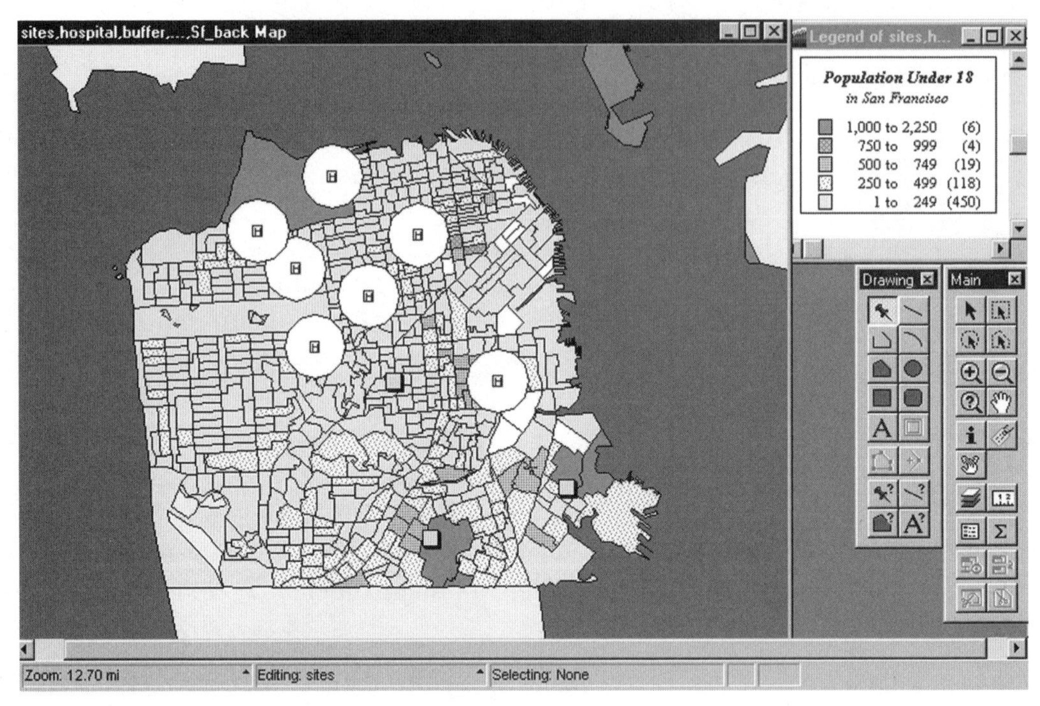

San Francisco map of three potential sites for pediatric clinics.

The yellow squares were placed on hot spots depicted by red on the map. The hot spot at the top of the map was not selected because of its proximity to two existing hospitals. In contrast, an area at the center of the map was selected, even though it did not show the peak range of concentration for the target population.

Assume you decide to investigate this location based on the number of block groups and the absence of hospitals in the surrounding area. The analysis could have been approached from a population density standpoint by dividing the number of people in the target population by block group area for each block group. Because MapInfo Professional provides an area function for such purposes, the thematic map for a population density analysis would have been just as easy to create as the one you are currently working with.

Radial Queries

At this stage you will identify the absolute numbers of the target population associated with each potential clinic site. Before you begin, select Map | Layer Control from the main menu bar, and set the Selectable option (under the pointer icon) for the *sites* layer to Off. If Editable is checked for the *sites* layer, you need to remove the check mark before you can alter the Selectable option. The Layer Control dialog box is shown in the following illustration.

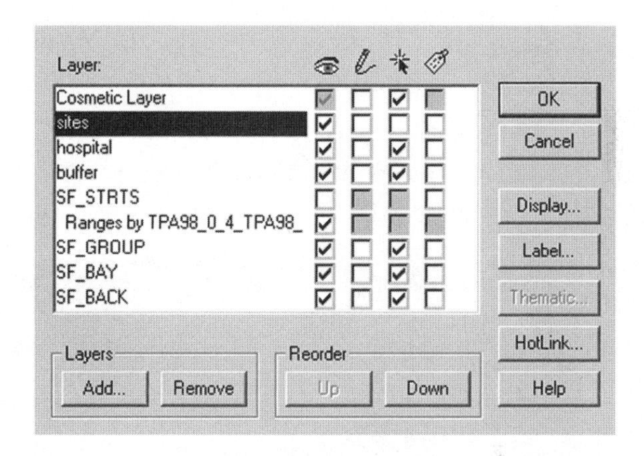

Layer Control dialog box showing Selectable option turned off for the sites layer.

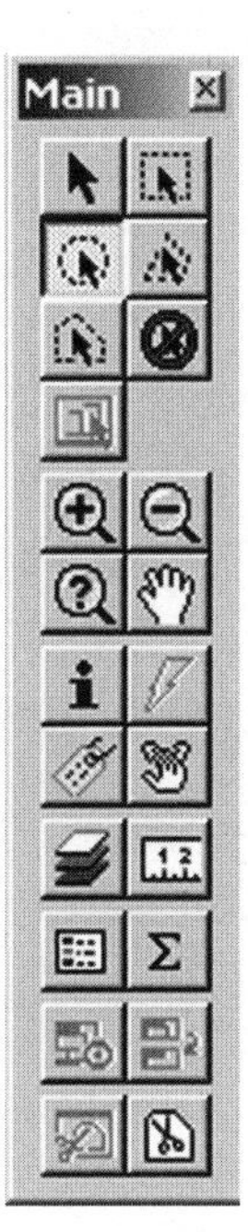

To calculate the number of people under 18 years of age, you first need to select the block groups within the one-mile trade area. This is done using the Radius Select tool on the main button pad, shown at left. Tutorial 2-12, which follows, takes you through the process of selecting block groups.

Main button pad.

▼ TUTORIAL 2-12: SELECTING BLOCK GROUPS

1. Click on the Radius Select tool (second from the top on the left in the previous illustration) from the Main toolbar.
2. Place the cursor on one of the sites, and click on this spot on the map.
3. While holding the left mouse button down, start to expand the ring. A radius indicator will appear in the bottom left corner of the status bar in the MapInfo screen.
4. Expand the ring to a 1.00-mile radius, as indicated in the lower left corner of the screen.

You have now defined the trade area for the new site. The selected block groups on the map will contain a hatch pattern to differentiate them from other block groups. (See the lower right portion of the map in the following illustration.)

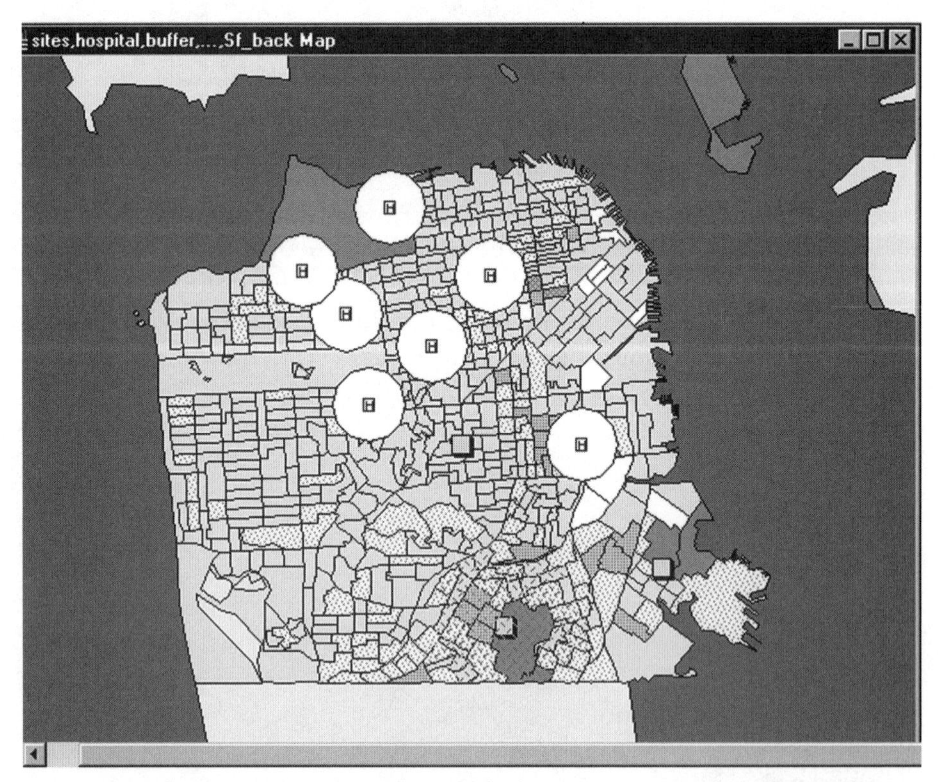

Hatch pattern showing selected block groups surrounding the southern-central site within a 1-mile radius.

You can zoom in on the hatched area to more closely view the selected block groups. To zoom in, perform tutorial 2-13, which follows.

▼ *TUTORIAL 2-13: ZOOMING IN ON BLOCK GROUPS*

1 Select the Zoom-in tool (plus sign in a magnifying glass) on the Main toolbar.

2 Click and drag a rectangle around the selected block groups. The map will zoom in to the selected location, as shown in the following illustration.

Results of a zoom-in on selected block groups surrounding a potential pediatric clinic site.

With the close-up view of the trade area, you could easily make adjustments to the area. However, for purposes of simplicity and consistency, use the block groups selected in the 1-mile rings as the trade areas. In a real-world analysis, adjustments to the trade area are virtually inevitable due to competition, terrain, street networks, and so forth. To return to the complete San Francisco area map, select Map | Previous View from the main menu bar.

Statistical Analysis

To obtain the count (total) of the target population, perform tutorial 2-14, which follows.

▼ *Tutorial 2-14: Obtaining Totals*

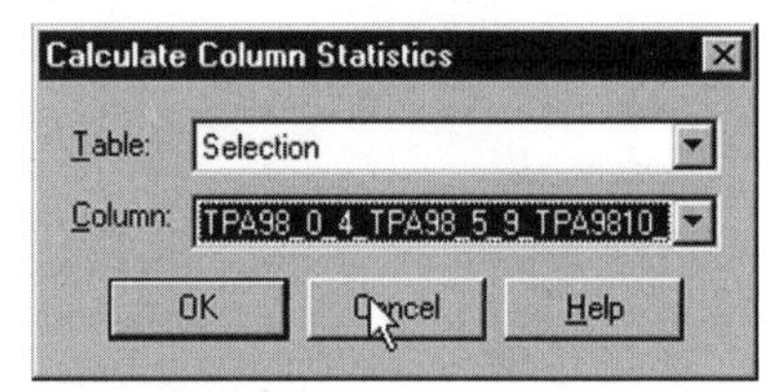

Calculate Column Statistics dialog box.

1. Select Query | Calculate Statistics from the menu bar. The resulting dialog box will prompt you to choose a table and a column. Select the Selection table as the basis for the statistics (shown in the illustration at right of the Calculate Column Statistics dialog box). If this option is not available, and the block groups are no longer highlighted (selected) on the map, use the Radial Select tool to again select the block groups within a 1-mile radius of the site.
2. Select the previously created column that contains the population under age 18.
3. After clicking on OK, the Statistics dialog box will appear.
4. You are interested in a sum of all persons under age 18 for each site. In this example, the site trade area contains 12,895 people in the target population.
5. Repeat the steps for radial queries and statistical analysis for each site to calculate the total target population in each potential site's trade area.
6. Assign an identification number to each site (e.g., 1, 2, 3).
7. From the Column Statistics dialog box, record the identification numbers and the target populations for each site on a piece of paper. The Column Statistics window is shown in the following illustration.

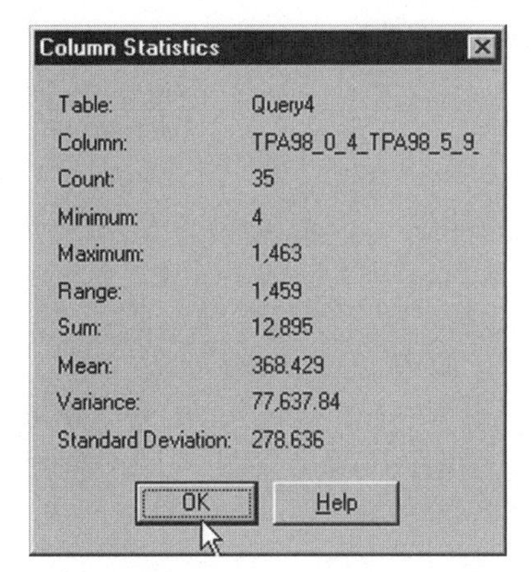

Column Statistics window.

Once you have recorded the appropriate statistics for all sites, return to the Layer Control dialog box (Map | Layer Control), and click on the Selectable option (under the arrow icon) for *sites*. At this point, you are ready to update the *sites* table. Tutorial 2-15, which follows, takes you through this process.

▼ *TUTORIAL 2-15: UPDATING A TABLE*

1 From the Main toolbar, select the Info Tool (*i* button). This tool accesses a window that contains information on a selected object. The Info Tool window can also be used for updating data.

2 Select a site by clicking on it. The Info Tool dialog box will appear. Other layers that are selectable will also be accessible via the Info Tool window. Select the *sites* layer by clicking on the word *sites* in the Info Tool window, shown in the illustration at right.

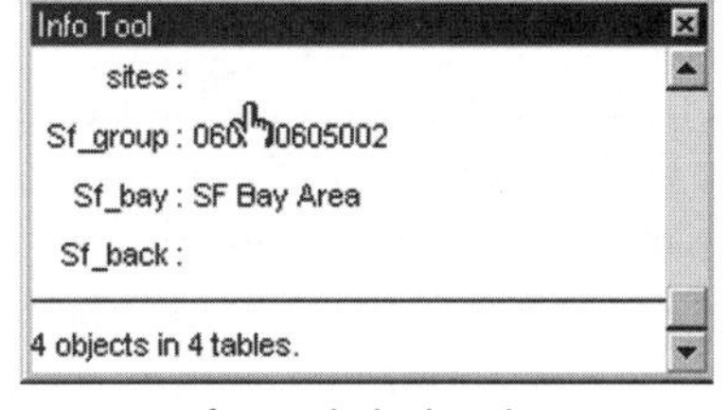

Info Tool dialog box.

3 In the blank information box that appears, key in the identification number you assigned to the site, as well as its corresponding trade area potential (sum of target population).

4 Repeat steps 2 and 3 for all sites until the trade area information has been updated for each site, as shown in the following illustration.

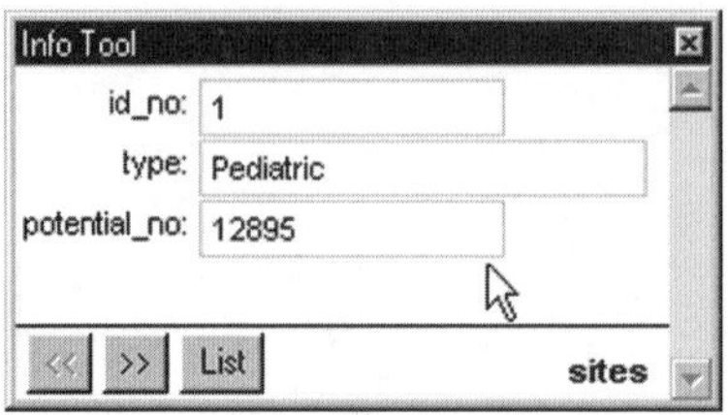

Info Tool dialog box showing identification number assigned to site and trade area potential.

You are now ready to prepare a hard copy of the map for presentation to county officials. Tutorial 2-16, which follows, takes you through this process.

▼ Tutorial 2-16: Preparing Hard Copy

1 Select Window | New Layout Window from the menu bar.

2 In the resulting dialog box, select "One Frame for Window" for the Map window, which should be the default option. The map and legend will appear in separate frames on the new Layout window. The New Layout Window dialog box is shown in the illustration at right.

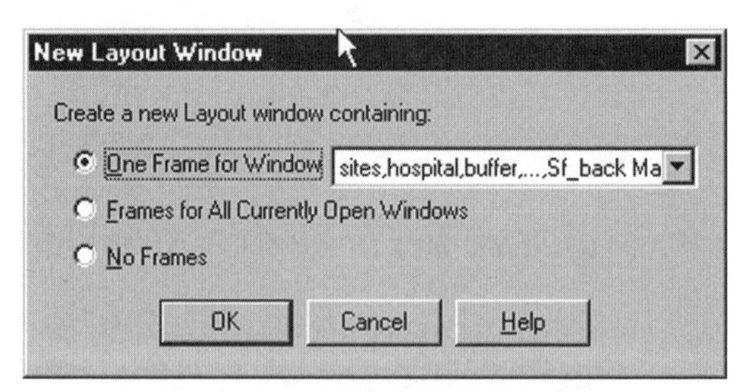

New Layout Window dialog box.

3 If you cannot see the entire layout, or the map appears in portrait mode in the Layout window, you need to make some adjustments. From the Layout option on the main menu, change the zoom using Change Zoom until you can better see the layout, or select View Entire Layout. The Layout Zoom dialog box is shown in the illustration at right.

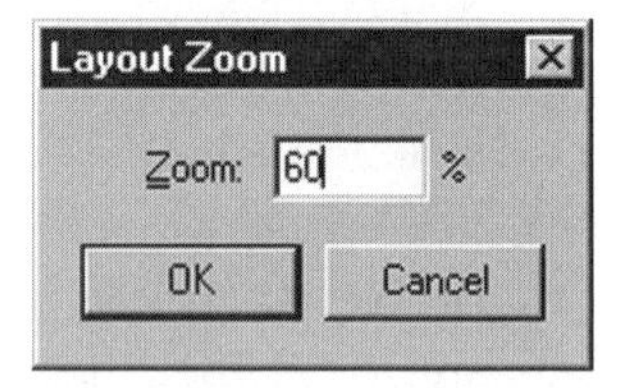

Layout Zoom dialog box.

4 To change the map to landscape mode, click on File | Page Setup from the main menu bar. Select Landscape to change the page orientation.

5 To resize the map in the Layout window, click on a corner of the map and drag it to fill the white background (which shows the printable area of the Layout window), leaving some room at the top for the output title.

6 Place the legend window in the bottom left corner of the map, as shown in the following illustration.

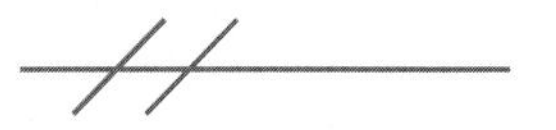

To add a title to the map, perform tutorial 2-17, which follows.

▼ TUTORIAL 2-17: ADDING A TITLE TO A MAP

1 Click on the Text Style tool (A? button) on the Drawing toolbar to set the format for the title. A font size of 24 points or larger is recommended.

2 Click on the Text tool (A button) on the Drawing toolbar, and then click on the Layout window where you wish the title to appear.

3 Key in text for the title. If you need to edit the text, double click on the title you have created and a dialog box will appear. You can make changes to the text or its appearance in this dialog box. The following illustration depicts the Layout window created for potential pediatric clinic sites.

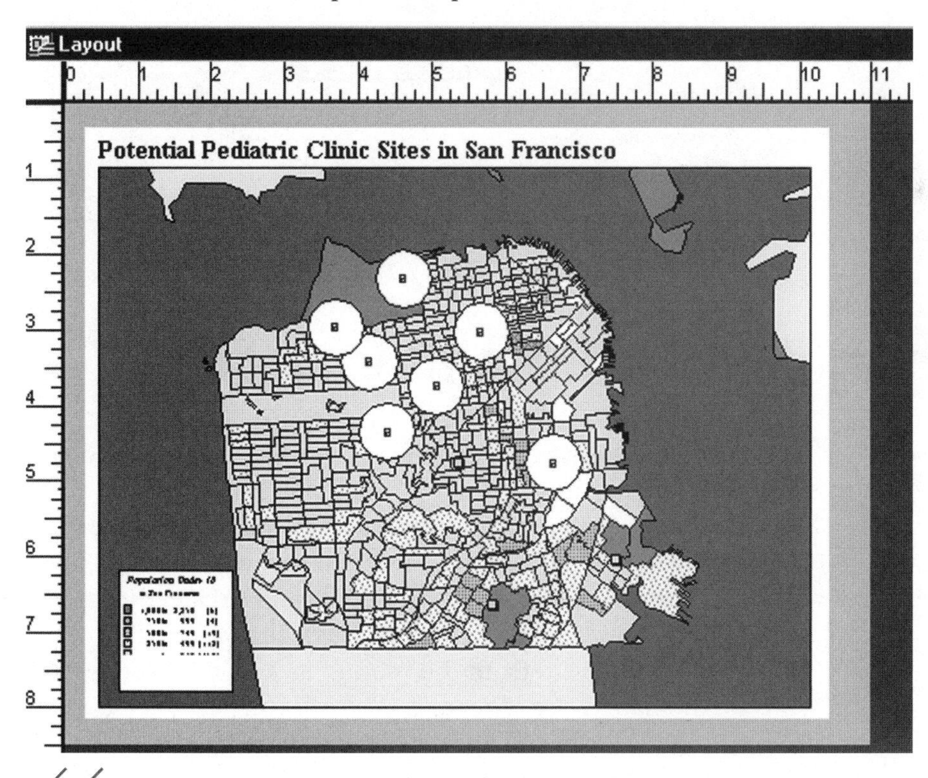

Map layout for hardcopy output of potential pediatric clinic sites.

In the event you wish to change the map at a later time for presentation purposes, it would be wise to save the current session as a workspace so that you can quickly return to the present Map and Layout windows. Opening a saved workspace will open all files that were open at the time the workspace was saved, as well as the thematic maps, labels, cosmetic objects, query results, and window settings. To save the workspace, select File | Save Workspace from the menu bar. The Save Workspace window is shown in the following illustration.

Save Workspace window.

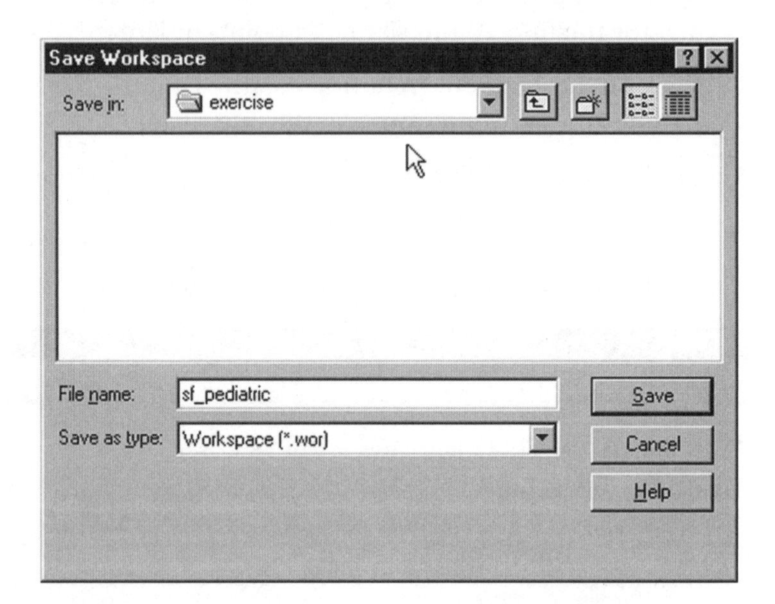

Identifying Potential Geriatric Clinic Sites

The location analysis for geriatric clinic sites can be conducted in the same manner as the analysis of pediatric clinic sites. The target population for geriatric clinics has already been defined as the 55-and-over age group, and the buffers have been created around existing hospitals. At this point, you would create the thematic map for a visual scan of potential geriatric clinic sites. Select Map | Create Thematic Map from the menu bar. Click on Ranges as the type of thematic map. Create an expression for the population over age 55 by adding all age categories 55 and above.

You are encouraged to follow the same steps used in the location analysis of potential pediatric clinic sites. You could execute this analysis now, or after you have studied subsequent chapters. The following illustration shows the final results of the analysis aimed at identifying four potential geriatric clinic sites.

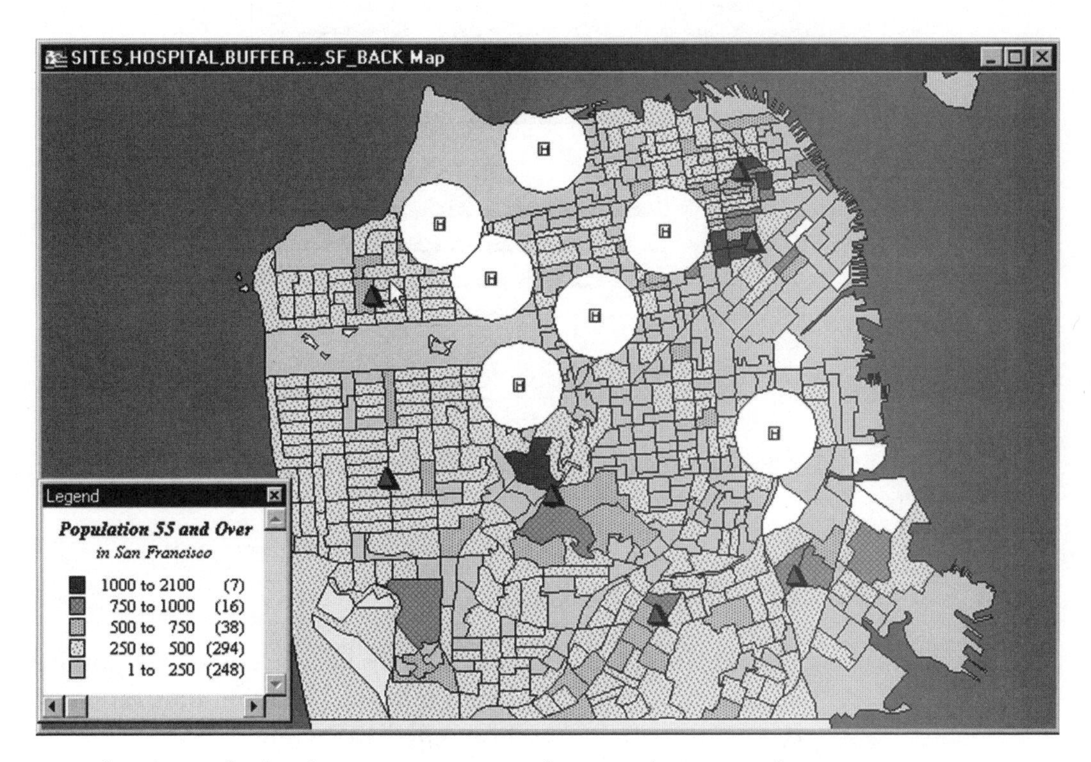

Completed map for hard-copy presentation of potential geriatric clinic sites.

Assume now that you need to make recommendations regarding the optimum sites for pediatric clinics in San Francisco. Recommendations will derive from selecting the sites with the largest target population(s) within respective trade areas. To create a table containing this information, perform tutorial 2-18, which follows.

▼ TUTORIAL 2-18: CREATING A QUERY RESULTS TABLE

1 Select Query | SQL Query from the main menu bar.

2 Build the query as shown in the following illustration of the SQL Select dialog box. First, in the "From Tables" section, select *sites*. Next, in the Order by Col-

umns section, select *potential_no*. As shown in the illustration, key in a space and *desc* (for descending) after *potential_no* to sort by the highest number of potential target customers to the lowest.

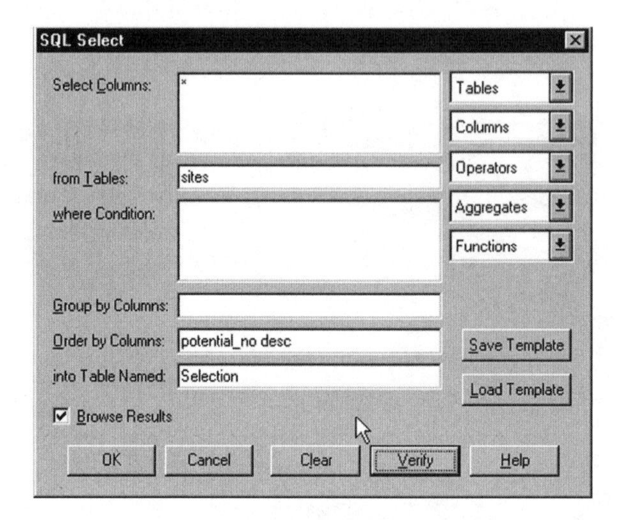

SQL Select dialog box showing query definition.

3 Verify that the Browse Results option is checked so that you can view the query results.

The previous query will list the sites by type and target population in descending order. As shown in the following illustration, site 1 is the optimum location for a pediatric clinic.

Browser showing site rankings by target populations.

Query6 Browser

id_no	type	potential_no
1	Pediatric	12,895
3	Pediatric	8,661
2	Pediatric	8,541

Select an optimum site by clicking on the box adjacent to the *id_no* column. The site should be highlighted on the map. Site 1 will be the only one selected, as shown in the following illustration.

Site 1 selected in the Browser window.

Query6 Browser

id_no	type	potential_no
1	Pediatric	12,895
3	Pediatric	8,661
2	Pediatric	8,541

In order to take a closer look at site 1, zoom in to the map. Access the Layer Control dialog box to return visibility to the *Streets* layer. This will facilitate examination of the site relative to the transportation network. As seen in the following illustration, site 3 is located near the San Francisco County border at the intersection of Mansell, Brazil, and Persia streets.

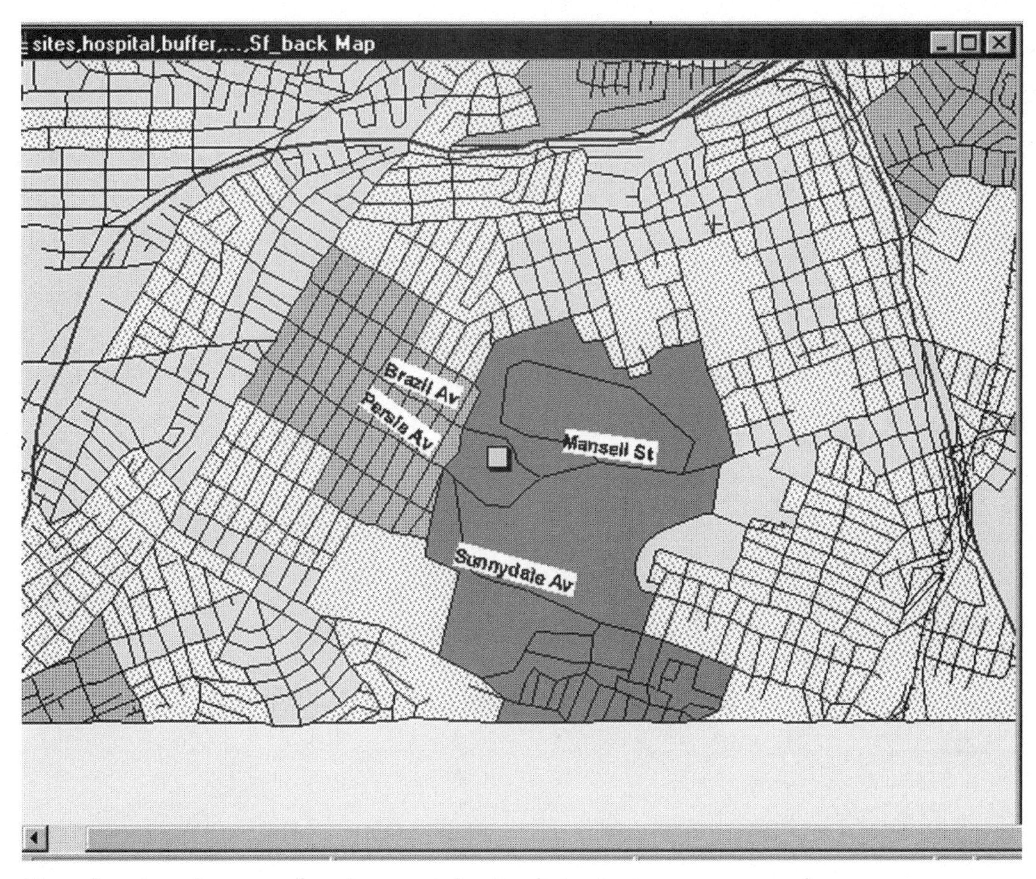

Map showing close-up of optimum pediatric clinic site in street network.

Summary

At this juncture in the series of tutorials, you can offer alternative sites for pediatric clinics to county planners. More importantly, you have just completed a typical GIS analysis using MapInfo Professional. If you struggled through the tutorials, do not worry. The rest of the book provides detailed information, tutorials, and exercises that will help answer all of your questions about MapInfo functions.

CHAPTER 3

MANAGING THE MAPINFO PROFESSIONAL ENVIRONMENT

MANAGING AND NAVIGATING WINDOWS are part of using any software in the Windows environment. File and table manipulation and maintenance are essential to database management. Data organization is also a key issue that can profoundly affect the usability of data in MapInfo Professional. MapInfo allows you to join, add, and delete data within tables as the need arises. Although many data vendors provide information in a format that insulates you from design considerations, it is often necessary to change file structures in MapInfo. This chapter covers a variety of topics related to getting started with MapInfo, managing windows and files, and special topics such as changing file structures.

Start-up and Preference Settings

If you have just installed MapInfo Professional, or have not turned off the Quick Start dialog box, the dialog will appear as shown in the following illustration when you start MapInfo. The first option, Restore Previous Session, will restore the files, themes, and windows that were open when you last exited MapInfo.

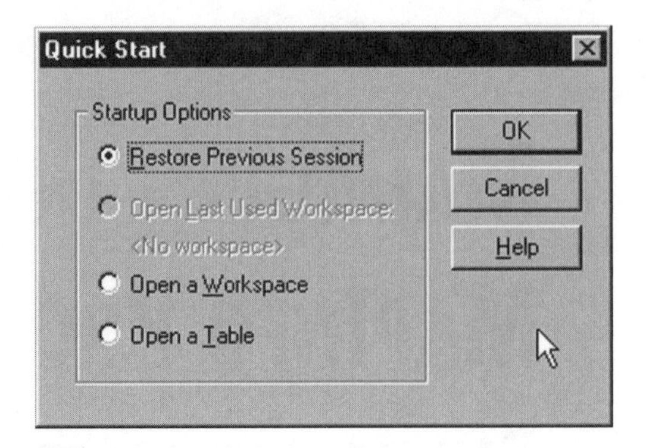

Quick Start dialog box.

The second option, Open Last Used Workspace, will open the last workspace you saved. The Open a Workspace option will allow you to select any previously saved workspace, and Open a Table will allow you to open a saved table. If you select Cancel at the bottom of the dialog box, you can perform any of these operations from the main menu.

When you initially open MapInfo Professional, it is a good idea to set preferences. Preference settings determine the way many MapInfo functions globally operate. For example, Preferences influence the way selected objects appear, which makes editing geographic objects easier. Other Preference settings affect the way MapInfo looks and operates at start-up. The functions affected are System, Map Window, Legend Window (new to MapInfo Professional 5.0), Start-up, Address Matching, and Directories. To access the Preferences options, select Option | Preferences from the main menu bar. The Preferences dialog box is shown at left in the following illustration.

The System Settings Preferences dialog box is shown at right in the following illustration. Although you can change any object within this window, the setup in the illustration is suggested. However, if you are printing to metric-sized paper, change the Paper and Layout units to centimeters or millimeters. You can also change the Paper and Layout settings to points or picas. Color Defaults should be set to Color if you are using a black-and-white monitor, but intend to print in color. You may wish to change the Number of Undo Objects. This is the number of commands or changes you can undo with the Edit | Undo command.

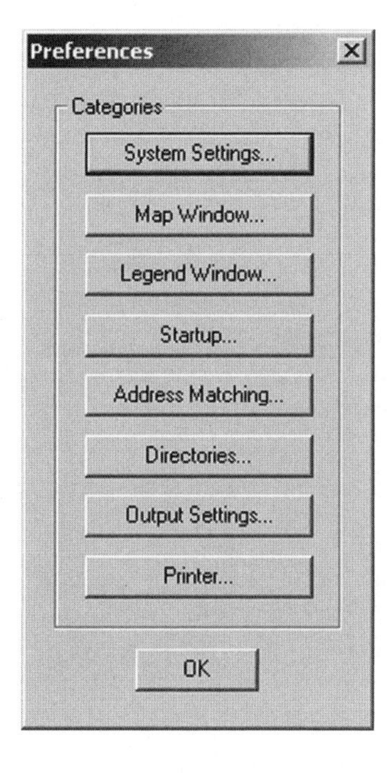

Preferences dialog box.

System Settings Preferences
Copy to Clipboard
Copy Text to Clipboard
Copy Bitmap to Clipboard
Copy Metafile to Clipboard
Color Defaults
Monitor Setting
Black and White
Color
Aspect Ratio Adjustment
Use Custom Screen Size:
Width: in Height: in
Paper and Layout Units: inches
Number of Undo Objects: 10
Pre-Version 4 Symbols
Display Using TrueType Font
Date Window for 2-digit Years
Turn date windowing off (use current century)
Set date window to:
Years entered as 00 - XX become 2000 - 20XX
Years entered as XX - 99 become 19XX - 1999
OK Cancel Help

System Settings Preferences dialog box.

The maximum number of undos allowed is 800. For optimal program speed and memory usage, setting a maximum of 10 undo objects is suggested. Date window settings of two-digit years allow you to use MapInfo's date window feature to convert two-digit years to four-digit years. This setting is a matter of user preference and should be set to suit your data.

Establishing date window settings is important if you perform date calculations on your data. If you choose to turn the date window off, MapInfo will assume the current century for all two-digit dates and turn off the date window function. If you choose "Set date window to:" MapInfo will prompt you to enter the maximum year for this century. For example, if you enter 30, MapInfo will assume all two-digit dates from 30 to 99 are 1930 to 1999 and those from 0 to 29 are 2000 to 2029.

As its name suggests, the Map Window Preferences dialog box (shown in the following illustration) changes features related to

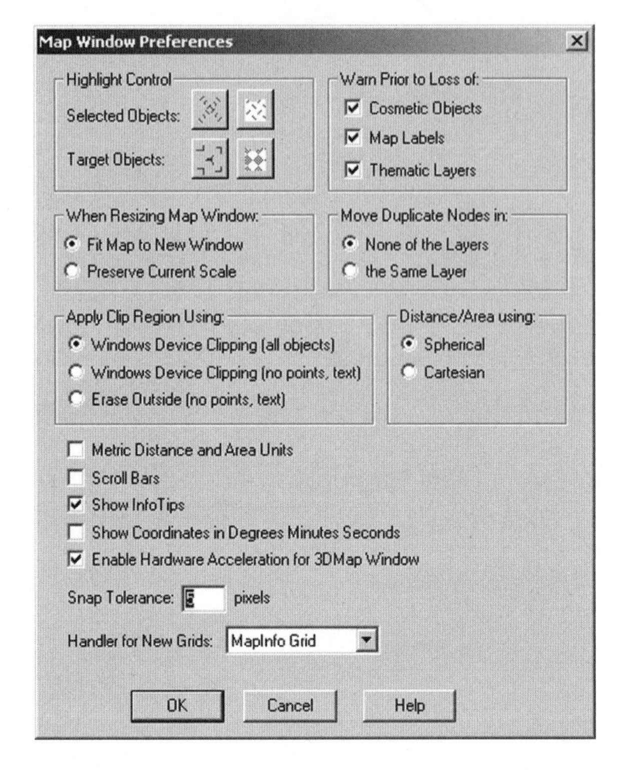

Map Window Preferences dialog box.

maps. One choice involves resizing a Map window. If you choose to fit a map to the size of the new window, you will retain the same map within the resized window. If you decide to preserve the current scale, you will see more of the map if you enlarge the window, and less of the map if you shrink the window.

After using MapInfo Professional for a while, you might want to experiment with changing the Highlight Control. Changing this option will make it easier for you to see selected objects and target objects.

If you are editing many polygons within the same layer, you may wish to change the "Move Duplicate Nodes in" to "the Same Layer." If you choose this option, moving a portion of a polygon will move any corresponding objects sharing the same node.

Snap Tolerance should be set at 5 or more pixels if you want objects that are close by to snap to a node when editing lines or polygons. If you do not want nearby objects to automatically snap to the node, select 1, 2, or 3 as the snap tolerance. Select Metric Distance and Area Units if you wish to see distance displayed in kilometers instead of miles.

An X in the Scroll Bars option will make scroll bars appear on the Map window. Scroll bars on the sides of the map will allow you to pan (move around the map) using the scroll bars. You can also use the Pan tool (hand symbol) on the Main toolbar for moving around the map.

The Show InfoTips feature was added in version 4.5. InfoTips appear on the map window when you move the mouse over map objects. These tips will show the information in the first data field in the tables associated with the map objects in the Map window. The check box toggles InfoTips on and off in the map window. InfoTips are useful for quickly determining differences in your data while on the map, such as store number, elevation, and so forth.

MapInfo also allows you the option of showing coordinates in degrees, minutes, and seconds, which is especially helpful when your data is in this format, such as that from some GPS units. The "Distance/Area using" option allows users to choose either spherical or Cartesian calculations. Spherical is MapInfo's default and is best suited to calculating distances when using earth-bound geography. Choosing Cartesian calculates distance and area of objects projected onto a flat plane and is best suited for non-earth projections. It cannot be used with latitude/longitude projections.

Choose the Apply Clip Region Using option that best suits your needs when using the Map Set Clip Region and the Map | Clip Region On/Off commands. The choices include "Windows Device Clipping (all objects)," "Windows Device Clipping (no points, text)," and "Erase Outside (no points, text)." You may want to experiment to determine the best option for your applications.

MapInfo Professional allows you to toggle warnings on and off. These warnings alert you when the loss of cosmetic objects, labels, or themes is imminent. It is recommended you use the warnings to prevent unintentional loss of work.

A general Legend window is a new feature of MapInfo Professional 5.0, which makes available a general cartographic legend for listing each feature on a map. The Legend Window Preferences dialog box, shown in the following illustration, allows you to set the appearance for the cartographic legend, and the Title Pattern changes the way a layer is referred to in the cartographic legend.

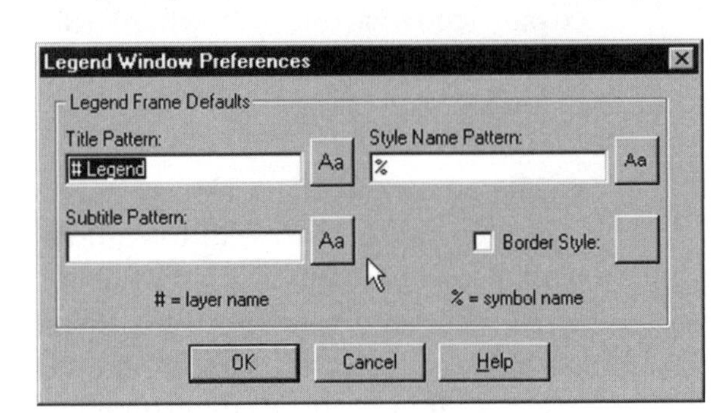

Legend Window Preferences dialog box.

The pound sign (#) is a placeholder for the layer name. The default layer reference in the Legend Window is # *Legend*, which would be *States Legend* for the *States* layer. You can change the default layer reference to *Legend of* # (e.g., *Legend of States*), or remove the word *Legend* entirely (e.g., *States*).

Entering the pound sign (#) or other text in the Title Pattern will give *every layer* a title. If you do not wish each layer to have a title, you can delete the text in the Title Pattern, as well as the pound sign. In the Create Legend dialog box, discussed in Chapter 5, you can override the default options when creating a cartographic legend for a map.

The Style Name Pattern option determines the way each layer is described. The default method (indicated by the percent sign, %) is to place the layer type next to the symbol, line, or region style depiction. For example, the default legend description for a point layer would be the global symbol representing the point on the map followed by the word *point.* If you do not wish to have the layer type referenced in the Legend Window, remove the percent sign from the Style Name Pattern. You can place the pound sign (layer name placeholder) in the Style Name Pattern if you wish the layer name to appear next to the global symbol, line, or region depiction of a layer.

For the Title Pattern, Subtitle Pattern, and Style Name Pattern, you can set the default text that will appear in the Legend window. Finally, MapInfo Professional allows you to set the border style that will appear around the legend, or to select no border. The border chosen will also affect the Legend window for any thematic map.

The Startup Preferences are the most pertinent to the subsequent discussion of MapInfo's file management capabilities. If you frequently return to the last analysis you were working on, select the Save MAPINFOW.WOR when Exiting MapInfo option in the Startup Preferences dialog box, shown in the following illustration. This option will save the MapInfo desktop when you exit. Using this option ensures that you will always have a backup copy of your last session. When you choose this option, you will be able to retrieve the last session by issuing the File | Open Workspace command and selecting *MAPINFOW.WOR*.

Startup Preferences dialog box.

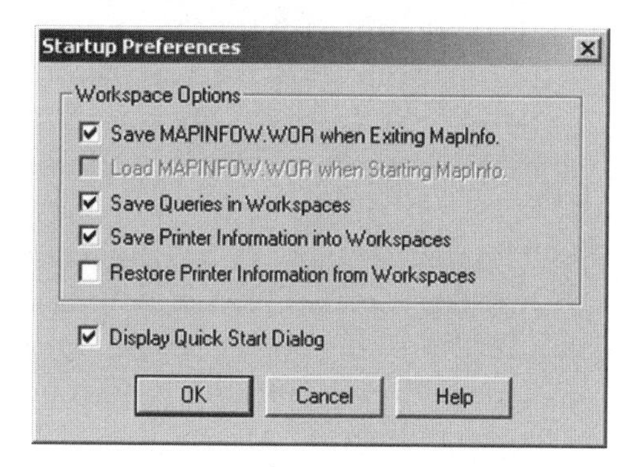

Note that you do not have to select the Load MAPINFOW.WOR start option if you select Save MAPINFOW.WOR. In fact, the Load option will slow you down, unless you typically want to revisit your last analysis.

The Save Queries in Workspaces option is a powerful feature that allows you to save temporary tables created by queries, as well as any map layers associated with queries, in your workspaces for use at a later time. Prior to version 4.5, MapInfo did not save these temporary tables, which required users to save additional tables if they wanted to return to the exact same analysis. This feature has some limitations. Some map layers based on queries are not saved, such as those created by a query of a query (subquery). For critical data or complex maps with layers based on queries, it is still best to save temporary tables as new tables to ensure the integrity of your workspace.

The Save Printer Information into Workspaces option saves all settings for each layout window, including those that override any Default Printer options (discussed later). The Restore Printer Options from Workspaces setting allows users to toggle whether to use the printer options set when saving a workspace or MapInfo's Default Printer options. These features are important because saved printer settings in MapInfo workspaces do not work with all printers. In some cases, even when these options are both selected, MapInfo's printer drivers are not able to implement those saved in a workspace, and in such cases the program typically reverts to Windows' default printer options.

If you do not wish for the Quick Start dialog to appear when opening MapInfo Professional, remove the check mark from the box beside the Display Quick Start Dialog option. All options in this dialog box can be duplicated using the File menu.

The next option on the Preferences dialog box is Address Matching. For most address matching operations, you will want to select the "Numbers before street name" option under House Numbers. The Address Matching Preferences dialog box is shown in the following illustration.

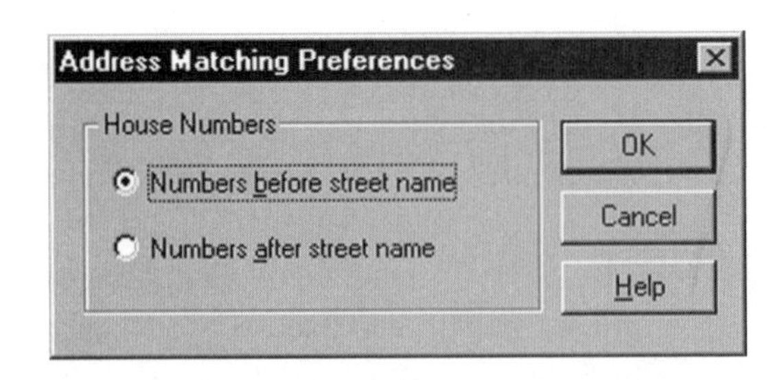

Address Matching Preferences dialog box.

The Directory Preferences dialog box, shown in the following illustration, allows users to specify default file paths. In the Initial Directories for File Dialogs field, you are allowed to set the default path for opening and saving Tables, Workspaces, MapBasic programs, Import Files, and ODBC QSL Queries, Theme Templates,

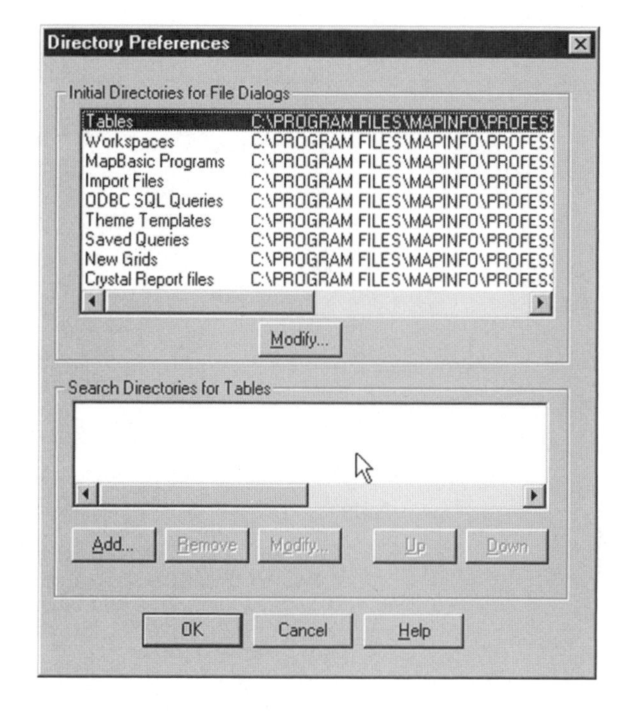

Directory Preferences dialog box.

and Saved Queries. The default path will be the MapInfo directory or one of its subdirectories. You should change this path if you frequently go to the same directories for the same file type. To change the path, highlight the appropriate file type (e.g., workspace) and select Modify to redirect the path.

The Search Directories for Tables field allows you to specify additional paths for opening tables, workspaces, and so forth. Select Modify to change an existing path, Add to create an additional search path, or Remove to delete a search path. If you use other directories to store data, MapBasic programs, workspaces, and so on, you may wish to specify the most commonly used directories as additional search paths. You can change the order by selecting the path and using the Up button to move a path higher in importance, or the Down button to make it search the path at a lower priority.

The Output Preferences dialog box, shown in the following illustration, allows users to have more control over their outputs, both electronic and hard copy. The first option concerns the display of raster images. If you use raster images, you may want to adjust the defaults, depending on your image. The choice of Output

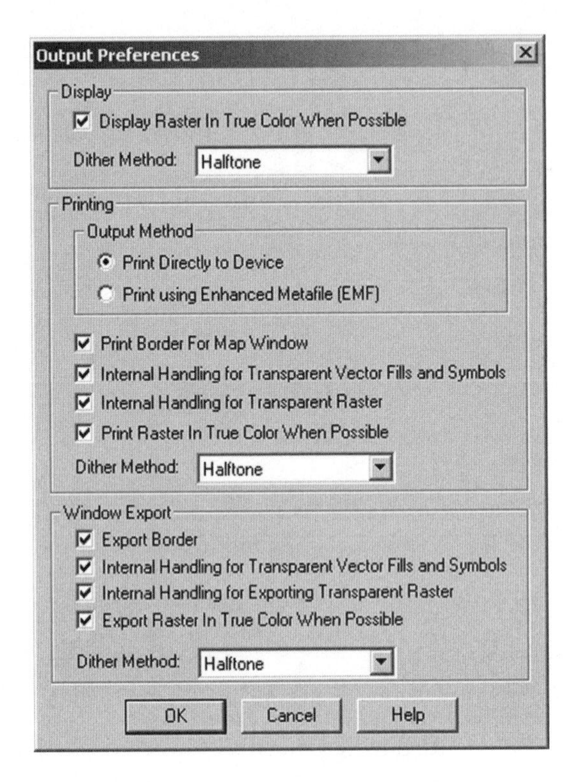

Output Preferences dialog box.

Method allows users to print directly to a printer (the default), using MapInfo's previous printing process, or to print to a specified printer by first creating an enhanced metafile (EMF) and sending that to the printer. Printing to an EMF may improve the quality of your printed output.

Toggling the Print Border For Map Window option allows users to turn on or off the border around a map window when printing. When checked, the options for Internal Handling of transparent vector fills, symbols, and raster images allow MapInfo to handle these items when printing. If unchecked, the print device handles these items, which could lead to unexpected results in your output. The Print Raster In True Colors When Possible option allows MapInfo to output raster images in true color. The next set of options concerns exporting windows. These options operate the same in exporting windows as the print options previously discussed.

The final option in the Preferences dialog box, Printer, is a new feature that allows users to choose either the Windows default printer or to specify a default printer for MapInfo to use, as well as a default page orientation. However, MapInfo's default printer settings do not work with all printers. As discussed previously, in some cases MapInfo's printer drivers are not able to implement some or all of the options defined in this dialog box. In such cases, MapInfo typically reverts to Windows' default printer options. The Printer Preferences dialog box is shown in the following illustration.

Printer Preferences dialog box.

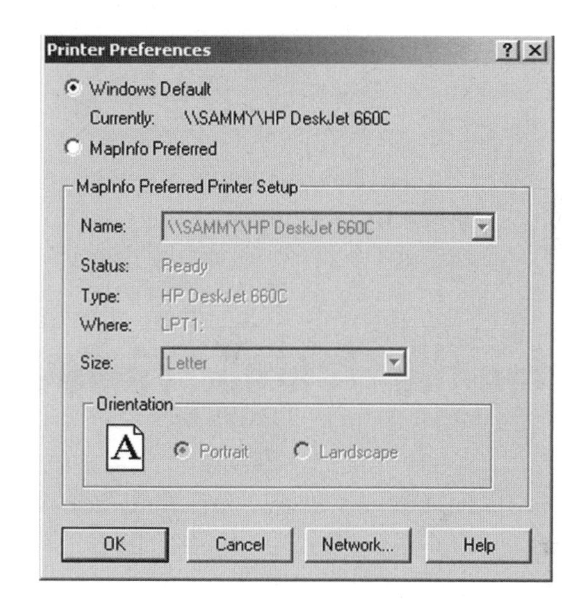

The MapWizard

New with the release of MapInfo Professional 6.0 is a MapWizard designed to help users more easily map data. To access the MapWizard, select Tools | Mapping Wizard Tool | Run Mapping Wizard Tool. The Wizard will take you step by step through the process of creating a map. It is not, however, as easy to use as wizards included in some other Windows packages. You must understand the basic concept of mapping and how MapInfo works. The first step involves opening the data for your analysis. You must know the file type of the data and its location. In step 2, the wizard prompts you to open map files relating to your data, and then creates a map based on these files.

Step 3, Analyze, allows you to perform four different actions: Create Points (geocode), Create Thematic Map, Modify Thematic Map, and Create Graph. When selecting an action, the window to the right of that will allow users to select options relating to the action, such as selecting a table to geocode. Finally, step 4 allows users to output their data. Action choices under Output include Create Layout Window, Print Window, Save Window to File, and Generate Report. Again, the window on the right changes with each action selected and lists the various options allowed for the selected action.

New users may find the MapWizard helpful, once they understand basic terminology. More experienced users may find the wizard does not improve their performance at all or actually slows them down. In addition, much of the powerful capabilities of MapInfo (such as creating spatial queries to analyze disparate data) are not available in the MapWizard.

Managing the Overall Windows Environment

MapInfo Professional looks and feels like most Windows software packages. Included are window options such as tile, cascade, and minimize/maximize. In addition, you can open multiple windows.

Multiple Windows

As previously mentioned, multiple windows can be opened simultaneously. These windows can be of the same type, such as several Map or Browser windows, and can even be the same window, such as two Browsers of the same table or two Map windows containing the same tables. To open additional Map, Browser, Graph, Layout, or Redistrict windows, select Window from the main menu bar, and then select the additional type of window you wish to open.

Multiple Map windows are useful when you wish to see a large-scale view and a zoomed-in view of the same general area at the same time. They are also useful when you wish to compare two maps of the same general area: one with selected data shaded, and the other without shading. In addition, multiple Map windows allow you to incorporate several maps into a single Layout window, and to print them on the same page. Because Map and Browser windows are linked, it is often easier to edit data by having both open. For example, data selected in a Map window are also selected in a Browser window, and vice versa. This ability allows you to see the data you select on a map, or to select a map object and easily find its associated data in the Browser window.

MapInfo Professional allows users to clone a Map window. To clone a Map window, select Map | Clone View from the main menu bar. All tables, themes, labels, and drawings of the map you are cloning will be copied into a new Map window. A more in-depth discussion of cloning windows is found in Chapter 5.

Map and Layout windows are also closely linked. In a Layout window, you can print multiple windows on a single page, as well as add titles, legends, graphics, and more to your output. The size and shape of the Map window is directly related to the size and shape of the map depicted in the Layout window. When you zoom in on part of the map in the Map window, the map will also change in the Layout window.

However, maximizing or minimizing the Map window does not affect the map in the Layout window. A Map window that has been maximized prior to creating a Layout window will be placed on the layout scaled to the size of the original map window. You can change the active window by clicking anywhere on the desired window, or by accessing the Window menu option and then choosing the appropriate window from the list of open windows at the bottom of the menu option.

MapInfo allows you to open numerous files concurrently. This feature can result in many open windows because each table has an associated Map and/or Browser window. For this reason, it is suggested that you close the files you no longer need, using the File | Close Table command. In addition, when you have completed an analysis, save the *workspace* (discussed later in the chapter), and then select File | Close All to clean up the MapInfo Professional application and desktop.

TIP: *MapInfo Professional often fails to release Windows resources upon completion of tasks, and maintains copies of all open tables in memory. Therefore, the longer a MapInfo session, the more Windows resources are held and unavailable to applications. To release Windows resources and avoid a lock-up, restarting MapInfo periodically and rebooting your machine after an intensive session are recommended. Given the processor speed and memory of most of today's desktop computers, this is becoming less of a consideration.*

Menu Bar Dependence on Active Window

The MapInfo Professional menu bar will change with the "active window." If a Layout window is the active window, a Layout menu option will appear on the bar, and similarly for Map, Browser, and Graph windows. MapInfo allows you to access the Table and Query menu options, regardless of which window is active. To change the menu bar, you need to select the window you wish to alter, making it the active window. The next section covers navigation through multiple windows.

TIP: *A shortcut to accessing the menu specific to the active window is to click the right mouse button. For example, if a Map window is the active window, a right button click will activate a pop-up window of the Map menu. For those accustomed to using the right mouse click to stop MapInfo from drawing, this action can now be accomplished with the <Esc> key.*

Basic Windows Functions

Like most Windows-based programs, MapInfo Professional allows you to manipulate open windows. You can maximize and minimize each window, as well as cascade and tile all open windows. The following illustration shows a disorganized desktop.

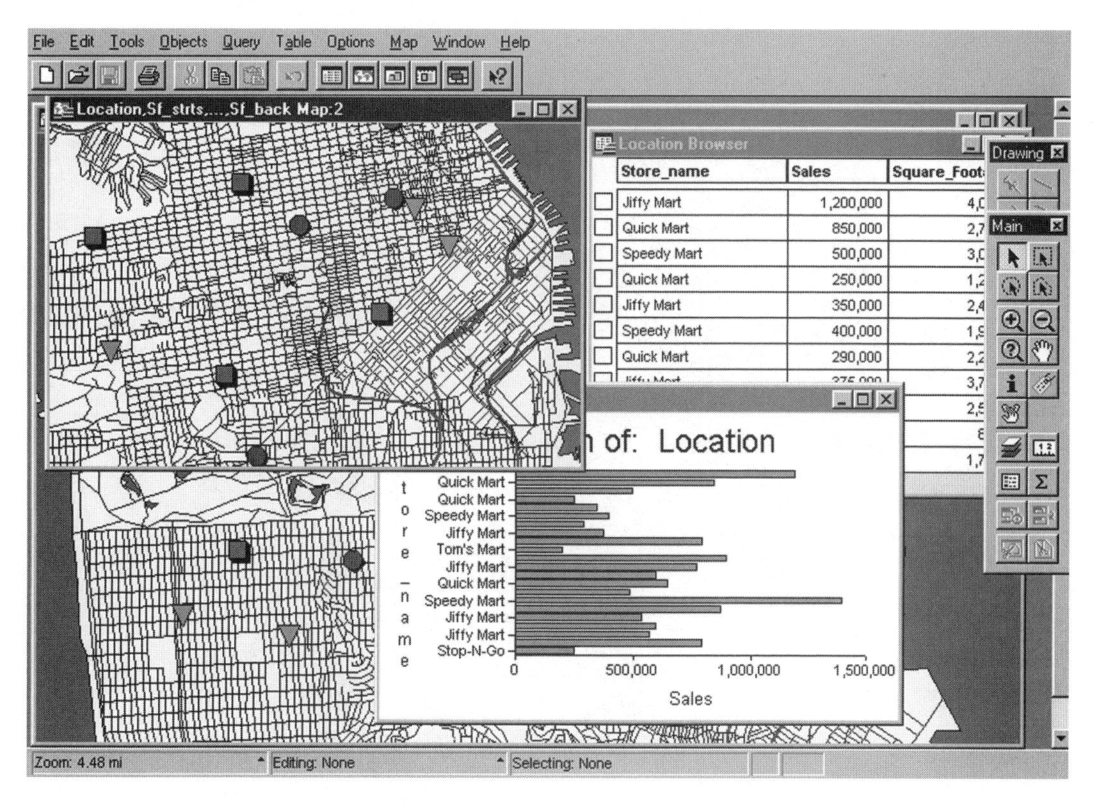

A messy desktop.

Tile Windows

A Browser, two Map windows, a Graph window, and a Legend window appear on the desktop depicted in the previous illustration. You could move, size, and arrange these windows however you wish. To size a window, grab a corner or side with the mouse and drag it to the desired size. To move a window, click on the window and drag it to the location desired. In this way, you can (however laboriously) organize your desktop.

An easier method is to select Window | Tile Windows from the menu bar. Tile Windows will automatically arrange and size all windows so that they can be viewed simultaneously. The following illustration shows the results of using the tile option to arrange the example desktop.

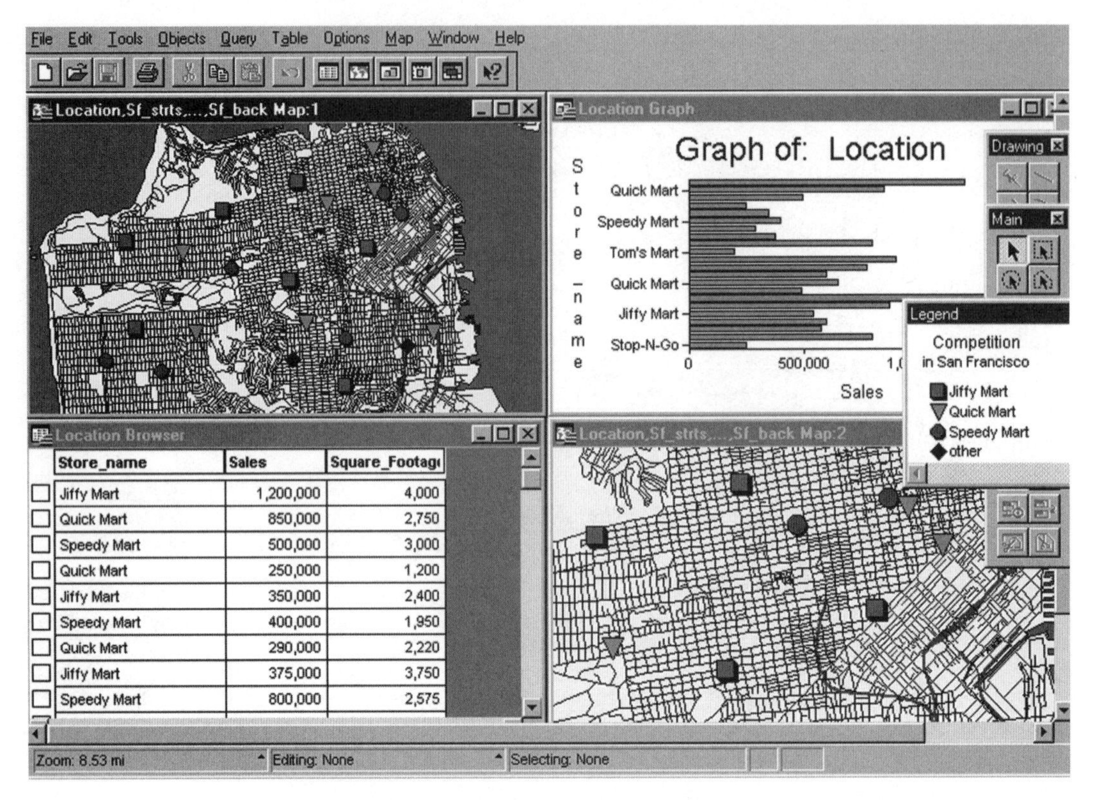

Desktop after using Tile Windows.

As shown in the previous illustration, the Tile Windows command does not affect the Legend window. The Legend window changes with the active map or graph, and thus is not affected by the Tile Windows command. Tile Windows is also a good way to set up the desktop for a Layout window. The layout appearing in the following illustration was created by selecting Window | New Layout Window, followed by the Frames for All Currently Open Windows option. Once the layout has been created, you can move and size the frames as desired for your printed output.

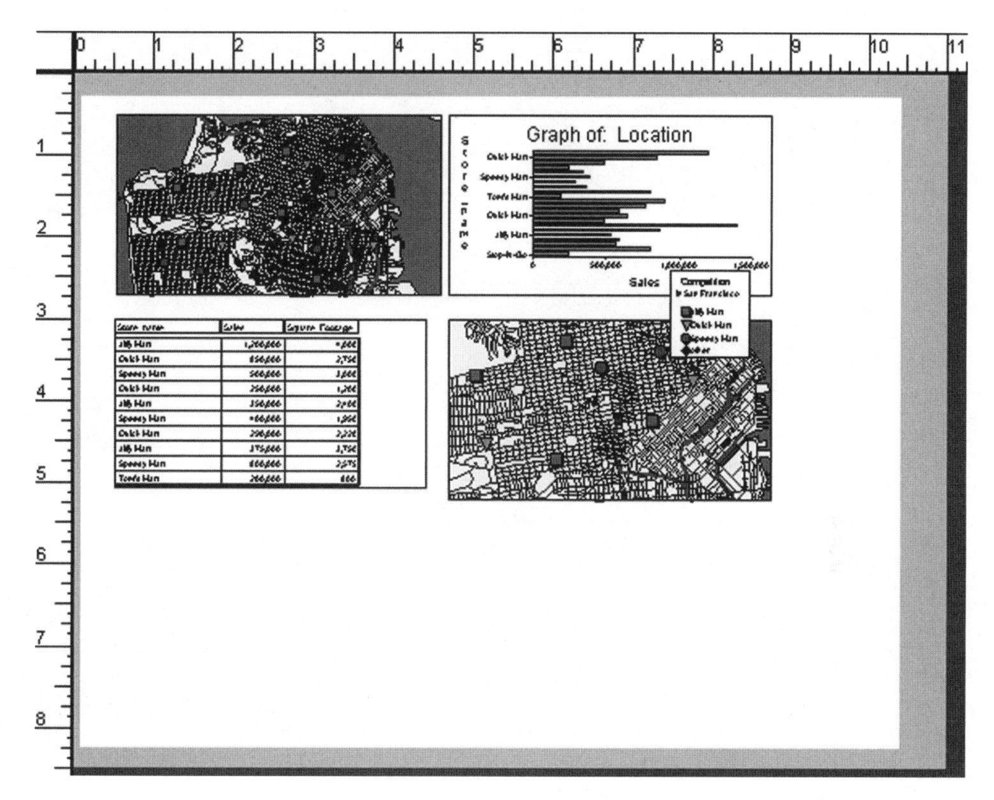

Layout window created with Frames for All Currently Open Windows option.

Cascade Windows

Cascade is another option for organizing the MapInfo Professional desktop. The following illustration shows the same messy desktop after selecting Window | Cascade Windows. Cascade will diagonally line up all open Map, Graph, Browser, and Layout windows on the desktop. The active window will be visible; all others will be lined up behind the active window, with only the description bar showing. In this mode, it is easy to switch between windows by clicking on the description bar of the window you wish to view.

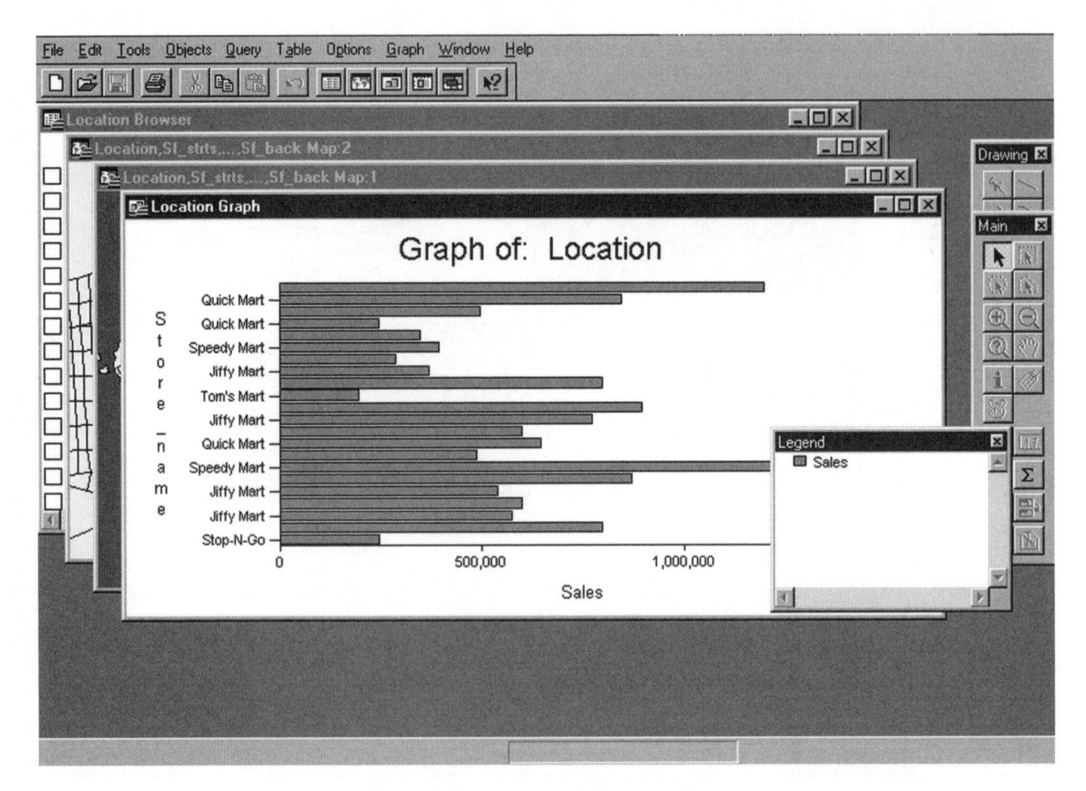

Desktop after selecting Cascade Windows.

Maximize and Minimize

MapInfo Professional also allows you to maximize, minimize, or close windows by using the buttons in the upper right corner of nearly every window. The rectangle button in the center maximizes the active window, making it expand to fill up the entire MapInfo desktop. The minus button minimizes the active window, transforming the window into an iconized title bar at the bottom of the screen.

To reopen the window, double click on the appropriate title bar, or select the maximize or restore window button on the title bar. To close a window, select the button containing the X. However, if you close a window accidentally, you will have to recreate the window. In the following illustration, all open windows on the example desktop are minimized.

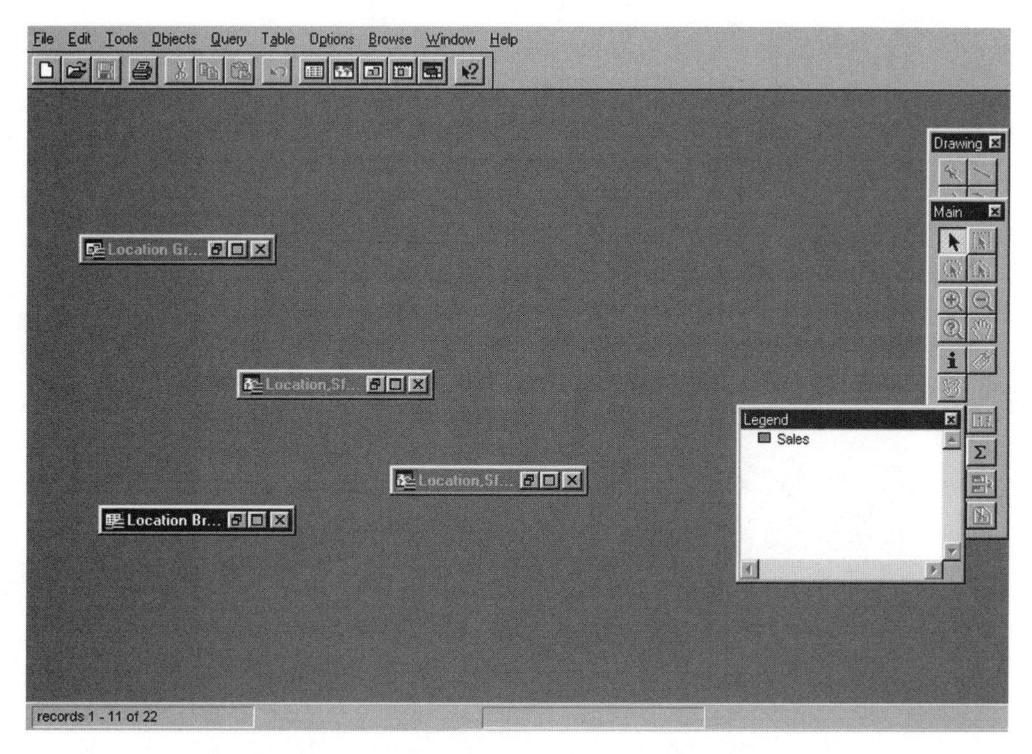

Minimized windows.

To organize all of the minimized windows, select Window | Arrange Icons; all minimized windows will be arranged at the bottom of the MapInfo Professional desktop, as shown in the following illustration. Minimizing allows you to maintain a clean desktop, while maintaining all of your queries, browsers, graphs, and maps so that you do not have to recreate those you may need later.

The only windows lacking maximize and minimize options are the button pads and the Theme Legend window. The only view options for these windows are open or closed. If you have closed these windows and wish to reopen them, select Options | Toolbars, and Option | Show Theme Legend, respectively. Note that a Theme Legend no longer appears automatically in its own window, but is instead included in the general Map Legend. If you wish to see the Theme Legend in its own window, select Option | Show Theme Legend from the main menu.

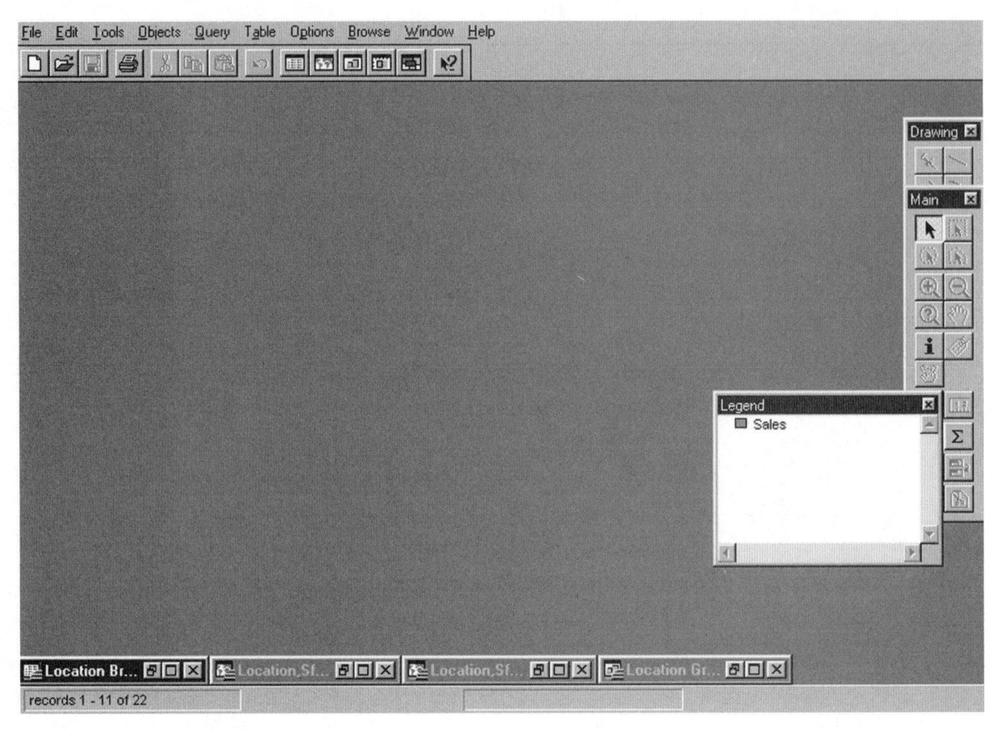

Minimized windows after selecting Window | Arrange Icons.

Managing Files

The outward appearance of MapInfo Professional data is simply files residing in directories. File manipulation is the process by which you choose the files to work with and how they will appear in your application. File manipulation in MapInfo is consistent with most common Windows-based applications. For example, to open a file, select File | Open Table from the MapInfo main menu bar, or select the open folder icon from the Standard toolbar.

The default location of the Standard toolbar is "docked" under the main menu, shown in the following illustration. Like other toolbars in MapInfo, the Standard toolbar can be docked under the main menu or "floated." The Main, ODBC, and Drawing toolbars float by default. MapInfo also refers to toolbars as "button pads." To dock a toolbar, click and drag it to the top of the screen, and it will appear under the main menu. To float a docked toolbar, click on part of the toolbar (not on a tool button) and drag it to the MapInfo desktop. You can change the size and shape of a floating toolbar by clicking on the edge of the toolbar; when the sizing arrows appear, drag the toolbar to the desired size.

NOTE: *"Tool tips" have been incorporated in MapInfo Professional. Tool tips give the name of the tool or text describing the tool's function. Tool tips appear automatically when you position the cursor over a tool button. (Do not click, or the tool will be activated.)*

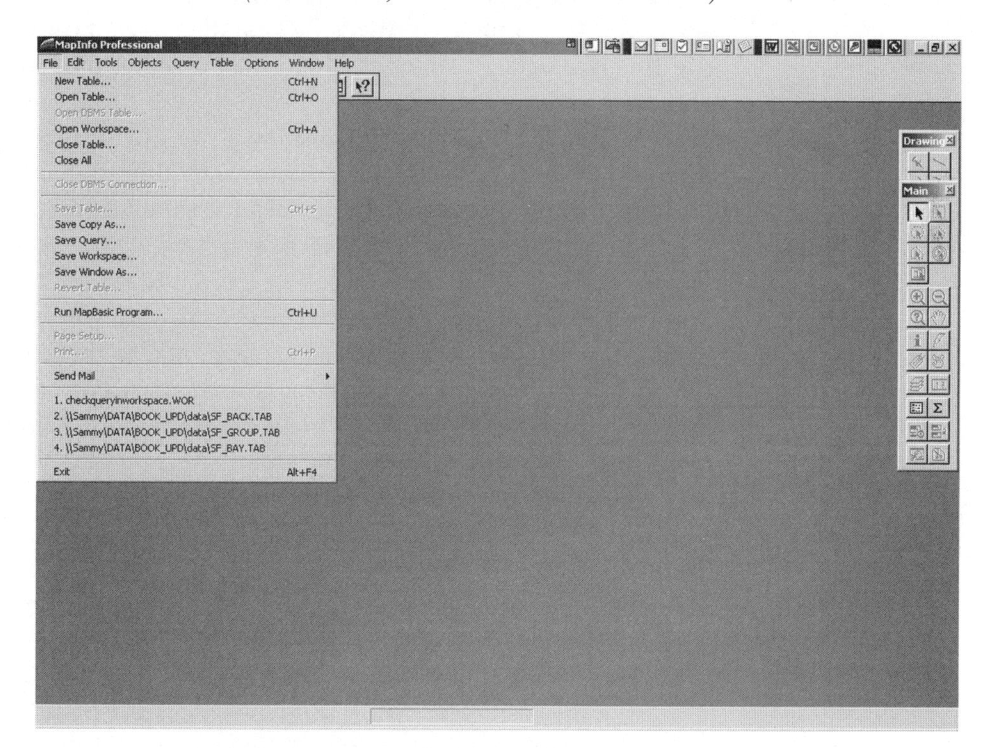

MapInfo Professional main menu.

Opening a file in MapInfo is referred to as opening a table. The following are file structure types you can open directly within MapInfo Professional.

- MapInfo **.tab*
- Dbase **.dbf*
- Delimited ASCII **.txt*
- Lotus 123 **.wk1*, **.wks*, **.wk3*, and **.wk4*

- Microsoft Excel **.xls*
- Raster **.bil, *.tif, *.grc, *.bmp, *.gif, *.tga, *.jpg,* and **.pcx*
- Grid Image **.mig* and **.grd*
- Microsoft Access Database **.mdb*

Whereas most of these file types contain data (or images, in the case of raster files), the MapInfo file type (*.tab*) contains information about associated data (*.dat*), map (*.map*), index (*.ind*), and id (*.id*) files. Opening a *.tab* file will open all associated files. Each of the file types has a specific function. For example, the *.dat* file contains data accessible by a Browser, the *.map* file contains graphic information and will display a map of the data, the *.ind* file tracks the indexes associated with the table to make querying the data faster, and the *.id* file tracks information associated with the map.

MapInfo Professional will also create a *.tab* file when opening a different file type (such as an *.xls* or *.dbf* file). In this case, the *.tab* file will keep track of the type of information contained in the associated data file. The next time you wish to access this information, you will be able to directly open the *.tab* file. MapInfo will locate the data associated with the *.tab* file.

MapInfo Professional's ability to directly access and use other file formats saves you time because you do not have to import files. This ability also saves disk space because you do not need two versions (i.e., file formats) of the same file saved on your computer. The only problem with using file types such as ASCII, Excel, and Lotus 1-2-3 files is that the data and structure of these files cannot be modified within MapInfo.

Because MapInfo uses spreadsheet file formats, you can specify a range of data to use when opening the file, rather than the entire spreadsheet. This feature is handy when you have titles, totals, graphs, and so forth in your worksheet rather than the data alone. Tables based on MapInfo, Dbase, and Microsoft Access files can be modified within MapInfo.

If you attempt to open a table that is already open in MapInfo Professional, the program will open another copy of the table. When you attempt to open a file in use by another program, MapInfo will display an error message and the file will not be opened.

NOTE: *When you close one copy of a table duplicated on the display, both copies are closed.*

As noted previously, the MapInfo Professional main menu changes depending on the active window. All menu options allow you to manipulate the active window in some manner. Note, however, that the Table and Query menus are always accessible. Consequently, tabular data can be modified or queried at any time.

When performing an analysis using MapInfo Professional, you will typically have many different files open at the same time. MapInfo keeps track of these tables in separate layers. The layering capability allows information to be integrated and complex analysis to be performed. In essence, the layering capability allows MapInfo to tell a story, either pictorially or through data integration.

At a minimum, an average user will have street, boundary, point (e.g., customer), and data files (e.g., demographics) open at the same time. In Chapter 2, block groups, potential sites, hospitals, demographics, and streets were used, as well as a land and water backdrop to give the map presentation more color.

When opening mappable tables (i.e., tables with an associated *.map* file), MapInfo will place the tables onto the same map as long as they are approximately in the same geographic area. If they are not in the same geographic area, MapInfo will open a separate Map window for the newly opened file, provided the Preferred View is set to automatic. (The Preferred View options are located at the bottom right corner of the Open Table dialog box.) Other Preferred View options, as summarized in the following list, will produce different results.

- *Automatic.* Opens table and determines the appropriate display (current mapper, new mapper, or browser).

- *Browser.* Opens the file as a browser. (If the file contains graphical data, you can manually add the data to the map later.)
- *Current Mapper.* Adds the table to the current Map window (if it contains graphical data), regardless of whether the file is in close proximity to the existing map.
- *New Mapper.* Opens the table in a new Map window (if it contains graphical data).
- *No View.* Opens the table without displaying the data.

You can change the appearance of a map by accessing the Layer Control dialog box through the Map menu. Chapter 5 focuses on mapping data and on the Layer Control dialog box. The Layer Control function is mentioned briefly here, however, because it concerns file manipulation.

You can hide or show layers on a map by toggling the check box under the visibility (eye) icon. If the box contains a check mark, the layer will be visible on the map. You can add files to a current map by selecting Add at the bottom of the Layer Control dialog box, and then selecting the file to be added.

Data: The Fuel for the Analysis Engine

Data are the fuel necessary to run any analysis engine, and MapInfo Professional is no exception. This means that an analysis using MapInfo is only as reliable as the information you use.

Internal Data

Internal data refer to proprietary data belonging to an individual or an organization. In the past, internal data were typically stored on paper or in someone's mind. Such data were not easily accessible or transferable throughout an organization. Businesses and public sector organizations are always seeking ways of developing competitive advantages or improving efficiency. With the advent of point-of-sale (POS) systems and in-depth marketing research to address competitive issues, more and more internal data are being computerized. Examples of internal data include customer information, store location information, and unique geographies, such as trade areas, sales territories, or bird migration patterns.

Computerizing internal data is typically a laborious process. Getting information such as trade areas or other unique geographies into computerized form can be an even more tedious chore. Options to help you computerize information include university departments, temporary agencies, and scanners. Temporary employees can be hired to input basic and repetitive information, such as survey results.

Scanners can also be employed for this task if survey forms are set up to facilitate this process. University students can help with digitizing information (transforming hardcopy maps into graphics files) or creating unique geographies (through aggregating or disaggregating available information) with the use of GIS software. With MapInfo Professional's raster imaging capabilities, scanners also allow photographs and floor plans to be incorporated into your electronic data.

External Data

External data refer to publicly or commercially available information. Much commercially available data are already in a computerized format that can be imported or used directly by MapInfo. Example commercial data sets are boundaries such as census tracts, block groups, states, and zip codes. Boundary data sets also include corresponding information relating to boundaries, such as census data, lifestyle market segmentation information, psychographics, market research data, business mailing lists, and electronic street files.

Commercial data providers are numerous. The two largest data providers in the industry at present are Claritas and Equifax/National Decision Systems (NDS). Some providers offer specialized data sets, such as RL Polk (automotive registration information), MRI and Simmons (marketing research and expenditure estimates), ABI (business/industry competition lists), American Digital Cartography Inc. (large-area maps and tabular data), and

Geographic Data Technology (GDT; street files and traffic counts).

MapInfo has strategic alliances with Oracle, Informix, and Microsoft, as well as many partners who resell MapInfo and related data. Many MapInfo partners have developed custom, add-on applications for MapInfo. Although many providers may offer data in MapInfo format, data purchased from MapInfo and MapInfo partners (i.e., resellers) ensure that purchased data sets are optimized for MapInfo and meet MapInfo Corporation's data quality standards.

Relative Versus Absolute Accuracy

An attribute of graphic data that should be considered in creating internal data or purchasing external data is the level of accuracy. "Relative accuracy" refers to placing objects (such as customer residences, places of business, boundary lines, and streets) at the approximate location. "Absolute accuracy" means that an object is placed in a precise geographical position.

Address matching is an example of relative accuracy because locations are approximated according to address. From the address ranges of a street network, you know that 201 E. Main lies on the southeast corner of Main and Second Avenue. Address matching in GIS software assigns the coordinates of the intersection to the customer residing at that address.

In the same example, absolute accuracy would require knowing that the customer's home is 50 feet from the centerline of Main, and 40 feet from the centerline of Second Avenue. Coordinates defined in this manner provide absolute accuracy in siting the location. Tax parcels maintained by a city or county government usually require absolute accuracy to ensure that the proper tax is levied against a property owner. Locating customers for a customer profiling or site selection analysis would most likely require relative accuracy.

Purchasing Considerations

The type of data you require will depend on the type of analysis you wish to perform. Budget constraints will also play a part in data acquisition. For example, computerizing all of your data currently stored in hardcopy format may not be cost effective. Digitizing data (transforming hardcopy maps into digital graphics files) is time intensive, and inputting infrequently accessed data is not cost effective.

Data providers offer vastly different pricing structures for the same data. Shop various vendors and ask for price quotes. Do not be afraid to haggle; at least one of the vendors may be willing to sell below list price. Cost will also be affected by the number of users requiring access to the data. Typically, the more users with access, the higher the cost. Network licenses for data can also be acquired.

The type of hardware and software you are running can also affect the price and availability of data. MapInfo data providers are numerous, whereas another GIS program that is not as widely distributed will not have nearly as many data options. Data may not be as readily available for DOS- and UNIX-based programs as for Windows-based programs.

The final cost issue can be summarized as relative versus absolute accuracy. Relatively accurate data are typically less expensive than absolutely accurate data.

Workspaces

Workspaces in MapInfo Professional are much like your physical desktop. If you took a snapshot of your desktop at any point in time, you could recreate its appearance very easily by placing the pencils, papers, and books in the same position as they appear in the photograph. Workspaces operate on this principle.

Saving a workspace with the File | Save Workspace command will allow you to recreate the MapInfo desktop as it appeared when you saved it. Workspaces track information on which files are open, the themes applied to each map layer (i.e., thematic maps that have been created), the layer settings applied to each layer (such as visibility or customized appearance), labels created, and the windows (e.g., maps, layouts, and browsers) that were open.

In addition, workspaces can capture queries as well as browsers, map layers, and graphs based on queries if the Options | Preferences | Startup | Save Queries in Workspaces option is selected. However, MapInfo's ability to capture temporary tables and map layers created from queries is not guaranteed, especially for queries based on other queries. For this reason, when creating complicated or time-critical maps, it is best if you save temporary tables as new tables using the File | Save Copy As command, and then create maps and themes based on the newly saved tables.

Workspaces can also retain print settings if Options | Preferences | Startup | Save Printer Information into Workspaces is checked. You must check Restore Printer Information from Workspaces as well if you wish to automatically recall print settings saved in workspaces. MapInfo's saved print settings, however, do not always restore correctly when opening a workspace, because of compatibility issues between MapInfo and some printer drivers.

In addition, workspace text can actually be used as functioning code in MapBasic applications. This allows you to configure the display the way you want and to use the underlying code (saved in the *.wor* file) in MapBasic programs. (See Chapter 12 for more information on this feature.)

MapBasic applications, functioning code in MapInfo Professional tracks labels and objects drawn on the cosmetic layer in workspaces. Opening a workspace will open all associated files and windows, recreate any drawings on the cosmetic layer, and apply theme, layer, and label settings.

TIP: *An additional benefit of workspaces is that they can be used for presentation purposes. In MapInfo Professional, you can save a map or a layout as a bitmap file, but bitmap files per se do not lend themselves to presentations. Instead, you can open a series of saved workspaces to effec-*

tively present your ideas. Workspaces also allow you to answer questions as they are asked by demonstrating your answer on the screen.

Modifying File Structures

It is relatively easy to change the structure of an existing table (as long as the data file is *not* an Excel, Lotus, or ASCII file) or create a new table. If you wish to change the structure of an ASCII, Excel, or Lotus 1-2-3 table, you must select the File | Save As option from the main menu bar and save the table in MapInfo, Dbase, or Microsoft Access format under a different name, or return to the application that created the file to make adjustments.

Changing data in MapInfo Professional will change the underlying data in the MapInfo *.dat* or *.dbf* files. Data cannot be changed in Excel, Lotus 1-2-3, or ASCII text files opened in MapInfo. MapInfo does not work like a conventional database in that changing information does not automatically save the changes, nor does MapInfo support a read-consistent snapshot of the database among multiple users. Basically, you keep a database fresh by periodically issuing the File | Save Table command to update the file. MapInfo stores a copy of all open files in memory, which allows you to save or reject changes during a session.

When you exit MapInfo Professional, however, you are asked if you want to save changes in the tables you have modified. If you do not choose to save the changes, they are lost when you exit MapInfo. You can select the File | Save Copy As option to store both the original file and the changed file. Be careful not to save (File | Save Table command) the original file when you intend, rather, to create a second version, or you will overwrite the original table.

When using the File | Save Copy As command, you must give the file a new name, or save the file to a different directory location. Make sure you are keeping track of your file versions in order to recall which file contains the original data, and which file contains the changed data. If you have made changes to data during a MapInfo session, and you wish to return to the original file, select

File | Revert to restore the original data. Remember that all changes will be lost.

The ability to create a new table in MapInfo Professional is very handy. Recalling the clinic example in Chapter 2, three new tables were created for the analysis: *buffer rings*, *hospitals*, and *new sites*. Different methods were used to create each of the three files. The *buffer rings* table was created after buffers were created around the hospitals. First, the Map | Save Cosmetic Objects option was executed and the buffer rings were saved to a new layer. The *hospitals* table was created by issuing a query based on another table (*landmark*). Query results were saved as a new table via the File | Save Copy As option. Finally, the File | New Table command was used to create from scratch a table for the identified sites.

As you can see, there are many reasons for creating a new table, such as saving labels or objects drawn on the cosmetic layer, creating a new data set from an existing data set, or defining new data. When creating a new table through File | New Table, MapInfo will ask if you want to open a browser, add the new table to the current map, or create a new map. Once you have made your selection(s), select Create. In the resulting New Table Structure dialog box, you can define the table's columns (fields).

Database Version Backups

File version backups are very important in database applications. There are many ways you could lose data in MapInfo Professional, such as making a modification to a table that causes data loss. It is recommended that you store copies of original data on floppy disk, employ a networked drive that is regularly backed up, or maintain some other type of storage device.

File version backup is a slightly different issue in MapInfo. This is because MapInfo differs from many other database applications in that changes are permanent only when the table has been saved. In most other database applications, the database is overwritten as changes are made to the data. As noted previously, it is often necessary to return to an earlier version. Regardless of the reason for returning to an earlier version, the original files should be accessible to prevent regenerating all of the work that went into creating the initial database. It is frequently useful to save the file in various states as new information is added.

Version backup is not necessary for all databases. Temporary tables that were easily created and are not reusable should not be tracked. Location databases that require a lot of time and money to create are good candidates for version backup tracking.

For database backup control, maintaining a copy of the original database on floppy disk or tape backup is recommended so that you can always return to the starting point. Next, for accessibility and depending on disk space, you may wish to keep copies of the database in various states (labeled *ver1*, *ver2*, and so on) on either floppy disk or in a subdirectory of the original file, or both.

Database backup control also serves as an audit trail to allow you to return to all previous versions to ensure that data are correct and up to date. If the data are not correct or updated, you do not have to return to the beginning, but only to the point where the data have been corrupted. Tracking file backups will save a lot of headaches in the event something goes wrong with your database or system.

Table Maintenance

Modifying the structure of a table is often necessary. Sometimes you need to add more spaces to character fields, or change a number from an integer to a decimal. Other situations require that you add columns of information to a table. All of these operations can be performed through the Table | Maintenance | Table Structure command. (The Modify Table Structure dialog box is shown in the following illustration.) Remember that any changes to the *structure* of a file are immediate and irreversible once OK has been selected.

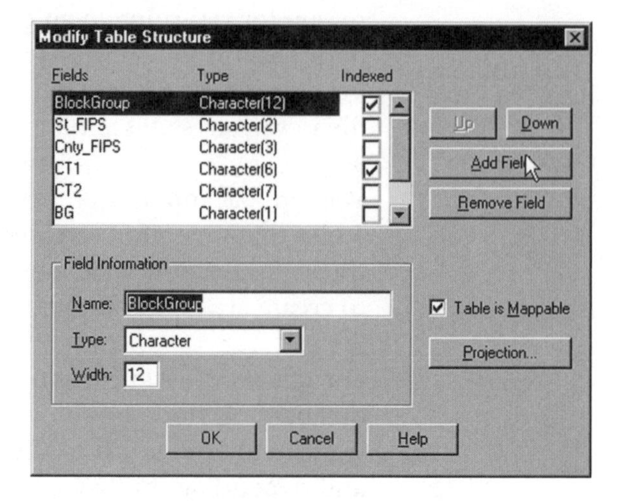

Modify Table Structure dialog box.

In contrast, changes to the data *must* be saved. As previously mentioned, you cannot modify the table structure of ASCII, Excel, or Lotus 1-2-3 files within MapInfo. In addition, MapInfo cannot modify the structure of a read-only table. If you wish to modify a read-only table, you will need to use the File | Save Copy As command, or change the file's properties within Windows Explorer.

NOTE: *MapInfo Professional does not support the Memo field from Dbase tables, or the Memo, OLE, and Replication ID fields from Microsoft Access tables. If you modify the structure or pack these types of tables within MapInfo, permanent data loss of these fields will result.*

Adding Fields

To add a new field to the table, select the Add Field option. This action will create a new field, placing a temporary name in the Name box under Field Information. The temporary name will indicate how many fields are currently in the table. For large tables, this is important because MapInfo Professional limits the number of fields in a table to 254. Although 254 fields sounds like a large number, in practice it is not. Running up against this limitation is common when using large demographic files and databases of proprietary information such as store location attributes.

Enter a descriptive name for the new field in the Name box. MapInfo allows alphanumeric names. Whereas underscores are allowed, spaces are not. Because you are not limited to the typical Dbase field name length, the name can be more descriptive. However, field names cannot begin with a number. If a field begins with a number, MapInfo will place an underscore at the beginning of the field name. The Type and Width boxes will automatically be filled in with the same type and width as that of the last field added, a useful feature if you are entering the same data types over and over. After entering the field name, select the data type for the field. MapInfo Professional's data types and respective limitations follow.

- *Character.* Stores up to 254 alphanumeric characters.
- *Integer.* Stores whole numbers between –2 billion and +2 billion.
- *Small Integer.* Stores whole numbers between –32,768 and +32,768.
- *Float.* Stores numbers in floating point decimal form.
- *Decimal.* Stores decimal numbers. [You must indicate the fixed number of places behind the decimal point to track, as well as the width of the field. Be sure to include spaces for the decimal point and the sign for the number (+ or –) when defining column width.]
- *Date.* Stores data in date format.
- *Logical.* Stores data in T (for True) and F (for False) format.

Be careful to choose the correct data type when first defining a field. For example, you will most likely want to select Character as the data type for zip code fields. If you do not select Character, you will lose leading zeroes in the data, such as in zip codes for the eastern United States (e.g., 04598). You should not select Small Integer for any field that may contain data over 10,000. If you do, you run the risk of losing data once the 32,768 mark has been reached.

Next, avoid choosing a data type that is too large for the field. For instance, if you select Float as the data type for all numbers, your file will be very large, even though the numbers contained in the data set are small. MapInfo Professional will reserve space for the large numbers associated with the float data type, whether or not the space is used.

Once the data type has been selected, indicate whether the field will be indexed. If you plan to frequently query the field, or join the table to other tables through the field, you will want to index the field to speed up the process.

Removing Fields

To delete a field from the field list, highlight the field, and then select Remove Field. The field will be permanently deleted from the table once the changes have been accepted by pressing the <Enter> key or clicking on OK. If you wish to be able to access the information in this field again, either save a copy of the table before editing the file structure or do not delete this field. Because it is very easy to modify tables, and especially to delete columns, maintaining backup copies of the original table is recommended so that you will always have this information available and are able to return to your starting place.

Reordering Columns (Fields)

If you are unhappy with a table's column order, you can reorder the field list by highlighting a field and then selecting the Up or Down button until the field is positioned where you want it. Changing the column order is somewhat clumsy. When working with a large file, you may want to consider exporting the table to another file format and reordering the columns in another application.

You could also leave the column order as is, and then reorder them while viewing the data in a Browser or through SQL queries. The only place that column order cannot be changed without modifying the table structure is the Info window. Data will appear in this window in the same order it appears in the underlying table.

Changing Existing Fields

You can change the name of an existing field by highlighting the field and retyping the field name in the Name box. Changing the type of data in an existing column is also relatively easy. Highlight the column to be changed, and then change the data type or the field width. If you shorten the field width, or change the data type (e.g., from an integer to a small integer), data can be lost. Be sure you allow enough space to track all of your data.

Table Is Mappable Option

The Table Is Mappable option will be selected if the table has corresponding graphic objects. Any MapInfo Professional table can be mapped if it contains coordinates (latitude and longitude entries). If you remove the X from this box, the table is no longer mappable. All geographic objects associated with this table will be lost, and the file will no longer appear in the Map window. Sometimes this is desirable, especially when working with point objects where you have updated latitude and longitude coordinates, and you want the map objects to reflect the changes.

Executing Changes

All modifications can be carried out by accessing the Modify Structure dialog box. If you click on the OK button prior to making all desired changes, the changes you have made up to that point will be applied. Remember, these changes are permanent. You will have to reaccess the Modify Structure dialog box to make any further changes.

Because constant reaccess of the dialog box can slow down the process of modifying a table's structure, click on OK only when you have made all intended changes. MapInfo Professional defaults to OK in this dialog box, so be careful when pressing the <Enter> key; you may find your changes applied before you want them applied. Selecting the Cancel button will revert the table to the state it was in when you entered the Modify Structure dialog box.

If you are attempting to modify a read-only MapInfo table (e.g., from a CD-ROM), or to modify an Excel, Lotus, or ASCII file, only the field list, type of data, and index information will appear in the Modify Table Structure dialog box. You will not be able to modify the field names, change data types, or modify the structure in any other manner. The Table Is Mappable option and an option to Change or Add indexes will appear as these are stored in separate files and thus tracked separately from the data by the *.tab* file.

Data Import and Export

Importing data to MapInfo Professional's file format does not work like most other Windows applications. File formats that can be used directly by MapInfo have already been discussed. The following file formats, however, require an Import command to convert them to MapInfo format.

- MapInfo Interchange (*.mif*)
- AutoCAD (*.dxf*)
- MapInfo DOS (*.mbi*)
- MapInfo DOS (*.mmi*)
- MapInfo DOS image (*.img*)

Most of the options under the Table | Import command are related to importing files from the MapInfo for DOS format. (The Import File dialog box is shown in the following illustration.) These include *.mbi*, *.mmi*, and *.img* files. AutoCAD *.dxf* files are drawings that can be imported and used in MapInfo. MapInfo Interchange files (*.mif*) are ASCII files that contain graphic data. Every *.mif* file has an associated *.mid* file containing textual data. Both *.mif* and *.mid* files can easily be edited with any text editor. The *.mid* and *.mif* file types are also commonly used by many third-party application programs.

Import File dialog box.

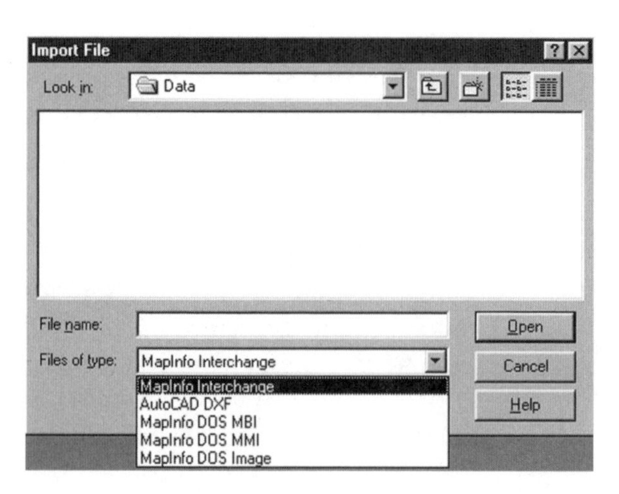

Although you can import from MapInfo's DOS version, you cannot export to this file format. When you select the Table | Export command, you can export data to the following file formats.

- MapInfo Interchange (*.mif*)
- Delimited ASCII (*.txt*)
- AutoCAD DXF (*.dxf*)
- Dbase (*.dbf*)

Exporting to Delimited ASCII or Dbase allows a MapInfo file to be used in many other applications, such as Excel, Lotus 1-2-3, SPSS, Access, Dbase, and FoxPro, among others. Exporting to an AutoCAD *.dxf* file format allows a MapInfo image to be altered within AutoCAD.

Another export option, saving a window as a graphic/picture file, is not accessible from the Table | Export command. To save a window as a graphic/picture file, verify that it is the current window, and then select the File | Save Window As command. Available formats include **.bmp, *.wmf, *.emf, *.jpg, *.png, *.tif,* and **.psd.*

In addition to importing and exporting data, you can save tables in MapInfo Professional, MapInfo version 2.x, and Dbase formats. MapInfo format is the default. To perform this operation, select File | Save Copy As, and select the desired format.

Saving data to a MapInfo format is useful when you are attempting to modify a read-only file or wish to change the file structure of an ASCII-, Lotus-, or Excel-based table. Saving a file as a MapInfo version 2.x table allows the file to be used in an older Windows version of MapInfo. The Dbase file format allows the file to be shared with many other applications, yet be modified within MapInfo.

Additional import and export option capabilities can be accessed through MapInfo Professional's Tool menu option. For instance, one available add-in option under the Tool menu is the import/export utility for ArcInfo file formats. OLE and ODBC capabilities provide other forms of data access. (OLE is discussed in Chapter

11.) The Tool menu option in MapInfo's main menu allows you to extend MapInfo's capabilities with add-in applications such as a legend manager, scale bar, grid maker, and Crystal Reports. (The Tool Manager is discussed in more detail in Chapter 11.)

Exercise 3-1, which follows, is aimed at helping you understand file manipulation and maintenance in the MapInfo world. In this exercise, you will use files from the companion CD-ROM to practice file manipulation and other file-related tasks.

■ EXERCISE 3-1: FILE MANIPULATION AND MAINTENANCE

1 Open the *STATES*, *ST_ATTR*, and *CITY_1K* tables in the *samples* directory on the companion CD-ROM. These tables are shown in the following illustration.

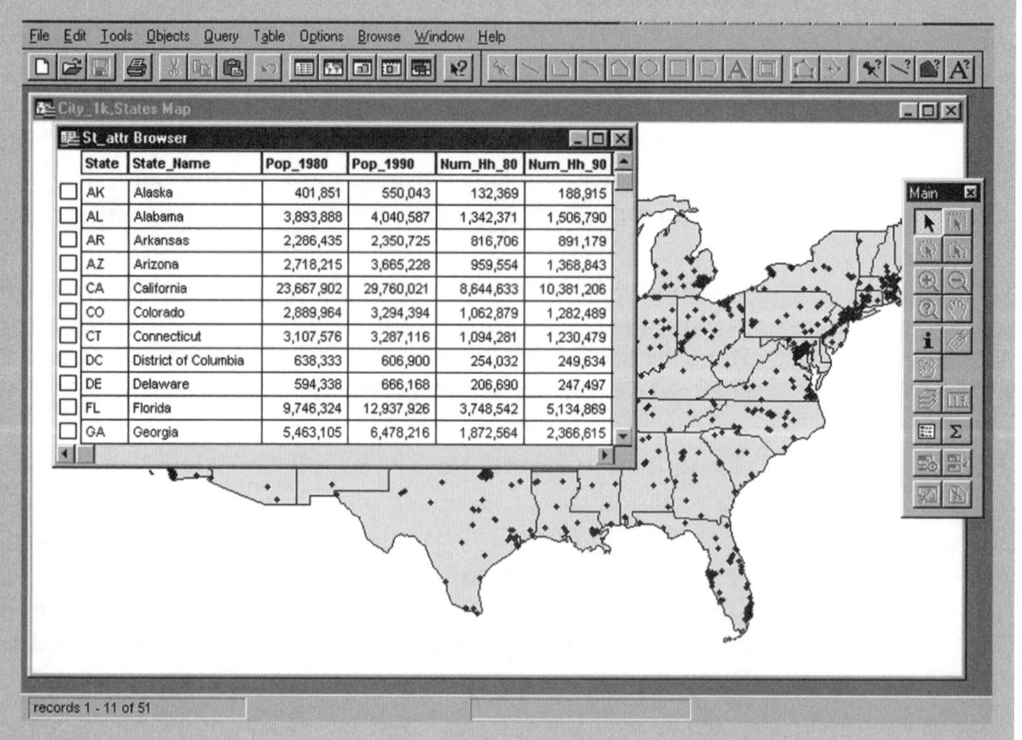

Opened tables.

2 Select Window | New Browser Window and browse the *STATES* table. The STATES table browser is shown in the following illustration.

States Browser

State_Name	State	FIPS_Code	Pop_1980	Pop_1990	Num_Hh_80	Num_Hh_90	Med_Inc_80	BPI_1990	Pop_
Alabama	AL	01	3,893,888	4,040,587	1,342,371	1,506,790	21,714	1.4017	2,
Alaska	AK	02	401,851	550,043	132,369	188,915	34,130	0.2446	
Arizona	AZ	04	2,718,215	3,665,228	959,554	1,368,843	25,218	1.3875	3,
Arkansas	AR	05	2,286,435	2,350,725	816,706	891,179	20,361	0.8111	1,
California	CA	06	23,667,902	29,760,021	8,644,633	10,381,206	33,342	12.9833	27,
Colorado	CO	08	2,889,964	3,294,394	1,062,879	1,282,489	28,558	1.3332	2,
Connecticut	CT	09	3,107,576	3,287,116	1,094,281	1,230,479	36,961	1.5841	2,
Delaware	DE	10	594,338	666,168	206,690	247,497	28,887	0.2919	
District Of Columbia	DC	11	638,333	606,900	254,032	249,634	26,962	0.2639	
Florida	FL	12	9,746,324	12,937,926	3,748,542	5,134,869	25,914	5.4385	10,
Georgia	GA	13	5,463,105	6,478,216	1,872,564	2,366,615	26,342	2.4956	4,
Hawaii	HI	15	964,691	1,108,229	294,934	356,267	34,997	0.5198	
Idaho	ID	16	943,935	1,006,749	324,889	360,723	24,475	0.3470	
Illinois	IL	17	11,426,518	11,430,602	4,046,638	4,202,240	31,119	4.7604	9,
Indiana	IN	18	5,490,224	5,544,159	1,928,375	2,065,355	25,982	2.0413	3,
Iowa	IA	19	2,913,808	2,776,755	1,053,107	1,064,325	24,699	1.0364	1,
Kansas	KS	20	2,363,679	2,477,574	873,336	944,726	26,557	0.9721	1,
Kentucky	KY	21	3,660,777	3,685,296	1,263,102	1,379,782	21,454	1.2793	1,
Louisiana	LA	22	4,205,900	4,219,973	1,413,394	1,499,269	23,167	1.4799	2,
Maine	ME	23	1,124,660	1,227,928	395,474	465,312	25,652	0.5006	

STATES table browser.

3 Select Window | New Layout Window and select the map to be shown in the layout (see following illustration).

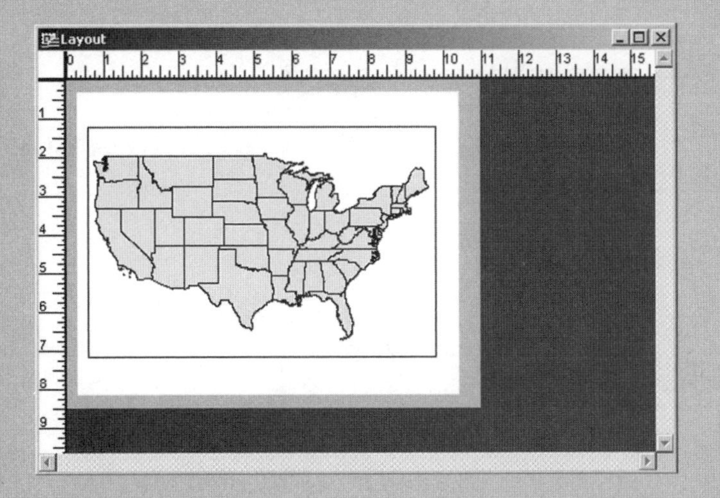

Map in layout.

4 Save the workspace as *STATE.WOR* (File | Save Workspace) to the *exercise* directory.

5 Close all files (File | Close All).

6 Open the *STATES* workspace (File | Open Workspace).

7 Create a new address table based on the structural components provided in table 3-1 (which follows), and name the file *Sales*. First, select File | New

Table. Select Open New Browser and deselect Open New Mapper in the New Table dialog box, and then select Create. Key in the Name, Type, and Width (see table 3-1 and the following illustration) for each field in the New Table Structure dialog box. After you have entered all fields, select Create. An empty browser for the new table will appear.

Table 3-1: Input to Address Table

Field	Entry
Name	Character 10
Address	Character 20
City	Character 20
State	Character 2
Phone	Character 13
Items_Sold	Integer
Total_Sales	Integer

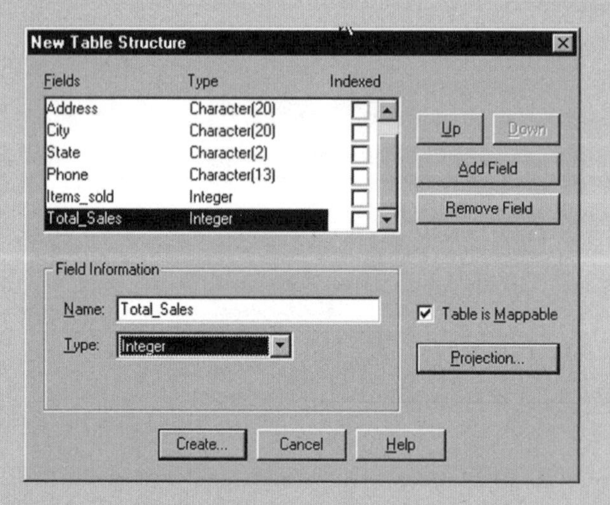

Setting up table structure.

8 Enter in the new table the records shown in table 3-2, which follows. Select Edit | New Row, with the browser for the new table as the active window. (See also the following illustration.)

Table 3-2: Records to Be Entered in the New Table

Name	Address	City	State	Phone	Items Sold	Total Sales
Tom Jones	1235 New Haven Dr.	Las Vegas	NV	221-8965	5	77.89
Sam Spade	444 Main St.	Las Vegas	NV	555-8965	14	198.78
Grace Allen	879 1st Ave.	Las Vegas	NV	986-7786	22	254.89

SALES Browser

Name	Address	City	State	Phone	Items_Sold	Total_Sales
Tom Jones	1235 New Haven D	Las Vegas	NV	221-8965	5	77
Sam Spade	444 Main St	Las Vegas	NV	555-8965	14	198
Grace Alle	879 1st Ave	Las Vegas	NV	986-7786	22	254

Records entered in Sales browser.

9 As you have likely noted, some of the information you entered into the new table was truncated. You need to change the table's structure (select Table | Maintenance | Table Structure). The Modify Table Structure dialog box is shown in the illustration at right (showing the step 12 field change). Change the width of the Name field to 20, and change the *Total_Sales* field to decimal 11,2.

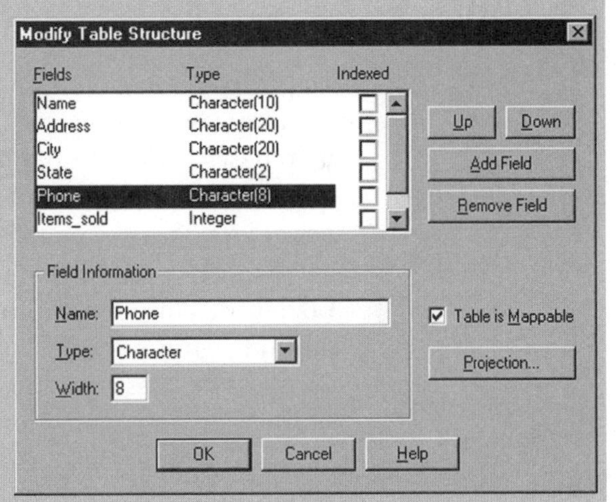

Modify Table Structure dialog box.

10 Add a column for *Sales Representative* to the table.

11 Delete the *Items Sold* column.

12 Shorten the phone number field width to 8 characters.

13 Export the *Sales* table to a Dbase (*.dbf*) file using Table | Export. The Export Table to File dialog box is shown in the following illustration.

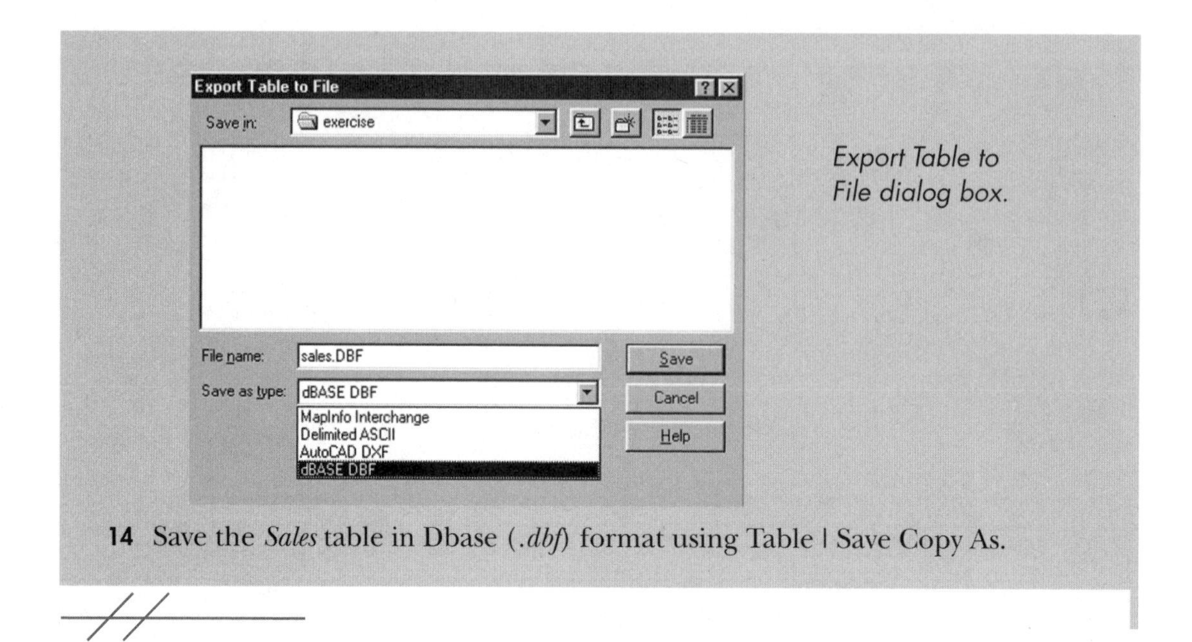

Export Table to File dialog box.

14 Save the *Sales* table in Dbase (*.dbf*) format using Table | Save Copy As.

Summary

MapInfo Professional allows you to affect the way the program operates by setting preferences, such as where the program searches for files, and whether or not it retrieves your last work session. With the workspace facility, you can save an entire analysis or project, thereby enabling you to return to it later.

Like most Windows applications, MapInfo allows the user to open multiple windows, and arrange windows with standard commands such as Tile, Cascade, Maximize, and Minimize. File manipulation within MapInfo Professional is relatively painless, partially because MapInfo offers many easy methods of sharing data with other applications. Many file formats can be used by MapInfo without prior data importing, especially given the ODBC functionalities. MapInfo also provides many export and "save as" options that permit you to use MapInfo data in other applications.

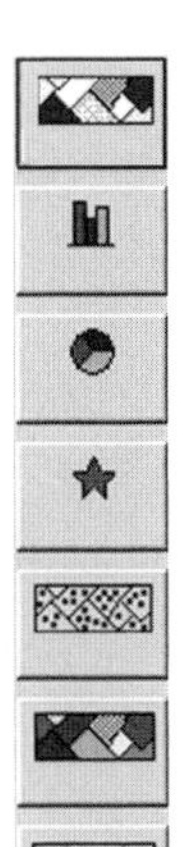

CHAPTER 4

QUERIES AND BROWSING

ONE OF THE BASIC FUNCTIONS PROVIDED BY GIS SOFTWARE is the ability to extract specific data records or map objects from a map linked to a database. The selected subsets of data can then be manipulated in many of the same ways as an entire MapInfo table (data file), referred to as a *base* table.

MapInfo Professional creates a temporary table to store the records for each selection (data subset) you make. The temporary tables are named *Query* by default. The first selection results will be displayed in table *Query1,* the second selection results will be displayed in table *Query2,* and so on.

To create a permanent table from any temporary selection table, you can select the File | Save Copy As command. As seen in subsequent chapters, you will also be able to browse, map, graph, and edit any temporary table. A temporary table may also be used to create additional selections. In addition, most temporary tables are saved along with maps, graphs, and browsers when saving a workspace. MapInfo also allows you to save query templates to access the same query at a later time.

MapInfo Professional provides many commands and tools for making selections. The first methods explored involve selecting data from the screen by using several tools on the Main toolbar.

Next, finding data using MapInfo's Find features is examined. In addition, selecting data through the Select and SQL Select dialog boxes are discussed, as well as ways of customizing the look of data in the Browser window.

Data Searches

The sections that follow explore Browser window basics, Map window basics, and selection tools associated with data searches. The series of tutorials that runs throughout this chapter begins in this section.

Browser Window Basics

The Browser window allows you to view data in the traditional row and column format found in many popular spreadsheet and database software packages. Each column contains a different piece of information about a record, often referred to as an *attribute.* Each row in the Browser window contains all of the information about a particular record in the database.

Per tutorial 4-1, which follows, open the *States* table (*samples* directory on the companion CD-ROM) and examine its Browser window. Because the tools and functions examined here are best explored interactively, you are encouraged to carry out the command sequences and other instructions as they are presented in the tutorials.

▼ *TUTORIAL 4-1: OPENING A TABLE AND EXAMINING ITS BROWSER*

1 Select File | Open Table from the Main menu bar. This accesses the Open Table dialog box, shown in the following illustration.

2 Select the *States.tab* file name. Click on the Open button.

3 Select Window | New Browser Window to view the data.

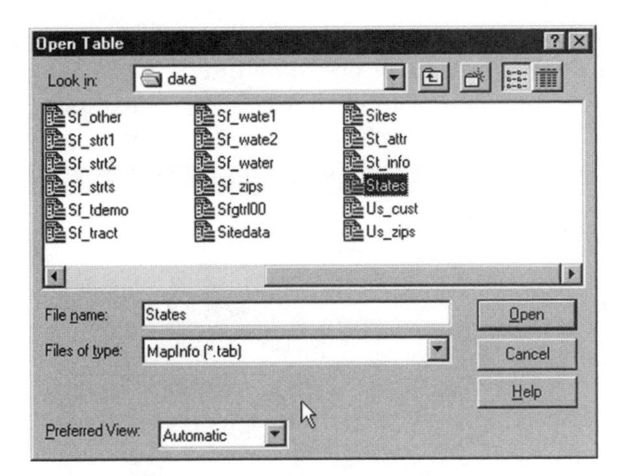

Open Table dialog box.

4 In the Browser window of the *States* table, shown in the following illustration, you see that every row of data represents a state in the United States. The following columns or attributes describe each state in the table: *State, State_Name, FIPS_Code, Pop_1980, Pop_1990, Num_Hh_80,* and *Num_Hh_90.*

States Browser

State_Name	State	FIPS_Code	Pop_1980	Pop_1990	Num_Hh_80	Num_Hh_90	Me
Alabama	AL	01	3,893,888	4,040,587	1,342,371	1,506,7[illegible]	
Alaska	AK	02	401,851	550,043	132,369	188,[illegible]	
Arizona	AZ	04	2,718,215	3,665,228	959,554	1,368,843	
Arkansas	AR	05	2,286,435	2,350,725	816,706	891,179	
California	CA	06	23,667,902	29,760,021	8,644,633	10,381,206	
Colorado	CO	08	2,889,964	3,294,394	1,062,879	1,282,489	
Connecticut	CT	09	3,107,576	3,287,116	1,094,281	1,230,479	

Browser window for States table.

You can use the horizontal scroll bar to view additional attributes for a particular state, and the vertical scroll bars to view information in other rows (i.e., states), or to resize the Browser window to view more (or fewer) rows or columns.

Map Window Basics

Although the Map window will be explored in greater depth in subsequent chapters, it is discussed briefly here to examine its relationship to the Browser window. Open the *City_1k* table (*samples* directory on the companion CD-ROM). The table will be added to the opened *States* table. (If you are viewing a browser, click on the

Map window to see both tables displayed. The *City_1k* table contains the 1,000 largest cities in the United States. The Map window of the two tables is shown in the following illustration.

Map window of States and City_1k.

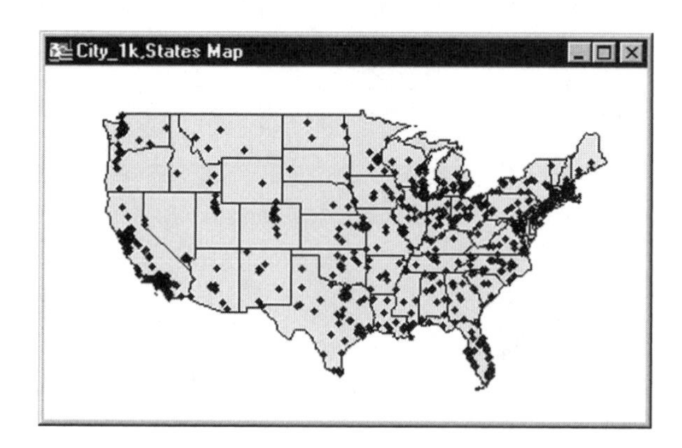

Selection Tools

MapInfo Professional provides several ways to select or query data objects, which are reviewed in this chapter. The first selection methods covered are the toolbar selection tools that allow you to manually select graphical data objects. You would use these manual selection tools when the data you wish to further analyze are located in the same area on a map. Later in the chapter, methods of querying for data objects based on their attribute characteristics are examined.

Making data object selections using manual or querying techniques is fundamental for creating meaningful analyses. Once you have selected a data subset, you can then display the subset in various MapInfo windows. For example, if you are working with a database of all clients who have made purchases in a particular store, you may want to analyze only those clients who have purchased over $1,000 worth of merchandise during the past year. In this case, you would select the clients meeting the "minimum amount" criterion and then display them in either a Map or Browser window.

When using the manual selection tools, you will usually select only a few data objects, or a group of data objects located in close proximity to one another. For example, you may want to select all clients within a 1-mile radius of a particular store for a special mailing. Manual selection tools and how they work are examined in the following section.

Select Tool

The Select tool, shown at left, allows you to select one or more objects. Use this tool to select either a single object or a limited number of objects. For instance, to select the states of Colorado and Texas, perform tutorial 4-2, which follows.

▼ TUTORIAL 4-2: USING THE SELECT TOOL

1 Select the Select tool from the Main toolbar.
2 Click on the state of Texas in the Map window.
3 Hold down the <Shift> key and click on the state of Colorado.
4 The states of Texas and Colorado will be highlighted in the Map window, as shown in the illustration at right.

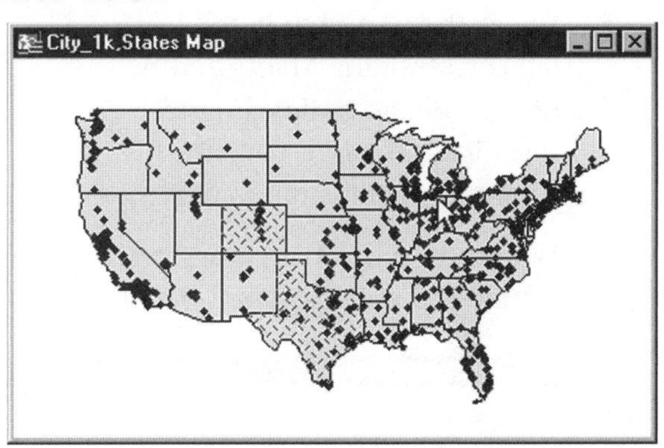

Select tool map of Colorado and Texas.

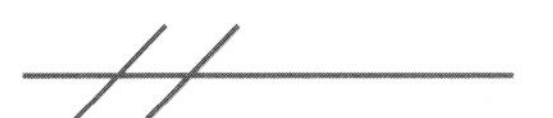

Radius Select Tool

The Radius Select tool, shown at left, allows you to select all objects that fall within a certain radius. This selection tool is appropriate when you want to find all objects that fall within a certain radius of another object. All selectable objects in the topmost map layer will be included in the selection. For example, to find the cities that fall within a 280-mile radius of the northeast corner of Colorado, perform tutorial 4-3, which follows.

NOTE: *You may have to reorder layers or toggle the selectability option for layers higher in the hierarchy to get desired results from the select tools. Only the topmost layer (with the selectable option turned on) in the area you click on using the select tools will be selected.*

▼ TUTORIAL 4-3: USING THE RADIUS SELECT TOOL

1 Select the Radius Select tool from the Main toolbar.

2 Position the cursor at the northeast corner of Colorado.

3 Press the mouse button and hold it down.

4 Drag the mouse away from the northeast corner of Colorado.

5 Check the status bar in the lower left corner of the MapInfo window to see the radius distance the selection tool is covering, as shown in the illustration at right.

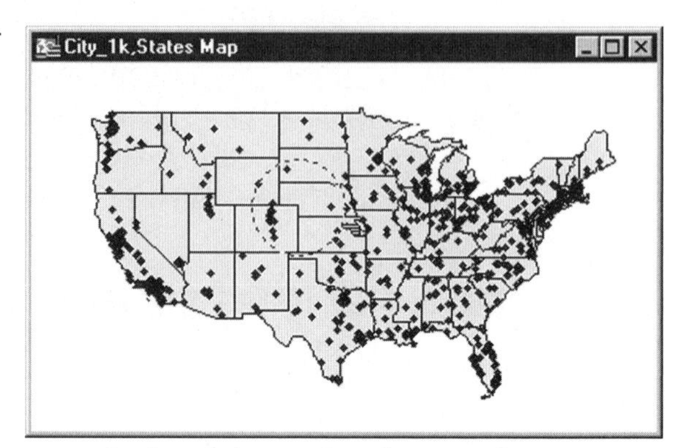

Radius Select tool making a selection.

6 When the status bar shows 280 miles, release the mouse button. The cities within 280 miles will be selected, as shown in the following illustration.

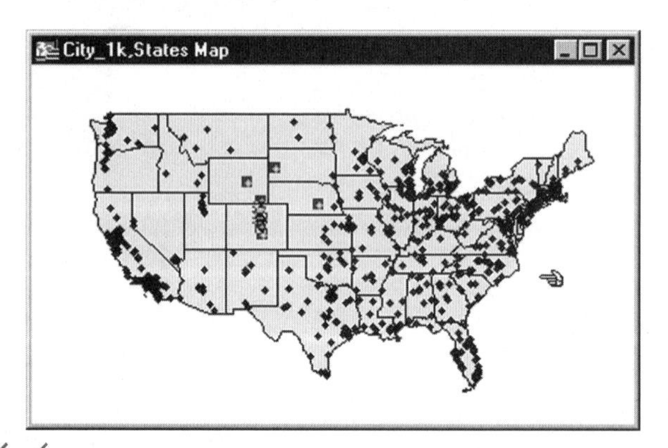

Radius Select tool result map.

Marquee Select Tool

The Marquee Select tool, shown at left, operates very much like the Radius Select tool except that instead of using a circle it uses a rectangle. This tool can be used to select all cities that fall in the eastern half of the United States, as demonstrated in tutorial 4-4, which follows.

▼ TUTORIAL 4-4: USING THE MARQUEE SELECT TOOL

1 Select the Marquee Select tool from the Main toolbar.

2 Position the cursor at the upper right of the state of Maine.

3 Press the mouse button and hold it down.

4 Drag the mouse to the lower left until the rectangle covers the eastern half of the United States, and release the mouse.

5 All cities in the eastern half of the United States will be selected, as shown in the illustration at right.

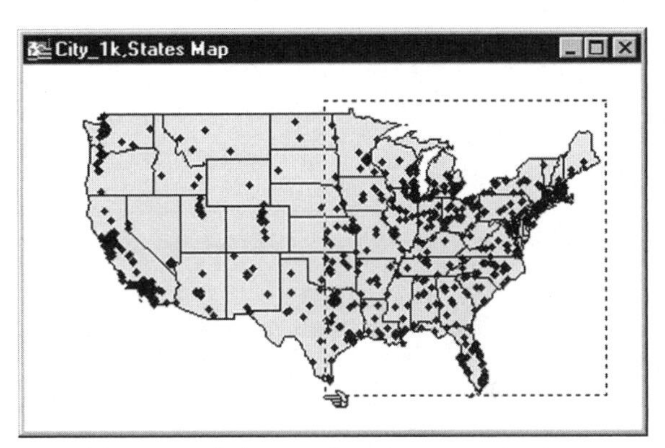

Marquee Select tool making a selection.

Polygon Select Tool

The Polygon Select tool, shown at left, allows you to draw a polygon on a map to select objects. This tool is useful when you wish to select objects and a radial or marquee tool is inappropriate, such as when defining trade areas or other irregular boundaries. For example, you can use this tool to select the cities in the Northwest, as demonstrated in tutorial 4-5, which follows.

▼ TUTORIAL 4-5: USING THE POLYGON SELECT TOOL

1 Select the Polygon Select tool from the Main toolbar.

2 Position the cursor in the northwestern corner of Washington State.

3 Click and draw a line down the Oregon coast to define the first side of the polygon, as shown in the following illustration.

4 Click and draw the next line across the bottom of Oregon and Idaho and a little into Wyoming (as shown).

5 Click again on the southeastern corner of Idaho and draw a diagonal line to the northeastern corner of Idaho (as shown; be sure to avoid any cities in Montana!).

6 Click and draw a line back to the original point in the northwestern corner of Washington State (as shown).

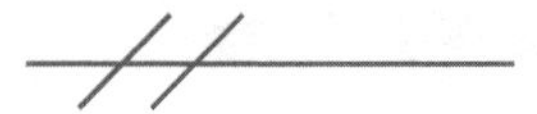

As shown in the following illustration, all cities in the three Pacific Northwestern states are selected.

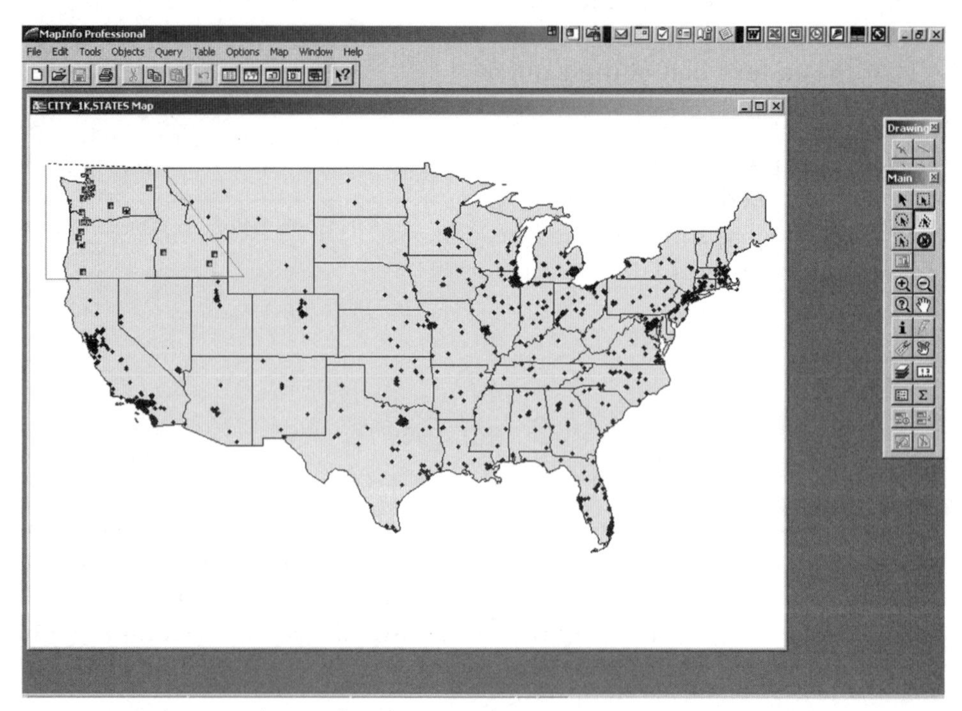

Polygon Select tool making a selection.

Boundary Select Tool

The Boundary Select tool, shown at left, allows you to select all objects that fall within the selected boundary or polygon object. For example, this tool is appropriate if you want to view all customers in a sales territory, work orders within a zip code, or outages in a phone exchange. Let's use the Boundary Select tool to select all cities within the state of Texas, as demonstrated in tutorial 4-6, which follows.

▼ TUTORIAL 4-6: USING THE BOUNDARY SELECT TOOL

1 Select the Boundary Select tool from the Main toolbar.

2 Click on the state of Texas.

3 All cities in the state of Texas will be selected, as shown in the following illustration.

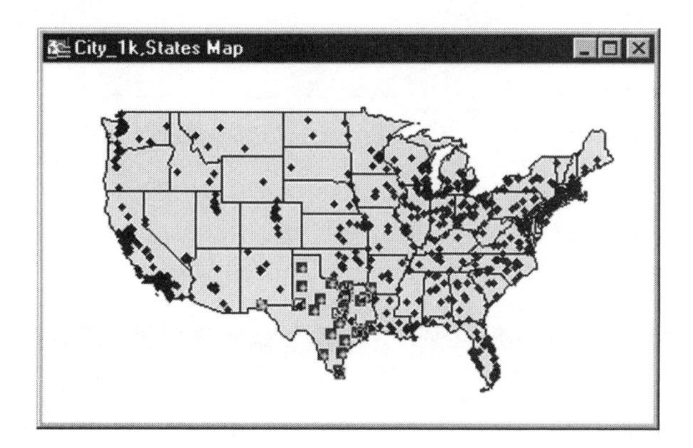

Boundary Select tool example.

Four selection tools for selecting map objects have been reviewed: Select, Radius Select, Marquee Select, and Boundary Select. Determining which selection tool to use may be confusing to a new user. Ultimately, the use of these tools will be driven by the application. As you work with MapInfo Professional, your choice of selection tools will become second nature.

Relating Map Selections to the Browser Window

Thus far, the entire browser table has been viewed. Typically, you will want to view and analyze the attributes of only the objects you have selected. Once you have made a selection from the Map window, you can then display a Browser window of the selected objects. For example, to browse selected map objects, perform tutorial 4-7, which follows.

▼ TUTORIAL 4-7: BROWSING SELECTED MAP OBJECTS

1 Select the objects on the map that you wish to analyze.

2 Select Window | New Browser Window from the Main menu bar. This accesses the Browse Table dialog box, shown in the illustration at right.

3 Select the Selection option in the table list box, shown at right, and click on the OK button.

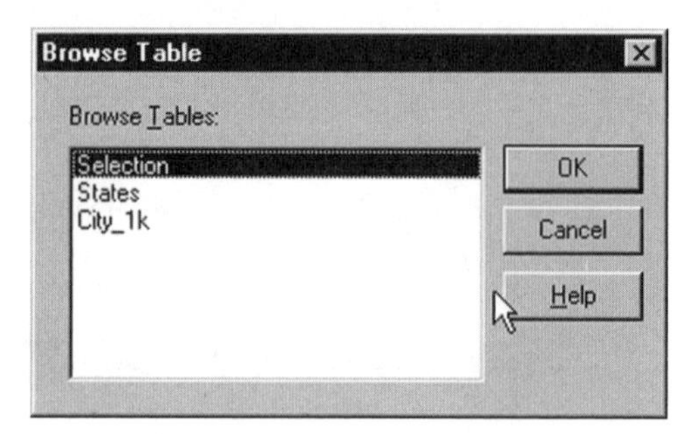

Browse Table dialog box.

Rows in the Browser table are limited to objects selected in the Map window, as shown in the following illustration.

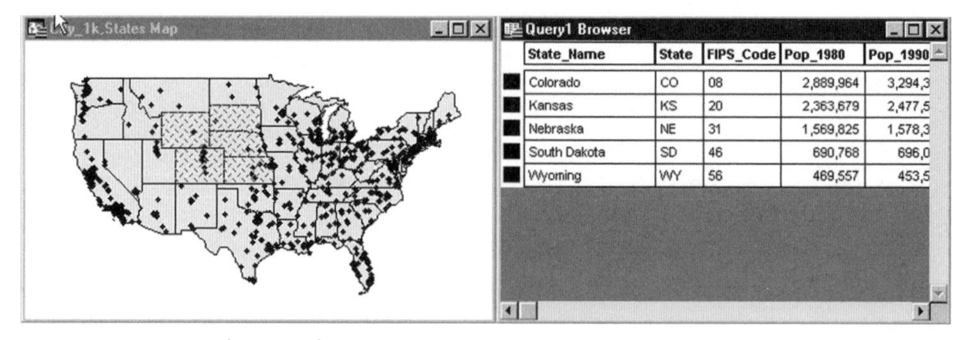

Map selection shown in a Browser window.

Find Functions

The sections that follow explore the Find and Find Selection options. The difference between these two options is explained, and tutorials for practice in using them are included.

Find

Occasionally you will want to select a certain map object you cannot visually locate. Find is a useful function for locating such objects. The object will be located in both Map and Browser windows. When MapInfo Professional finds the requested object on the map, the object is marked with a symbol. MapInfo will also scroll any open Browser windows to the requested object.

NOTE: *You can use the Find command only on tables that contain an indexed field.*

In tutorial 4-8, which follows, use the *States* table to find the state of Nevada.

▼ TUTORIAL 4-8: USING THE FIND COMMAND

1 Select Query | Find from the Main menu bar.

2 Set the Find dialog box to search the *States* table for objects in the *State_Name* column, as shown in the illustration at right. Click on the OK button.

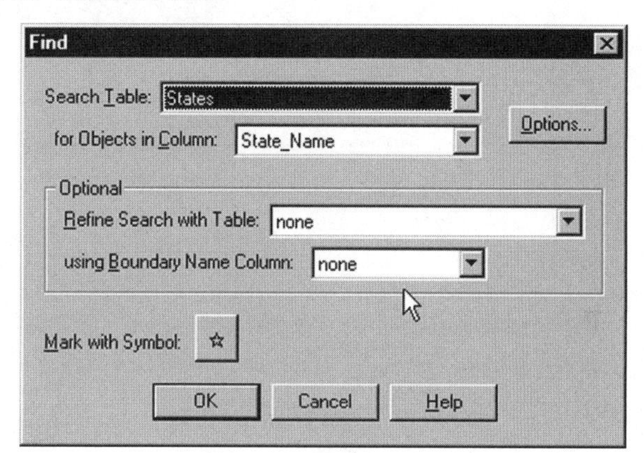

Find dialog box for specifications.

3 Key in *nevada* for *State_Name* in the Find dialog box as shown in the following illustration. Click on the OK button.

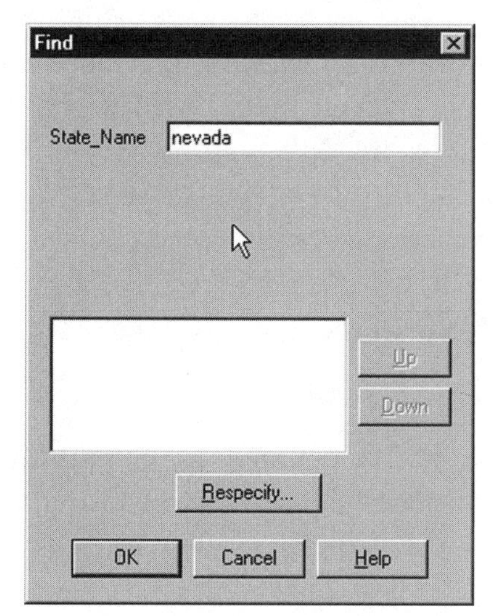

Find dialog box for items.

4 A star will appear in the center of the state of Nevada in the Map window. The *States* Browser window will display Nevada in the first row. These are shown in the following illustration.

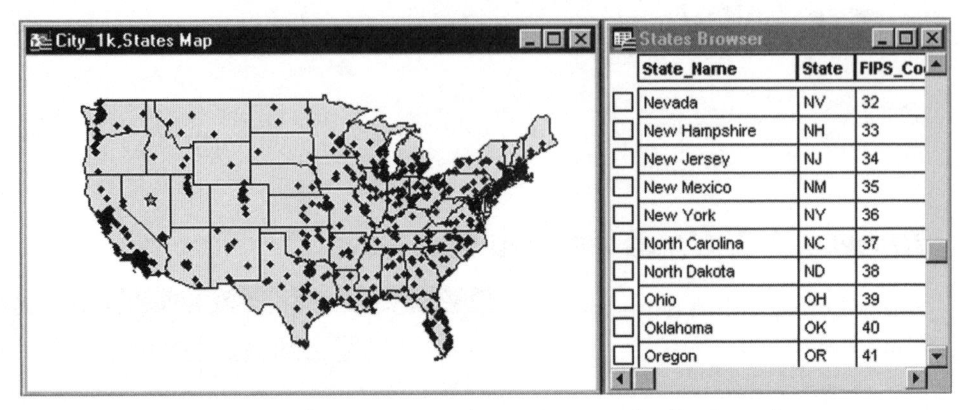

Map and Browser windows showing the Find results.

Find Selection

The Find Selection option automatically finds and displays the selection in all windows. When you select a map object and return to a Browser window, the Find Selection option locates the item in the Browser window. Find Selection also works in the other direction. If you select a row in the Browser window and select Find Selection when on the Map window, the map will center on the object and show the selected item. With these bidirectional features, tabular selects work with maps, and map selects work with tables. For example, to find the state of Oregon, perform tutorial 4-9, which follows.

▼ *TUTORIAL 4-9: USING THE FIND SELECTION COMMAND*

1 Click on the state of Oregon in the Map window.

2 Return to the Browser window and select Query | Find Selection from the Main menu bar. The Browser window will place the row containing data for the state of Oregon at the top of the Browser window, as shown in the following illustration.

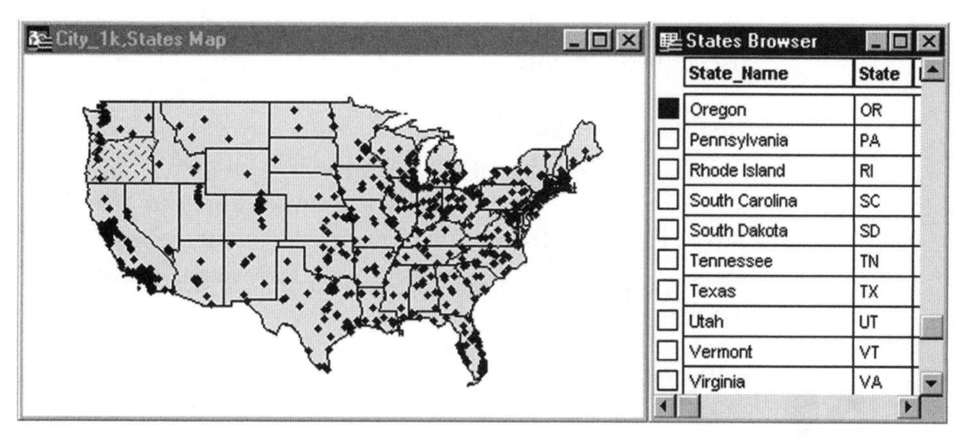

Find Selection results.

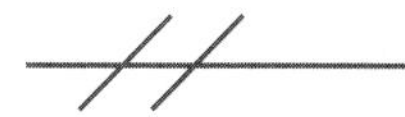

Info Tool

The Info tool, shown at left, allows you to easily examine the attributes of features on a map. With a simple mouse click, you can see all data associated with the state of Texas, as demonstrated in tutorial 4-10, which follows.

▼ *TUTORIAL 4-10: USING THE INFO TOOL*

1 Select the Info tool from the Main toolbar.

2 Select the state of Texas on the *States* Map window. The Info Tool window, shown in the illustration at right, displays the attributes for the state of Texas.

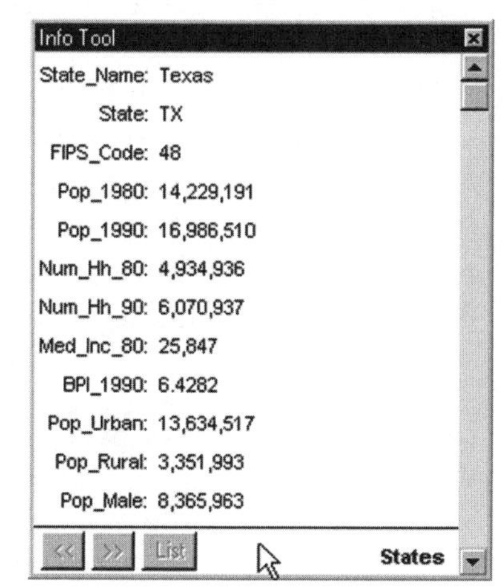

Info Tool window.

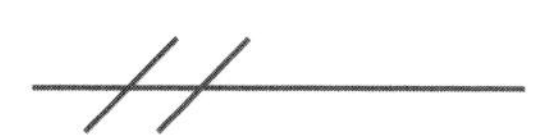

The Info Tool window can be enlarged to simultaneously view more attributes. The scroll bar at the side of the window (shown in the previous illustration) allows you to move up and down the list of data attributes.

If you select a location that contains more than one layer (table) of information, the Info Tool window will display all layer names. You can then move through the layers and the attributes in the layers for the information that interests you, using the arrow buttons (shown in the previous illustration) at the bottom of the Info Tool window.

TIP: *The Info tool not only displays data but provides a neat shortcut for editing data.*

Query Vehicles: Select Versus SQL Select

Thus far, the discussion has focused on selection tools. Although these tools are powerful in their own right, MapInfo Professional has many other ways of interacting with your database, such as *queries.*

Queries are statements that allow you to analyze the information in a database, and they are usually built using simple dialog boxes. MapInfo displays in maps and browser windows the results of queries built using these dialog boxes.

MapInfo offers two different dialogs for querying subsets of information from database tables. The first dialog (Select) allows you to issue user-friendly, albeit somewhat limited, queries on a single table. The second dialog (SQL Select) allows you to build more complex queries, based on multiple tables, for retrieving only the columns of data you wish to see, in the manner you wish to see them.

Select

The Select dialog allows you to build simple queries to select records and objects from a single table. The items that meet your criteria can be displayed in a Map or Browser window.

To become familiar with what the Select functions allow, you will first work through an example Select query, and then examine the features of the Select dialog in more detail. To find all states in the *States* table with a population greater than 5 million, perform tutorial 4-11, which follows.

▼ TUTORIAL 4-11: USING THE SELECT FEATURE

1 Select Query | Select from the Main menu bar.

2 Select *States* from the Select records in the Table pull-down menu.

3 Type in the expression *POP_1990 > 5000000* in the Select dialog, shown in the illustration at right. Click on the OK button.

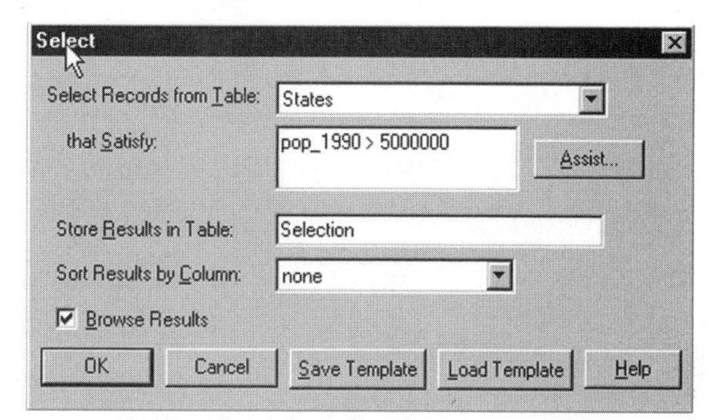

Select dialog for selecting States with 1990 population > 5000000.

4 The results should show a Browser window containing the rows of *States* data for 15 states. The Map window will display the 15 states selected, as shown in the following illustration.

NOTE: *You can easily tell the number of items selected while in a resulting Browser window. The number of rows of data for a given browser appears on the bottom left corner of the MapInfo window on the status bar. The number of rows equates to the number of selected items.*

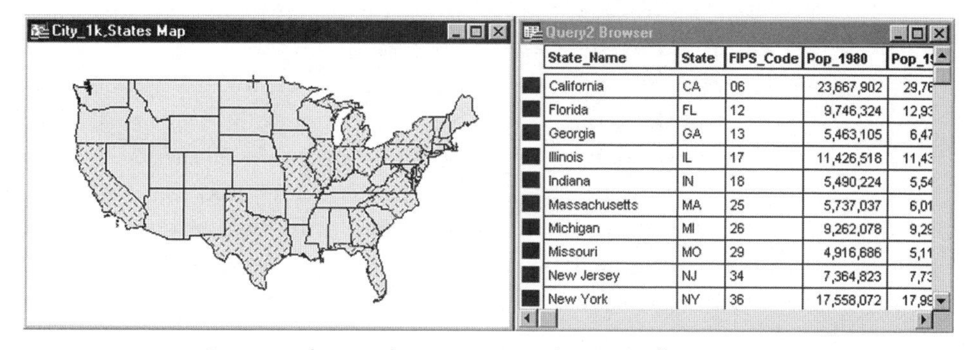

State_Name	State	FIPS_Code	Pop_1980	Pop_1!
California	CA	06	23,667,902	29,76
Florida	FL	12	9,746,324	12,93
Georgia	GA	13	5,463,105	6,47
Illinois	IL	17	11,426,518	11,43
Indiana	IN	18	5,490,224	5,54
Massachusetts	MA	25	5,737,037	6,01
Michigan	MI	26	9,262,078	9,29
Missouri	MO	29	4,916,686	5,11
New Jersey	NJ	34	7,364,823	7,73
New York	NY	36	17,558,072	17,99

States with population greater than 5 million.

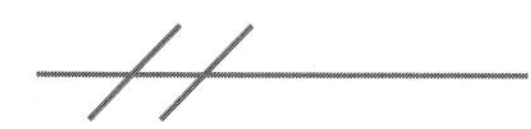

The Select dialog offers several features to make it easier for you to formulate the selection you wish to make. Results are stored in the default table named *Selection*, but you can change the table name. Giving your selections meaningful names can be particularly useful during a work session when you are issuing several queries. Without meaningful names, you could forget which query contains the results you want, and thus find yourself rebuilding queries.

You must, however, remember to change the name of the resultant table with each query if you change MapInfo's default, as MapInfo retains the user-assigned name and will continue to use it until changed back to MapInfo's default or the user changes it. The dialog box also offers a pull-down menu allowing you to select a column in the table by which you want to sort results. You also have the option of displaying the results of the selection in a new Browser window.

MapInfo Professional allows you to save the query template, using the Save Template button on the Select dialog box for later retrieval. Choose a descriptive name for your query so that it will be easy to find the next time you wish to use the template. To retrieve a saved query template, simply select Load Template at the bottom of the Select dialog box, and select the MapInfo query (**.qry*) file to be loaded.

Expression Dialog Box

The Expression dialog box is displayed when you select the Assist button on the Query | Select dialog box. The dialog contains three pull-down menus to make it easier for you to formulate expressions or criteria for selecting data: Columns, Operators, and Functions. All of these pull-down menus can be used for assistance in building expressions. First, let's take a quick look at the items contained in each of the pull-down menus. An example expression will be built using the Expression dialog, and additional example expressions for selecting data are reviewed.

Columns

The Columns pull-down menu on the Expression dialog, show in the following illustration, will display a list of all column names in the table. You can scroll up and down the list to select the required column name.

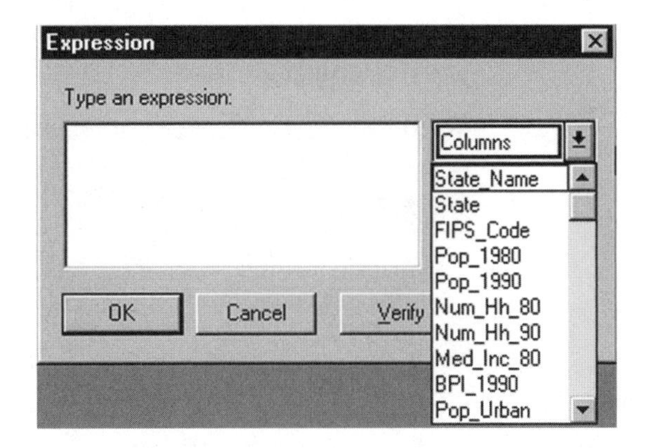

Expression dialog with Columns pull-down menu.

As you build an expression using a column from a table, you can either type in the column name or use the column pull-down list. Because all column names appear in the pull-down list, you do not have to remember the exact name or spelling of all attributes in data tables.

Operators

MapInfo Professional supports many operators in the Operators menu, shown in the following illustration. An operator is a symbol that tells the software to perform a mathematical or logical manipulation that relates the data you want to see to the attributes by which they can be selected. There are three classes of operators in MapInfo: arithmetic, logical, and geographical.

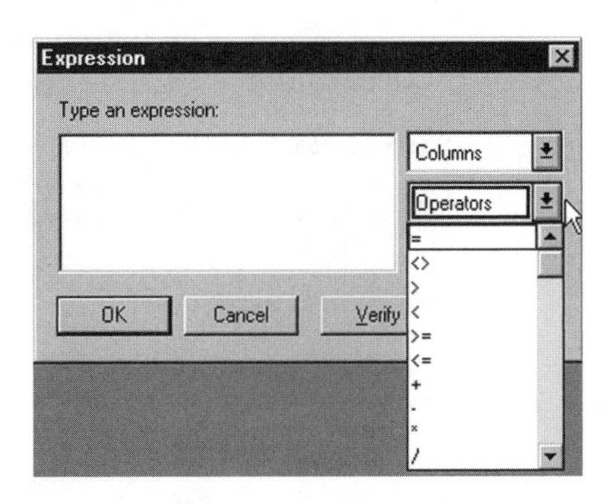

Expression dialog with Operators pull-down menu.

Arithmetic and logical operators are listed in tables 4-1 and 4-2, which follow, with notes to help you determine when and how to use them. Spatial or geographical operators (e.g., *contains*, *contains entire*, *within*, *entirely within*, and *intersects*), because of their greater complexity, are defined separately in a later section.

Table 4-1: Arithmetic Operators

Operator	Meaning
+	Addition
-	Subtraction
*	Multiplication
/	Division
^	Exponentiation
( )	Precedence

Table 4-2: Logical Operators

Operator	Meaning
=	Equal to
<>	Not equal to
<	Less than
>	Greater than
>=	Greater than or equal to
<=	Less than or equal to

Table 4-3, which follows, lists connectors used to connect two query conditions via the rules of logic.

Table 4-3: Connectors

Connector	Meaning
AND	For the features to be selected, the logical expressions on both sides of AND must be true.
OR	For the features to be selected, the logical expression on one or both sides of OR must be true.
NOT	Negate the condition.

Table 4-4, which follows, lists LIKE operators used for pattern matching.

Table 4-4: LIKE Operators

Operator	Meaning
_ (underscore)	Represents a single space.
%	Represents any number of spaces or characters.

By combining the operators of tables 4-1 through 4-4 and the columns in your database, you can construct the expression or criteria for selecting the data you want. It is possible to get the answer to a complex question by building a simple expression, and then building another expression based on the answer to the first query. However, it is more efficient to use connectors to build a single longer query for the results you want. The following are a few example expressions using the operators previously described to answer specific questions.

- Select all states with more than 5 million households, or total sales greater than $100,000,000. See the expression shown in the following illustration.

Expression selecting states with over 5 million households or sales greater than $100 million.

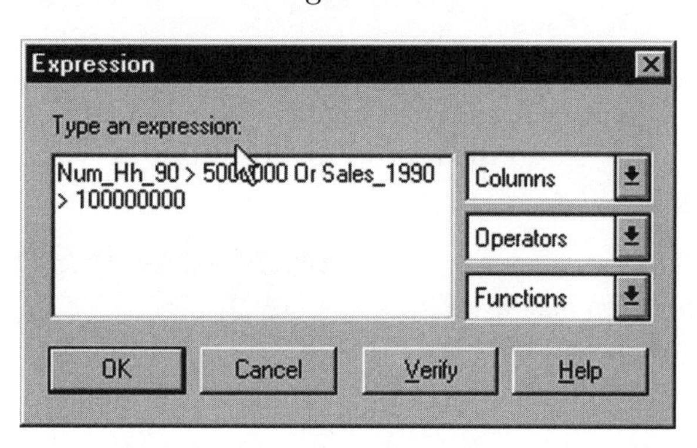

- Select all states whose population increased more than 10% between 1980 and 1990. See the expression shown in the following illustration.

Expression selecting states with greater than 10% population increase in the 1980s.

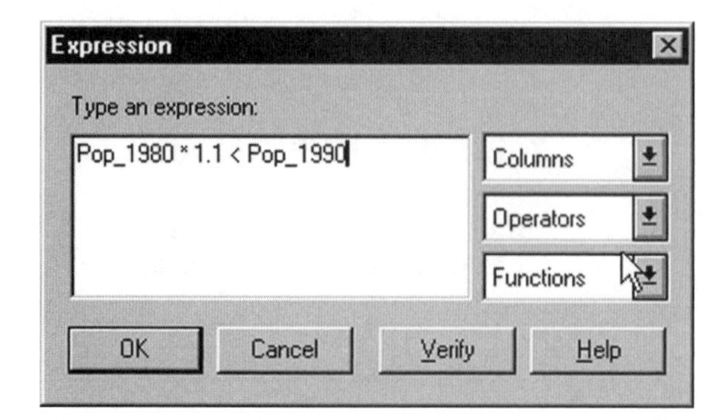

- Select all cities in the *City_1k* table that end with the word *City*. See the expression shown in the following illustration.

Expression selecting cities whose names contain City.

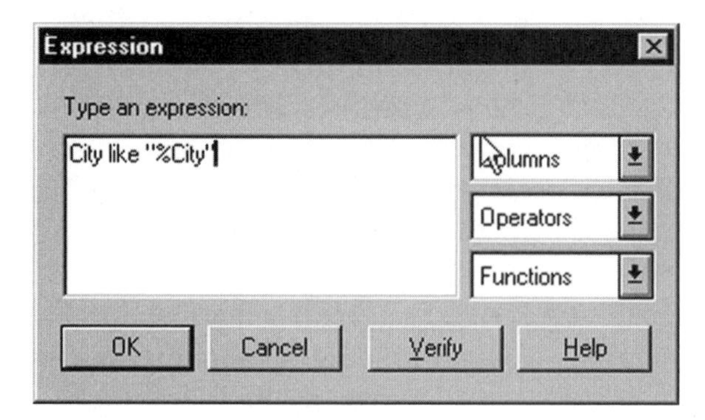

An expression can also include one or more function calls. MapInfo Professional supports a large number of functions, as shown in the following illustration.

Expression dialog with Functions pull-down menu.

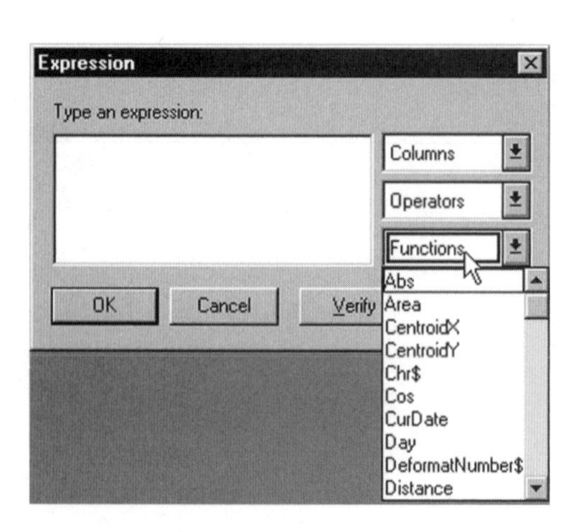

Tables 4-5 through 4-8, which follow, list each of the functions (and a short description of what the function does) for, respectively, math functions, date functions, string functions, and functions that return geographical calculations. In these tables, *num* is any numeric expression, *str* is any string expression, and *obj* is any object expression.

Table 4-5: Math Functions

Function	Effect
Abs(num)	Returns the absolute value of a number.
Cos(num)	Returns the cosine of a number; *num* is in radians.
Int(num)	Returns the integer (whole number) portion of a number.
Maximum(num, num)	Returns the larger of two numbers.
Minimum(num, num)	Returns the smaller of two numbers.
Round(num1, num2)	Returns a number (*num1*) rounded off to the nearest value of *num2*.
Sin(num)	Returns the sine of a number; *num* is in radians.
Tan(num)	Returns the tangent of a number; *num* is in radians.

Table 4-6: Date Functions

Function	Effect
CurDate( )	Returns the current date.
Day(date)	Returns the day of the month (1-31) portion of the date.
Month(date)	Returns the month (1-12) portion of the date.
Weekday(date)	Returns the weekday (1-7) portion of the date; 1 represents Sunday.
Year(date)	Returns the year portion (e.g., 1994) of the date.

Table 4-7: String Functions

Function	Effect
Chr$(num)	Returns a character that corresponds to a character code. For example, Chr$(65) returns the *A* character.
DeformatNumber$(str)	Reverses the effect of the *FormatNumber$* function; that is, it returns a string that does not include thousands separators.

Table 4-7: String Functions

Function	Effect
Format$(num, str)	Returns a string representing a formatted number. Example: *Format$(12345.678, "$,#.##")* returns $12,345.68.
FormatNumber$(num)	Returns a string representing a number formatted with thousands separators. Although it is simpler to use than *Format$*, it is also less powerful.
InStr(num, str1, str2)	Searches the string *str1* starting at character position *num*, and looks for an occurrence of the string *str1*. Returns the position where *str2* was found, or zero if not found. To start search at beginning, use a *num* value of one (1).
LCase$(str)	Returns a lowercase version of the string *str.*
Left$(str, num)	Returns the first *num* characters of the string *str.*
Len(str)	Returns the number of characters in a string.
LTrim$(str)	Trims any spaces from the start of *str* and returns result.
Mid$(str, num1, num2)	Returns a portion of the string *str* starting at character position *num1* and extending for *num2* characters.
Proper$(str)	Returns a string with capitalization (first letter of each word capitalized).
Right$(str, num)	Returns the last *num* characters of the string *str.*
RTrim$(str)	Trims spaces from the end of *str* and returns result.
Str$(expr)	Returns a string approximation of an expression.
UCase$(str)	Returns an uppercase (all capitalized) version of *str.*
Val(str)	Returns the numeric value of the string. For example, *Val("18")* returns the number 18.

Table 4-8: Functions That Return Geographical Calculations

Function	Effect
Area(obj, str)	Returns the area of the object. The *str* parameter specifies an area unit name, such as *sq. mi* or *sq. km.*
CentroidX(obj)	Returns the *x* coordinate of the object's centroid.
CentroidY(obj)	Returns the *y* coordinate of the object's centroid.
Distance(num_x, num_y, num_x2, num_y2, str)	Returns the distance between two locations. The first two parameters specify the *x* and *y* value of the start location. The next two parameters specify the *x* and *y* value of the end location. The *str* parameter is a distance unit name, such as *mi* or *km.*

Table 4-8: Functions That Return Geographical Calculations

Function	Effect
ObjectLen(obj, str)	Returns the length of the object. The *str* value specifies a distance unit name, such as *mi* or *km*. Only line, polyline, and arc objects have non-zero lengths.
Perimeter(obj, str)	Returns the perimeter of the object. The *str* value specifies a distance unit name, such as *mi* or *km*. Only region, ellipse, and rectangle objects have non-zero perimeters.

The functions that return geographical calculations also include *CartesianArea*, *CartesianDistance*, *CartesianObjectLen*, and *Cartesian-Distance* to perform these calculations using the Cartesian (non-earth, x/y) coordinate system, and *SphericalArea*, *SphericalDistance*, *SphericalObjectLen*, and *SphericalDistance* to perform geographic calculations using the earth-bound coordinate systems. Using some of the previous functions, you can calculate or format the results you want. Examples of using these functions follow.

- Select all states with an area of less than 50,000 square miles. See the expression shown in the following illustration.

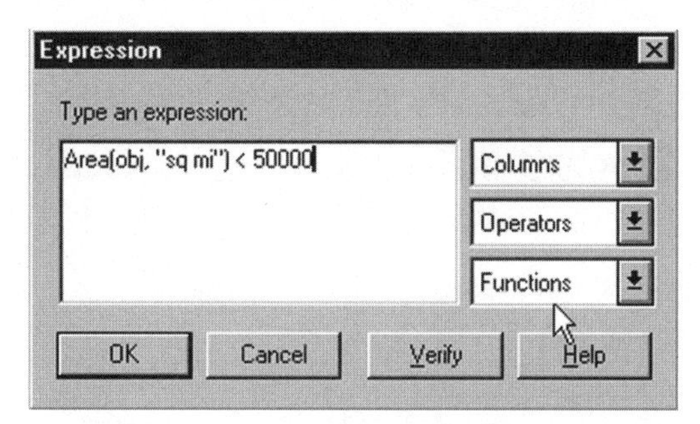

Expression dialog for finding small states.

- Select cities at longitude less than -95.0 (western U.S.). See the expression shown in the following illustration.

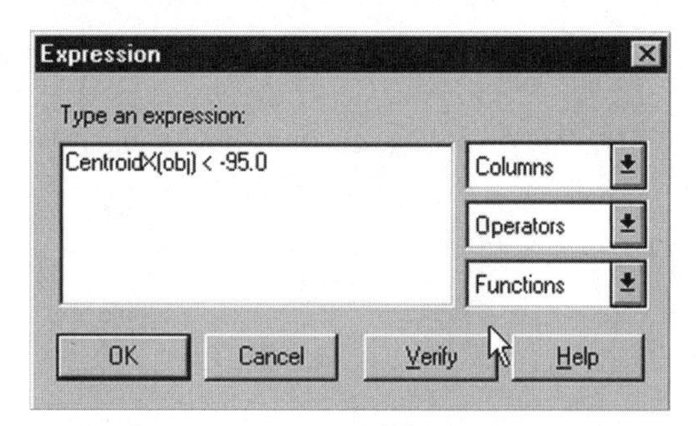

Expression dialog for finding western cities.

Building Expressions

Using the Column, Operator, and Functions pull-down lists makes building expressions easier. In tutorial 4-12, which follows, you will use the Expression dialog for building a Select statement to find the states with a population density greater than 100 people per square mile.

▼ TUTORIAL 4-12: BUILDING A SELECT STATEMENT USING THE EXPRESSION DIALOG

1 Select the Query | Select Menu bar option.

2 Set the States table as the table to select records from.

3 Click on the Assist button.

4 Use the pull-down menus to build the expression shown in the illustration at right. Select *Pop_1990* from the Columns pull-down menu.

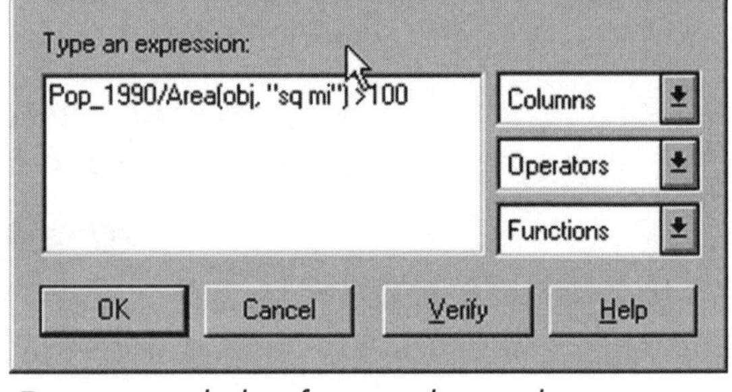

Expression dialog for population density greater than 100 per square mile.

5 Select / from the Operators pull-down menu.

6 Select Area from the Functions pull-down menu.

7 Select > from the Operators pull-down menu.

8 Key in *100.*

NOTE: *As an alternative to using the pull-down menus (operator, column, and function), you can simply type the expression in the text box. Many MapInfo users build expressions with both key-ins and pull-down menu selections. When keying in an expression, be sure that your column names are correct, or your query will not process. An easy way to check the query statement is to use the Verify button on the bottom of the dialog box.*

9 Click on the OK button. This will access the Select dialog box, shown in the illustration at right.

10 Click on the OK button on the Select dialog.

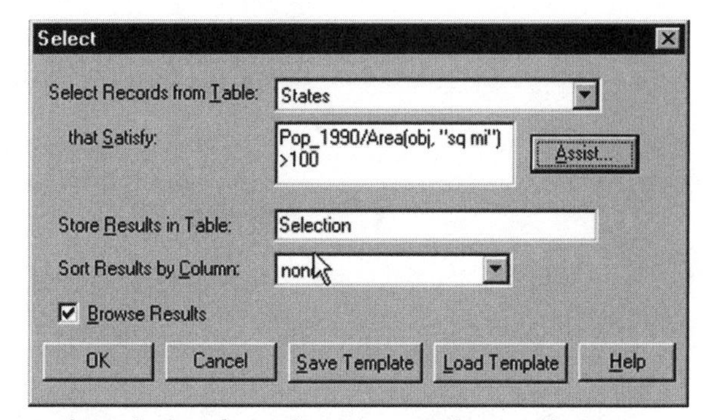

Select dialog for selecting population density greater than 100 per square mile.

The Browser window shows the 22 states with a population density of greater than 100 people per square mile. The 22 states are also highlighted in the Map window, as shown in the following illustration.

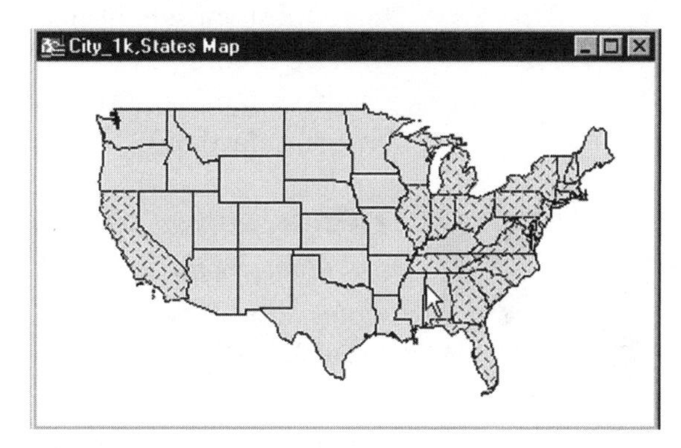

Map showing densely populated states.

Additional examples of how the Select statement method can be used are explored in tutorial 4-13, which follows. The first selection can be executed using the *States* table found on the companion CD-ROM. The remaining two selections are examples of the type of query you can carry out using Select statements.

▼ TUTORIAL 4-13: FURTHER SELECT STATEMENT METHODS

1 Select the states with an area of less than 50,000 square miles and with a population greater than 1,000,000.

2 Select all sales records for 1993.

3 Select all sales records for the month of March.

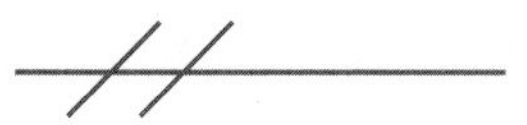

The Query | Select dialog box will retain the last query issued. This allows you to easily modify your last query, or issue the same query again.

SQL Select

The SQL Select dialog allows you to build more complex queries than the Select dialog. SQL stands for Structured Query Language and is usually pronounced "sequel." SQL, finalized by an ANSI committee in 1981, is an official standard relational language for databases.

The SQL Select dialog is based on the SQL standard. The main keywords for selecting information are *select, from, where, order by,* and *group by.* These keywords are assembled and issued to MapInfo to request subsets of information from tables. If you have worked with SQL before, you will find MapInfo Professional's implementation very familiar, but somewhat limited (see the sidebar "MapInfo and Standard SQL" in this chapter for further discussion).

TIP: *You can type SQL statements directly on the MapBasic window. For users that are familiar with SQL, his may be faster than using the query dialog boxes. To access the MapBasic window, select Options | Show MapBasic Window from the Main menu.*

Every query formulated with the Select dialog can also be formulated with the SQL Select dialog, shown in the following illustration. The SQL Select dialog is a bit more complicated, but with powerful features to provide greater control over data selection.

The SQL Select dialog provides the same Columns, Operators, and Functions pull-down menu items as the Select dialog. The pull-down menus perform the same as those on the Query | Select dialog box previously discussed. The SQL Select dialog contains two additional pull-down menus for selecting Tables and Aggregates.

The Tables pull-down menu item gives you a list of all currently open tables. The list will contain both base tables and temporary tables (query tables you have created using Select or SQL Select).

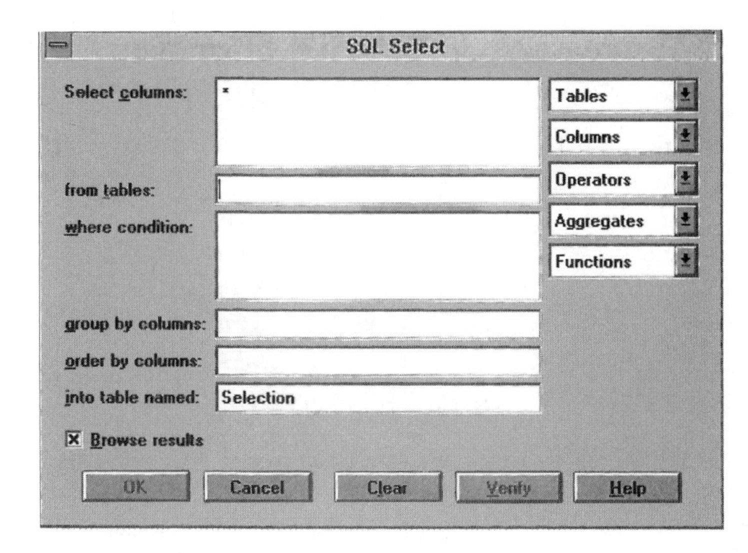

SQL Select dialog.

Aggregate functions are discussed in the "Group By Clause" section that follows.

Where Clause

The *Where* clause specifies the qualifications to use for selecting the desired data. This clause is constructed the same way clauses are constructed in the Expression dialog, discussed previously in the section on the Select dialog. As seen in the following, more complex expressions can be built with the SQL Select *Where* clause. In addition, when creating a query based on more than one table, the *Where* clause allows you to specify how the tables are related.

Group By Clause

The *Group By Columns* clause is used to display summary information about *groups of rows* that contain the same values in one or more fields. For example, assume you are working with a sales report table. The rows in the table contain fields for salesperson, product, amount, and client, among others. Assume also that you want the average sales figure for each salesperson. In this case, you would use a *Group By* clause to group the data by salesperson (salesperson field). MapInfo Professional supports the following six aggregate functions that are used for summarizing information with the *Group By* clause.

Table 4-9: Functions Used with the Group By Clause

Function	Effect
Avg(column_name)	Calculates an average for the values in the specified column in the query result group.
count(*)	Counts the number of rows in the query result group.
Max(column_name)	Determines the maximum value for the specified column in the query result group.
Min(column_name)	Determines the minimum value for the specified column in the query result group.
Sum(column_name)	Calculates a total for all values in the specified column in the query result group.
WtAvg(column_name, column_name)	Calculates a weighted average for the values in the first specified column, weighted by values in the second column, in the query result group.

For example, using the *Group By* clause, you can find the average population of the 1,000 largest cities by state. Tutorial 4-14, which follows, takes you through this process.

▼ TUTORIAL 4-14: QUERYING USING A GROUP BY CLAUSE

1 Select the SQL Select option from the Query menu.

2 Use the pull-down menus to build the query shown in the illustration at right. With the cursor in the "from Tables" text box, select *City_1k* from the Tables pull-down menu.

3 With the cursor in the Select Columns text box, select *State* from the Columns pull-down menu.

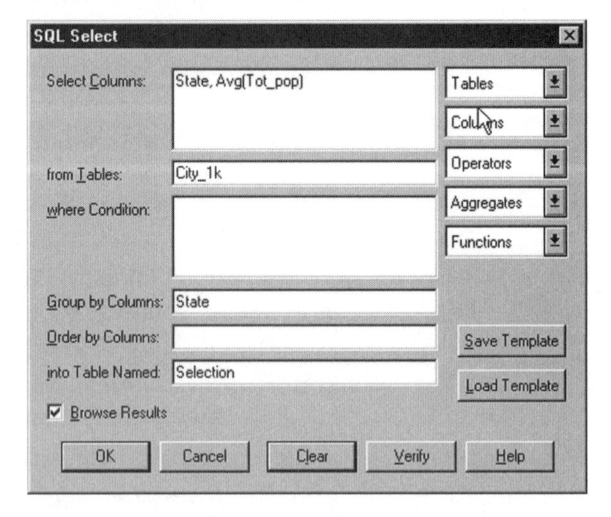

SQL Select dialog of Group By *clause.*

4 Verify that the cursor is in the Select Columns text box, and select *Avg* from the Aggregates pull-down menu.

5 Select *Tot_pop* from the Columns pull-down menu.

6 Move the cursor to the Group by Columns text box, and select *State* from the Columns pull-down menu.

NOTE: *As in the case of the Expression dialog box discussed previously, you can type in the required fields, use the pull-down menus, or a combination of both input methods.*

7 Click on the OK button. The resulting Browser for the *Group By* clause is shown at right.

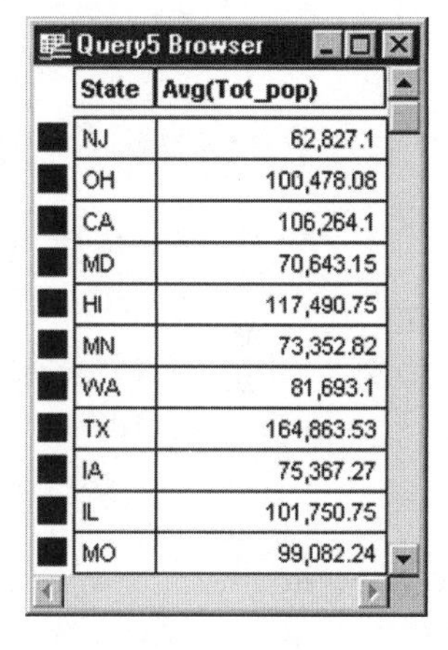

Query5 Browser

State	Avg(Tot_pop)
NJ	62,827.1
OH	100,478.08
CA	106,264.1
MD	70,643.15
HI	117,490.75
MN	73,352.82
WA	81,693.1
TX	164,863.53
IA	75,367.27
IL	101,750.75
MO	99,082.24

Resulting Browser window for Group By clause.

The resulting Browser window, shown in the previous illustration, displays each state name, followed by the average population of the cities in the state.

TIP: *The SQL Select dialog retains the settings from the previous SQL Select command. Use the Clear button to start with a fresh dialog.*

The *Group By* clause can answer many summary types of questions, such as the following:

- How many customers are located in each county in California?
- What is the average household income, by zip code area, in Missouri?
- Which cities in West Virginia have the largest populations?
- How many sales of product X, by state, were recorded in the Northeast this month?

Order By Clause

The *Order By* clause returns the requested data sorted in the specified order. You may wish to order data in alphabetical order from A to Z, or from the largest to smallest value. Frequently, ordering the SQL Select statement results makes analysis of the information easier. For instance, to analyze the 1,000 largest cities in the United States by population, let's sort the table by population, as demonstrated in tutorial 4-15, which follows.

▼ TUTORIAL 4-15: QUERYING USING AN ORDER BY CLAUSE

1 Select the SQL Select option from the Query menu.

2 Build the query shown at right. Use the pull-down menus or type in the query. If you are uneasy about using the pull-down menus, refer to the steps in the example in the previous "Group By Clause" section.

3 Click on the OK button. The result is displayed in the Browser window, shown in the following illustration.

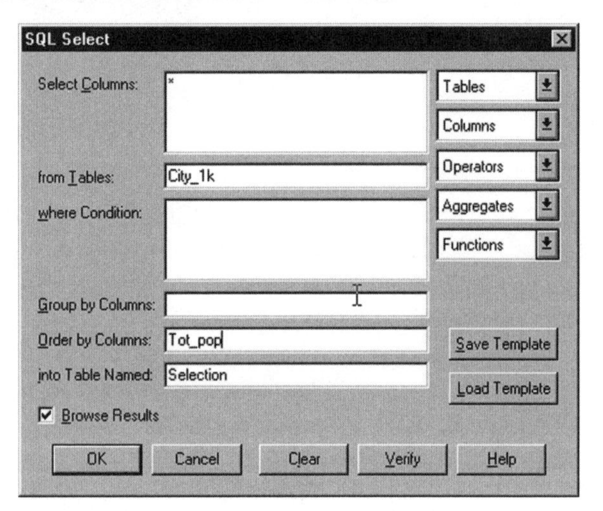

SQL Select dialog using Order By.

Query6 Browser

City	State	Tot_hu	Tot_pop
North Brunswick Township	NJ	12,186	31,287
Fairborn	OH	13,288	31,300
Danville	CA	11,466	31,306
Chillum	MD	12,093	31,309
El Centro	CA	10,180	31,384
Waipahu	HI	7,739	31,435
Mankato	MN	11,688	31,477
Gilroy	CA	9,767	31,487
Longview	WA	13,441	31,499
Pleasant Hill	CA	13,653	31,585
Sherman	TX	14,261	31,601

Browser window of largest cities sorted by population in ascending order.

North Brunswick Township is the city with the smallest population, followed by Fairborn. Upon scrolling to the bottom of the Browser window, you see that New York has the largest population.

To examine the cities from the largest to the smallest, sort the table in descending order, so that those at the top of the list have the largest populations. The SQL Select statement for this would include the keyword *desc*, as follows:

```
Select * from CITY_1K order by tot_pop desc
```

In the SQL Select dialog box, select the text boxes and select or key in the statements shown in table 4-10, which follows.

Table 4-10: First Settings for the SQL Select Dialog Box

Text Box	Statement
Select Columns	*
from Tables	CITY_1K
Order by Columns	tot_pop desc

Sorting data by more than one column is sometimes desirable. If you wanted to analyze large cities in each state, you would sort by state and then sort by population. The resulting table will contain cities organized by state, with the largest cities listed first. The query for this data sort follows:

```
Select * from CITY_1K order by State, tot_pop desc
```

Table 4-11, which follows, lists the text box selections and statement entries associated with this query.

Table 4-11: Second Settings for the SQL Dialog Box

Text Box	Statement
Select Columns	*
from Tables	CITY_1K
Order by Columns	state, tot_pop desc

Joining Databases

In the desktop mapping environment, data are typically stored in several different tables, such as geographical data obtained from a commercial vendor and company databases containing diverse information. SQL Select statements allow you to join the information from various tables into a single results table.

To join two tables, a data column in the first table must match a data column in the second table. The matching column is often referred to as a *key* column. To build an SQL Select statement that joins the *STATE* table with the *City_1k* table, perform tutorial 4-16, which follows.

▼ Tutorial 4-16: Joining Database Tables

1 Select the SQL Select option from the Query menu.

2 Use the pull-down menus. Select the *States* table first, and then the *City_1k* table.

3 Once you have selected the tables, MapInfo automatically sets the *Where* condition portion of the SQL Select dialog box to *States.State=City_1k.State*. In this instance, MapInfo sets the *Where* condition based on *State*, the matching column. This is the join you want, shown in the illustration at right.

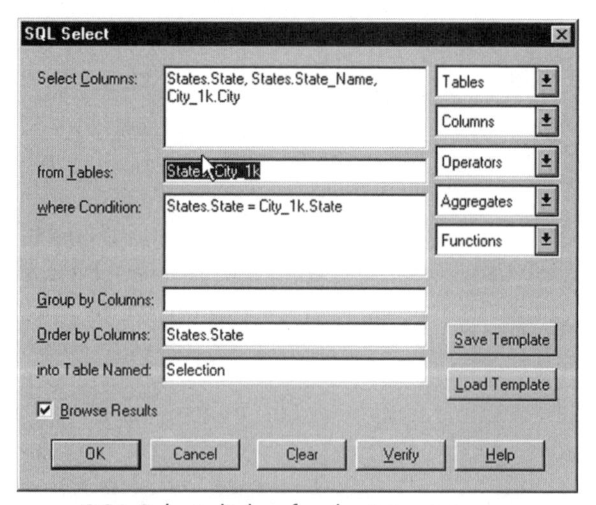

SQL Select dialog for the Join statement.

NOTE: *If MapInfo's "guess" about the join condition is not what you want, you would replace the* Where *condition of the SQL Select dialog box with the desired condition.*

4 Click on the OK button. In the resulting table you can see the *State* and *State_Name* columns from the *States* table, followed by the *City* column from the *City_1k* table.

5 Set the Order by Columns clause to *States.State*. The resulting Browser window of joined data is shown in the following illustration.

NOTE: *When querying multiple tables, you will need to specify both the table and column names in the query. For example, when querying both the* States *and the* St_attr *tables, you would need to specify* States.State *to query the* States *column from the* States *table. When you use the pull-down list of columns, MapInfo will automatically place both the table and column name in the query.*

Query8 Browser

State	State_Name	City
AK	Alaska	Anchorage
AL	Alabama	Prichard
AL	Alabama	Florence
AL	Alabama	Hoover
AL	Alabama	Gadsden
AL	Alabama	Decatur
AL	Alabama	Dothan
AL	Alabama	Tuscaloosa
AL	Alabama	Huntsville
AL	Alabama	Montgomery
AL	Alabama	Mobile

Resulting Browser window showing joined data.

The order of the tables named in the From Tables box of the SQL Select statement is important. The resulting table will contain only the map objects from the first table listed in the "from Tables" section of the query.

When you join two tables, the number of rows in the results table will depend on how well the two tables match. For example, if you join a table of *CITY* records to a table of *States* records, your results table will either contain all rows from the *CITY* table or only those that have a matching state record. The *States* rows may not match due to foreign city records or data entry errors.

TIP: *To optimize your SQL statements, index the columns of data on which you will commonly be basing selections or joins. To index a data column, select Table | Maintenance | Table Structure from the Main menu bar. Check the index box of each data attribute or data column you*

want to index. Remember that indexing too many columns will defeat the purpose of indexing and could actually increase processing time.

NOTE: *The sidebar that follows contains valuable information regarding the differences and similarities between MapInfo and standard SQL. Readers are encouraged to examine this information.*

MapInfo and Standard SQL

If you are a newcomer to MapInfo from the world of databases, or if you are a mapping professional who has become recently educated about relational databases, you might be tempted to characterize the package as a relational database with geographic functions and operations.

For certain, the similarities between MapInfo's table management and various RDBMS functions are plentiful. The TABLE:MAINTENANCE operation, for example, positions the user for easy table management, whereas MapInfo's TABLE:SQL SELECT positions the user with expectations of full SQL functionality. Be aware, however, that there are many differences between MapInfo's SQL implementation and ANSI standards.

Perhaps the most quickly identifiable differences surface with Delete and Update operations. Delete, a common SQL command, is not even found among MapInfo's standard menus. The Update performs within significant limitations. The Standard MO (mode of operation) for most commercial RDBMS packages is to enable users to specify *Where* clauses at the end of their Update statements. For example, one might expect the statement *Update Stores Set Priority=1 Where Sales>1000000* to function cleanly. A *Where* clause, though, cannot be found anywhere in MapInfo's *TABLE:UPDATE COLUMN* dialog. To achieve the desired update, one must instead query out the stores where sales are greater than *1000000* and then update the resultant query set.

MapInfo does address some of these syntactical limitations to a limited extent in MapBasic (the PC application development environment). Delete and Updates can be performed with *Where* clauses. However, these *Where* clauses are a bit hampered; that is, they are not SQL Where clauses in the ordinary sense because they can only support equivalence tests with row *ids*. To move around quickly with Update and Delete processes, users should become familiar with a two-step process: query on the target records and then delete or update the queried records.

A second readily apparent difference in MapInfo's SQL implementation is the lack of some keywords. For example, if you want to retrieve nonredundant fields from a table, you cannot specify DISTINCT or UNIQUE as a function upon the column. Past RDBMS users might be thinking *Select Unique(Customer_Name) from Customers*, but in MapInfo this needs to executed using a *Group By* clause (e.g., *Select Customer_Name from Customers Group By Customer_Name*).

Similarly, if you want to detect records with columns that have not been assigned values, no NULL function is available. In this case, one needs to query in a manner that identifies null properties (e.g., *Select Customer_Name from Customers where Address =""*).

One area that can at first cause chagrin with MapInfo is subqueries. If one types in *subqueries* in the Help Index for MapInfo or MapBasic, it can appear that subqueries are unsupported. Fortunately, this is misleading: subqueries can be used successfully in both environments. For example, if you want to acquire information about retailers located in the same city as your higher-value customers, you can issue:

```
Select * from Retailers where City in (Select City from
   Customers where value>100000)
```

Note that the data in this type of query could be returned through a simple join as well (e.g., *Select * from Retailers, Customers where Retailers.City=Customers.City and Customers>100000*), However, (1) the amount of columns in the returning data set is different, and (2) the "join" does not break out the logic as well as the subquery.

If you come to MapInfo from the relational world, you might also note that this discussion is of queries, not "views." Views are logical representations of physical tables in an RDBMS. The nearest equivalent in MapInfo is the concept of a "template," in which users can open queries constructed from previous sessions.

The absence of views reflects a relatively lean data dictionary in MapInfo that is different from the overhead maintained for packages such as Oracle, Informix, Access, and SQL Server. These packages set up the tables for administrative and optimization practices to a degree that MapInfo does not attempt. Of course, in fairness to MapInfo, it is also right to point out that MapInfo sets up tables for geographic operations and efficient processing that the RDBMS packages do not attempt!

You will find MapInfo very nimble and capable of providing many aspects of the better-known commercial database systems. It is remarkably flexible with respect to its ability to connect with Oracle, Informix, Access, and other packages. However, keep in mind that MapInfo does not fully replicate ANSI standard SQL and relational operations by itself.

Saving Queries

Query statements can be saved for later use. In the SQL Select dialog box, Save Template and Load Template work the same as in the Select dialog box (previously discussed). To save a query statement for later use, click on the Save Template button on the SQL Select dialog box, and input a descriptive name for the query template. To retrieve a saved query, click on Load Template from the SQL Select dialog box, and select the saved query you wish to use. The chosen query will populate the SQL Select dialog box with the saved query statement.

Subqueries

A subquery is a *Select* statement placed inside the *Where* clause of the SQL statement. You might say it is a "Select within a Select." MapInfo Professional first evaluates the subquery and then uses the results of the subquery to evaluate the main SQL statement. Subqueries are more complicated than any of the queries reviewed thus far. Knowledge of how to build SQL statements is required to effectively use subqueries. The following examples are intended as a sampling of how subqueries are structured. For detailed information on building SQL statements, numerous references are available, such as *A Guide to the SQL Standard* by C. J. Date (Reading, MA: Addison-Wesley Publishing Company, Inc., 1987).

Suppose you want to select all states where the population is greater than the national state average for 1990. Because you do not know the average population, the first (or inner) query will calculate the average. Then the main (or outer) query will select all states with a population greater than the average. To begin working with subqueries, perform tutorial 4-17, which follows.

▼ *TUTORIAL 4-17: WORKING WITH SUBQUERIES*

1 Select the SQL Select option from the Query menu.

2 Use the pull-down menus to build the following *Select* statement. Table 4-12, which follows, provides the text box selections and statement entries associated with this subquery. The illustration that follows the table shows the SQL Select dialog box for this subquery.

```
Select * from STATES where pop_1990 > (Select Avg(pop_1990)
   From STATES)
```

Table 4-12: Third Settings for the SQL Select Dialog Box

Text Box	Statement
Select Columns	*
from Tables	STATES
where Condition	pop_1990 > (Select Avg(pop_1990) From STATES)

SQL Select dialog box for the subquery.

3 Press the OK button and review the results, shown in the following illustration.

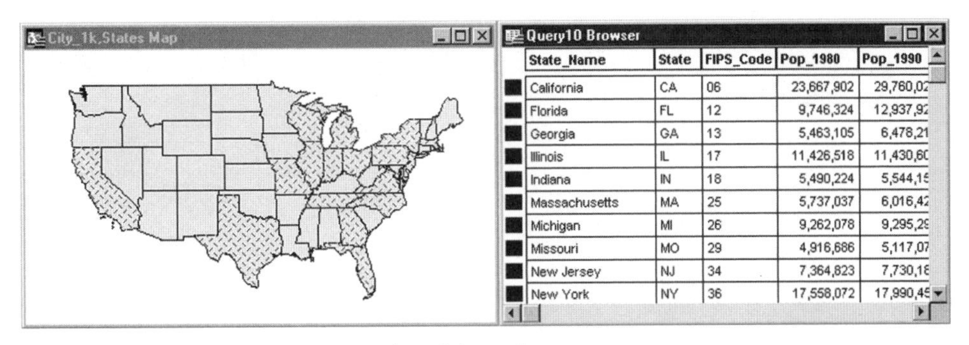

Browser showing results of the subquery.

Another example of how you could use a subquery is to locate all states east of Colorado. The query to answer this question follows.

Table 4-13 provides the text box selections and statement entries associated with this subquery.

```
Select * from STATES where CentroidX(obj) > (Select CentroidX(obj)
  from STATES where State = "CO")
```

Table 4-13: Fourth Settings for the SQL Select Dialog Box

Text Box	Statement
Select Columns	*
from Tables	STATES
where Condition	where CentroidX(obj) > (Select CentroidX(obj) from STATES where State = "CO")

The previous query becomes more complex upon requesting the states east of Colorado with a population greater than 5 million. Adding another qualification at the end of the previous query results in the following query. Table 4-14 provides the text box selections and statement entries associated with this subquery.

```
Select * from STATES where CentroidX(obj) > (Select CentroidX(obj)
  from STATES where State = "CO" And Pop_1990 > 5000000)
```

Table 4-14: Fifth Settings for the SQL Select dialog Box

Text Box	Statement
Select Columns	*
from Tables	STATES
where Condition	CentroidX(obj) > (Select CentroidX(obj) from STATES where State = "CO" And Pop_1990 > 5000000)

TIP: *To reduce processing time when using subselects (in subqueries), make your inner SQL statement as specific as possible.*

Spatial Queries

A spatial query helps you locate geographic objects in relation to other geographic objects. Spatial data are classified as points, lines, areas, or surfaces. By using the join capability together with a geographic operator, you can join data that are not specifically related by a common data attribute (i.e., column).

Earlier discussion covered the Boundary Select tool from the Main toolbar. This selection provides the same results as a very simple spatial query. If you want to find all cities in a state, or customers in a zip code, the Boundary Select tool can easily provide such results, assuming you know where the subject boundary is located. To review how to build a spatial query, let's take a simple example and build a query that finds all cities located in Texas. Tutorial 4-18, which follows, takes you through this process.

▼ *TUTORIAL 4-18: PERFORMING SPATIAL QUERIES*

1 Select the SQL Select option from the Query menu.

2 Use the pull-down menus to build the following *Select* statement. Table 4-15 provides the text box selections and statement entries associated with this query. The illustration that follows the table shows the SQL Select dialog box for this query.

```
Select * from STATES where obj intersects (Select obj from
   STATES where state = "TX")
```

Table 4-15: Sixth Settings for the SQL Select Dialog Box

Text Box	Statement
Select Columns	*
from Tables	STATES
where Condition	obj intersects (Select obj from STATES where state = "TX")

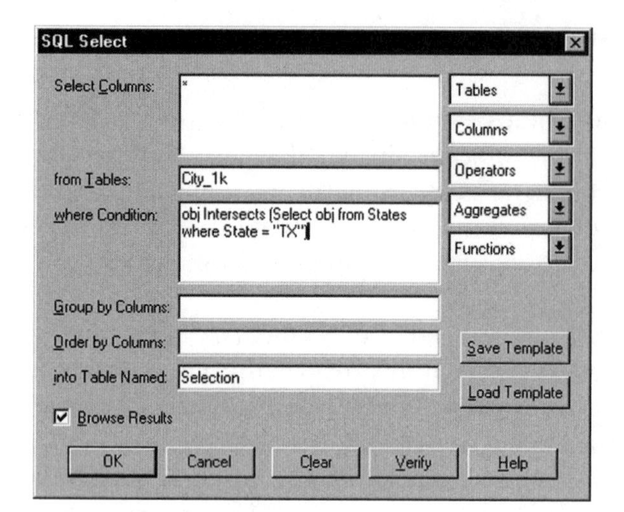

SQL Select dialog box for the spatial query.

3 Click on the OK button and review the results, shown in the following illustration.

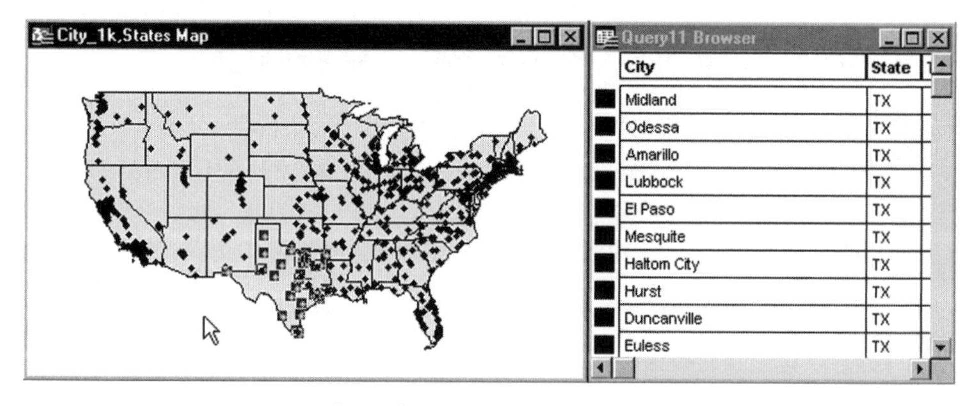

Browser window of cities in Texas.

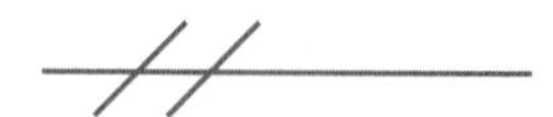

Geographic Operators

Spatial queries will use one of several geographic operations in the SQL Select statement. Table 4-16, which follows, summarizes the five geographic operators.

Table 4-16: Geographic Operators Used with the SQL Select Statement

Operator	Purpose
Contains	Object A *Contains* object B if B's centroid is anywhere within A's boundary.
Contains Entire	Object A *Contains Entire* object B if B's boundary is entirely within A's boundary.
Within	Object A is *Within* object B if its centroid is inside B's boundary.
Entirely Within	Object A is *Entirely Within* object B if A's boundary is entirely within B's boundary.
Intersects	Object A *Intersects* object B if the objects have at least one point in common or if one of them is entirely within the other.

Using NOT with Geographic Operators

It is sometimes easier to formulate a query using the graphical operators for the data you do not want. In these cases, use the *NOT* operator to change a query of the data you do not want into a query of the data you want. Suppose you want to find all cities not in the state of Texas. The query would be changed to the following:

```
Select * from STATES where NOT obj intersects (Select obj from STATES
  where state = "TX")
```

Table 4-17, which follows, provides the SQL Select dialog box settings used with the *NOT* operator.

Table 4-17: SQL Select Dialog Box Settings Used with a NOT Operator

Text Box	Statement
Select Columns	*
from Tables	STATES
where Condition	NOT obj intersects (Select obj from STATES where state = "TX")

TIP: *Graphical joins can be time consuming. If you have columns of data for creating the join, use the columns instead of the graphical relationship for faster processing.*

Examples of SQL Select Queries

All SQL types previously described can be used in conjunction with each other to answer common business questions. Many companies have data collected at the zip code level, but wish to aggregate data at the county level for decision making. The SQL Select statement for this situation follows:

```
Select TX_CNTY.County, TX_ZIP.ZIP_CODE from TX_CNTY,TX_ZIP where
  TX_CNTY.obj Contains TX_ZIP.obj
```

Table 4-18, which follows, provides the SQL Select dialog box settings used with this aggregator.

Table 4-18: SQL Select Dialog Box Settings Used with an Aggregator

Text Box	Statement
Select Columns	TX_CNTY.County, TX_ZIP.ZIP_CODE
from Tables	TX_CNTY, TX_ZIP
where Condition	TX_CNTY.obj Contains TX_ZIP.obj

Assume you live in the city of Washington, DC, and wish to travel to a city within a 200-mile radius. You know that Washington is located at -77.016167, 38.805050. You can find all cities within 200 miles of Washington and display that distance, along with the city names, via the following query:

```
Select City, Distance(CentroidX(obj), CentroidY(obj), -77.016167,
  38.805050, "mi") "dist" From CITY_1K Where Distance(CentroidX(obj),
  centroidy(obj), -77.016167, 38.805050, "mi") < 200.0
```

Table 4-19, which follows, provides the SQL Select dialog box settings used with this radius query.

Table 4-19: SQL Select Dialog Box Settings Used with a Radius Query

Text Box	Statement
Select Columns	Select City, Distance(CentroidX(obj), CentroidY(obj),-77.016167, 38.805050, "mi") "dist"
from Tables	CITY_1K

Table 4-19: SQL Select Dialog Box Settings Used with a Radius Query

Text Box	Statement
where Condition	Distance(CentroidX(obj), CentroidY(obj), -77.016167, 38.805050, "mi") < 200.0

TIP: *Columns can also be referred to by their relative numbers. The first column is also known as* col1, *the second as* col2, *and so forth. With this shortcut, you could add an* Order by Col2 *clause to the preceding query to order the distances.*

You may occasionally wish to select all point features that fall within boundaries defined by certain characteristics. For example, to select all cities in states with populations in excess of 4 million, use the following query:

```
Select * from CITY_1K where obj within any (Select obj from STATES
  where Pop_1990 > 4000000)
```

Table 4-20, which follows, provides the SQL Select dialog box settings for this type of defined radius query.

Table 4-20: SQL Select Dialog Box Settings Used with a Defined Point-Boundary Query

Text Box	Statement
Select Columns	*
from Tables	CITY_1K
where Condition	obj within any (Select obj from STATES where Pop_1990 > 4000000)

TIP: *When a subselect is not used with* any, all, *or* in, *the subselect must return exactly one row of values.*

With the SQL Select dialog you can also use some of the MapBasic commands within the *Select* statements. The use of MapBasic commands adds spatial characteristics to MapInfo Professional's SQL that are not available in most SQL implementations. (MapBasic is introduced in Chapter 12.) At this juncture, a couple of example queries are analyzed to demonstrate the results you can obtain by using MapBasic commands. To select all cities that fall within a

100-mile radius of Washington, DC, you can use the *CreateCircle* command in the *Select* statement as follows:

```
Select * from CITY_1K where obj within CreateCircle(-77.016167,
  38.805050,100)
```

Table 4-21, which follows, provides the SQL Select dialog box settings used with this type of MapBasic command.

Table 4-21: SQL Select Dialog Box Settings Used with MapBasic Commands

Text Box	Statement
Select Columns	*
from Tables	CITY_1K
where Condition	obj within CreateCircle(-77.016167, 38.805050,100)

To determine how many miles you travel on streets between the two points, you can use a combination of a MapBasic command and the *sum* aggregation function. First, select all street segments to be traveled, and then issue the following *Select* statement:

```
Select Sum(ObjectLen(obj, "mi")) from Selection
```

Table 4-22, which follows, provides the SQL Select dialog box settings used with this type of *sum* function.

Table 4-22: SQL Select Dialog Box Settings Used with a MapBasic Command and the Sum Aggregation Function

Text Box	Statement
Select Columns	Sum(ObjectLen(obj, "mi"))
from Tables	Selection

Displaying Query Records

To this point, the focus has been on methods of selecting data objects. Once you have made data selections, you will often want to present them in either a Map or Browser window to analyze the data subselections. In the remainder of the chapter, the Browser window is further examined; review of the Map window appears in subsequent chapters.

Data selections can easily be displayed in a new Browser window by selecting Window | New Browser Window from the Main menu bar. MapInfo Professional then provides several ways of controlling the appearance of the Browser window. This is desirable for purposes of data analysis and visualization, as well as for printing.

The first controls to be reviewed are those you can work with in a Browser window already on the screen. The second set of control options are built into the SQL Select dialog box.

Browser Window Customization

The Browser window has some of the same column controls found in many of today's popular spreadsheet packages. The columns in the Browser window can be moved, and can be made larger or smaller. In addition, you can choose which columns to view.

You can change column widths to make your data easier to view. By positioning the mouse pointer on the dividing line between attributes of the Browser window, the cursor changes shape to a line with arrows on both sides (see the following illustration). To resize the column, hold down the mouse button and drag the column to a new size.

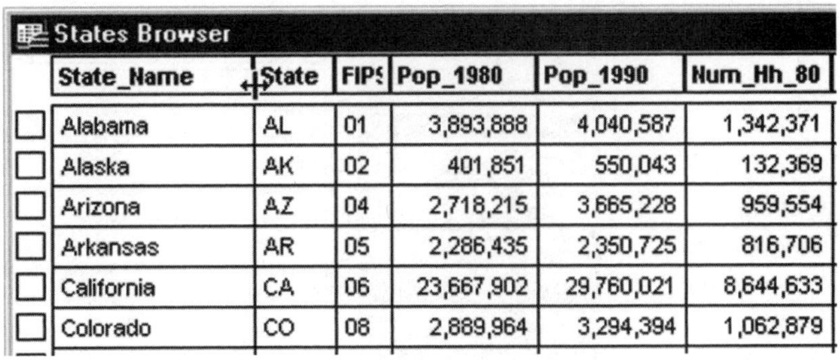

Cursor for resizing Browser window columns.

States Browser

State_Name	State	FIPS	Pop_1980	Pop_1990	Num_Hh_80
Alabama	AL	01	3,893,888	4,040,587	1,342,371
Alaska	AK	02	401,851	550,043	132,369
Arizona	AZ	04	2,718,215	3,665,228	959,554
Arkansas	AR	05	2,286,435	2,350,725	816,706
California	CA	06	23,667,902	29,760,021	8,644,633
Colorado	CO	08	2,889,964	3,294,394	1,062,879

The Browser window columns can also be moved around so that two data columns you are reviewing are located next to each other. Position the mouse pointer on the attribute name at the top of the Browser window. The cursor changes to the shape of a hand, as shown in the following illustration.

Cursor for moving Browser window columns.

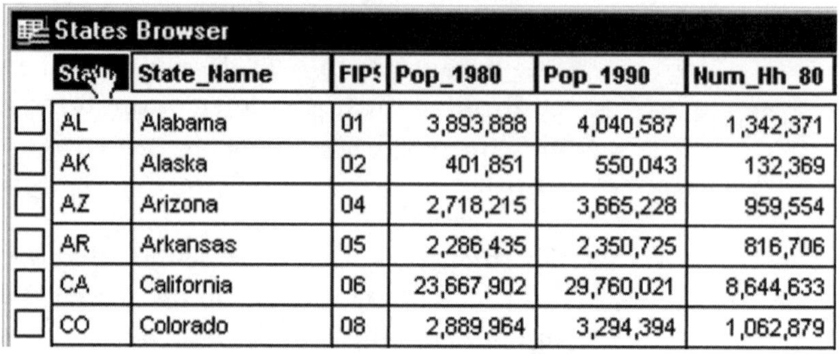

States Browser

	State	State_Name	FIPS	Pop_1980	Pop_1990	Num_Hh_80
☐	AL	Alabama	01	3,893,888	4,040,587	1,342,371
☐	AK	Alaska	02	401,851	550,043	132,369
☐	AZ	Arizona	04	2,718,215	3,665,228	959,554
☐	AR	Arkansas	05	2,286,435	2,350,725	816,706
☐	CA	California	06	23,667,902	29,760,021	8,644,633
☐	CO	Colorado	08	2,889,964	3,294,394	1,062,879

To place the column to the left or right of its current position, move the mouse so that the cursor is over the column name. Then, hold down the mouse button and drag the column in the desired direction. In addition to controlling the size and position of columns in the Browser window, you can also control the display of the grid lines and select which columns should be visible in the Browser window. When a Browser window is the active window, the Main menu bar contains an additional Browse menu item with the submenu items of Pick Fields and Options.

The Options menu item displays the dialog shown in the following illustration. This dialog allows you to remove the grid lines from the Browser window. Without the grid lines, the Browser window has an entirely new look.

Browse Options

☐ Show Grid Lines

OK Cancel

States Browser

	State_Name	FIPS	Pop_1980	Pop_1990	State	Num_Hh_80	Num_Hh_
☐	Alabama	01	3,893,888	4,040,587	AL	1,342,371	1,506,7
☐	Alaska	02	401,851	550,043	AK	132,369	188,
☐	Arizona	04	2,718,215	3,665,228	AZ	959,554	1,368,
☐	Arkansas	05	2,286,435	2,350,725	AR	816,706	891,1
☐	California	06	23,667,902	29,760,021	CA	8,644,633	10,381,
☐	Colorado	08	2,889,964	3,294,394	CO	1,062,879	1,282,4
☐	Connecticut	09	3,107,576	3,287,116	CT	1,094,281	1,230,4
☐	Delaware	10	594,338	666,168	DE	206,690	247,4
☐	District Of Columbi	11	638,333	606,900	DC	254,032	249,
☐	Florida	12	9,746,324	12,937,926	FL	3,748,542	5,134,
☐	Georgia	13	5,463,105	6,478,216	GA	1,872,564	2,366,

Browser Options window and the resulting Browser window.

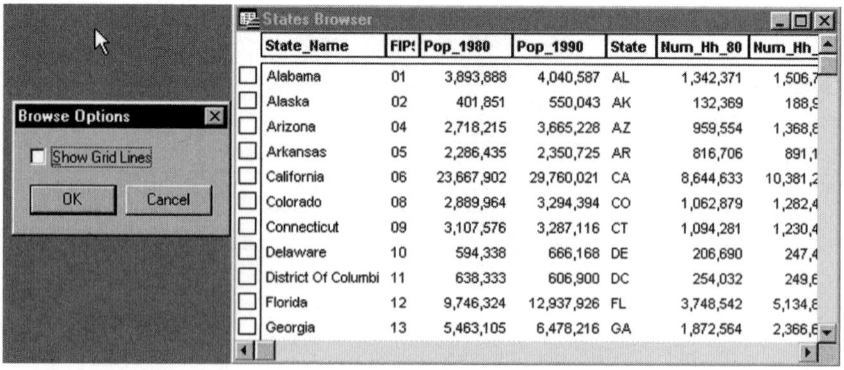

With the Pick Fields dialog, you can remove columns from the Browser window. For example, to remove all columns from the *States* table except *State_Name* and *1990_Pop*, perform tutorial 4-19, which follows.

▼ TUTORIAL 4-19: CUSTOMIZING THE BROWSER WINDOW

1 Select the Pick Fields option from the Browse menu.

2 In the Pick Fields dialog box, shown in the illustration at right, remove all columns from the list of column names at the right, except the *State_Name* and *1990_Pop* columns.

3 Click on the OK button. The Browser window now contains only the *State_ Name* and *Pop_1990* columns, as shown in the following illustration.

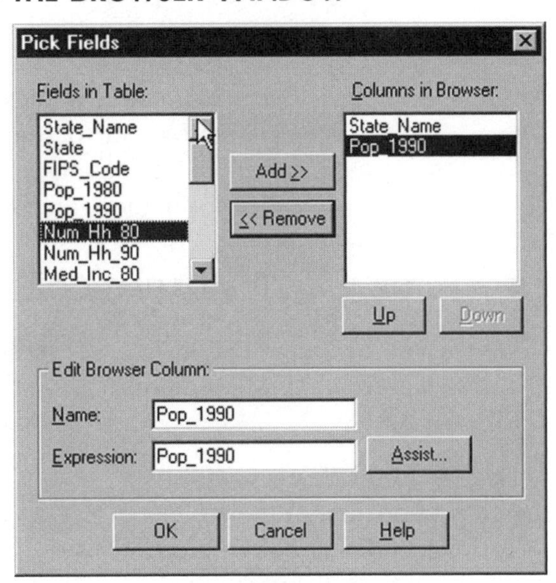

Pick Fields dialog box.

States Browser

State_Name	Pop_1990
Alabama	4,040,587
Alaska	550,043
Arizona	3,665,228
Arkansas	2,350,725
California	29,760,021
Colorado	3,294,394
Connecticut	3,287,116
Delaware	666,168

States *Browser window showing only the* State_Name *and* Pop_1990 *columns.*

SQL Select Browser Control Options

The previously discussed controls on the Browser window help customize the size and content of the Browser window. This section takes a look at a means of changing column names or controlling the format of data inside Browser cells. The SQL Select dialog can be used to create alias column names and format the data items.

Alias Column Names

The column names used in the Browser window are derived from the column names of the data table. Place alias names (in quotation marks) directly after the column name you wish to alias (e.g., *State_Name, "State Name"*), as shown in the following illustration. Because column names are often awkward and cryptic, using an alias allows you to rename the column with a more descriptive moniker. For example, to use aliases for columns while selecting state, population, and sales, perform tutorial 4-20, which follows.

▼ *TUTORIAL 4-20: ALIASING COLUMN NAMES*

1 Select the Query | SQL Select from the Main menu bar.

2 Fill in the Select Columns text portion of the SQL Select dialog box as shown at right by using the keyboard and the pull-down menus. If you cannot recall how to use the pull-down menus, refer to the steps in the "Group By Clause" section.

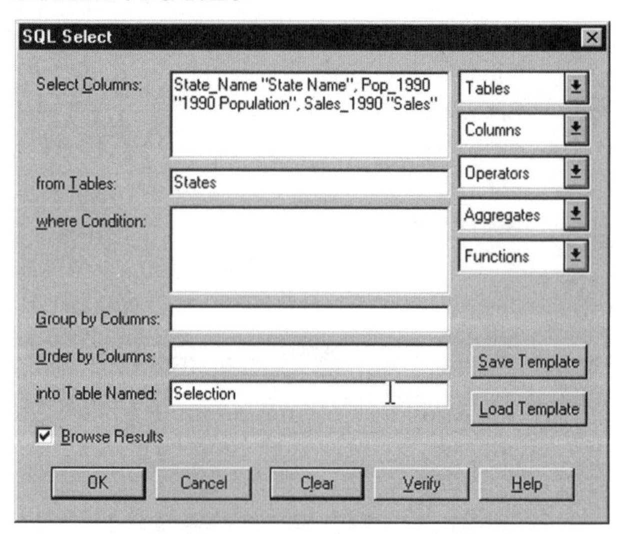

SQL Select dialog box example of column aliases.

3 Click on the OK button. The resulting Browser window is shown in the illustration at right.

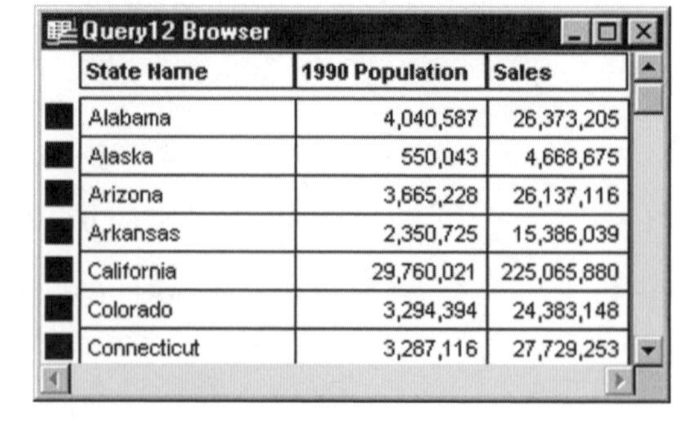

State Name	1990 Population	Sales
Alabama	4,040,587	26,373,205
Alaska	550,043	4,668,675
Arizona	3,665,228	26,137,116
Arkansas	2,350,725	15,386,039
California	29,760,021	225,065,880
Colorado	3,294,394	24,383,148
Connecticut	3,287,116	27,729,253

Resulting Browser window with alias column names.

Format Data Columns

In the Browser window you will notice that the data in the columns have been automatically formatted with commas. To make the data more aesthetically pleasing, you can change data output using the *Format$* function within the SQL Select dialog. Assume you wish to format the sales column in the *States* table with a dollar sign, comma, and decimal point. Tutorial 4-21, which follows, takes you through this process.

▼ TUTORIAL 4-21: FORMATTING DATA COLUMNS

1 From the Query Menu, select the SQL Select option.

2 Modify the SQL Select dialog using the keyboard and the pull-down menus to match the statement shown at right. This SQL Select statement will format the sales data column.

3 XClick on the OK button. The resulting Browser window is shown in the following illustration.

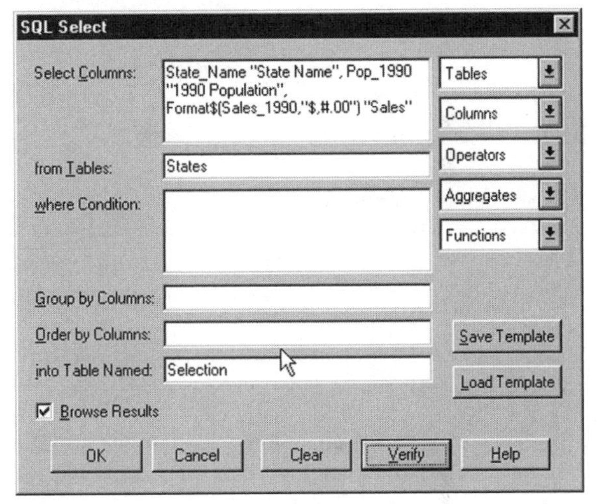

SQL Select dialog box example using the Format$( ) *function.*

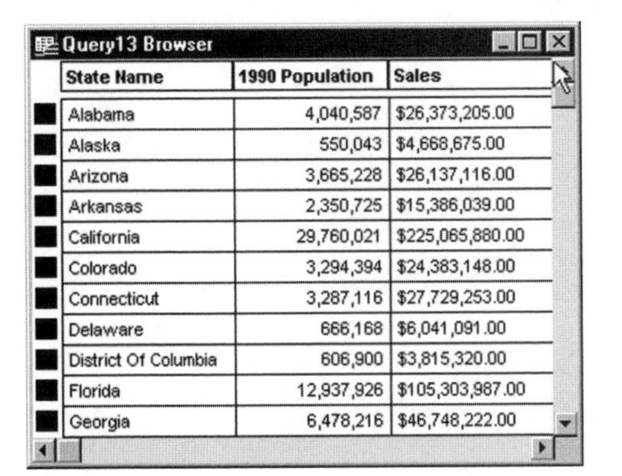

Query13 Browser

State Name	1990 Population	Sales
Alabama	4,040,587	$26,373,205.00
Alaska	550,043	$4,668,675.00
Arizona	3,665,228	$26,137,116.00
Arkansas	2,350,725	$15,386,039.00
California	29,760,021	$225,065,880.00
Colorado	3,294,394	$24,383,148.00
Connecticut	3,287,116	$27,729,253.00
Delaware	666,168	$6,041,091.00
District Of Columbia	606,900	$3,815,320.00
Florida	12,937,926	$105,303,987.00
Georgia	6,478,216	$46,748,222.00

Resulting Browser window with Format$( ) *function.*

TIP: *Issuing a File | Save Copy As command will save both the query results and browser formatting for later use. Another option is to save the workspace, which saves most query results within the workspace (queries based on temporary tables are not saved in workspaces). The final option,*

which is often preferable to using the File | Save Copy As command, is to save the query template, and reload and issue the query when needed again.

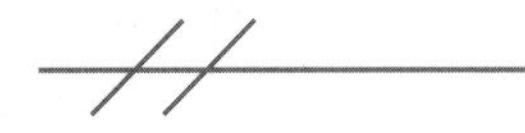

With the *Format$( )* function, the first parameter is the column name you format with the pattern string specified as the second parameter. The pattern string should include one or more special format characters, such as #, 0, %, the comma character, the period, or the semicolon; these characters control the appearance of your results. The pattern string can also include one or more cosmetic characters, such as the dollar sign, to make the results more attractive. Table 4-23, which follows, summarizes the format characters.

Table 4-23: Format Characters

Character	Purpose
#	The result will include one or more digits from the value. However, if the control string contains one or more pound (#) characters to the left of the decimal place, and the value is between zero and one, the formatted result string will not include a zero before the decimal place.
0	A digit place holder similar to the pound (#) character. However, if the control string contains one or more 0 characters to the left of the decimal place, and the value is between zero and one, the formatted result string will include a zero before the decimal place.
. (period)	The period is used in conjunction with the pound (#) character. If the pattern string includes a period character, the number of pound (#) characters to the right of the period will dictate the number of decimal places the resulting string will display.
, (comma)	If you include a comma before the first pound (#) character, the resulting string will include a comma every three digits to the left of the decimal place. Thus, the number 10 million appears as 10,000,000.
%	The result will represent the value multiplied by 100 (e.g., the value 0.75 will produce a result of 75%). If you wish to see the percentage sign in the result, place a backslash character (\) before the percentage sign (i.e., \%).
E+	The result will be formatted according to scientific notation (e.g., the value 1234 produces the result "1.234e+03"). If the exponent is positive, a plus sign will appear after the e. If the exponent is negative, such as in fractional numbers, the formatted results include a minus sign after the *e*.
E-	This string of control characters functions the same way as the *E+* string, except that the result will never show a plus sign following the *e*.

Table 4-23: Format Characters

Character	Purpose
; (semicolon)	By including a semicolon in your pattern string, you can specify one format for positive numbers and another for negative numbers. The semicolon should appear after the first set of format characters, and before the second set of format characters. The second set of format characters will apply to negative numbers. If you wish to see a minus sign in the results, you should include a dash (-) character in the second set of format characters. (See examples in table 4-24.)
\	When the backslash character appears in a pattern string, MapBasic does not perform any special processing for the character that follows the backslash. This lets you include special characters (e.g., %) in the results without causing the special formatting actions previously described.

Table 4-24, which follows, provides examples of format statements.

Table 4-24: Sample Format Statements

Statement	Effect
Format$(12345, ",#")	Returns "12,345"
Format$(12345, "$#")	Returns "$12345"
Format$(12345.678, "$,#.##;($,#.##)")	Returns "$12,345.68"
Format$(-12345.678, "$,#.##;($,#.##)")	Returns "($12,345.68)"
Format$(12345.6789, ",#.###")	Returns "12,345.679"
Format$(12345.6789, ",#.#")	Returns "12,345.7"
Format$(0.054321, "#.##%")	Returns "5.43%"
Format$(0.054321, "#.##\%")	Returns ".05%"
Format$(0.054321, "0.##\%")	Returns "0.05%"

Queries and Workspaces

As mentioned previously, query results derived from base tables will be saved in workspaces. Maps, graphs, and browsers based on these queries will also be saved when saving a workspace. However, query results derived from temporary tables will not be saved in workspaces.

Exercise 4-1, which follows, serves as a summary hands-on of the general processes covered in the tutorials of this chapter. In this exercise, you will perform a site analysis using various query techniques. A chapter summary follows this exercise.

■ EXERCISE 4-1: SITE ANALYSIS USING QUERY TECHNIQUES

Assume you operate a real estate office offering relocation services. A client requests detailed information on the locations of schools and other places of interest surrounding her prospective new home, 49 Laskie Street.

Opening the Applicable Tables

To prepare for a meeting with the client, you need to open the *Sf_strts* and *Sf_landm* tables (*samples* directory on the companion CD-ROM), and then geocode the target address.

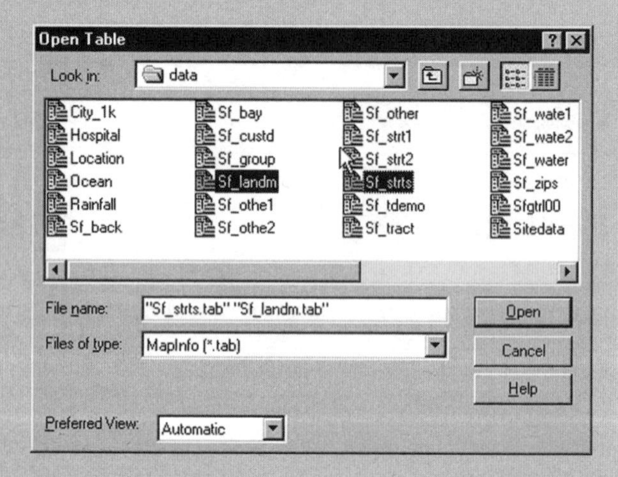

Open Table dialog box for Sf_strts *and* Sf_landm.

1 Select the File | Open Table option from the Main menu bar. This will access the Open Table dialog box, shown in the illustration at right.

2 Open the *Sf_strts.tab* and *Sf_landm.tab* tables. (Hold down the <Shift> key to select both tables.) Click on the OK button. The resulting Map window is shown in the following illustration.

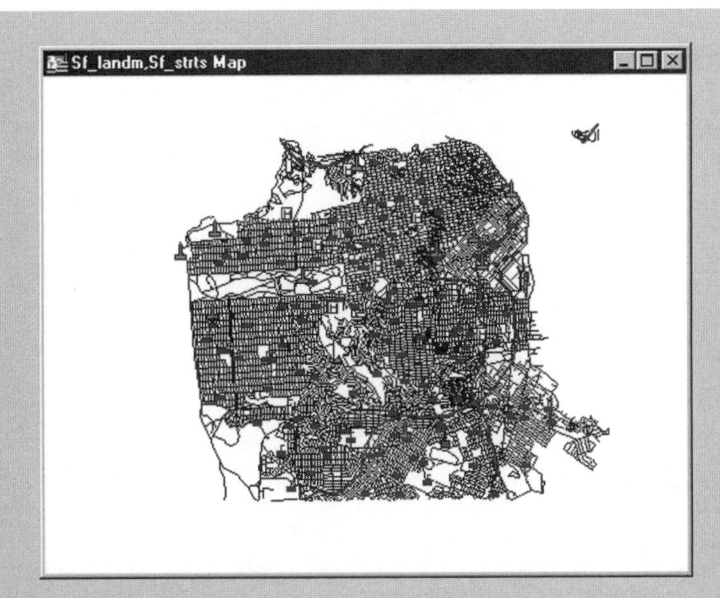

Map window for the Sf_strts *and* Sf_landm *tables.*

Displaying the Location on a Map

Next, assume you wish to display the client's property location on a map. MapInfo will locate the site through the Find command, as follows.

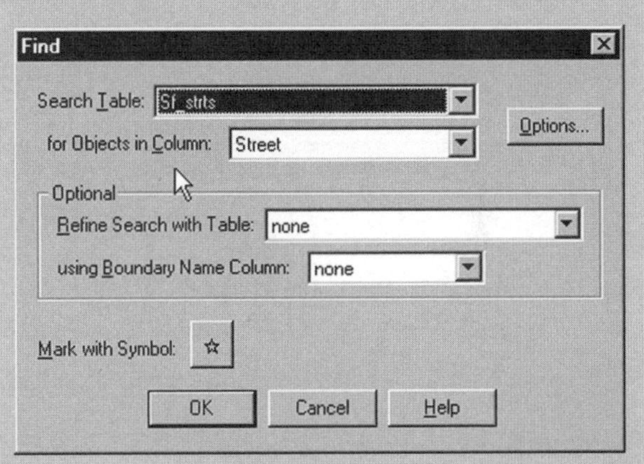

Find dialog box.

1 To prepare the Map window to easily visualize the house location, use the Zoom-in tool to zoom in on the map to a scale of approximately 1 mile.

2 Select the Query | Find option from the Main menu bar.

3 Set the Find dialog box, shown in the illustration above, to search the street column of the *Sf_streets* table. Click on the OK button.

4 Input the address of the client's new home in the Find dialog box, as shown in the following illustration. Click on the OK button.

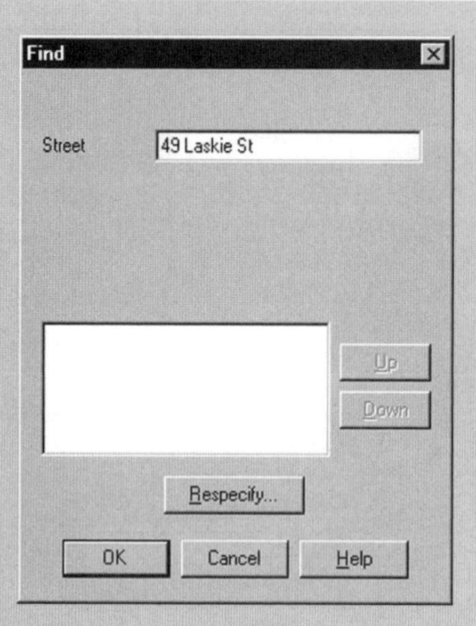

Inputting an address in the Find dialog box.

When MapInfo finds the 49 Laskie St. location, a new symbol is placed on the address, and the location appears at the center of the map screen, as shown in the following illustration. (If you are not zoomed in to the map when issuing the Find command, the map will not change.)

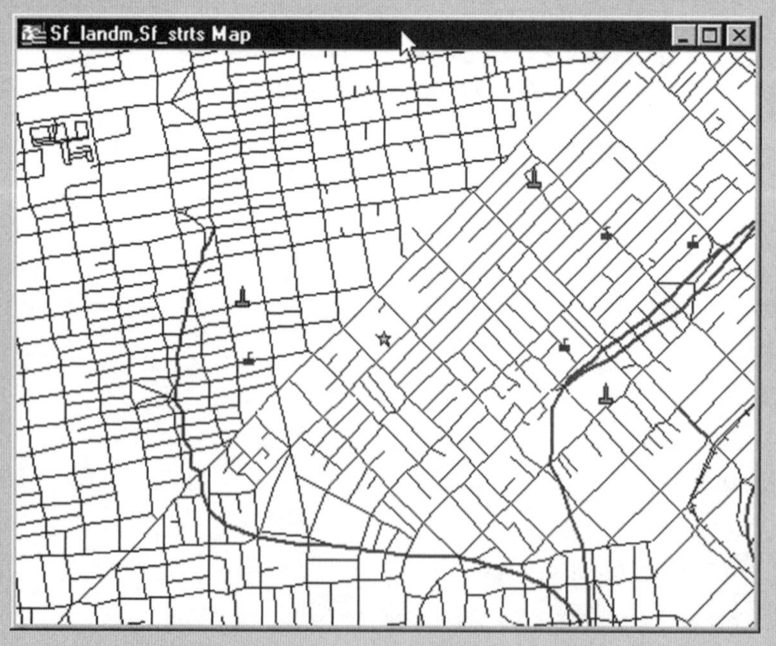

Map window zoomed in, showing location of 49 Laskie St.

Enhancing Screen Appearance and Saving the Table

At this point, you can enhance the appearance of your work and save the table in order to accelerate subsequent queries. Perform the following steps.

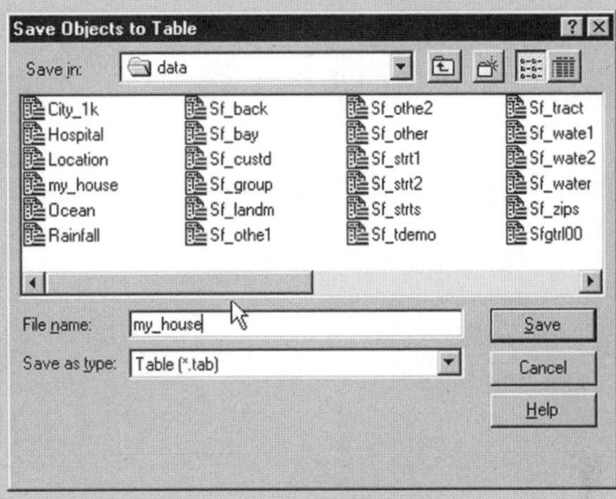

Save Objects to Table dialog box.

1. Select Map | Save Cosmetic Objects from the Main menu bar.
2. Set the File dialog to save the cosmetic point to a table named *my_house.tab* on your hard drive. Click on the OK button.
3. The objects in the cosmetic layer have now been saved to a table named *my_house*. Note that the Map window is now displaying the new table, as shown above.

Querying for Schools in the Neighborhood

You are now ready to issue a query locating schools in the neighborhood. The client prefers that her children attend schools no farther than 1 mile from home. You could use the Radius Select tool to make a selection circle with a 1-mile radius. However, this action would collect *all* points in the *Sf_landm* table rather than only the schools. Instead, you will use the SQL Select dialog box to select all school objects in the *Sf_landm* table that fall within a 1-mile radius of the client's prospective home.

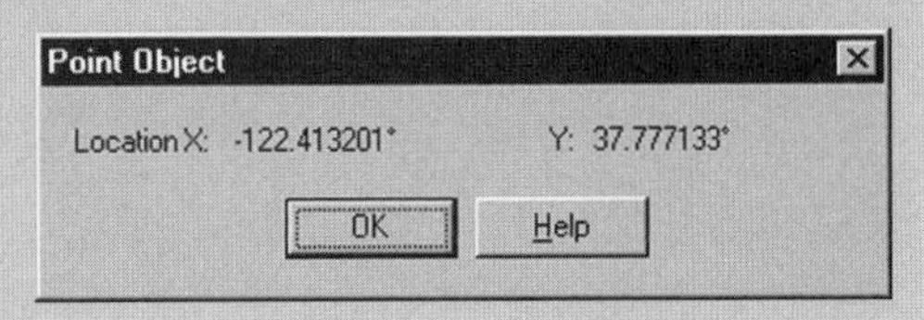

House point location.

1. First, you need to locate the exact (x,y) position of the home. Double click the left mouse button on the star positioned at the house location. MapInfo will display an object information box about the point, showing the exact latitude and longitude of the house address. Note that the house is located at X position -122.413201, and Y position 37.777133, as shown in the illustration above right.

2 Select Query | SQL Select from the Main menu bar.

3 Set the fields in the SQL Select dialog to find all school class points in the *Sf_landm* table that are within 1 mile of 49 Laskie St, as shown in the illustration at right.

4 Click on the OK button. The results of the query are shown in the illustration below.

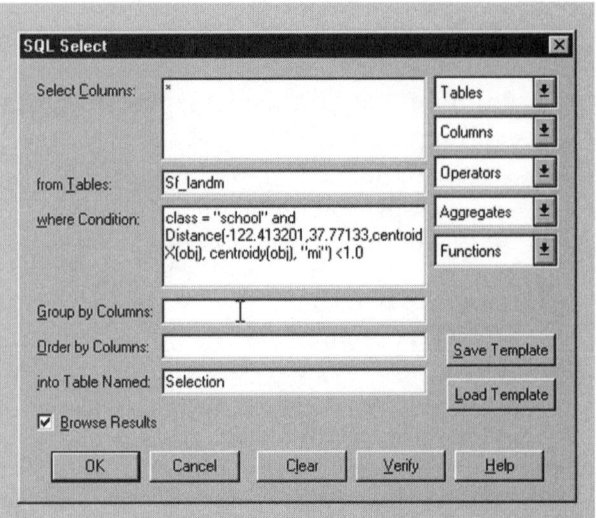

SQL Select for schools within 1 mile of 49 Laskie St.

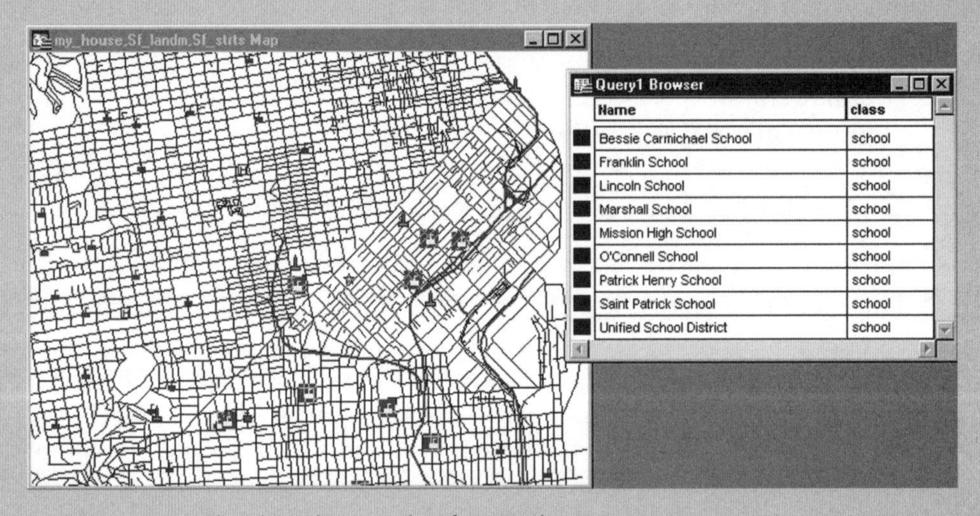

Schools within 1 mile of 49 Laskie St.

Querying for Points of Interest in the Neighborhood

Next, let's explore points of interest in the neighborhood. Rather than seeking all points in the *Sf_landm* table within a certain distance of the home (such as with the use of a Radius Select tool), you will query all points and report respective distances from the house in a table format.

1 Select Query | SQL Select from the Main menu bar.

2 Set the SQL Select dialog box to obtain all non-school points in the *Sf_landm* table. Select the name of the landmark, landmark class, and distance from 49 Laskie St. Alias the Distance Calculated column as *dist* (see illustration at right). In addition, order the list so that the landmarks closest to the house are at the top of the list (by *dist*).

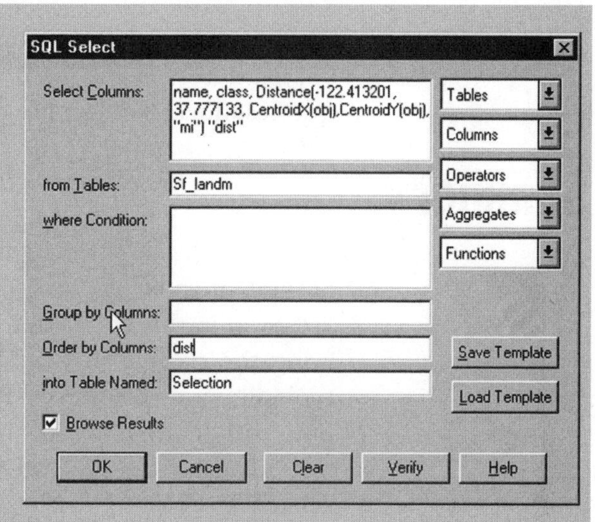

SQL Select of landmark points.

The resulting Browser window, shown in the following illustration, shows the landmarks and respective distances from the client's prospective home.

Query3 Browser

Name	class	dist
Unified School District	school	0.344794
San Francisco Opera House	building	0.370469
Bessie Carmichael School	school	0.448058
Old Mint Building	building	0.538987
Hall of Justice	building	0.56867
Saint Patrick School	school	0.610156
Lincoln School	school	0.80206
Marshall School	school	0.806388
Franklin School	school	0.897787
Redding School	school	0.900449
Green Hospital	hospital	1.01573
Lombard School	school	1.09429
Mission Dolores	church	1.13073
Gough School	school	1.22647
Patrick Henry School	school	1.22776
Mission High School	school	1.24736
Sanchez School	school	1.30236
Franklin Junior High School	school	1.34522
McKinley School	school	1.39621
O'Connell School	school	1.39744
Fremont School	school	1.46374

Browser window of landmarks and distances from 49 Laskie St.

This exercise provided an example of a MapInfo Professional application in the residential real estate market. Once you master queries, dozens of applications are possible.

Summary

Several means of querying MapInfo Professional data were reviewed in this chapter. You can perform custom analyses by querying the data and formatting it the way you wish. Many other tools allow you to query data. However, the power of desktop mapping is that the data and results relate to a map. You can analyze not only how the data are interrelated but how they are spatially related.

The Selection tool items on the Main toolbar are very useful for high-level data analysis. The purpose of the Select and SQL Select dialogs is to help you construct precision queries of data. Finally, the SQL Select dialog provides a great deal of control over the format of query results.

CHAPTER 5

DISPLAYING MAPS

THIS CHAPTER COVERS CHANGING AND MANIPULATING map displays. Discussion topics include how MapInfo Professional handles map data, as well as how the appearance of map data can be changed permanently or temporarily. MapInfo places each open table containing graphic data (with associated *.map* and *.id* files) in a separate layer. Each layer can be changed, edited, or manipulated independently of all other layers.

You can apply temporary or permanent changes to the appearance of each layer. You can also thematically map data associated with each layer, the topic of the next chapter.

Most of the chapter is organized as an exercise. To follow along in MapInfo Professional, access the following tables from the *samples* directory on the companion CD-ROM: *SF_STRTS*, *Location*, *SF_GROUP*, *SF_BAY*, and *SF_BACK*.

Layer Control Dialog Box

The Layer Control dialog box controls the order and appearance of the layers on a map. To access this dialog box, verify that a Map window is active. Next, select Map | Layer Control from the Main menu or the Layer tool from the Main toolbar (or right-click your mouse and then select Layer Control from the top of the pop-up menu). If you cannot access the Map menu, check to confirm that a Map window is the active window. Each Map window accesses a separate Layer Control dialog box and displays the layers (tables) associated with a particular map.

Working with the Layer Control Dialog Box

The Layer Control dialog box operates on a hierarchical principle: a layer appearing higher in the list will be drawn on top of the layers listed below it. When you create a thematic map (i.e., a map displaying a selected variable; see Chapter 6), the theme will appear above its base layer (the layer the theme is based on) in the Layer list. In the following illustration, the Location table is the base layer for two types of thematic maps: *Ranges by Sales* and *Ind. Value with Store_name.*

Table associated with two themes.

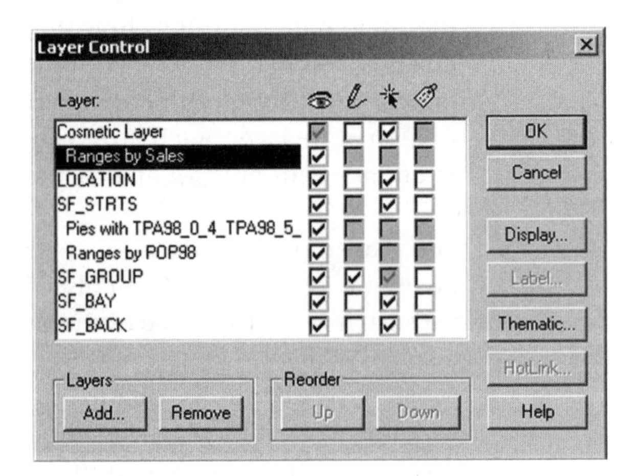

You can change, edit, or manipulate each layer independently of all other layers. The Layer Control dialog box lists the map layers in the same order as they appear in the Map window. The higher the layer in the order, the later the layer is drawn. In other words, the layer at the top in the Layer Control dialog box (the cosmetic layer) will be drawn last and on top of all other layers. Layers higher in the drawing order may hide from view details of layers below them.

The drawing order typically follows the order in which files were opened, although MapInfo Professional tends to place point data above region data. In addition, layers added later to the map tend to be placed at the top of the drawing order. You will typically need to alter the layer order to ensure your map looks the way you anticipate.

The following illustration shows the drawing order of the map layers in the Layer Control dialog box. The cosmetic layer (used for drawing on or annotating the map) will be drawn last.

Layer Control dialog box showing map layer order.

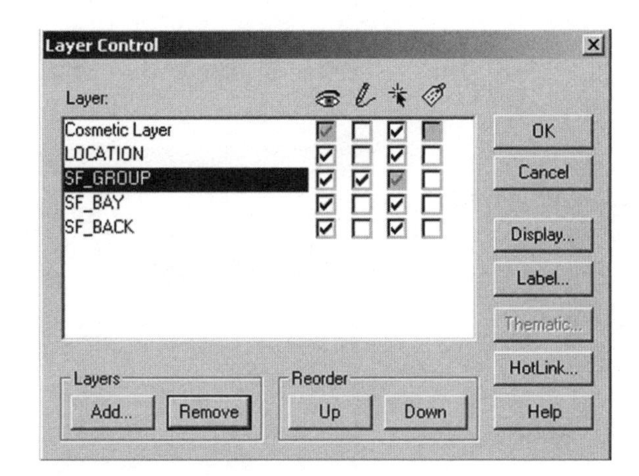

As shown in the following illustration, the point locations (*Location* table) are drawn on top of the block groups (*SF_GROUP*), which in turn are on top of San Francisco Bay (*SF_BAY*).

Map view showing example of layer priority.

If you change the order of the layers by moving the *Location* table below the block group table (*SF_GROUP*) and the *SF_BAY*, the map will change, as shown in the first of the following illustrations. The site locations are no longer visible because the block groups and bay area backdrop have been drawn on top of the sites, as shown in the second of the following illustrations.

Location layer moved down in layer order.

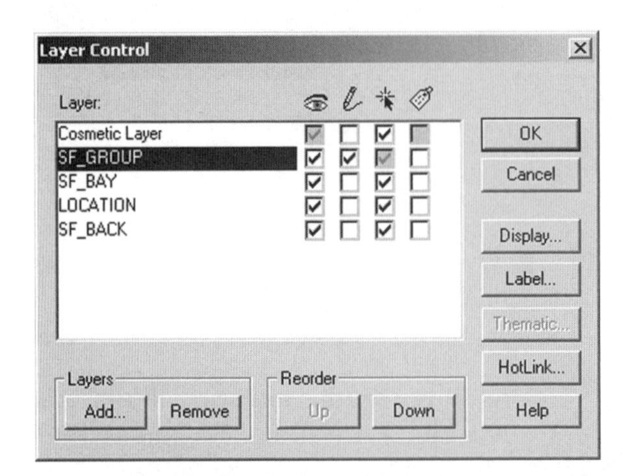

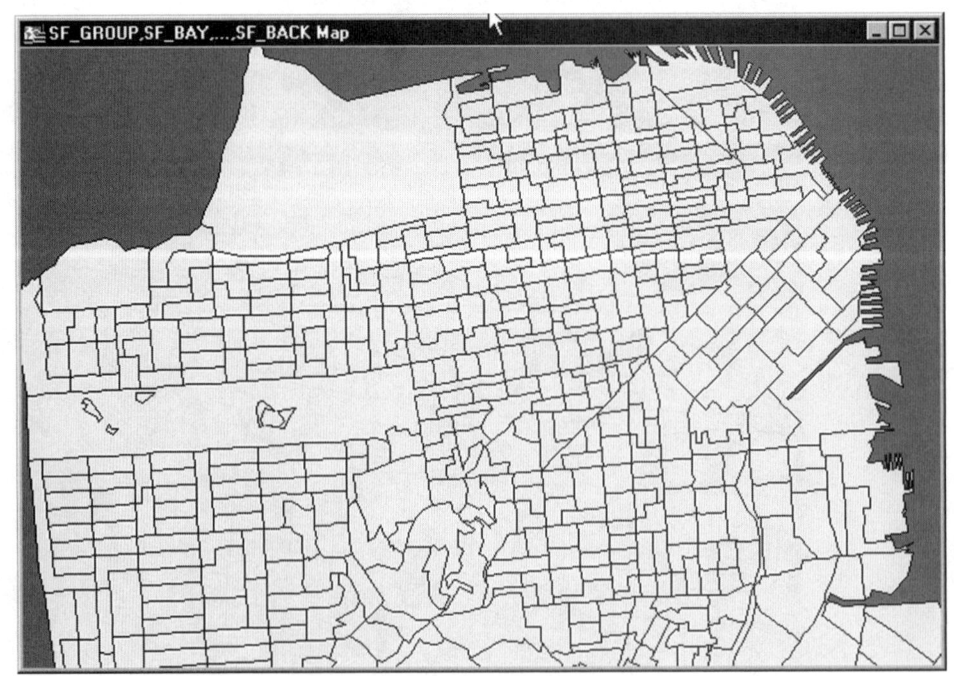

Results of changing the drawing order for the Location layer.

If you had moved the *Location* layer below the *SF_GROUP* layer only (instead of below both the *SF_GROUP* and the *SF_BAY* layers), you would have been able to see the features in the *Location* layer. This is because the *SF_GROUP* layer's default style did not have a background color or pattern, allowing you to see objects drawn below it. If you changed the fill pattern (discussed later in the chapter) for the *SF_GROUP* layer to a light green background, you would not be able to see the *Location* layer it if were placed below the *SF_GROUP* layer in the Layer Control dialog box.

A layer's position in the layer hierarchy also determines selection order. This is important when using the various selection tools and the Info tool. If all layers are selectable, when you click on the map to select an object, the object in the top layer in the drawing order will be selected. For example, assume a layer containing the *States* table was on top of a layer containing block groups for Arizona in the Layer list. In this case, clicking on part of Arizona with a selection tool would select the state of Arizona instead of a block group.

Layer Attributes

The columns next to the layer names—Visible (eye icon), Editable (pencil icon), Selectable (pointer icon), and Label (package label icon)—control the operation of each layer. These attributes are controlled by the check boxes under respective icons. To change attributes for a layer, highlight the layer and check or uncheck the appropriate boxes.

Visible

The Visible attribute determines whether a layer is visible on the map. If a layer is not visible, it can still be used in queries; that is, SQL Select or Select. A layer must be visible in order to select an object from the map (via the selection tools), or for the geographic objects associated with the table to be edited.

If you turn the visibility on or off for a layer with associated thematic layers, MapInfo will ask if you also wish to change the visibility for the themes based on this layer. To save time, consider turning visibility off for layers that take time to draw, such as detailed streets, until they are needed in your analysis or for printed output. However, with today's powerful desktop computers, redraw speed is becoming less of an issue. If you turn visibility off for a particular layer or layers to save drawing time, do not forget to turn layer visibility on again when you complete the analysis and wish for such layer(s) to appear in the output map.

Editable

Only one layer at a time can be editable. If a layer is editable, you can change the map (graphic) objects associated with the entire table, a subset of records, or an individual record. Making a layer editable allows its features to be edited, as in moving objects on the map, changing line segment length, and changing the color or style of an object. Making a layer editable will automatically check (and disable) the Selectable option, if it was not previously checked. Because thematic layers cannot be edited, all edits must be made to base tables.

A table does not need to be editable for you to change associated tabular data. You can change the tabular information in a Browser window or through the Table | Update Column command regardless of whether a layer is editable. When a Map window is the active window, the status bar at the bottom of the MapInfo window will indicate which layer, if any, is being edited.

Selectable

The Selectable attribute controls whether you can select objects on the map. If Selectable is not chosen, you cannot use the Info tool to view information on this layer from the Map window. It is often a good idea to turn off the selectability of all layers except the one you are analyzing before using the select tools to select objects from the map. This will ensure that objects from the intended layer are selected.

This suggestion is especially helpful if you only wish to add labels to some of the objects on a single layer. By turning the Selectable option off for the other layers, you are assured that only the layer you are attempting to label will in fact be labeled; otherwise, you have no guarantee as to what has been selected and thus labeled.

Label

Placing a check mark under the Label icon for a layer will automatically label all objects in that layer on the map (subject to constraints set under the Label Options dialog box, discussed later in this chapter). Labels are not part of the cosmetic layer, and are saved with other workspace information.

Settings Options and the Use of the Display Button

The Display button on the Layer Control dialog box allows you to change the look of the map, such as applying a background color for block groups; changing the line style (or color) of a highway layer to more closely resemble a highway; or changing the color, size, or style of the point symbols on the map. These changes are global, meaning that they affect all objects in the layer. However, these changes are not permanently saved.

If you close the table for which you have changed the display and then reopen it, the Display settings previously applied would be lost. These display settings, however, are tracked within a workspace. If you change the global display for a layer, and then save and reopen the same workspace, your display settings will reappear as you defined them.

In this discussion, references are made to *line*, *point*, and *region* data. Line data refer to streets, elevation contours, and other information stored as lines. Point data are defined by a table containing a single coordinate pair (latitude/longitude) for each record. Point data can include cities (e.g., the *City_1k* table), customer addresses, or competitor locations. Region data are defined by some type of closed shape, such as census tract, state, county, or country.

The following illustration (at left) shows the Display Options dialog box for the *SF_GROUP* layer. Assume that Style Override has

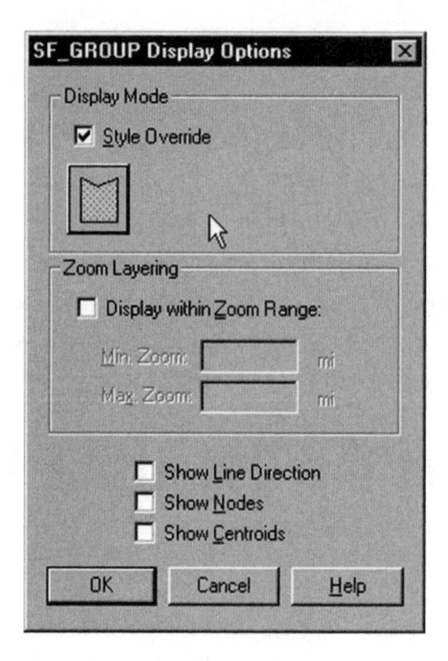

Display Options dialog box.

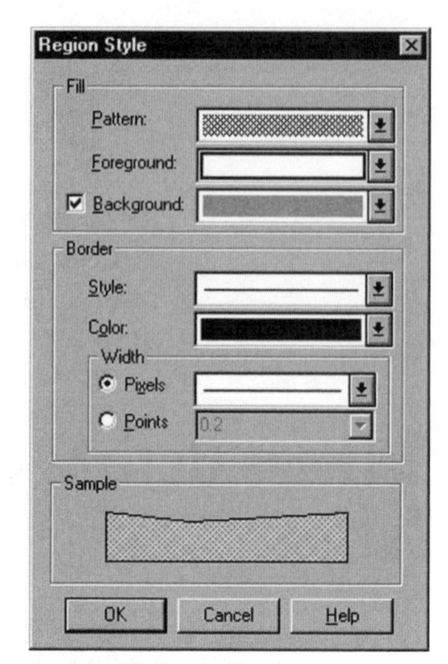

Region Style dialog box.

been selected to change the appearance of the block groups. The style box that displays depends on the type of data in the layer. Because this example uses regions, the Region Style dialog box (shown above right) will display. If your layer contains line or point data, different style dialog boxes will display. Click on the Region Style dialog box to access the Region Style dialog box. For region data, you can change the following elements.

- Foreground color
- Background color
- Boundary line width
- Fill pattern
- Boundary line style
- Boundary line color

For lines, you can change the following elements. The Line Style dialog box is shown in the following illustration.

- Line style

- Line color
- Line width

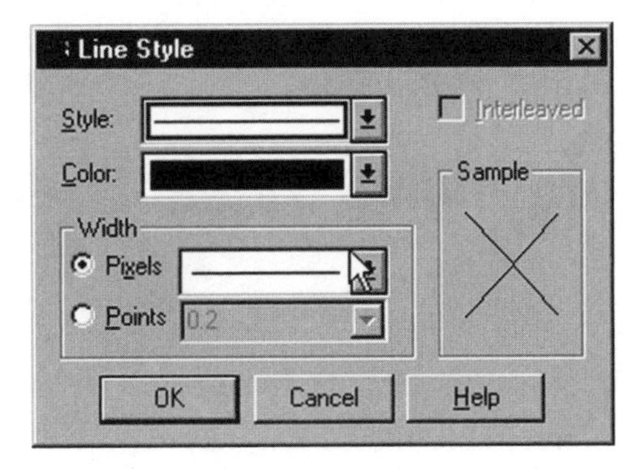

Line Style dialog box.

You can also change the following symbol features. The Symbol Style dialog box is shown in the following illustration.

- Font (or symbol set)
- Symbol style
- Rotation angle
- Effects
- Size
- Color
- Background appearance

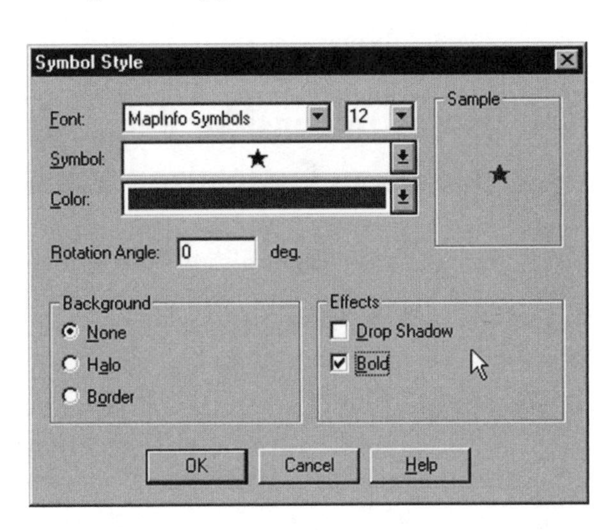

Symbol Style dialog box.

MapInfo Professional allows users to access TrueType fonts as symbol sets. These fonts have greatly extended MapInfo's symbol set capabilities. The ability to add a background or effect to a symbol allows you to further customize map symbols. The sample box displays how the symbol will appear on the map.

MapInfo Professional Symbol Sets

The symbols you choose to display for each point object on a map can greatly help your audience to instantly recognize what each point represents. For example, when you see a small airplane or a picnic table on a road map, you know that these symbols represent an airport and a park, respectively. These are examples of common symbols recognized by the general public. Many industries that rely on mapping have developed special symbol sets to represent industry-specific items.

Many of the most commonly used general mapping symbols ship with MapInfo Professional. You can add to this symbol set, or create an entirely new symbol file of point representations related to your field or discipline. With the release of 4.0, MapInfo began supporting three different mechanisms for symbol creation.

- MapInfo font file (*mapinfow.fnt*, the original mechanism)
- TrueType fonts
- Bitmaps

Issues to Consider Before Creating and Using Customized Symbols

Creating and maintaining custom symbols can be time consuming and expensive. Custom symbols have the benefit of being easily recognized. However, on the negative side, it can take considerable time to create an attractive symbol, and custom symbols usually must be maintained and/or periodically reworked. You can really go wild with symbols, and find yourself supporting multiple symbol sets for different regions, businesses, or applications. If you are working on a one-time project, you will probably wish to opt for the MapInfo basic symbol set.

If you create a custom symbol set for a GIS solution you are building and delivering to a client, you have additional issues to consider. Does that client use custom symbol sets other than the set you created? The client may also purchase additional data from other sources that use different symbol sets. You must ensure that clients using your custom symbols understand how to use, manage, and change the symbol sets they need in their environment. Thus, integration and distribution of symbol sets require careful planning and coordination.

A common mistake made when creating symbols is to make them too complex. The more intricate or detailed the symbol, the larger it must be displayed on the map for easy recognition. The map could easily become convoluted if all symbols must be placed at a point size of 48, thereby covering other map detail. The following illustration shows an example of a symbol that is too complex. The man with the fruit basket is so detailed that he must be displayed at a point size of 48 to allow the viewer to interpret and understand the symbol. Most symbols are easily recognized at a point size of 14 to 18. When displayed at the smaller size, the man with the fruit basket loses all detail.

Symbols should not be too complex.

Another problem is that symbols displayed at a large font size can be difficult to select using the Selection tool. For example, the large man displayed in the illustration must be selected at a point near the middle of his body. Selecting near his feet or head will not select the point. You can reduce selection problems if you always center new symbols you create and if you avoid excessively large font sizes.

MapInfo Professional Font Files

The sections that follow explore font file management and the creation of symbol sets. The starting point is the *mapinfow.fnt* file. Information on working with this feature under various versions of the software is provided.

Management

The *mapinfow.fnt* file is copied into the MapInfo application directory during the MapInfo installation process. When MapInfo Professional is opened, the program searches the application directory for the *mapinfow.fnt* file. If the *mapinfow.fnt* file is not found, it will then check the Windows directory for the file. MapInfo will initialize with whichever **.fnt* file it locates first.

If MapInfo fails to locate the *mapinfow.fnt* file, you will be presented with a file selection dialog box for selecting the **.fnt* file for the current MapInfo session. If you are constantly changing the symbol files in use, you may wish to use this method of initializing the files.

If you want to change the symbol file while you are still in MapInfo, consider investigating the MapBasic command for setting symbol files, or creating a single set of basic symbols for all of your needs. In MapInfo Professional, *mapinfow.fnt* file symbols are available under the Old MapInfo Symbols font title.

Creating Symbol Sets

If you are working with a MapInfo version prior to 4.0, the only mechanism available to you for modifying symbol sets is to create them. This mechanism allows you to create additional symbols with eight basic colors.

Two programs provided by MapInfo Corporation allow users to create additional symbols for the *mapinfow.fnt* file. The *mapfnt.exe* program was used in very early versions of the software, whereas the other, *symbol.mbx*, has been developed more recently.

The *mapfnt.exe* program is not shipped with MapInfo Professional. This program requires you to draw or graph out the symbol you wish to create. You must then create an ASCII file describing the symbol with coordinates. For example, 3> 30,2 -4,3 will draw a red line from the coordinates (30,2) to (-4,3). The *mapfnt.exe* program will read this ASCII description file and create a *mapinfow.fnt* file that is substituted for the MapInfo-provided file. Of course, this is a tedious, time-consuming process.

The *symbol.mbx* program is a MapBasic program included with MapInfo Professional that provides a Map window for drawing new logos. This program allows you to use the map drawing tools for creating new symbols. The program is difficult to work with to get consistently sized logos. In the Map window you are also not given control of which shape draws on top of what, thus causing additional frustrations. This program limits the complexity of a new symbol to 200 nodes.

TrueType Fonts

Beginning with version 4.0 of MapInfo Professional, you can use TrueType font file characters as symbol representations. Several TrueType font files included with MapInfo cover many mapping uses of symbols. There are also many additional TrueType font files available that have been developed by computer users over the years.

If you are unable to find a suitable TrueType font set that meets your needs, there are many tools available to help you create a custom font or edit an existing TrueType font. It should also be noted that TrueType fonts can only be displayed in two colors for both foreground and background.

Bitmaps

MapInfo Professional also permits you to save bitmap files that can be used as a symbol set. Predefined bitmap symbols are available under the Custom Symbols font title. MapInfo provides a set of bitmaps that are sized 64 by 64 pixels.

You can create your own bitmap files with one of the many bitmap editors available, store them in the *MapInfo\Professional\ CUSTSYMB* directory, and then access them through the Custom Symbols font. The Display Options dialog box also allows you to set ranges for zoom layering. Zoom layering enables you to set a distance range within which a layer will display.

The other options (Show Line Direction, Show Nodes, and Show Centroids) can be useful for various functions. Show Nodes is useful when editing map objects (such as trade areas or street segments). This option allows you to easily check on node connections, and to select nodes you wish to move. Show Centroids can be useful when working with regions and you need to associate a point with a region object. Show Line Direction can be useful when labeling line data; the label will follow the direction of the line if Rotate Label with Line is selected in the Label Options dialog box for the layer.

Label Button

The Label settings button allows you to choose the column or expression to use for a layer's labels, as well as change the label style for each layer. MapInfo Professional tracks labels by layer, meaning that labels (and edits made to labels) are saved with workspaces. Because MapInfo does not permit thematic maps to be labeled, when a thematic layer is highlighted in Layer Control, the Label option will be disabled. All label settings must be applied to the base layer.

To create labels, you can choose any column in the table, or an expression based on the columns in a table. This ability to label with an expression is very useful. For example, when labeling currency data, you can automatically add dollar signs to labels, or create a theme for a base layer with label values related to the theme.

For instance, suppose you have created a theme based on the block group layer to display population growth. In this case, you could label the block group table with *x.xx%* growth for each block group. MapInfo will automatically label numeric data with commas. (For more information on the *format$* command, see the

"Format Data Columns" section in Chapter 4.) To add dollar signs, use the following expression:

```
format$(column name, "$,#.##")
```

To add percent signs, key in:

```
format$(column name, "#.##%")
```

To label a coordinate pair for a location by latitude and longitude, key in:

```
str$(CentroidY(obj))+", "+str$(CentroidX(obj))
```

To create a multiple-line label for a layer, insert *Chr$(13)* where you wish the line break to occur. For example, assume you want to create a multiple-line label based on *sales, store_number, city,* and *state* columns to resemble the following format:

Sales
Store_number
City
State

In this case, you would enter the following formula:

```
Sales+Chr$(13)+Store_number+Chr$(13)+City+", "+State
```

In the Label Options dialog box, shown in the following illustration, you can also change label orientation with respect to an object, font style and size, label arrow, background, effects, and visibility, as well as label settings. To change a label's position with respect to the object, select the appropriate directional button under Position. By selecting the Rotated with Line Segments option, you can rotate labels to correspond with line segments, such as in the case of placing street labels parallel to (and in the direction of) a line segment. You can also set the offset from the object. If you select a large offset, lines connecting the object to the label will appear if you have chosen Simple or Arrow under Label Lines (in the Styles section).

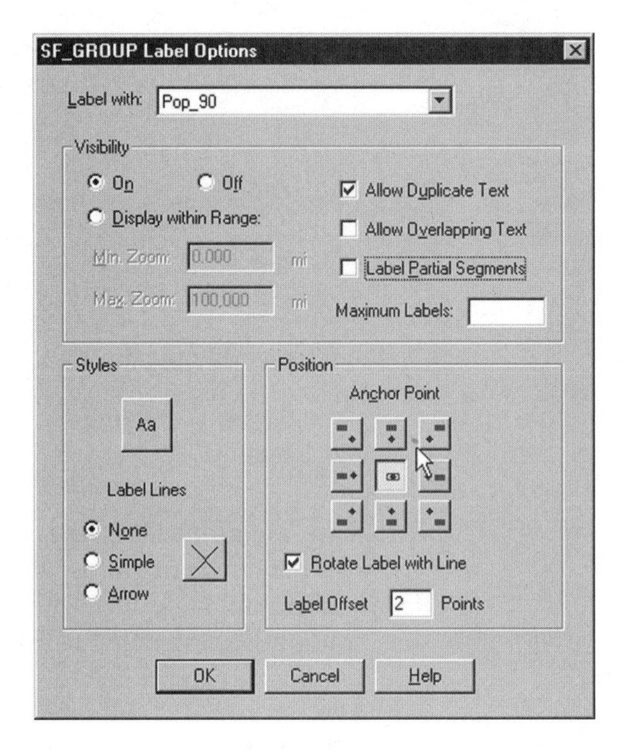

Label Options dialog box.

You can also set visibility on or off, as well as set a zoom range for the labels to be displayed. If the map range is greater than the maximum or less than the minimum zoom, the labels will not be displayed. You can choose whether to allow duplicate text or overlapping text, or whether to label partial line segments. The last option refers to polylines and whether they will be autolabeled if only a small portion of the line is visible. In addition, you can set the maximum number of labels.

MapInfo Professional allows many style options for labels. To change the style of a label, select the *Aa* button under the Styles section. In the Text Style dialog box, shown in the following illustration, you can set the font and font size for the label, as well as the text color. In addition, you can choose a background halo or box and its color. Selecting a background box or halo for your labels is recommended when your map contains a lot of color or detail that could detract from the label. Finally, you can choose from several effects, such as bold, italic, underline, all caps, shadow, and expanded text. All of these options allow you to create custom labels and enhance the appearance of your map.

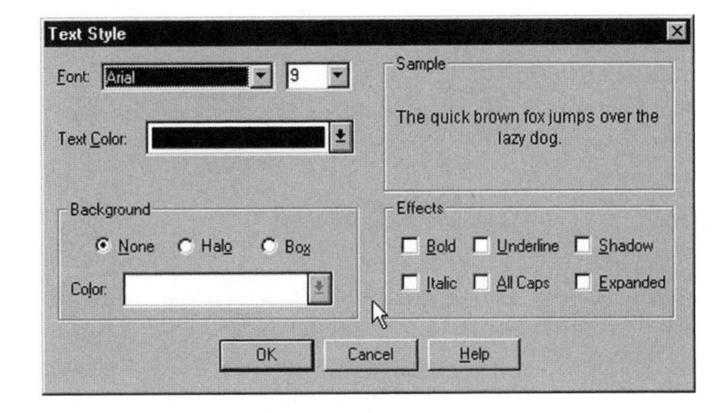

Text Style dialog box.

To label all objects on a map, check the Label option in the Layer Control dialog box. If this option is not available, access the Label Options dialog box (Label button in Layer Control dialog box) and set Visibility to on. While in the Label Options dialog box, you should determine whether you wish to allow duplicate or overlapping text. Select/deselect these options based on your needs.

To label specific items on the map, select the Label tool and then click on the map object you wish to label. Due to MapInfo's layer order system, the object you intended to label may not be labeled. In this case, you may want to turn off the Selectable option for all other layers, thereby leaving the layer you wish to label as the only selectable layer.

Labels can be edited from the Map window. To change the text or style of one label at a time, double click on the label in the Map window. You can also click and drag a label in the Map window to a new position. If you are moving labels on the Map window, Simple or Arrow under the Label Lines option on the Label Options dialog box should be selected to ensure that the audience for your map understands the objects to which the labels on the map refer.

Thematic Button

The Thematic settings button allows you to access the Modify Thematic dialog box associated with the highlighted thematic layer.

This dialog will allow you to change the appearance or the legend associated with a thematic layer. If a thematic layer is not highlighted, this option is disabled. (Thematic mapping is discussed in Chapter 6.)

HotLink Button

The HotLink button accesses the HotLink Options dialog box, which allows users to specify options relating to MapInfo's HotLink feature for a given layer. These options include URL/filename expression, file locations, what activates the HotLink, and saving options to table metadata. A mappable table must have a text column (field) that specifies the URLs or file names related to an object.

A layer must be active (i.e., selectable or editable) and have defined HotLink options to use the HotLink tool. The HotLink tool is enabled only for map windows with at least one active layer. Depending on how activation is defined under HotLink Options, the HotLink tool launches a URL or file associated with an active object when you click on the object or its label. (HotLinks are discussed in detail in Chapter 10.)

Layer Buttons

The Layer Add and Remove buttons do not open and close files. These options allow you to add and remove layers from the map as desired. It is often necessary to add layers to a map. If you open a map table that is not in the same general area as the rest of the map, and you want it to appear on the current map, you will have to add it separately. (Alternatively, you could have changed the Preferred View to Current Mapper on the Open Table dialog when you opened the table.) If you use the Create Points function, you will need to add the newly mappable table to the map. In addition, if you pack a table after deleting records, you will have to add it back to the map.

Selecting Add will activate the Add Layer dialog box, in which you can choose a layer to add to the map from a list of all open and mappable tables. You can also add another copy of a layer already appearing on the map; in which case, the newly added layer will revert to the default settings for the layer, despite any changes to the appearance you may have made to the map.

To remove a layer from the map, highlight the layer in the Layer Control dialog box and select Remove. Do not remove thematic layers, because you will have to recreate the themes. Once a layer has been removed, changes made to the display settings will be lost.

If you wish to access a theme again, or to retain your display settings, you should turn off visibility for such layers instead of removing them. In this way, you will not have to recreate your work.

Reorder Buttons

The Reorder Up and Down buttons allow you to change the order in which the layers are drawn and selected. The higher the layer is on the list, the later it is drawn on the map (to ensure it can be seen over the layers beneath it). Thematic maps can be moved up or down in layer order only by changing the order of the underlying table on which the theme is based. You can also change the order of some types of thematic maps related to the same base layer. MapInfo Professional also allows you to drag and drop a layer to a new position in the Layer list, instead of using the reorder buttons.

Cosmetic Layer

You can draw symbols, shapes, or text on the cosmetic layer, which is always the top layer on the map. Like an acetate layer overlaying a paper map, you can choose to leave it on the map or peel it off and start fresh. You can also make the drawings a part of the underlying map. All drawing tools (rectangle, symbol, text, and so on) can be used on the cosmetic layer to make finishing touches to your map.

The only Layer Control options affecting the cosmetic layer are Selectable and Editable. The cosmetic layer (or another layer) must be editable to add drawings to the map. Objects drawn on the cosmetic layer can be saved to either an existing layer or to a new layer using the Map | Save Cosmetic Objects command. Cosmetic objects can be edited and changed like any other mappable table by making the cosmetic layer editable in the Layer Control dialog box.

If you save cosmetic objects to an existing layer, new records will be appended to the existing table, although only the geographic object (invisible to users) will be updated. If you save to a new layer, MapInfo Professional will create a new table containing only an *ID* field. If you have saved the cosmetic objects as a new table, they will be added as a separate layer to the present map and stored with all other layers and settings when you save the workspace. The objects can be accessed later, edited, added to, and so on, like any other MapInfo table.

The cosmetic layer is useful for drawing new geographies or point objects, as well as for adding rings around point objects. When creating new objects, consider the following steps: draw a single object; save the object to a new layer; modify the structure of the new table; and make the layer Editable so that you are able to input information on each object as it is created. To save objects on the cosmetic layer, select Map | Save Cosmetic Objects. Select a current layer to add the objects to, or select New to create a new table for the cosmetic objects.

If you wish to clear all drawings from the cosmetic layer, select Map | Clear Cosmetic Objects. If you open a workspace file containing objects on the cosmetic layer, you can clear these objects with the Map | Clear Cosmetic Layer command. If you wish to permanently remove the cosmetic objects from the workspace file, you will have to re-save the workspace after clearing the cosmetic layer.

Legend Window

The ability to create a cartographic legend for a map, added with version 5.0, was a welcome addition to MapInfo, and has improved some with version 6.0. You can create as many different legends as you desire for a particular map, as well as include different features for each legend. In addition, thematic legends no longer automatically appear in their own window, but appear in the cartographic legend when already created, or in a new cartographic legend if one has not been created. To create a descriptive legend for a map, select Map | Create Legend. This option accesses the Create Legend Step 1 of 3 dialog box, shown in the following illustration.

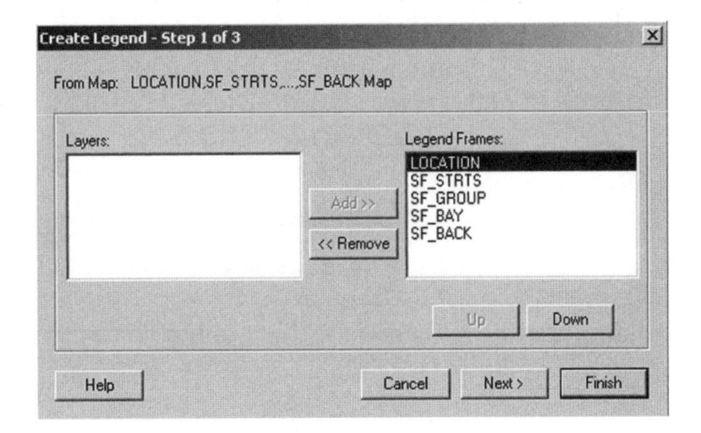

Create Legend Step 1 of 3 dialog box.

The Step 1 of 3 dialog box allows you to select which layers will appear in the legend. Assume that the *SF_BAY* layer's appearance in the legend is not desired. You can remove it from the Legend Frames box on the right, as shown in the following illustration. The Step 1 of 3 dialog box also allows you to determine the order for the layers in the Legend window by using the Up/Down buttons to reorder the layers. Make sure to order the layers the way you want them listed while in this dialog box. It is very difficult to reorder entries in the legend once created, as it must then be done by hand.

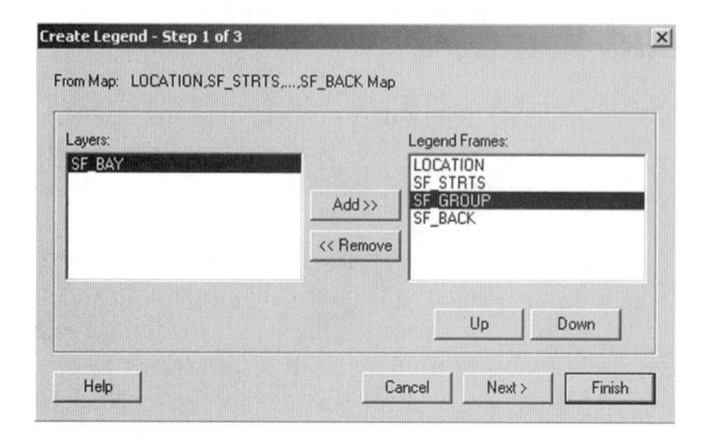

Create Legend Step 1 of 3 dialog box with SF_BAY removed from the legend.

The Step 2 of 3 dialog box, shown in the following illustration, allows you to define the look of the Legend window. You can choose a title for the Legend window in the Window Title box. This title is given to the Legend window; it is not a title for the legend. Other legend properties include whether scroll bars appear, and portrait versus landscape orientation.

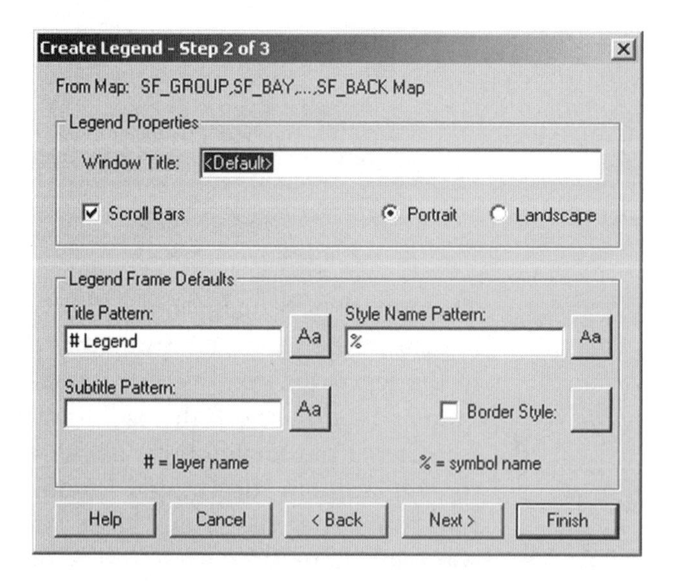

Step 2 of 3 dialog box.

Under Legend Frame Defaults, the user can define a title for each layer in the legend. The default is # *Legend,* which will place *layer name* and *Legend* as the title for each layer included in the legend. You can alter the legend title, or remove it completely by deleting # *Legend* from the box. The default Style Name Pattern is %, which will place the type of data (i.e., line, point, or region) next to the symbol representing a layer. To remove the type of data, delete the percent sign (%) from the Style Name Pattern box. As with labels and other text options, you can change the appearance of the title, subtitle, and name text by clicking on the appropriate text style button for the item you wish to alter.

The Border Style option determines the line style for the border of the legend. When you choose a border style on the Step 2 of 3 dialog box, it will place a border around the legend for *each* layer in the Legend window.

The Step 3 of 3 dialog box, shown in the following illustration, allows you to further customize each legend frame. This dialog box shows you the current title and subtitle for the legend frame selected on the left, and allows you to more easily change the text. The "Styles from" option offers users better control over the legend. By selecting unique map styles, a legend entry will be created for each unique map style within the table, preventing duplicate entries. Selecting Unique Values in Column and then selecting a column will create a legend entry for each different (i.e., unique) record in a column.

Step 3 of 3 dialog box.

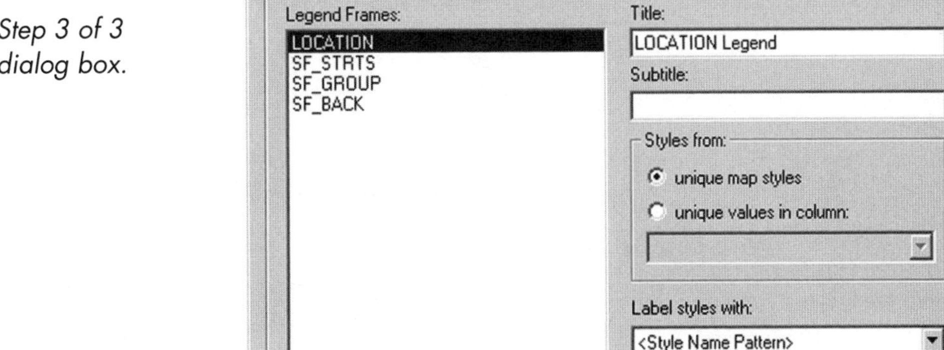

This is helpful if you have several styles and/or values within your table and wish to distinguish them in the legend by their different values. A legend entry will be created for duplicate symbols. Do not select columns with more than a few different values. Ten is a good maximum, so that MapInfo can handle the differences and the legend remains readable. The "Label styles with" option allows you to select a column for labeling the different values.

You can use expressions or data from other tables through joins as labels. The "Save frame settings to Metadata" option allows you to save the settings to the metadata table. If you select this option, the current settings will become the default for the Legend frame. The following illustration depicts the legend created for the *Location, SF_STRTS, SF_GROUP,* and *SF_BACK* layers in the related map windows by using the defaults and removing the *SF_BAY* table from the legend.

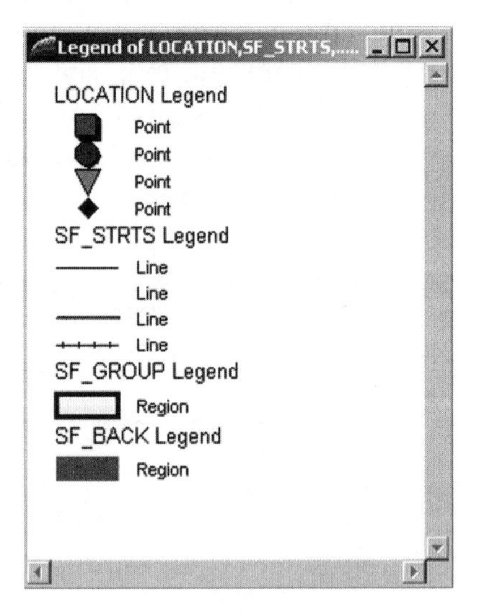

Standard legend created using default options.

As you can see, the Legend window created using the default options has a title for each layer. In addition, the label for each layer contains the data type. Note that the *Location* layer is depicted by several different symbols and the *SF_STRTS* layer consists of several types of lines. (The process for editing data in this manner is discussed later in this chapter, as well as in Chapter 7.) However, all data in the *Location* layer is labeled as *Point Location* in the legend.

In the Step 2 of 3 dialog box, you altered the Legend window by removing the Legend Title for each layer (which can be a bit excessive), as well as the data type from the Style Name Pattern. This is shown in the first of the following illustrations. In the Step 3 of 3 dialog box, the word *Legend* was added to the title field for the *Location* table (the first entry in the Legend), to give the legend a title. In addition, you selected "unique values in column" for the *Location* table, and chose the *storename* column for the unique label values. The results of modifying the default choices are shown in the second of the following illustrations.

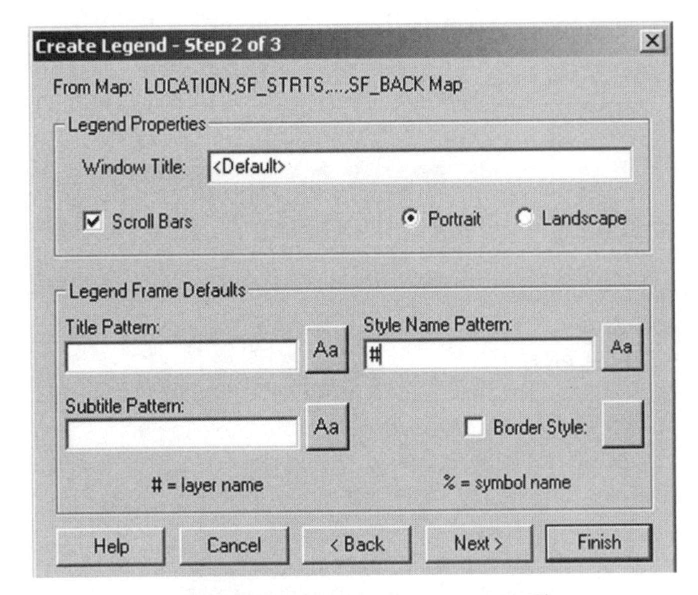

Step 2 of 3 dialog box modifying the default choices.

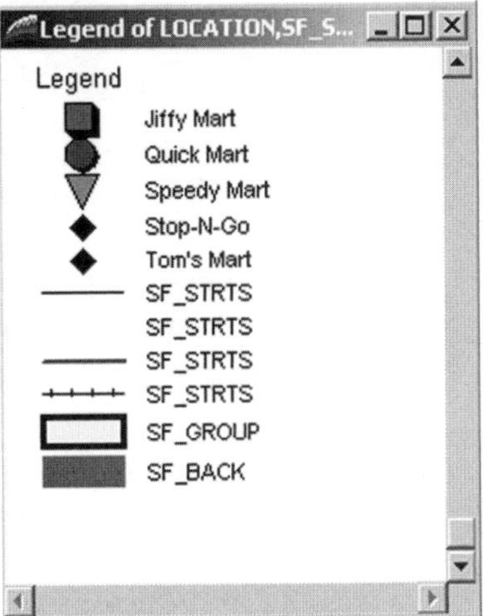

Results of modifying the default choices.

Once a legend is defined, it can be altered using the Legend option from the Main menu when a Legend window is active. Alternatively, you can double click on the Legend window to alter a portion of the legend. As noted previously, the Legend window will show all features for a particular layer, as in the case of the *Location and SF_STRTS* tables.

Although you can use the Step 3 dialog box to modify the *Location* table's legend, you will be required to manually edit the legend text for the *SF_STRTS* table to differentiate the various features for a given layer, as indicated in the following illustration. The reason is that there is not a field that adequately differentiates the styles in the *SF_STRTS* table. To edit the legend text, in the Legend window, double click on the layer you wish to edit.

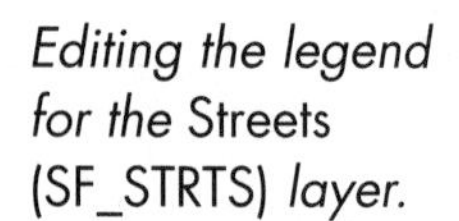

Editing the legend for the Streets (SF_STRTS) *layer.*

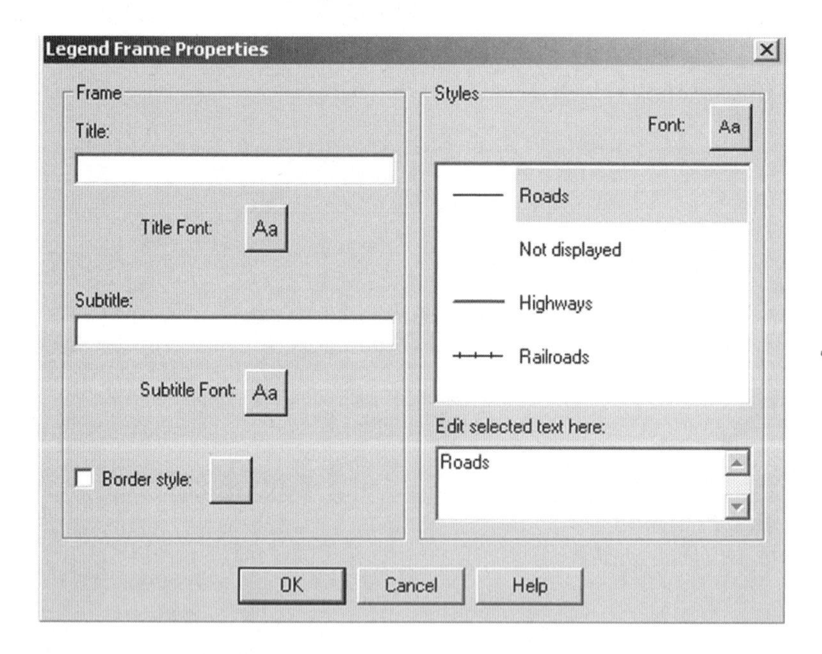

From the Legend Frame Properties dialog box, you can add or change the legend title for a layer, add a border style for the layer, or change the text describing the data. To change the text, in the Styles box, click on the item you wish to change, and type in the new text.

For the *Location, SF_STRTS, SF_GROUP,* and *SF_BAY* map, the final legend, shown in the following illustration, was created by accessing and customizing the text for each layer on the map. In addition, you can reposition items in the Legend window by clicking and dragging items to a new position, although aligning the legend entries manually tends to be very difficult.

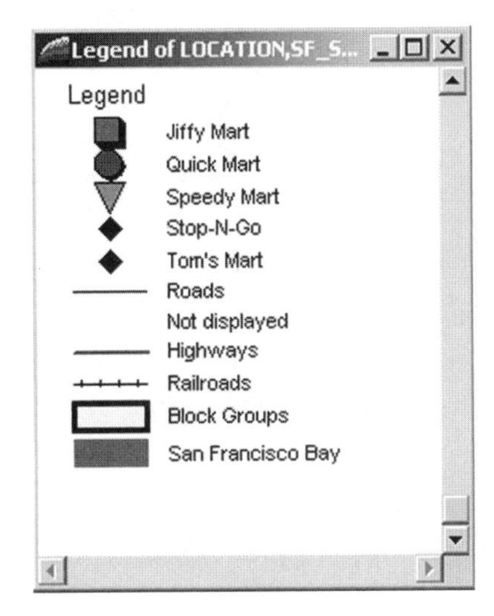

Final customized legend.

Animation Layer

The animation layer is intended for applications that track data in real time (i.e., the data are frequently updated). When activated, through rapid redraw, this layer allows you to see the position of data objects (such as a truck fleet) as they move by on the Map window. An example would be tracking trains in an interstate system. The animation layer allows rapid redraw regardless of map complexity because only the animation layer redraws. In effect, this layer becomes the top layer of the map, although it does not appear in the layer list in the Layer Control dialog box.

The animation layer can be created through MapBasic commands only. In addition, the animation layer will not be saved in a workspace. For greater detail and tips on using the animation layer, see Chapter 11.

Changing the Appearance of Individual Features

Changing the appearance of features on a map is often necessary, and MapInfo Professional provides you with several options. Some options permanently change the table, whereas others change only the display in the current analysis. Sometimes it is necessary to change an entire table, whereas in other cases you may wish to change the appearance of only a subset of the data.

As stated earlier, much of this chapter is organized as an exercise. If you choose to follow along on your computer, copy the *SF_GROUP* and *Location* tables (*samples* directory on the companion CD-ROM) to the *exercise* directory (to be able to edit the files) and open the *SF_BAY*, *SF_BACK*, and *SF_STRTS* files from the companion CD-ROM.

Global Changes

Global changes modify the way an entire table appears on a map. For example, assume a company's store locations are depicted as blue dots on the map, and you want to change them to purple to correspond to the company color scheme. This change would be global because the entire table is affected. To make global changes in MapInfo Professional, you can either change the display settings or edit the table.

Display setting changes are temporary. Such changes are in effect only during the current MapInfo session, and any subsequent session activating a workspace saved during the current session. The next time the file is opened (File | Open Table command), or if you display the file in a second Map window, it will revert to its default state (i.e., appear the same as before you made changes to the display settings).

The map shown in the following illustration demonstrates why global changes are often necessary. The map contains locations, streets, and block groups. However, the viewer is not able to distinguish the block groups from the streets in the map. By changing the appearance of the block groups in the map, you will be able to see both the block groups and streets at the same time. Even if the map were printed in color, the map shown in this illustration would still be mostly black and white. To avoid this, you can use a background color to make the map more pleasing to the eye. For block groups, you can select a light color (e.g., pale yellow) to give the map a little life while not overpowering it.

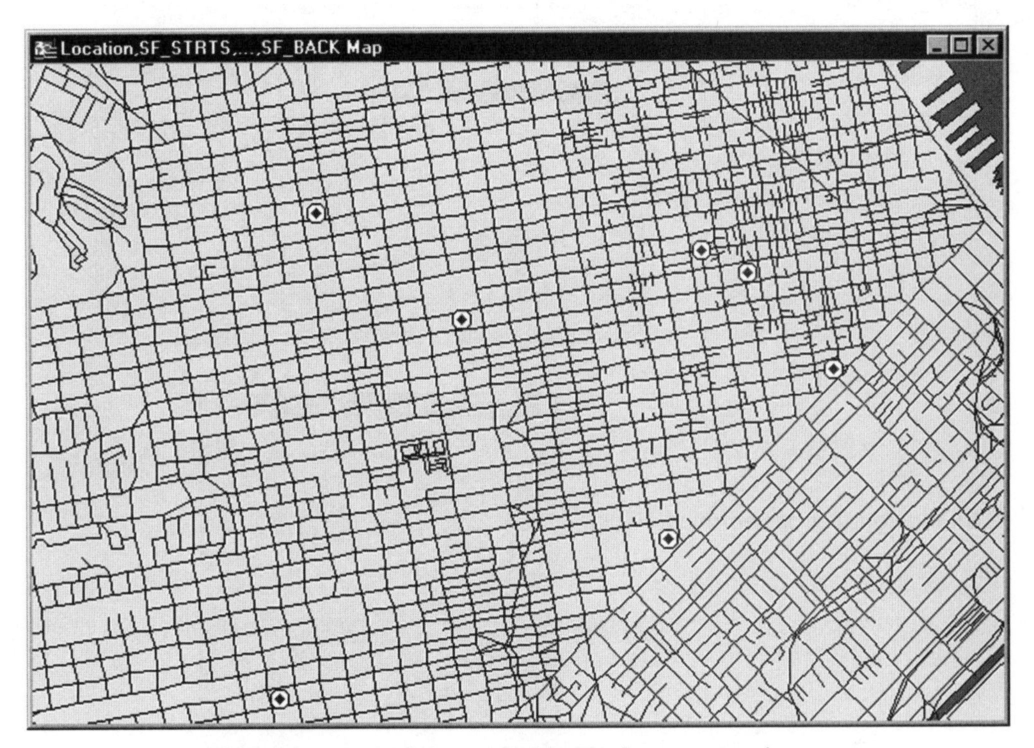

Map depicting indistinguishable block groups and streets.

To make a temporary change in the display settings for a layer, as shown in the first of the following illustrations, access the Layer Control dialog box (using the Map | Layer Control command). In Layer Control, select the layer you wish to change by clicking on that layer in the dialog. An example of changing layer appearance is shown in the second of the following illustrations.

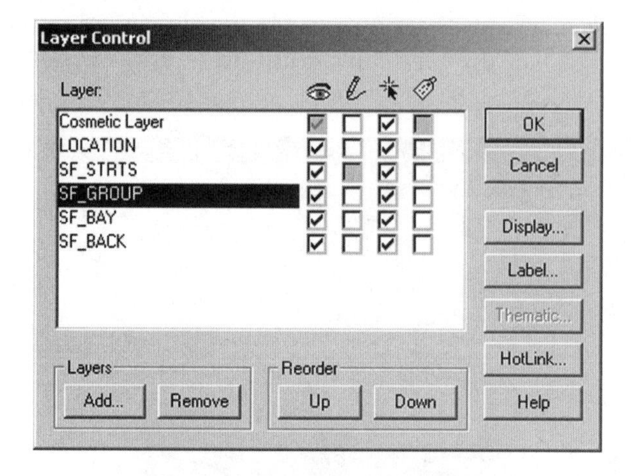

Changing display settings for a layer.

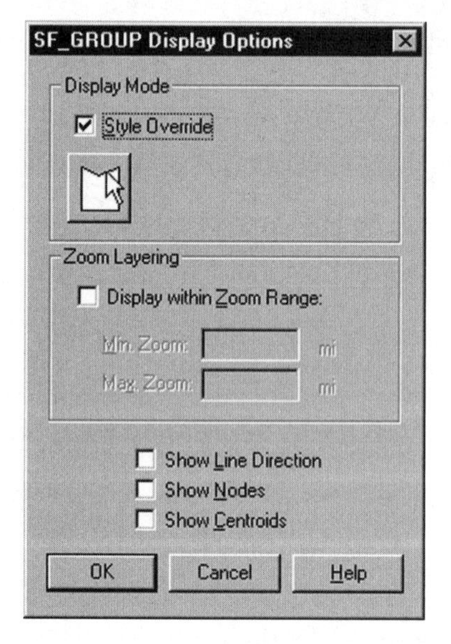

Changing layer appearance.

Once the layer is highlighted, select the Display button to access the Display Options dialog box. To change the layer appearance, select the Style Override box and click on the Style box below it.

A dialog box will appear, allowing you to customize the appearance of the layer. In this dialog, you can change the color, fill pattern, border style, and width for region data; line style and width for line data; and symbol size, color, and effects for point data. The sample style dialog box shown in the first of the following illustrations depicts how changes will appear on your map. Assume that the color, pattern, and border width for the block group layer are changed to be able to distinguish the block groups from the street layer, as shown in the second of the following illustrations.

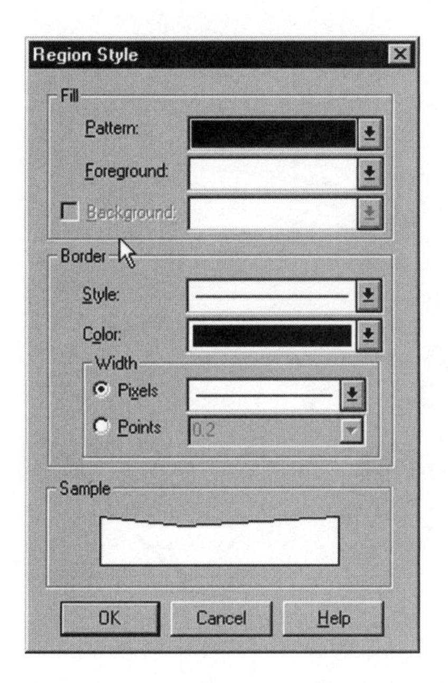

Sample field in Region Style dialog box.

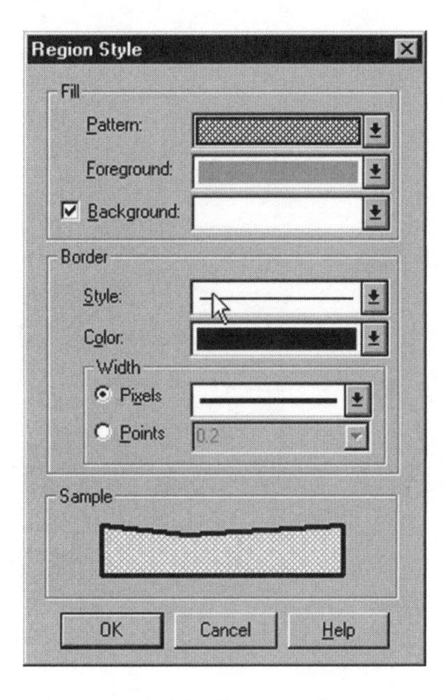

Changes to block group layer.

Once you click on OK to apply the changes to the map and dismiss the Layer Control dialog box, you will be able to easily identify the block groups (heavy lines) and the streets (lighter lines) above them. You could have used color as well as line width to show the difference between the two layers. Using lighter colors with larger line widths on lower layers will allow you to see darker colors on upper layers, or vice versa. For example, you could have made the block group border a thick yellow line, which would have allowed you to see the thin, black lines of the roads against the block group borders, as shown in the following illustration.

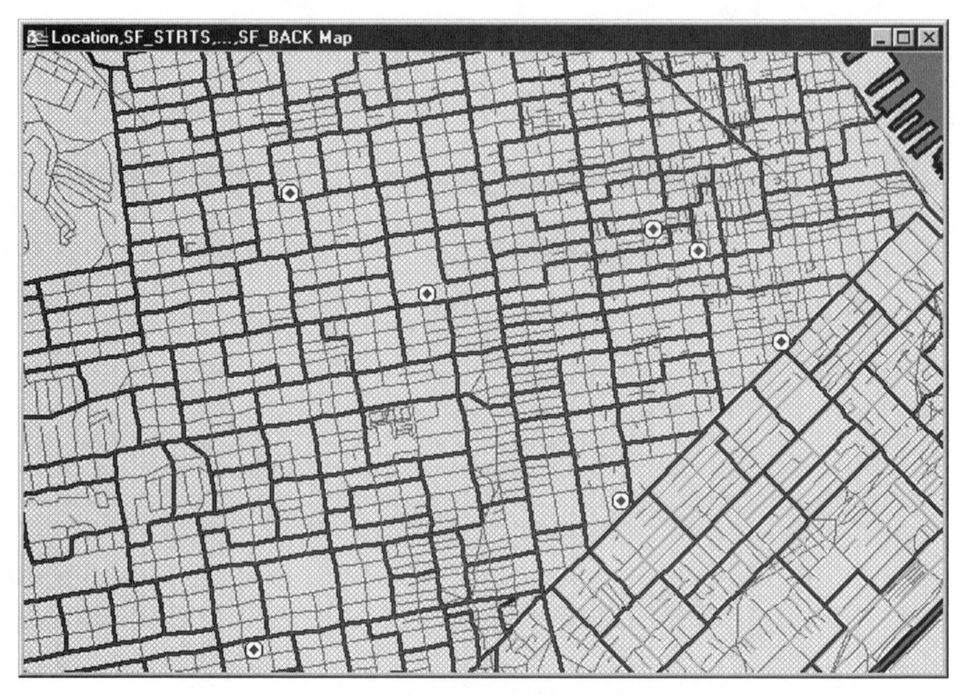

Block groups and streets made distinguishable.

The second option available for changing the global appearance of a layer is to edit the layer. This option will permanently change the layer when you save the table. (Whereas editing options are discussed in Chapter 7, this section attempts to explore alternative methods of changing map display appearance.) To make permanent global changes, access the Layer Control dialog box (Map | Layer Control), and make the layer you wish to change (*SF_GROUP*) editable, as shown in the following illustration.

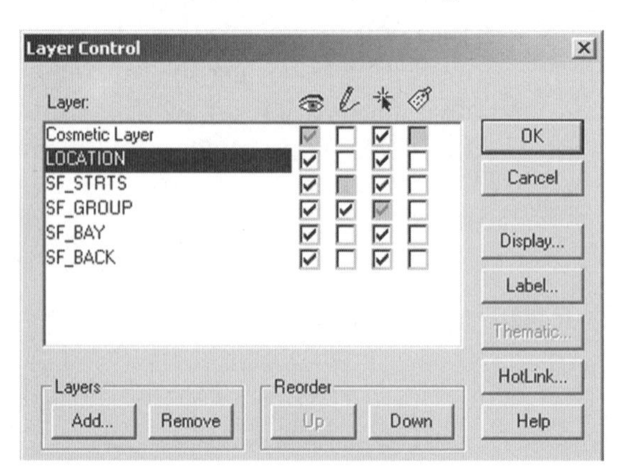

Changing attributes of a layer to editable.

Next, select all objects in the layer using either the Query | Select or Query | SQL Select commands, which access the Select dialog box, shown in the following illustration. Use Query | Select, and because you are interested in the appearance of the graphic objects only, deselect Browse Results.

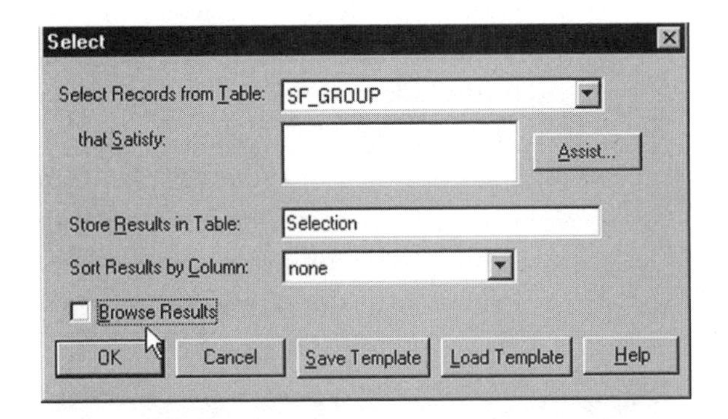

Select dialog box.

Once you have selected all objects in the layer, select Options from the Main menu bar. Verify that the Map window is the active window, or your changes will not take effect. This is one of the reasons the Browse Results option is turned off when issuing the Select command described previously. You are returned directly to the Map window to view the map objects selected as a result of the query. Select the appropriate data type from the Options menu and modify the features accordingly.

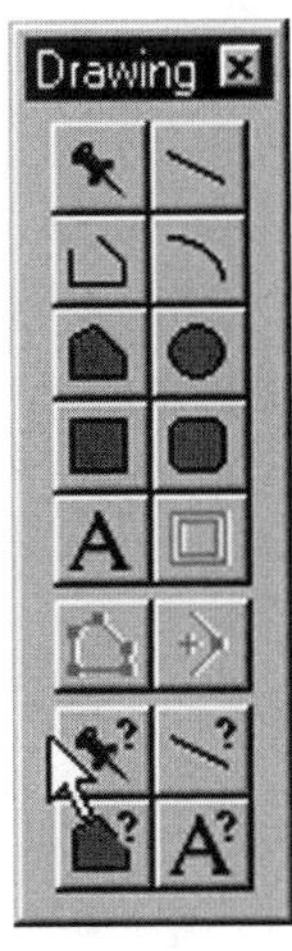

Drawing toolbar.

Because you are modifying the block group layer, you would select Options | Region Style to alter the appearance of the block groups. The corresponding Style dialog box activates to allow you to make changes to the layer's appearance. You can change the appearance of line, point, region, and text objects through Options style commands. You can also access the Style dialog box via the Drawing toolbar, shown at left.

The Region Style tool operates the same as Options | Region Style. The Symbol Style tool corresponds to the Option | Symbol Style command, the Line Style tool to the Option | Line Style command, and the Text Style tool to the Option | Text Style command. These tools are shown in the following illustrations.

Region Style tool.

Symbol Style tool.

Line Style tool.

Text Style tool.

You can access the Region Style dialog box through either the Options | Region Style command or the Region Style tool. Now, let's customize the appearance of the layer objects to suit the map, and click on the OK button to put the changes into effect. The appearance of the map has been modified to resemble the example created using the Display Options in the Layer Control dialog box. As shown in the following illustration, the results are the same: both maps have been modified so that both the block groups and the streets layers can be identified.

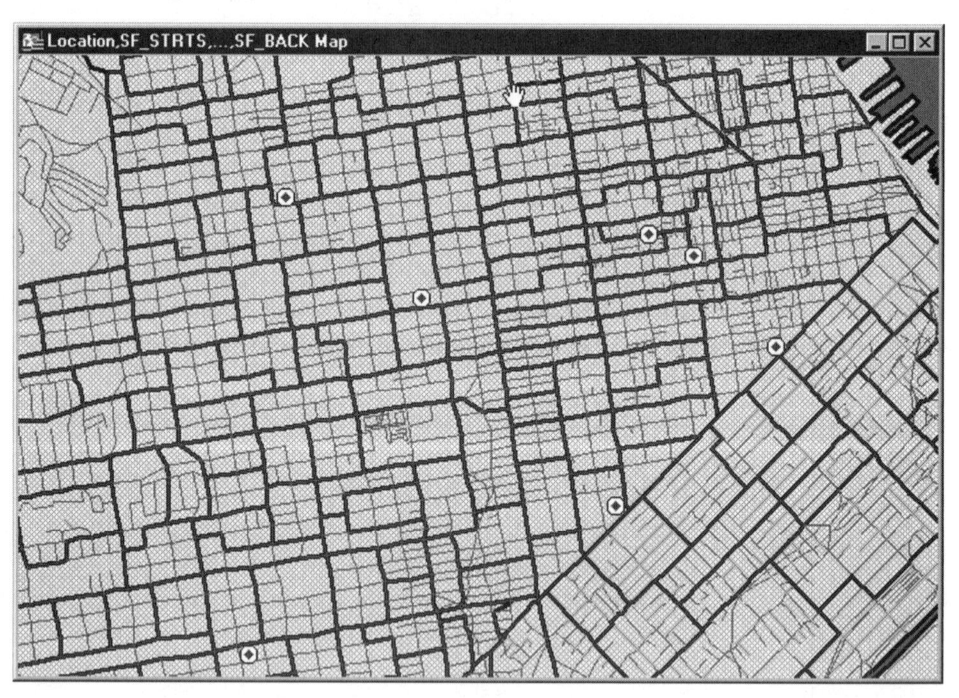

Changes made to map by editing the file.

Differences between the two methods pertain to the permanence of the changes. As stated previously, the changes made through the Display settings in the Layer Control dialog box are temporary and apply only to the map in which the changes are made, and are saved only by saving the current workspace. Editing the file, on the other hand, is permanent if you save the table. If you do not save the table, the changes will be lost once the file has been closed. Even if you have saved the workspace, the table will revert to its appearance prior to the changes (assuming the table was not saved separately).

The decision about which method to use should be based on the intended use of the file in later analyses. If you want the table to appear the same in most of your future analyses, edit the table, and make the changes permanent. If the change is specific to the current analysis only, change the Display settings.

Changing the Features of a Table One Object at a Time

Although the method of changing map features one object at a time is not viable for large-scale changes, it is useful when only a few changes are required. For example, assume you are examining six potential locations for a store. Assume also that you wish to show the six locations on a single map, but would like to distinguish the one you believe to be most advantageous compared to other potential sites.

Because all locations reside in the same MapInfo table, you might ask: How do I change the symbol for only one of the locations? You could make the cosmetic layer editable, and draw a symbol over the preferred location, or you could edit the file. Editing the file is recommended, to ensure that the preferred site is easily distinguishable in any subsequent maps you create. (Editing options are discussed in Chapter 7.)

To edit objects one at a time, under Layer Control, make the layer you wish to change editable. Select an object(s) on the map using the selection tools on the Main toolbar. If you are using the Select tool and wish to select more than one object, hold down the <Shift> key as you select each object. In the map shown in the following illustration, four locations are selected in order to change their symbols to reflect a different store chain, as shown in the following illustration.

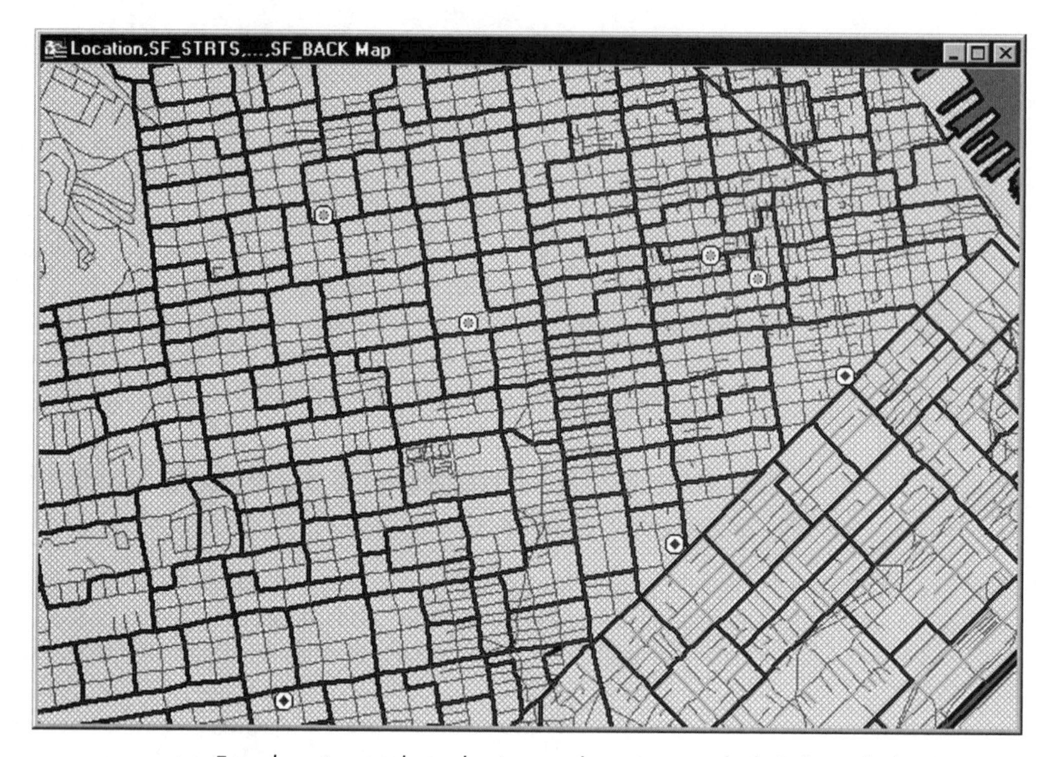

Four locations selected prior to changing symbol. Selected objects are lighter than other point objects.

Once the object(s) have been selected, select the appropriate option tool from the Drawing toolbar, or select Symbol Style from the Options menu. Selecting the Symbol Style tool (pushpin with question mark) activates the Symbol Style dialog box. The symbol has been changed from a red bull's-eye to a purple-shadowed box, as shown in the following illustrations.

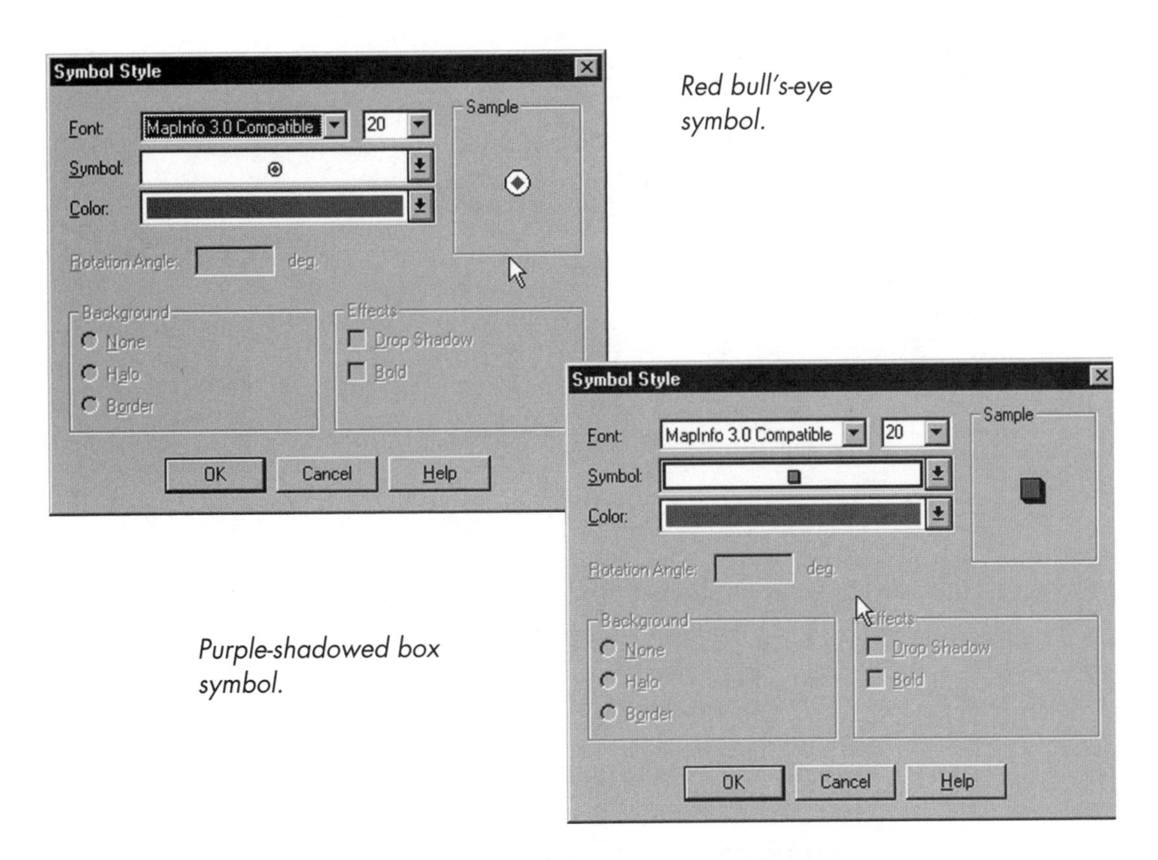

Red bull's-eye symbol.

Purple-shadowed box symbol.

The finished map is shown in the following illustration. Be sure to save the file, or your changes will be lost when the file is closed. Changing one or a few symbols at a time is fairly simple using this method, but this would not be the method of choice to change more than a handful. The following section discusses changing data subsets.

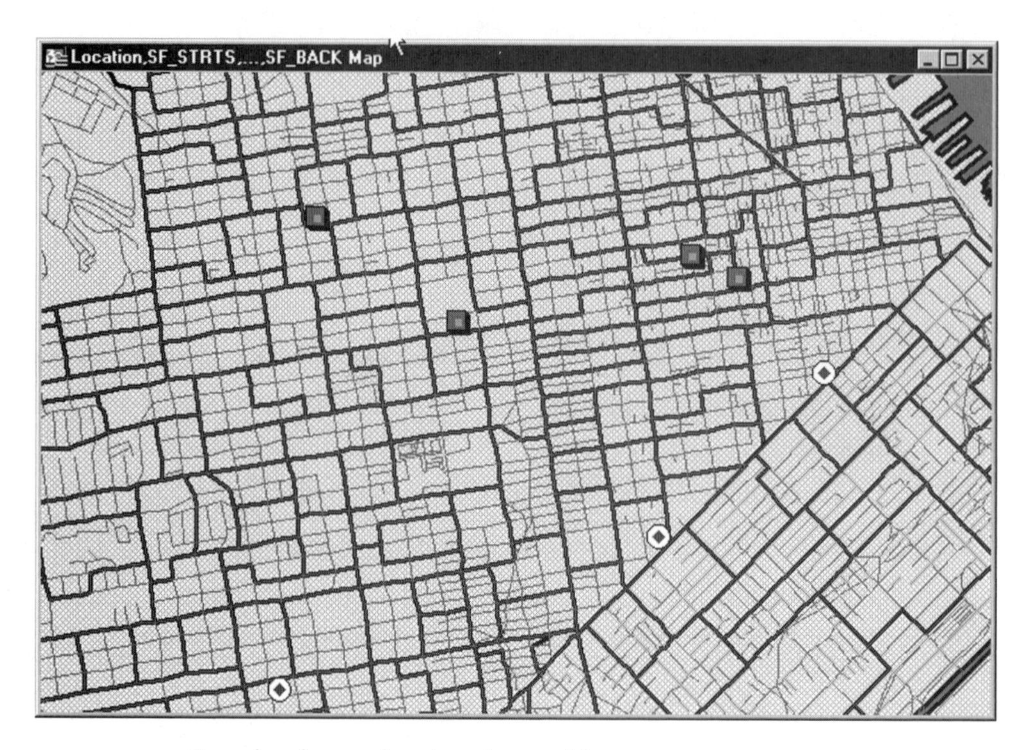

Completed map showing changed features.

Changing the Appearance of a Data Subset

Assume you want to distinguish highways from surface roads, even though both roadway types reside in the same table. Another example would be to differentiate within a single table among data regarding competitor companies. To distinguish among items within a table, you can either make changes to the table itself or create an individual-values thematic map to depict the different values.

Editing the table will result in permanent changes once the file is saved. (See Chapter 7 for details on editing.) Creating an individual-values thematic map will apply the differentiation to the present MapInfo session and any resultant workspaces. (See Chapter 6 for details on thematic maps.)

To edit a table, make the table editable in the Layer Control dialog (Map | Layer Control). This step is the same whether you wish to make global or individual changes, or to change only a subset of the data. Next, define a query that isolates the subset of data you wish to change. As an example, the symbol for different competitor companies used in the map shown in the following illustration will be changed.

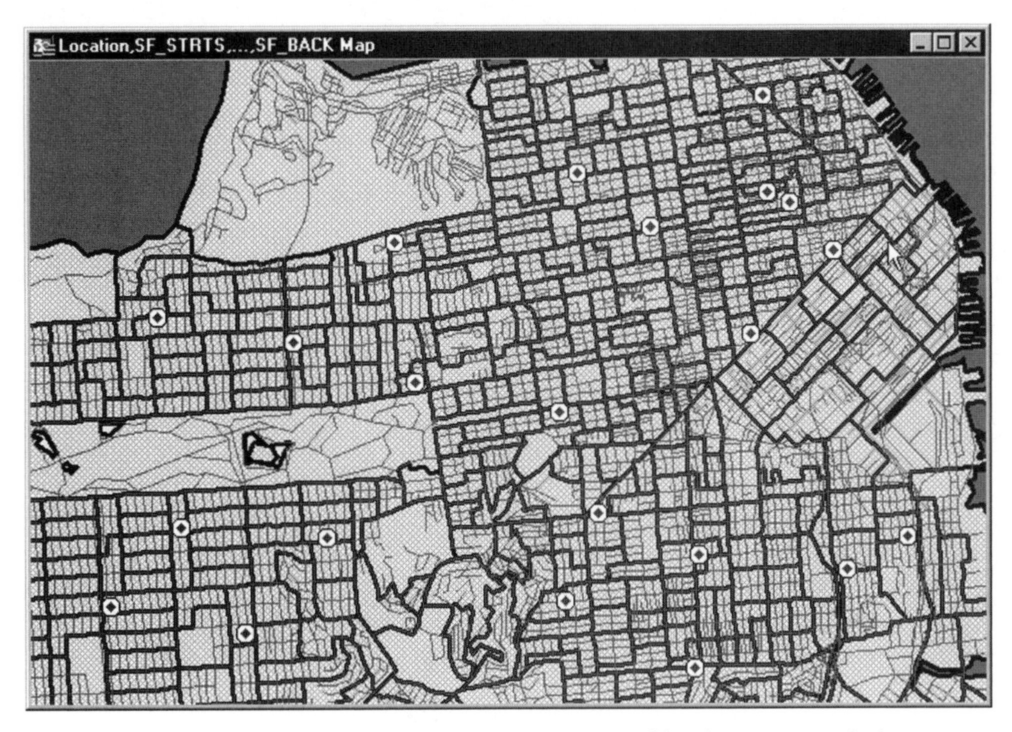

Map showing all competitors represented by the same symbol.

The easiest way to identify the different value types (e.g., competitor names) within a table is to issue a query based on the table. As shown in the following illustration, a query has been set up that will return each unique competitor name in the table, as well as the number of locations for each competitor.

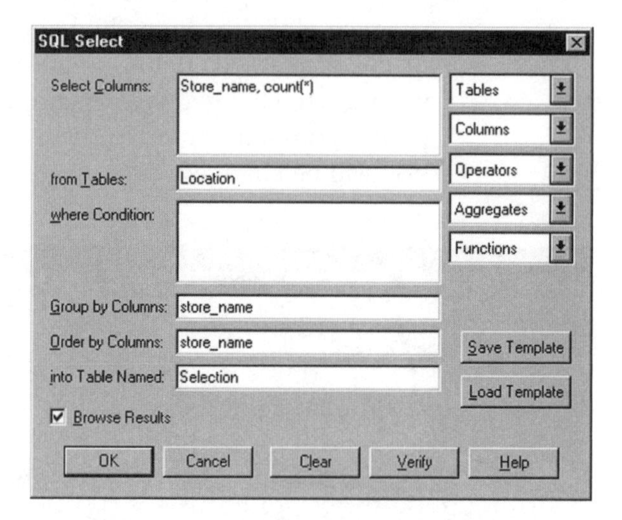

Query to identify unique competitor names and each competitor's locations.

The results of the query are shown in the following illustration. There are three major competitors (Jiffy Mart, Quick Mart, and Speedy Mart) in the area, and two minor competitors (Tom's Mart and Stop-N-Go). (Company names are fictitious.) Major and minor competitors are defined strictly on the basis of the number of outlets in the area.

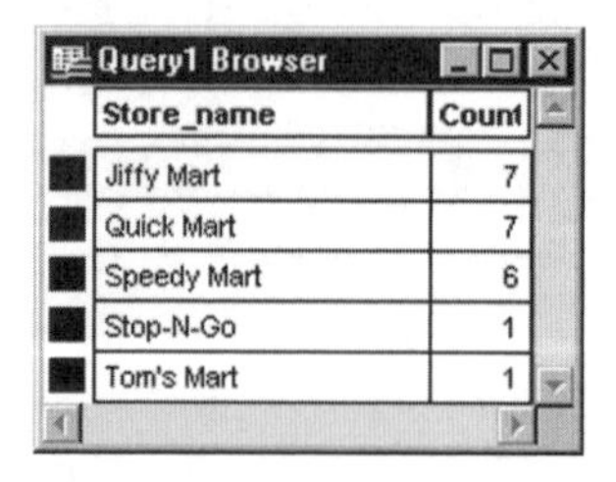
Query1 Browser

Store_name	Count
Jiffy Mart	7
Quick Mart	7
Speedy Mart	6
Stop-N-Go	1
Tom's Mart	1

Query results.

At this point, assume you wish to differentiate between the three major competitors using different symbols, and to combine the two minor competitors using the same symbol for both. To select a data subset, you have fashioned an SQL query to select all Jiffy Mart locations (see query definition in the following illustration). After running this query, all Jiffy Marts will be selected. Now you need to return to the Map window to change the symbols.

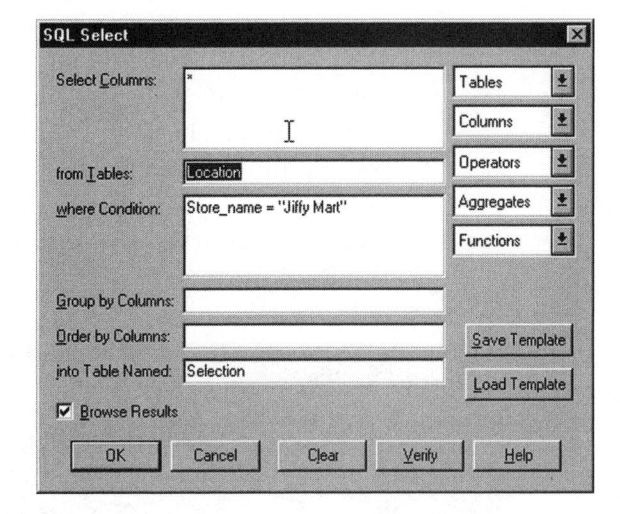

SQL query to select Jiffy Mart locations.

To change the symbol for Jiffy Mart, select either Option | Symbol Style or the Symbol Style tool from the Drawing toolbar. For the other major competitors, Quick Mart and Speedy Mart, issue the same query (changing the *Where* clause for each competitor), return to the Map window, and change the symbol style.

This selection process could also be accomplished with the more straightforward Query | Select command. Both types of queries will save the last query issued for easy modification. The selection statement shown in the previous illustration is depicted in the following illustration using Query | Select.

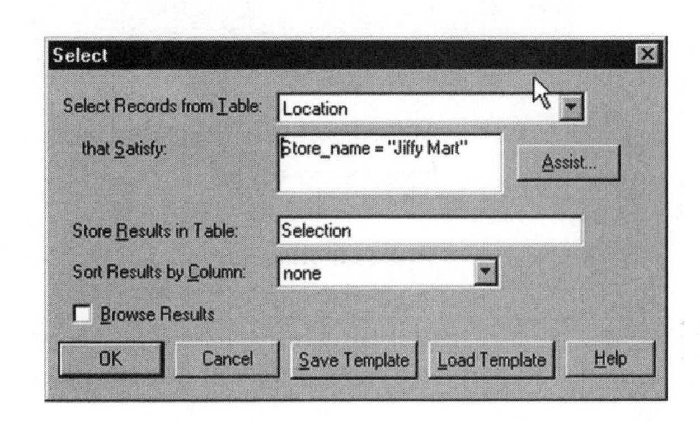

Selection statement using Query | Select.

If the symbol does not change when you have selected a new style, there are three things you need to check. First, access the Layer Control dialog box to ensure that the layer is indeed editable. If it is editable, verify that Style Override is not selected under Display Mode on the display dialog box for the layer. If the layer is editable and the default style has not been overridden, return to the Map window. The Map window must be the active window when you attempt to change the style, because styles can be changed only when the Map window is active. If all of these conditions are met, confirm that the objects to be changed are still selected. The results of the query are shown in the following illustration.

Map showing different symbols for major competitors, and the Location layer being edited.

You can also check the status bar to see which table is editable. Near the center of the status bar, the table being edited will be identified (e.g., *Editing: Cosmetic Layer*). If no table is being edited, the status bar will read *Editing: None.*

Now that the symbol for all major competitors has been changed, let's downplay the minor competition. Because there are only two, the minor competitors will be manually selected. They are easy to identify on the map because they are still depicted by the original bull's-eye symbol. You could have issued a query to select the minor competitors, but visual selection was just as easy in this case. Assume that with the Select tool (while holding down the <Shift> key), you selected both minor competitors, and then selected the Option | Symbol Style command to change their appearance. The resulting map is shown in the following illustration.

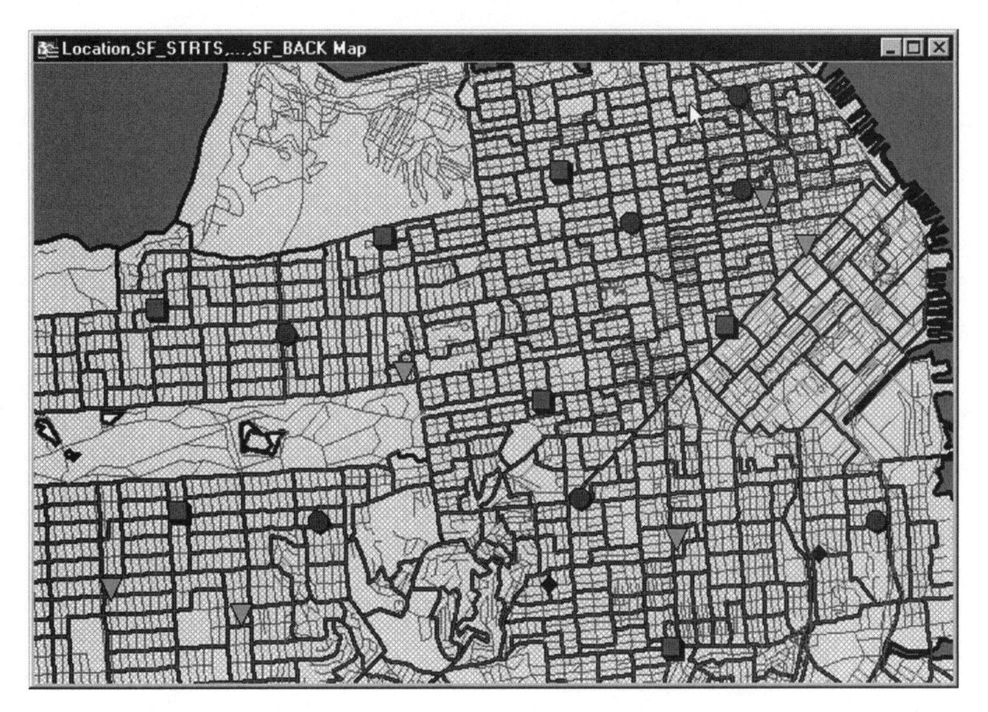

Map showing final results of symbol style changes.

The same results could have been accomplished with an individual-values thematic map. Although thematic maps are covered in depth in Chapter 6, a brief overview is provided here. To create a thematic map, select Map | Create Thematic Map from the Main menu bar. Select Individual as the type of thematic map to be created, and then select Point IndValue Default as the template.

To duplicate the map shown previously, the theme is based on the *Location* table, and *store_name* is selected as the basis for the theme. The symbol styles are modified until they resemble the styles defined previously for the competitors (when the file was edited).

As seen in the following illustration, the same symbol was used for both minor competitors. The legend was customized to group the minor competitors into an "Other" category by deselecting the Show this Layer option in the Customize Legend dialog box. When the two minor competitors were grouped, the Record Count option was turned off to ensure that incorrect information was not being depicted on the map.

Individual-values map showing same results as editing symbols with a MapInfo-created legend.

Pros and Cons of Editing Graphical Information

The advantage to permanently changing a file (by editing the geographic objects and saving the table) is that the table will always show the alterations (e.g., changed symbols) when you open it. You do not have to recreate the differentiation, as you would with an individual-values thematic map. You could also save the workspace containing the individual-values theme. However, to ensure that the map always appears the same in subsequent analyses, you would have to use the workspace saved with this theme as your starting place every time.

The major disadvantage to editing the table was addressed by MapInfo Professional 5.0 with the ability to create a standard legend for a map, which was not an option in previous versions of MapInfo. The decision to create an individual-values theme or edit a table is simple. If you need to include the same file in many analyses while differentiating between values, edit the table and save the changes. If you have created customized symbols that reflect competitors' logos, choose to edit the file directly and make the changes. If you do not believe you will access the changes in future analyses, create an individual-values thematic map.

Saving Changes

All methods discussed in this chapter for changing a display can be saved for use in other analyses. Changes made to the table by editing require that you issue the File | Save Table command, or save the table upon exiting MapInfo Professional. If these changes are not saved, the map features will revert to their appearance prior to any changes.

Appearance changes made using thematic maps and display settings will not be applied to the table the next time it is opened. However, you are able to return to the analysis (and the table's changed appearance) by saving the analysis to a workspace. Note that regardless of how the appearance of the objects was changed, you are always able to temporarily override the changes through global displays or new thematic maps. Exercise 5-1, which follows, provides you with practice in changing layer order and map appearance. A chapter summary follows the exercise.

■ EXERCISE 5-1: CHANGING LAYER ORDER AND MAP APPEARANCE

In this exercise you can experiment with several features discussed in the chapter.

1. Open the *SF_GROUP*, *SF_BACK*, *SF_BAY*, and *Location* tables (*samples* directory on the companion CD-ROM).
2. Access the Layer Control dialog box. Change the order of the layers (tables) from top to bottom as follows: *SF_GROUP*, *SF_BAY*, *Location*, and *SF_BACK*. Which layers are visible?
3. Access the Layer Control dialog box. Change the order of the layers (tables) from top to bottom as follows: *SF_GROUP*, *Location*, *SF_BAY*, and *SF_BACK*. Which layers are now visible?
4. Open the *SF_STRTS* table (*samples* directory), and wait for MapInfo to redraw the map.
5. Turn off Visibility for the *SF_STRTS* layer. Does the map redraw take less time than the previous redraw?
6. Turn off Visibility for the *SF_BAY* layer.
7. Remove the *SF_STRTS* layer.
8. Change the order of the layers from top to bottom as follows: *Location*, *SF_GROUP*, *SF_BAY*, *SF_BACK*.
9. Change the display for the *SF_GROUP* layer to a light green with dotted, dark green border lines.
10. Customize the labels for the *Location* table. Label this layer with the Sales column. Use the following currency format expression:

    ```
    format$(Sales,"$,#")
    ```

11. Change the label font to Times New Roman, 8 pt.
12. Add a light yellow background halo to the label.
13. Add bold and expanded effects to the label.
14. Change the label position to be above the object, with a 4-pt offset. Your map should resemble that shown in the following illustration.

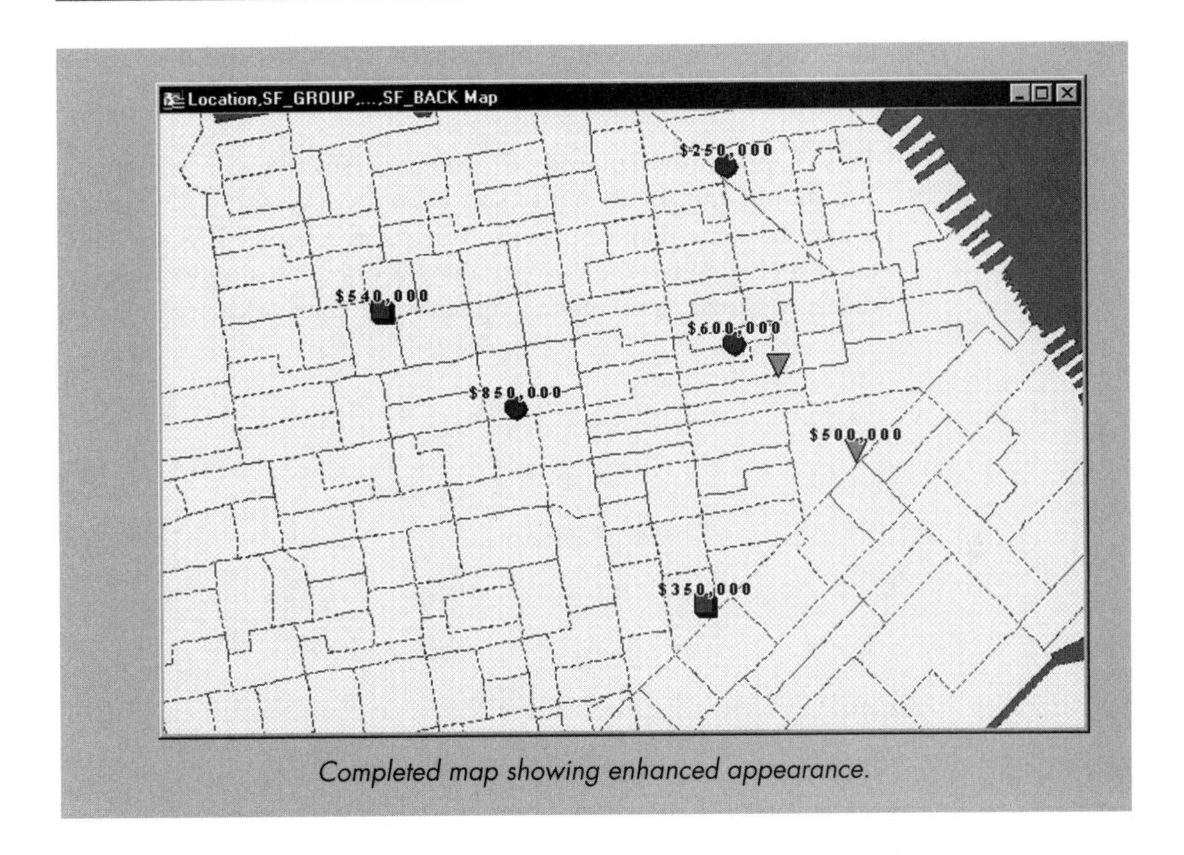

Completed map showing enhanced appearance.

Summary

Display options determine how objects appear on a map. These options are controlled through the Layer Control dialog box. In Layer Control, you can turn the visibility of a layer on or off, make it editable, allow map objects to be selected from the map, or label all objects. From this dialog box, you can also temporarily change the appearance of map features (e.g., add color to a block group table or change the symbol for a table), change the appearance of labels, or change a thematic layer. In addition, you can add tables not currently displayed on the map, or remove unnecessary layers.

The cosmetic layer enables you to add drawings to a map. Cosmetic objects that are not saved to a table (either new or existing) are lost when you close MapInfo Professional. However, if you save the workspace, MapInfo will recreate the objects on the cosmetic layer, along with all other objects, such as themes and display settings.

This chapter also explored permanent versus temporary changes. You should edit the file and make changes permanent if you want to use the same map display in subsequent sessions. If you wish to make a change for the current analysis only, you should alter the display or create a thematic map, and save the workspace.

CHAPTER 6

CREATING THEMATIC MAPS

AS MENTIONED PREVIOUSLY, GIS SOFTWARE PROVIDES YOU with the ability to integrate graphic and tabular information. GIS can also be used as a powerful means of communication through thematic maps, and MapInfo Professional has one of the strongest thematic tool sets of any desktop GIS program. These built-in tools let you depict many pages of tabular data on a single easily understood map.

Thematic mapping is the process of overlaying tabular data on geographical objects in a map. Thematic maps are also known as *themes.* Examples of thematic maps or themes are shading census tracts according to median income ranges, and graduating location symbol sizes by sales ranges.

To map data, a table must either contain geographies or a field (e.g., zip code or block group name) that will link the data to a geographic object or feature. If the data to be mapped are not contained in the same table as the graphic data, the name of the geography must be exactly the same in the data table as in the graphic table. To ensure that the two are in fact the same, you may want to open a browser for both tables, compare the geographic references side by side, and then edit the data and change the file structure as needed. You may wish to work with a copy or back up the files before making changes, to ensure data integrity.

Basic Thematic Map Types

MapInfo Professional allows you to choose from the following types (and combinations) of thematic maps: range, pie, chart, graduated symbol, dot density, individual values, and grid. Line, point, and region data were discussed in Chapter 5. The same terminology is used in this chapter. Recall that line data refer to streets, contours, and other types of information stored as lines. A table containing a single latitude and longitude coordinate pair for each record is considered point data. Examples of point data include store locations, customer addresses, and elevation points for mountain peaks. Region data refer to an area (or polygon), such as a block group, zip code, county, or state.

Ranges

Ranges allow you to map (or shade) classes of data by color or fill pattern. You can explore the same data in different ways through diverse ranging methods. Range themes can be created for symbol, line, or region data, although the data must be contained in a single column or in an expression to be mapped. MapInfo Professional will create a temporary column in the table if you wish to map an expression or information from another table. The data used to create a range map must be numeric.

An example of a thematic map containing region and range data is median income ranges (e.g., < $15,000, $15,000-$30,000, $30,000-$45,000, $45,000-$60,000, > $60,000) for a particular county by block group. In the following illustration, a data range thematic map of San Francisco County depicts point data (store locations of competitors by sales range) and region data (block groups by 1998 population range).

Pie Charts

Pie charts are useful when the data you wish to map are stored in several columns. Age and income ranges are good examples of data to map with pie charts. If your data table contains columns

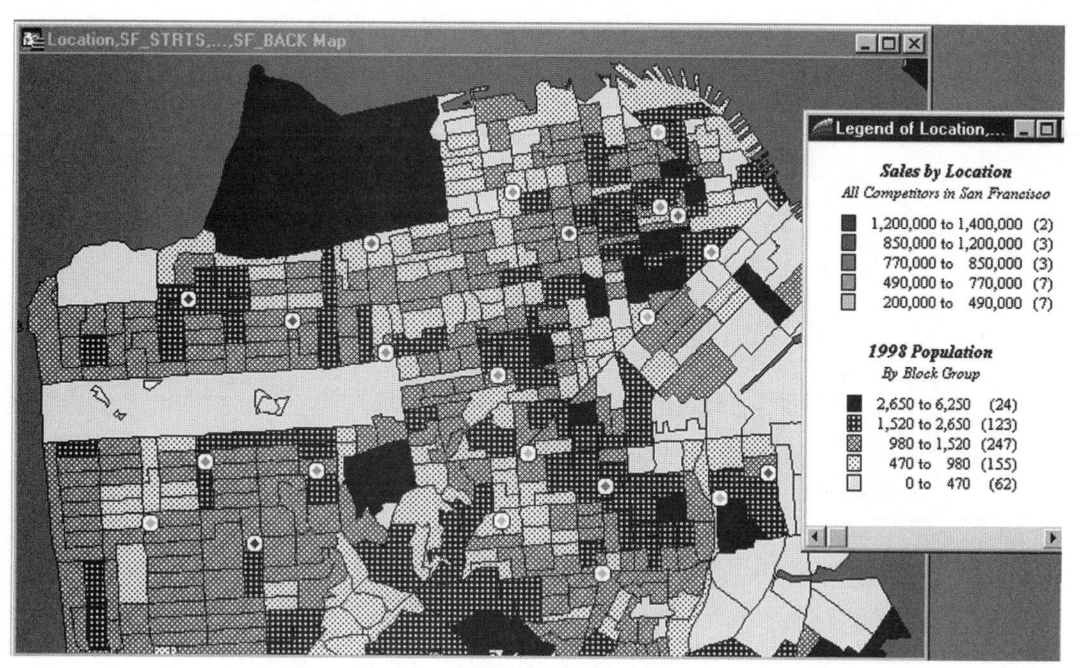

Data range thematic map using region and point data.

for numbers of people aged 55-59, 60-64, 65-69, and 70+, you could create a pie chart thematic map by census tract, showing the number of people that fall in each age group within each tract. The size of the pie chart could be graduated to show the total population age 55 and over within each geography.

Pie charts work best with three to five data categories. More categories make the "pie slices" difficult to interpret, especially in small geographies. For instance, pies do not work well if you are using small geographies, such as block groups, when viewing (analyzing) a large area in which the pies would overlap. Pies are best used to thematically map region data, as shown in the following illustration, because it might be difficult to track which layer the pie was attached to for point and line data. Pie chart maps will map numeric data only.

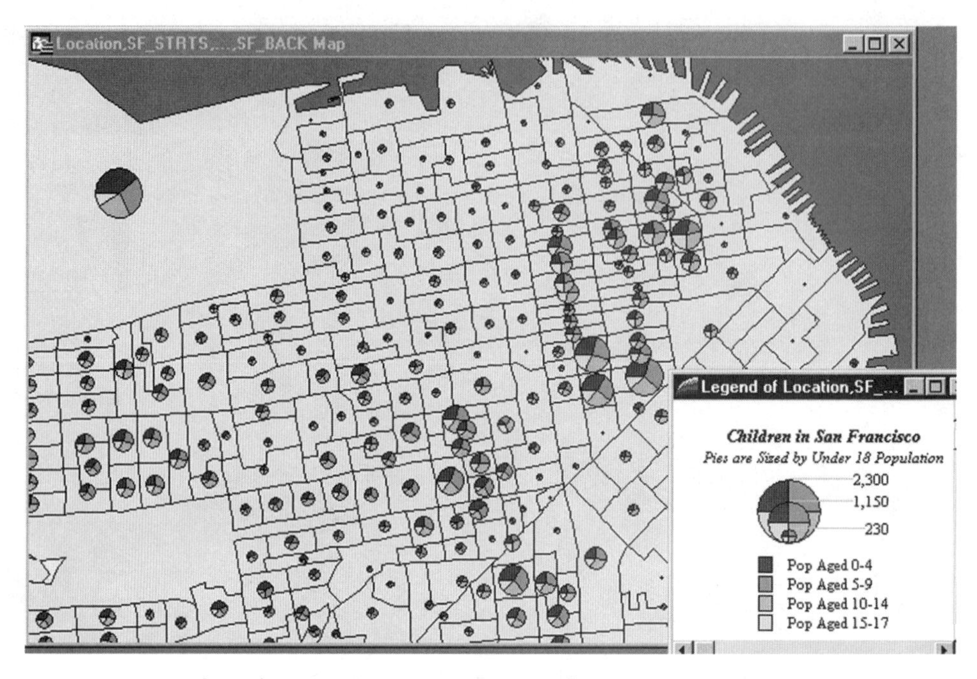

Pie chart thematic map using region data.

Bar Charts

Bar chart themes are similar to pie chart maps in that they map data contained in several columns. Like pie charts, bar charts are best used with region data instead of lines or point data, as shown in the first of the following illustrations. Again, bar charts will map only numeric data.

Graduated Symbols

In a graduated symbols thematic map, a symbol graduated by size is used to display information for both point and region data, as shown in the second of the following illustrations. The symbol is sized by the amount of the selected field or expression related to a point or contained within a geography. MapInfo allows symbols for positive and negative values to be graduated for this type of theme. Like ranges, pie charts, and bar charts, graduated symbols only work with numeric data.

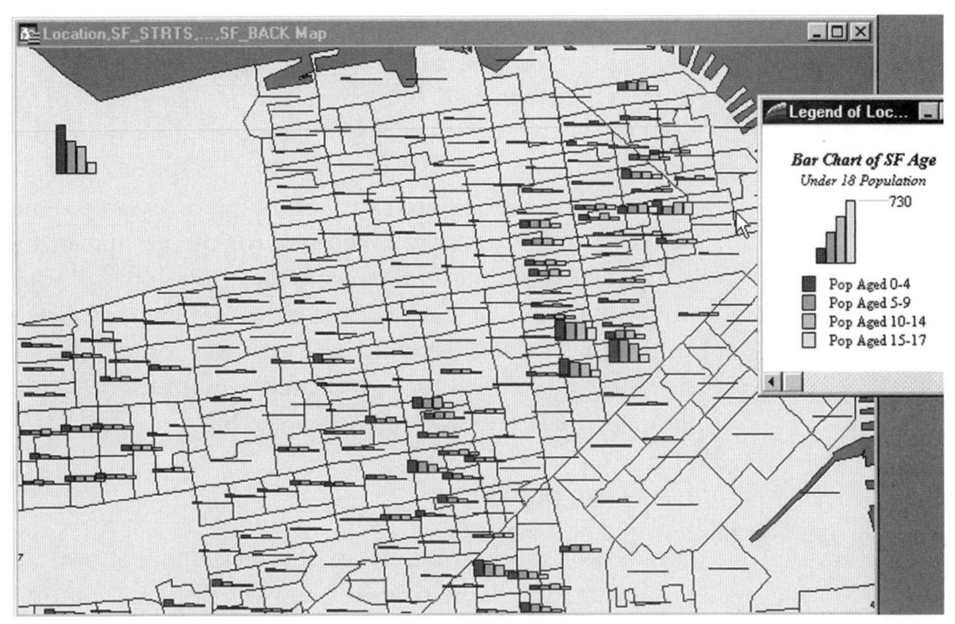

Bar chart thematic map using region data.

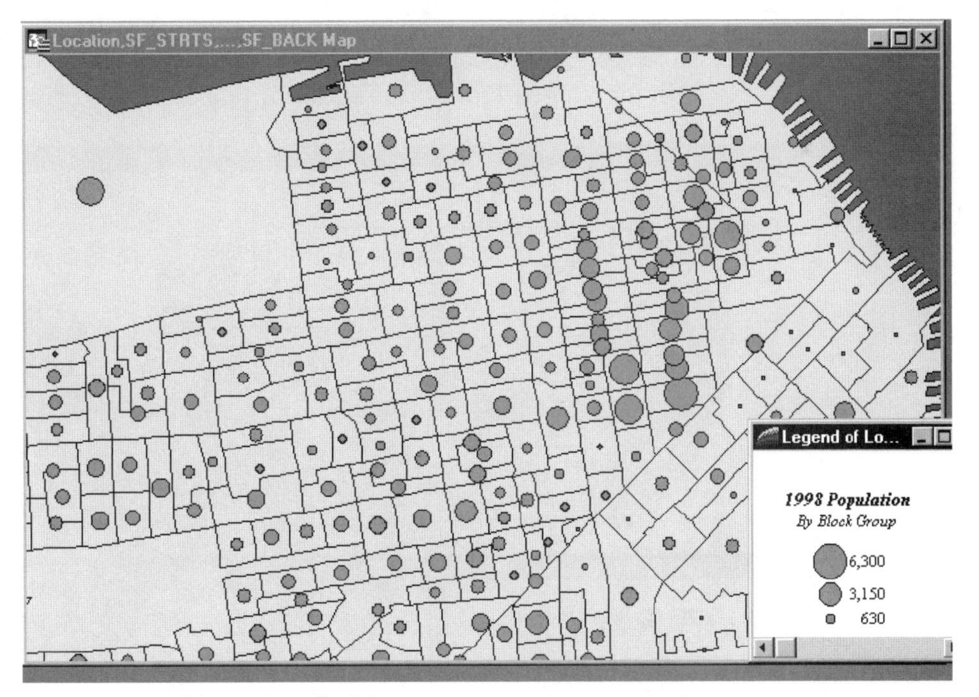

Graduated symbol thematic map using region data.

Other types of thematic maps use graduated size to help convey a message, such as symbol size in range maps, pie chart size in pie themes, and stacked bar chart size in bar chart maps. You may wish to choose another type of theme to achieve greater flexibility.

For example, a graduated symbol theme takes precedence over symbol style; that is, the symbol chosen for the graduated theme will be displayed over other themes using the symbol style. If you have edited the symbols in a table or created an individual values theme (see "Individual Values" section) and wish to preserve the symbols, consider this alternative to a graduated symbol theme: create a ranges thematic map, choose Auto Spread by Size, and apply size only.

Dot Density

Dot density maps are used for region data, as shown in the following illustration. This type of map will display randomly dispersed dots within a region. The dots depict the amount of the selected data each region contains by the number of dots placed within the region. Dot density themes can be created only for numeric data. To prevent confusion, it is suggested that you note the following in a theme's legend: dots are randomly placed in geographies.

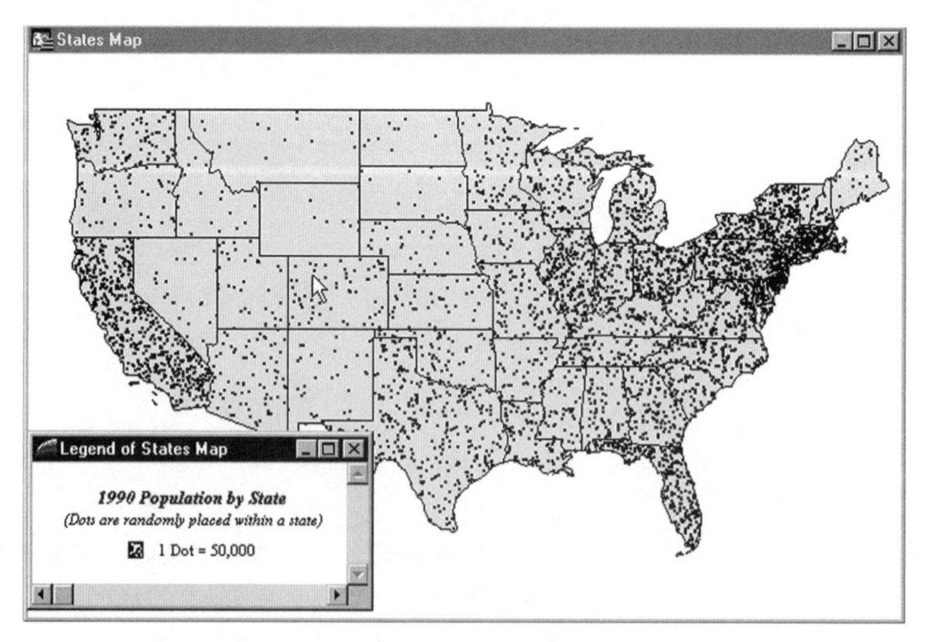

Dot density thematic map using region data.

Individual Values

An individual values thematic map is the only theme type available for mapping text values, although you can also map numeric values with this type of theme. MapInfo Professional will evaluate the data in the selected column and assign a color for each unique value in the column. You can change the symbol, region or line style, and color for each individual value. This type of thematic map, an example of which is shown in the following illustration, works equally well with line, point, or region data.

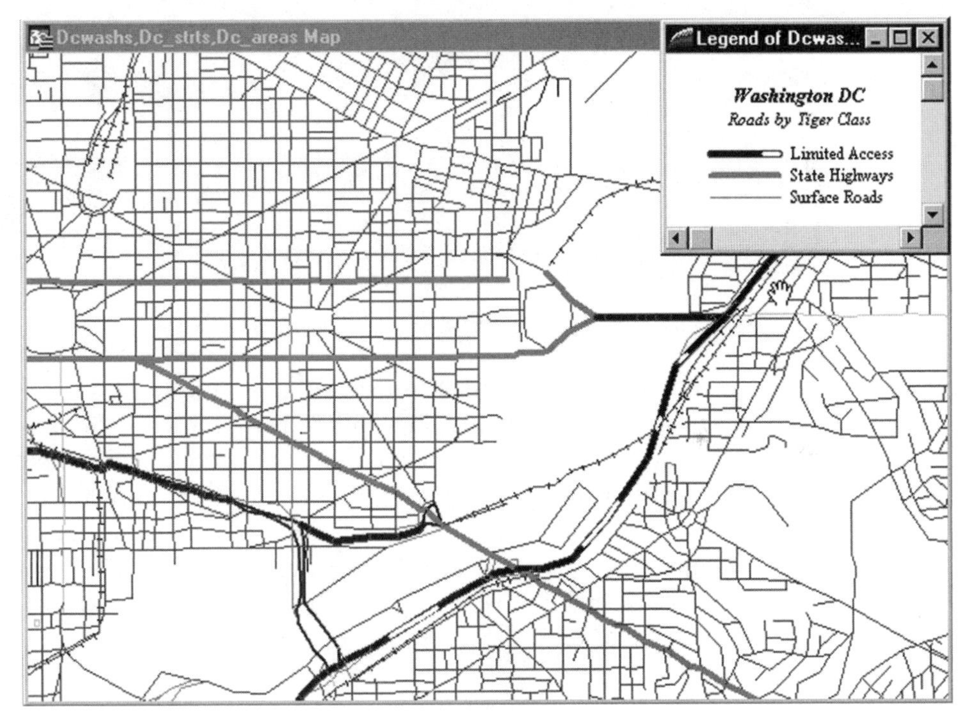

Individual values theme using line data.

Examples of cases where you may wish to map by individual values are coloring a floodplain area by soil types or mapping a market area by competitors. This type of theme, an example of which is shown in the following illustration, is limited to a data set with a small number of possible values. For instance, you could easily map six or seven different values, but the map would be confusing if you attempted to map the sales for 250 individual stores. The latter type of map is better suited to a ranges theme.

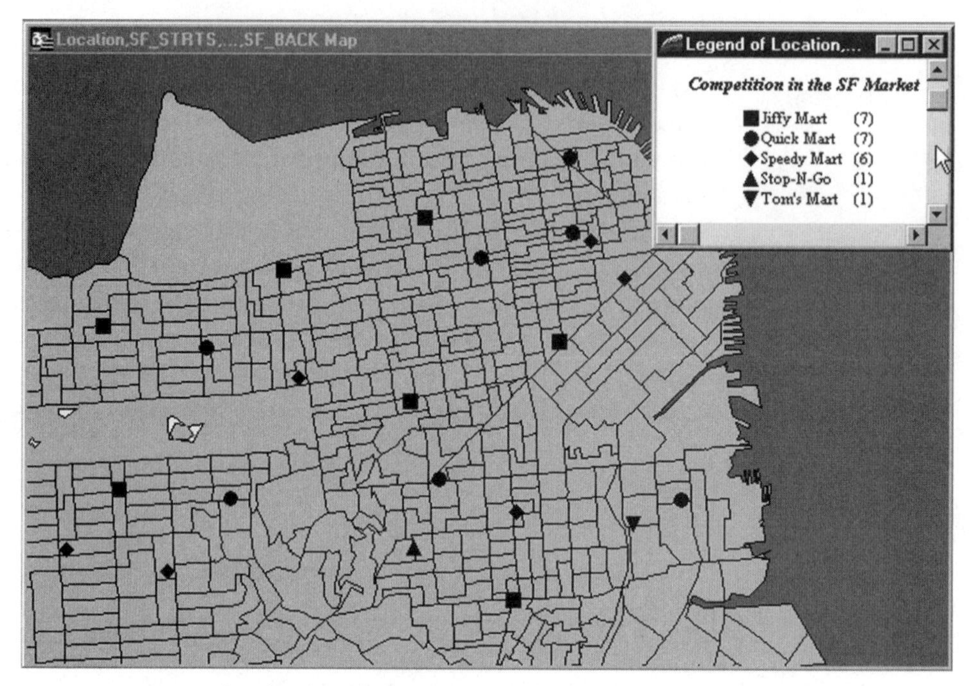

Example of individual values theme using point data.

The following illustration shows an example of an individual values theme depicting the 152 census tracts in the San Francisco area. MapInfo has reused colors, or used colors so similar that differentiating between the hues is nearly impossible. This is an example of a data set containing too many values to create a meaningful individual values thematic map. The legend is also confusing because it would be four to five times as long if stretched out to show all the individual values.

Grid Themes

Grid themes are a new addition to MapInfo Professional. This type of theme displays data as continuous color variations across a map. The map is created by the program interpolating point data from a source table, and then generating a grid file based on the interpolation and displayed as a raster image on the map. The data is projected onto grid cells, with values based on the point data that falls within each cell and estimated values for cells with missing data. Grid themes work best with point data that have

Example of individual values theme depicting too many values differentiated by color.

measured values at given locations, such as sales, elevation, temperature, and rainfall information. Through its continuous shading, this type of map allows you to infer a value where no information is available.

Grid themes are different from the other types of themes discussed so far. Instead of just creating a temporary layer like the other themes, a grid thematic map creates a raster image depicting a selected variable, saves the image as a new file, and adds it to the map. A grid theme is stored in a separate table (specified by the user), and can be opened independently in another analysis. It can, however, still be modified as a theme, using Map | Modify Thematic Map, if the base table is open.

In addition, in other types of themes, if the data on which the theme is based change, the thematic map will reflect those changes. However, with grid themes, because a new table is generated and added to the map, the data on which the theme is based is no longer tied to the original file. If the data in the original file are updated, you will need to recreate any grid theme based on the data.

The last difference is in how the themes are added to the map. In other types of themes, a thematic map appears directly above (in the layer list in the Layer Control dialog box) the table on which the theme is based. In grid themes, because a new file is created, the newly created table typically appears low in the layer order, because it is a raster image (based on MapInfo Professional's hierarchy). To view the results of a grid thematic map, you will often have to access Layer Control and reorder the layers. An example of this type of theme is shown in the following illustration.

Example of grid theme depicting average March temperatures across the United States.

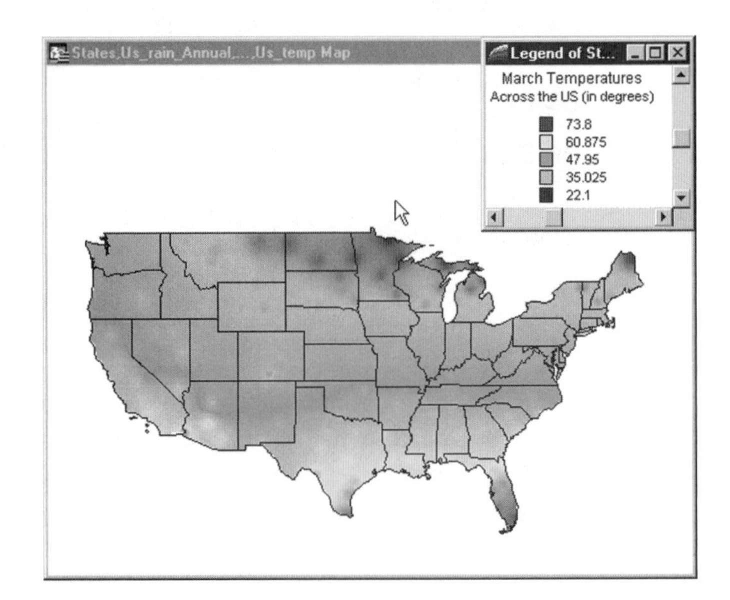

Combinations of Themes

Combinations of themes allow you to explore more than one dimension of data on a single map. A compelling example is a range map depicting various classes of median income by block group, overlaid with a pie chart map representing the number of households that fall into various income categories. With this map, a viewer can easily identify areas of high mean income with large numbers of high income households versus areas where mean income has been driven upward by a few wealthy households. In addition, by graduating the pies, you can determine the relative number of households within each block group.

To demonstrate a combination map, three themes have been created from a retail sales competitor database: the size of the symbol indicates size of the competitor's location, symbol color reflects sales range for each location, and the symbol shape indicates competitor identity (name) for each location. The combination map was created using a range map for square footage and applying the size attribute only, a range map for sales and applying the color attribute only, and an individual values map to create a symbol for each competitor and deselecting the color option. These options/selections are discussed later in this chapter.

NOTE: *When creating two or more themes based on a single layer (such as the retail table in the previous example), you may have to reorder the thematic layers (in the Layer Control dialog box) to achieve the desired results.*

Because of MapInfo's hierarchical layer structure, the individual values map was created first so that the other attributes would be applied to the symbols created by the individual values theme. As shown in the following illustration, the locations are set against a backdrop of 1998 population by block group, which was also created as a range theme.

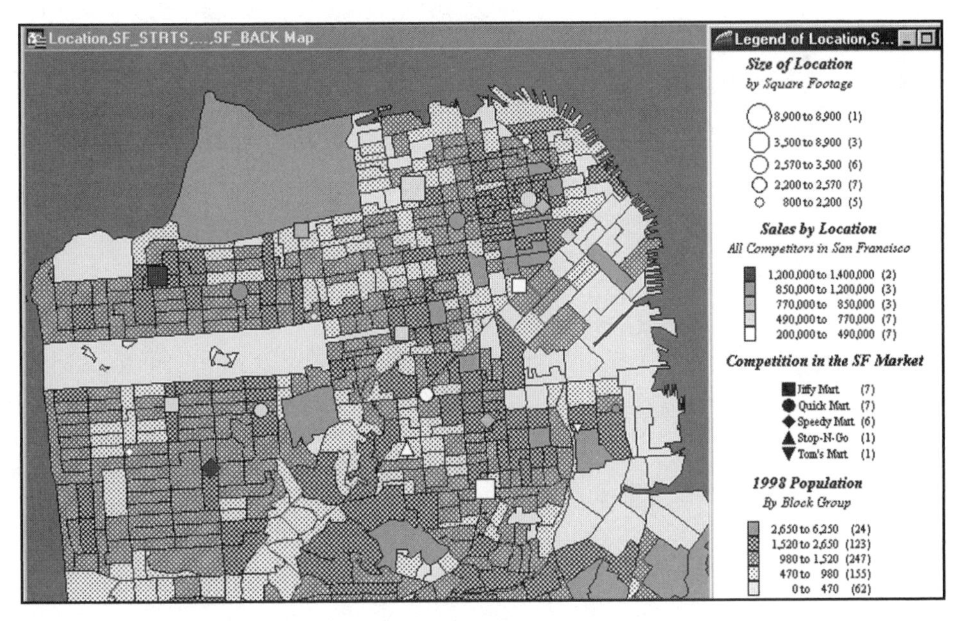

Combination map showing locations of competitors against backdrop of 1998 population by block group.

A different type of combination map is shown in the following illustration. In this map, the block groups are shaded by 1998 population. A pie chart theme showing proportions of minors (persons under age 18) by age group was placed on top of the population theme. This map is useful if you wish to display relative concentrations of both the general population and minors.

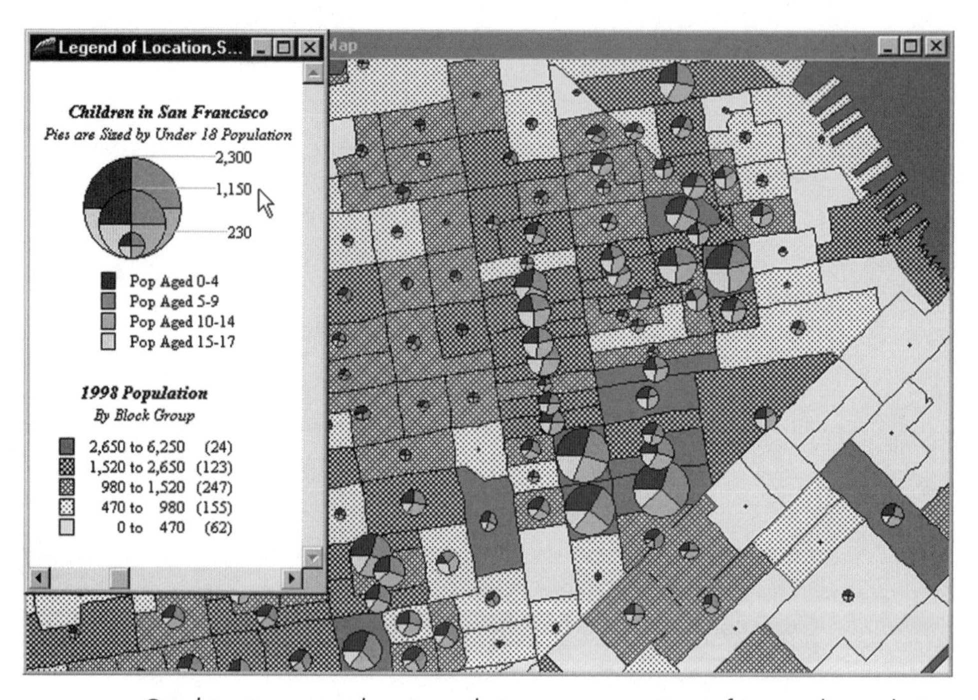

Combination map showing relative concentrations of general population and minors.

Although thematic maps (especially theme combinations) are very powerful, avoid creating too many themes on a single map. For example, a retail business could map the locations of both its stores and competitors' stores colored by sales (range map), sized by the square footage of the location (range map), with symbol shapes determined by competitor name (individual values). Under the point data, the retailer could show the market by block group, with the block groups shaded by median income (ranges), dotted with population under age 18 (dot density), and containing pie charts showing population groups under age 18.

Although this example may be an extreme situation, a number of interrelated factors can be effectively depicted on a single map. In this case, however, the following illustration shows that the seven themes created for the example produce a very confusing visual experience.

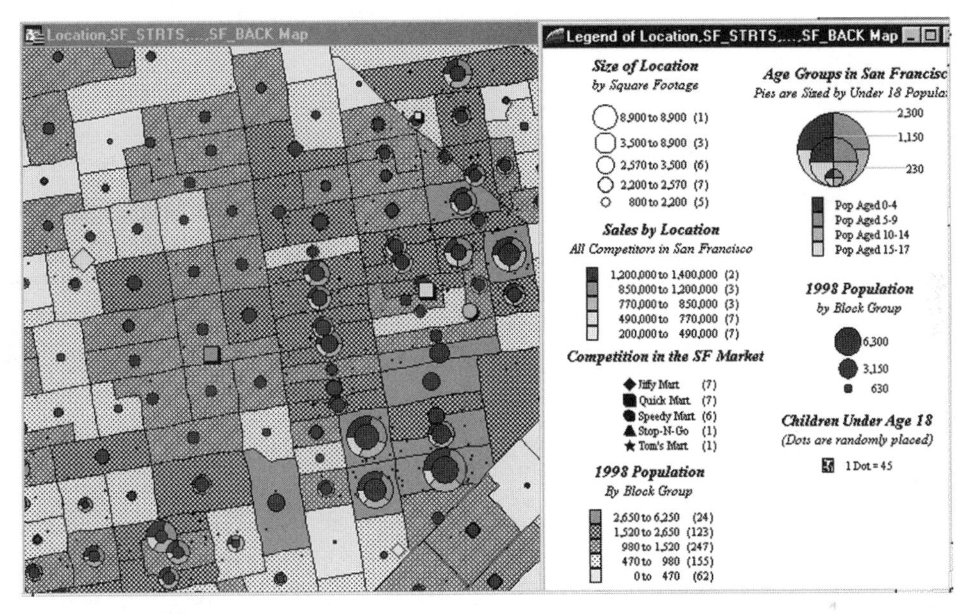

Example of confusing map based on seven themes.

As previously suggested, a large number of themes can produce a confusing map. Maps tend to become confusing with more than two themes, although you can map up to four themes in a coherent fashion. Keep the map's intended audience in mind when deciding on the number of themes for your map.

It may be easier to get your point across with a series of thematic maps, rather than attempting to communicate several messages with a single map. In fact, multiple themes, each depicting a different theme, can be effectively integrated onto a single printed page (rather than a single map) with MapInfo Professional's layout tools (see Chapter 9). A series of thematic maps can also be saved as separate workspaces for incorporation into a presentation. In this way, you can answer questions concerning the map and its supporting data on the screen as they develop.

Classification Techniques

MapInfo Professional provides the following methods for creating value ranges for range thematic mapping. An example of these methods is shown in the illustration that follows.

- Equal count
- Natural break
- Quantile
- Equal ranges
- Standard deviation
- Custom

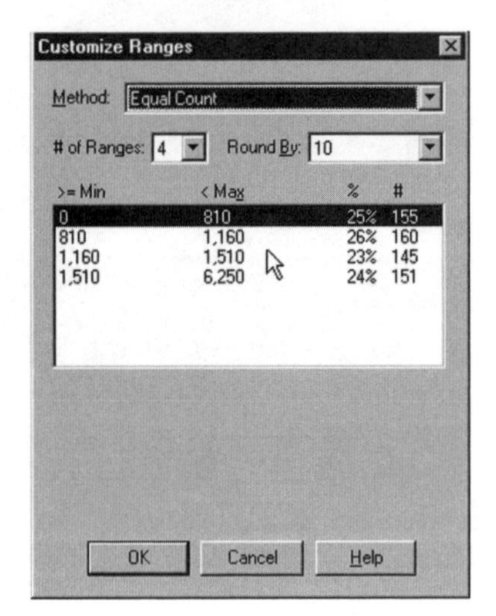

Example of ranging methods.

Data are broken down into ranges via one of the previously listed ranging methods. MapInfo allows you to choose the number of ranges. The minimum and maximum values for each range are displayed within the Customize Ranges dialog box, shown in the previous illustration. The data values that fall within each range are greater than or equal to the minimum value, and less than the maximum value. There is no data overlap for the built-in ranging methods.

When creating custom ranges, verify that you have included all values in the ranges, and that you have not created overlapping ranges. Ranging methods are demonstrated in the following by using block group maps showing 1998 population organized into five ranges. Ranges are depicted by different fill patterns in block groups on the map. Pay close attention to the legends on the maps depicting different ranging methods.

Equal Count

The equal count method, an example of which is shown in the following illustration, breaks data into ranges, resulting in a fairly even number of cases placed in each range. When developing customized MapBasic programs for thematic maps, you may wish to select the equal count method to ensure that the map will always create a broad distribution of the data.

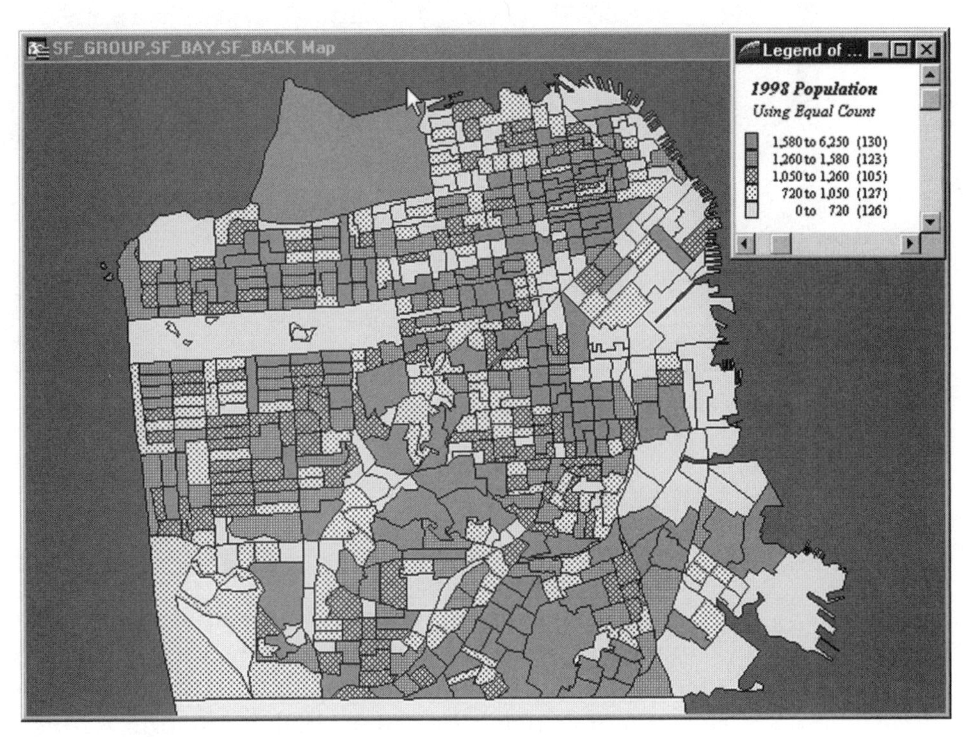

Example of equal count method.

Equal Ranges

Equal ranges, an example of which is shown in the following illustration, will break the data into ranges of equal size (e.g., 1-400, 400-800, 800-1200, and so on). MapInfo Professional takes the highest value and subtracts the lowest value, and then divides the difference by the number of ranges to determine the range boundaries.

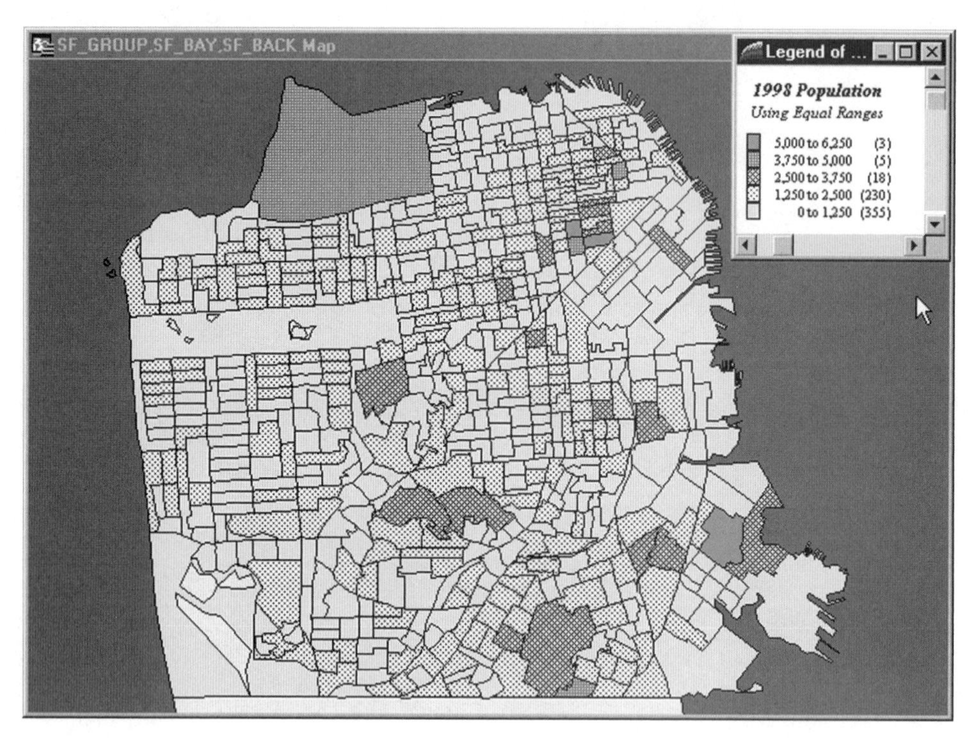

Example of equal ranges method.

Natural Break

The natural break method, an example of which is shown in the following illustration, minimizes the difference between the individual data values and the average value on a per range basis. This method will provide a more accurate representation of the data distribution than the other built-in MapInfo functions. This method attempts to ensure that the ranges are represented by respective averages and that data values within each range are close to the average for the range (and thus close to each other).

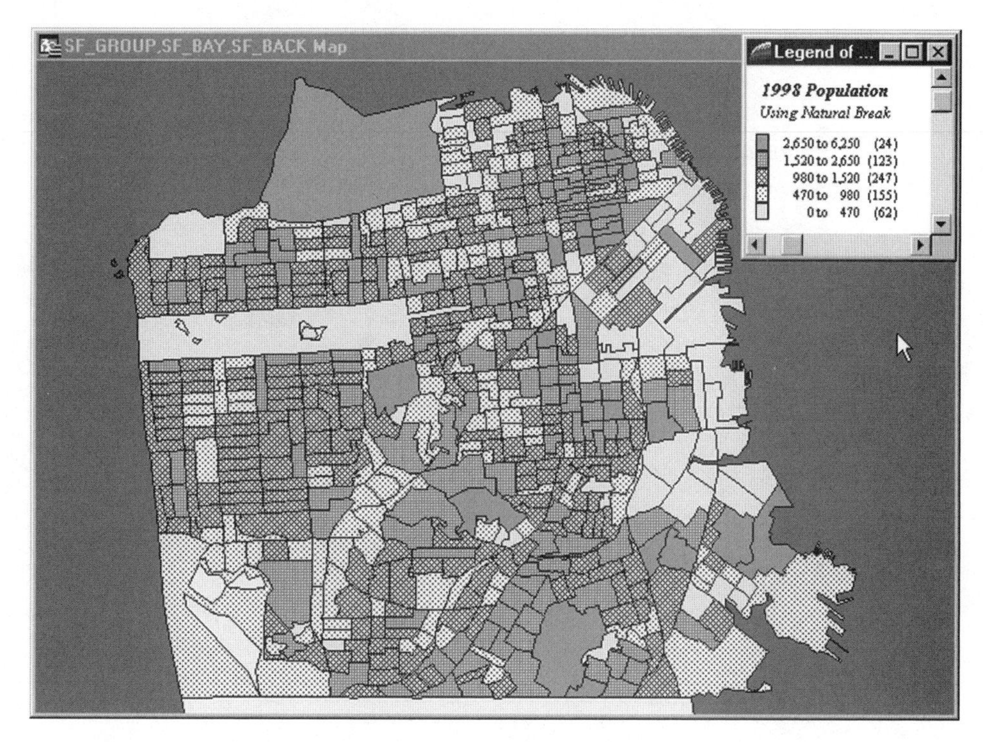

Example of natural break method.

Standard Deviation

The standard deviation method, an example of which is shown in the following illustration, of creating ranges breaks the middle range at the mean of all data values for the selected field or expression. The ranges above and below the middle range are one standard deviation above or below the mean, and two standard deviations above or below the mean, respectively. This method of ranging is useful for seeking outliers in the data.

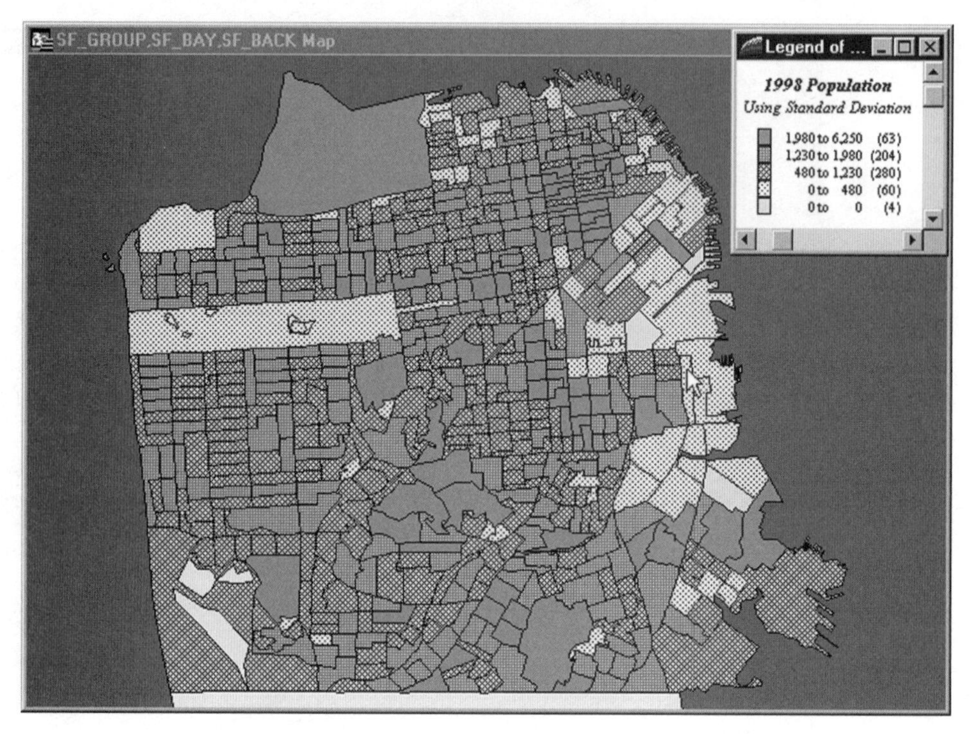

Example of standard deviation method.

Quantile

Selecting the quantile method will distribute a selected variable across another variable or expression. In this way you can create ranges that reflect the distribution of the thematic variable across a segment of the data set. For example, you can quantile a base population by the over-55 population to show how the over-55 population is distributed across a geographic area. In the following illustration, quantiles are used to distribute 1998 population over the population under age 18.

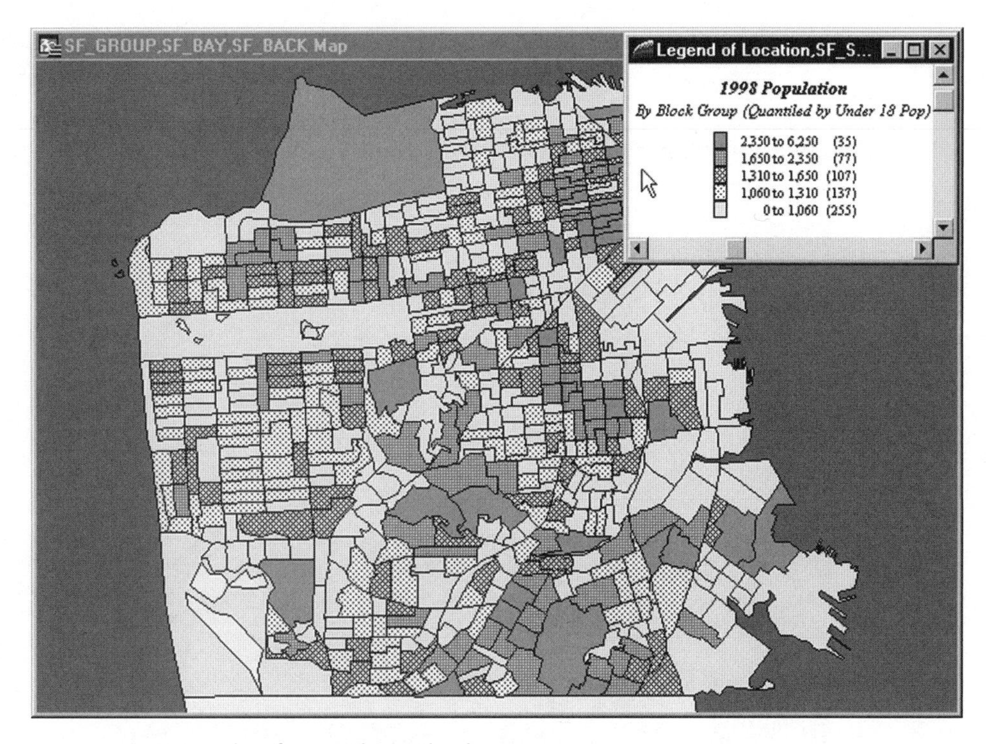

Example of quantile method.

Noting the field used to quantile the information in the theme's legend is recommended to assist the audience in understanding what the theme represents.

Custom

Selecting custom as the ranging method allows you to select and input the minimum and maximum values for each range. This choice will often best meet your needs, especially if you are producing several different thematic maps, but require continuity of ranges between the maps. The custom method is also useful when trying to isolate information, such as income ranges, that meet your analysis criteria. A Custom Ranges section will appear at the bottom of the Customize Ranges dialog box (shown in the first of the following illustrations) to allow you to fill in the minimum and maximum for the user-defined ranges. An example of the use of the custom ranging method is shown in second of the following illustrations.

Customize Ranges dialog box.

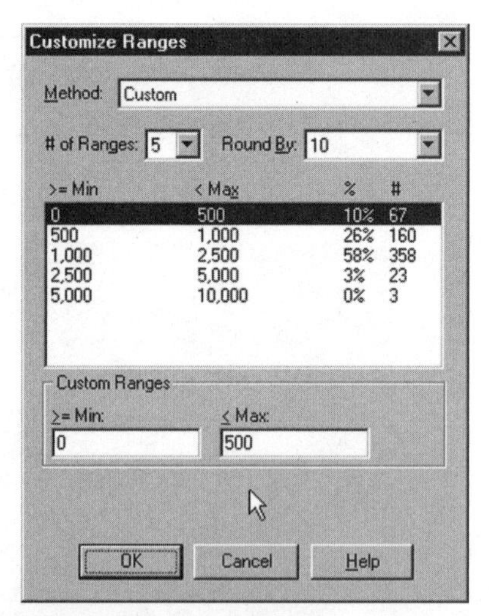

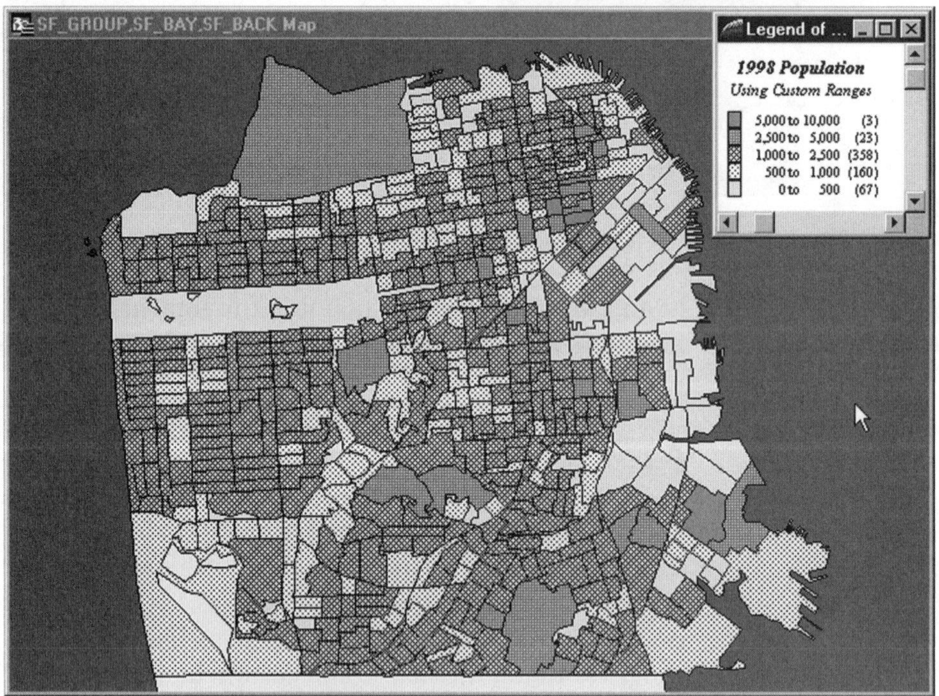

Example of custom method.

Creating Themes

Creating a thematic map is fairly straightforward. First, the Map window must be the active window. Next, select Map | Create Thematic Map. A series of dialog boxes will lead you step by step through the creation process. Selecting the Next button will take you to the next step in the process. If you wish to return to the previous step, in the Step 3 of 3 dialog, select the Back button prior to clicking on OK. The first dialog box, Step 1 of 3, is shown in the following illustration.

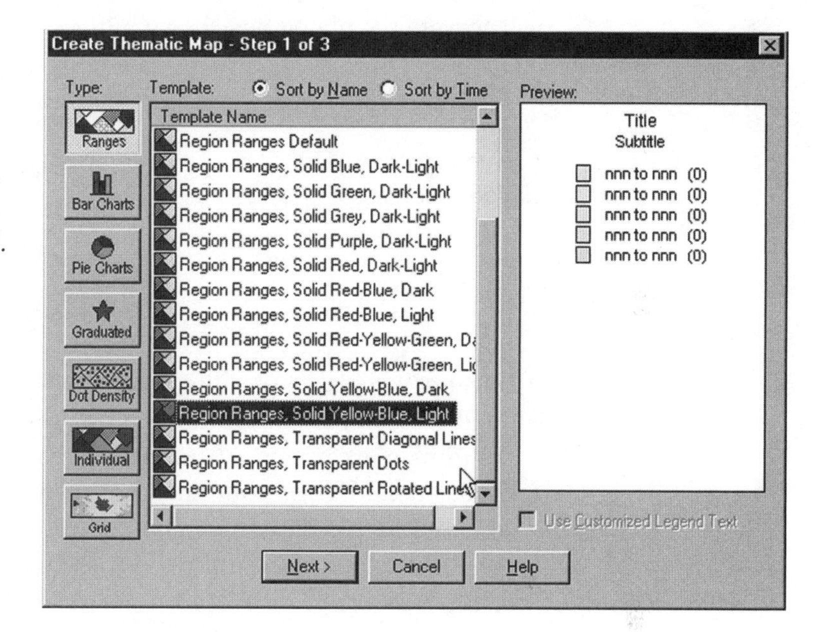

Create Thematic Map dialog box (Step 1 of 3).

Step 1

The Step 1 dialog box prompts you to select a type of thematic map from the buttons on the left of the dialog box. As discussed previously, bar charts, pie charts, and dot density work best with region data. Graduated themes work best with point or region data, grid themes work best with point data, and ranges and individual themes work equally well with region, point, or line data.

Once you have selected a type of theme, you will need to choose a template for the type of data (region, line, point) on which your theme will be based, as well as the color scheme, symbol, or line type (depending on the data) you wish to display. The preview box on the right side of the Step 1 dialog will preview the legend of the theme. If you do not like the appearance of the chosen template once you have completed the three steps, you can always alter the appearance of the theme.

Step 2

The Step 2 dialog box varies, depending on the type of thematic map you select. Selecting Ranges, Individual, Graduated, or Dot Density will activate the dialog box shown in the following illustration.

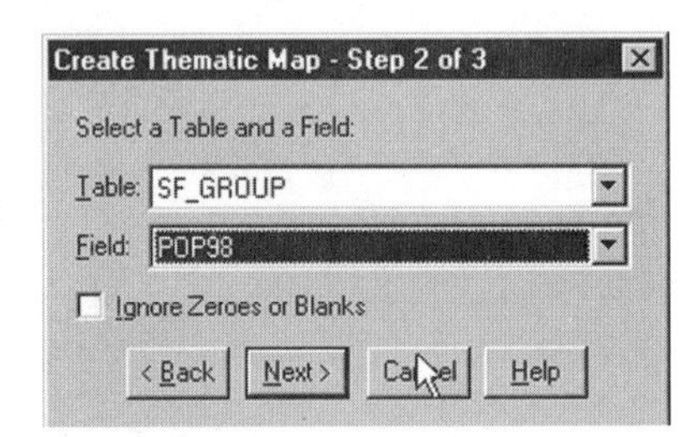

Create Thematic Map dialog box (Step 2 of 3).

The Step 2 dialog box allows you to map a column within the named table, an expression based on columns within the table, or a column or expression from another table (by joining the two tables). In addition, you can choose to ignore zeros or blanks when creating your theme. This option is helpful when you have zeros within your data and do not wish to modify data ranges or settings to remove or isolate the zeros. Tutorial 6-1, which follows, provides you with practice in mapping an expression.

▼ TUTORIAL 6-1: MAPPING AN EXPRESSION

1 To map an expression, select Expression from the Field list in the dialog box shown in the previous illustration. (Expression is located at the bottom of the Field list.)

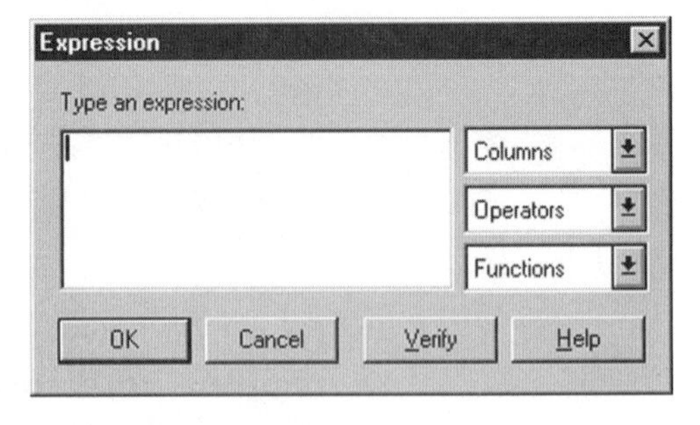

Expression dialog box.

NOTE: *The resulting Expression dialog box, shown in the above right illustration, is the same dialog used to create expressions throughout MapInfo Professional (see Chapter 4 for a detailed discussion of creating expressions).*

2 Scroll through the available columns and select the column(s) on which you wish to base the expression, as shown in the illustration at right.

3 Insert operators, functions, and constants where appropriate. Selecting the Verify button before clicking on the OK button is recommended to ensure that the syntax of your expression is correct.

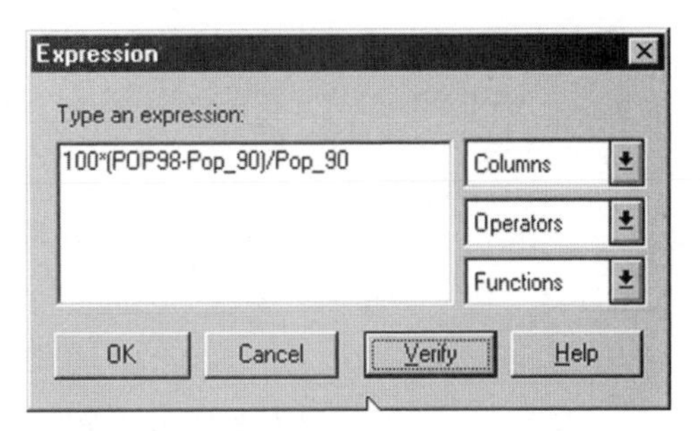

Example expression for percent population growth between 1990 and 1998.

If you wish to retrieve information from another table, perform tutorial 6-2, which follows.

▼ TUTORIAL 6-2: RETRIEVING INFORMATION FROM ANOTHER TABLE

1 Select Join from the available field list in the Step 2 of 3 dialog box. If there are multiple fields in your table, keep scrolling; Join and Expression should be at the bottom of the list.

2 Selecting Join will activate the Update Column for Thematic dialog box, shown in the illustration at right.

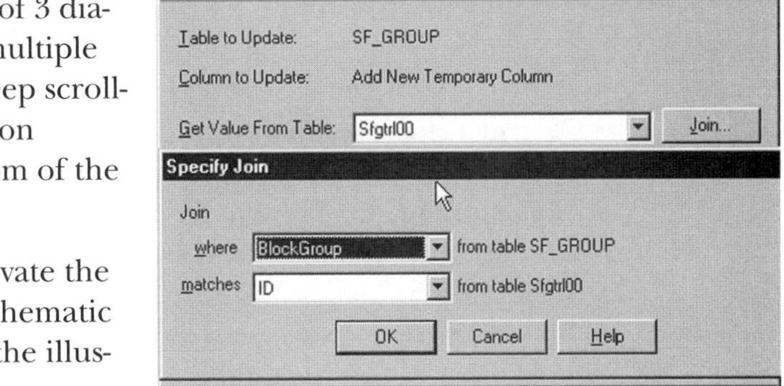

Update Column for Thematic dialog box.

3 Select the table from which you wish to retrieve information and then select the Join button to access the Specify Join dialog box.

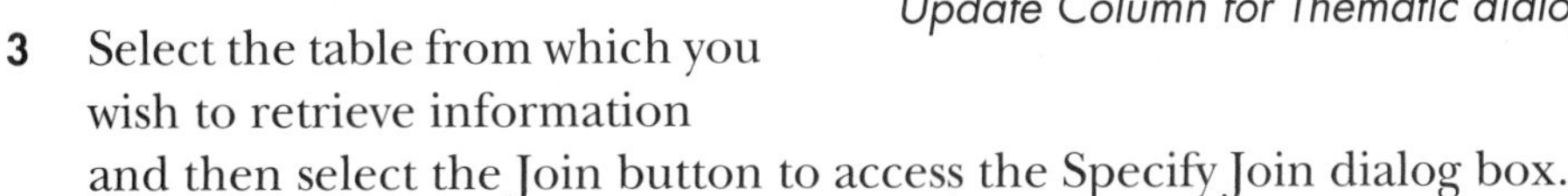

4 Select the associated fields in each table to tell MapInfo how the files are related. The two files can be joined by block group name (*ID* in the *Sfgtrl00* file is the name of the block group). The content must be identical for MapInfo to join the tables. In this case, the content of the two fields are identical, although the field names are different.

5 After the join criteria have been specified, select the Calculate method, such as Value, and then a field or an expression to be mapped, as indicated in the illustration at right.

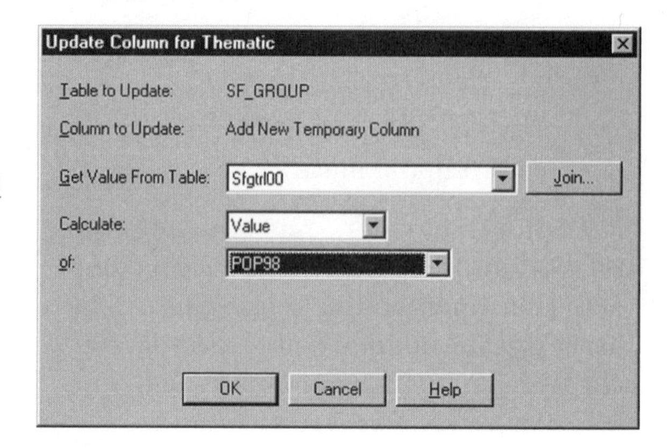

Update Column for Thematic dialog box showing Calculate method and POP98 field.

Joining tables will place a temporary column in the table you are mapping. Thus, you will be able to access the column throughout your analysis for querying, mapping, browsing, or creating expressions. Because joins are saved in workspaces, themes based on columns from other tables will be available in subsequent sessions (provided you save the workspace).

If you wish to build a thematic map based on an expression created from columns residing in separate tables, join the tables and then select a column or expression from the second table. Next, click on OK to return to the Step 2 of 3 dialog box. Now both columns are available in the base table. (If you need to access another field from the second table, select Join again, and then the additional field. At this point, click on OK to return to the Step 2 of 3 dialog box.) To proceed, select Expression from the list of fields and then build your expression as usual. This process is demonstrated in the following tutorial. The Step 2 of 3 dialog box for grid themes is shown in the above right illustration.

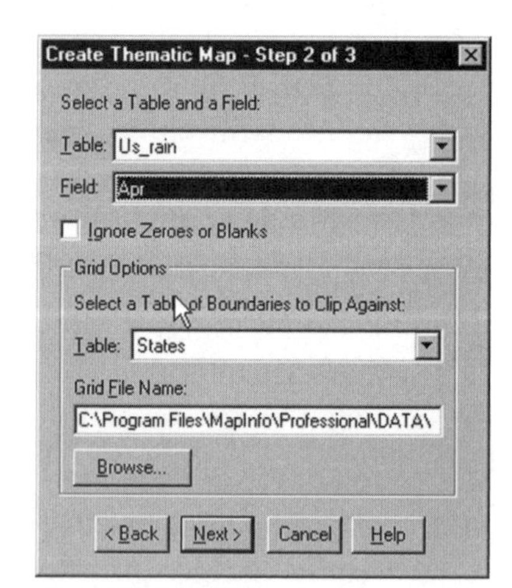

Step 2 of 3 dialog box for grid themes.

The Step 2 of 3 dialog box for grid themes is slightly different. Like other themes, you will choose the table and the field (or expression) on which to base the theme. However, you will have to select an output directory for the new raster image file created. The file is named in the following form: *original file name_field name.* If the original table were named *US_Rain* and the field mapped were April, the new table would be named *US_Rain_April.*

In addition, you can specify a boundary file the new image will be clipped against. In the previous grid example, the grid was clipped against the states boundary file to keep the raster image consistent with the shape of the United States. In tutorial 6-3, which follows, a thematic map is created depicting the number of people per hazardous waste site by state.

▼ TUTORIAL 6-3: CREATING A THEMATIC MAP OF POPULATION PER HAZARDOUS WASTE SITE

1. Select Map | Create Thematic Map.
2. Select Ranges as the type of map to be created, and Region Ranges Default as the template. Click on Next to move to the Step 2 of 3 dialog box.
3. Select the *States* table (*samples* directory on the companion CD-ROM) as the base for the theme. Click on Join to activate the Join dialog box.

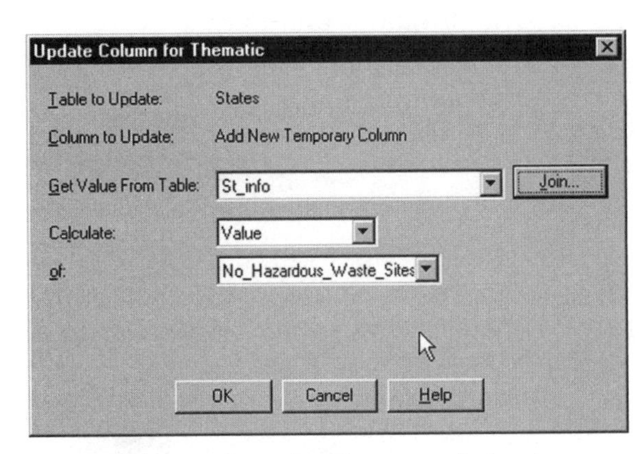

Update Column for Thematic dialog box.

4. In the Update Column for Thematic dialog box, shown in the above right illustration, select *St_info* as the table to Get Value From and *No_Hazardous_Waste_Sites* as the field.
5. Select Join to define the relationship between the two tables. In the Specify Join dialog box, join the tables where *state* matches *state,* as shown in the following illustration.

6 Click on OK to accept the join, and then select OK in the Update Column for Thematic dialog box to return to the Step 2 of 3 dialog box.

7 In the Step 2 of 3 dialog box, select Expression. In the Expression dialog box, as shown in the second illustration at right, key in the following: *Pop_1990/No_Hazardous_Waste_Sites.*

8 Click on OK to accept the expression. Click on Next to access the Step 3 of 3 dialog box.

9 Select OK in the Step 3 of 3 dialog box to accept the default values. Your map should resemble that shown in the following illustration.

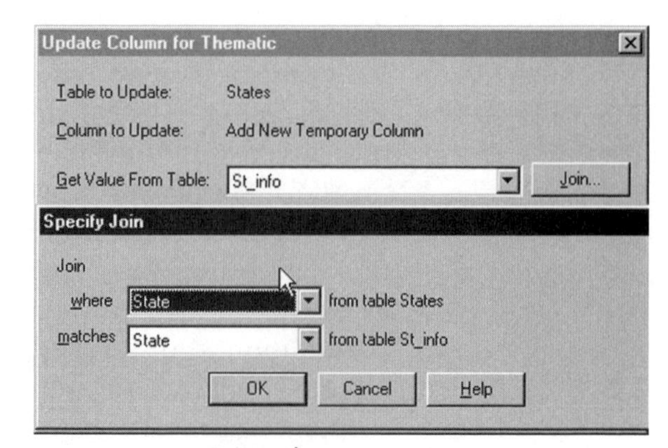

Example expression.

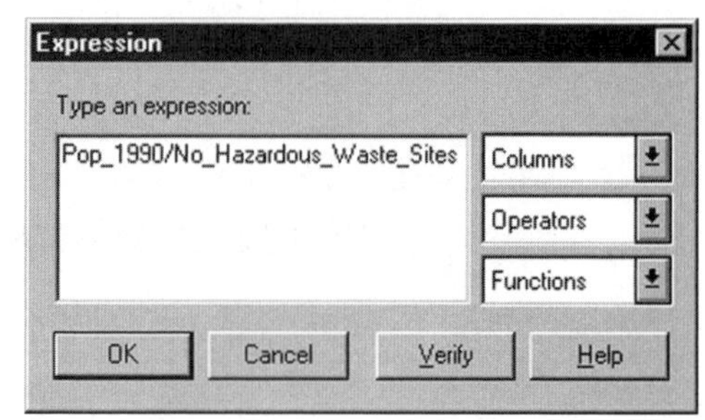

Specify Join dialog box.

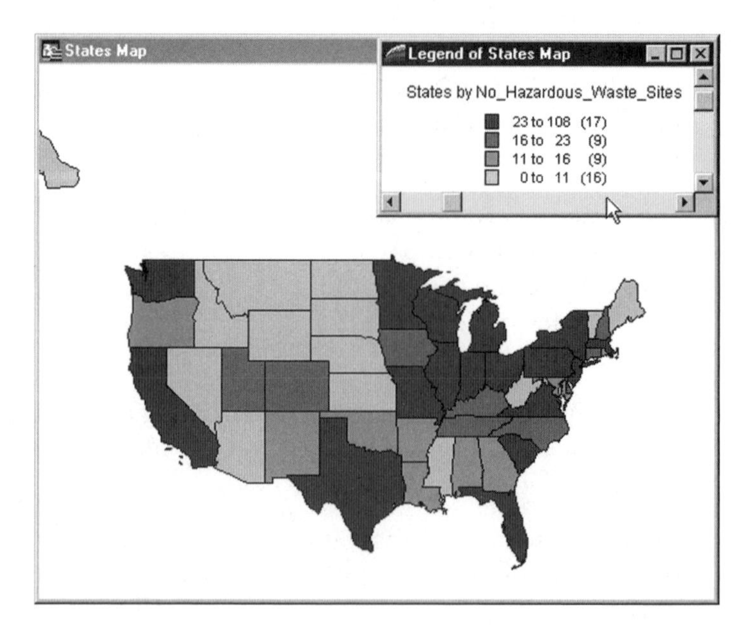

Thematic map showing population per hazardous waste site by state.

Step 3

Step 3 of creating a thematic map will activate a dialog box (shown in the following illustration) that allows you to change different aspects of the theme, depending on the type of map chosen. Using the Ranges, Styles, and Legend buttons on this dialog box to modify the look of the thematic map is the topic of the next section.

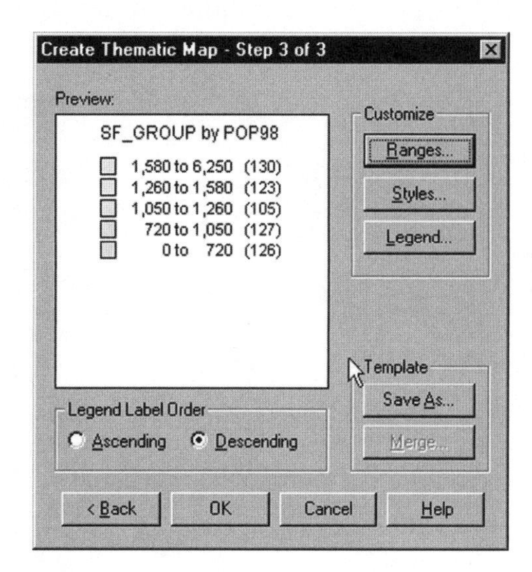

Create Thematic Map dialog box (Step 3 of 3).

Refining Map Appearance

You can modify the appearance of a theme (ranges/settings, styles, or legend) at any point in a MapInfo session. The options available for modification depend on the type of theme. If you wish to change the thematic map type (e.g., from a ranges map to a dot density or graduated map) or the column or expression the map is based on, you will need to create a new thematic map.

If you update or change the underlying tabular data the theme is based on, MapInfo will automatically update the map. For example, assume you are working with a theme based on sales information. When you increase the sales data by 20 percent, the thematic map will reflect the change. This is not true for grid themes, however. Because a new raster image file is created and added to the map when you create a grid theme, changing information in the original table has no effect on the grid theme.

You can modify the appearance of the map as Step 3 of the thematic map creation process, or click on OK in the dialog box to view the appearance of your chosen template settings on the map. If you do not modify the map's appearance as Step 3, you can select Map | Modify Thematic Map from the Main menu bar, or Thematic from the Layer Control dialog box, to change theme appearance. In the Layer Control dialog, shown in the following illustration, you must highlight (select) a thematic layer to make the Thematic option available.

Thematic layer selected in Layer Control dialog box.

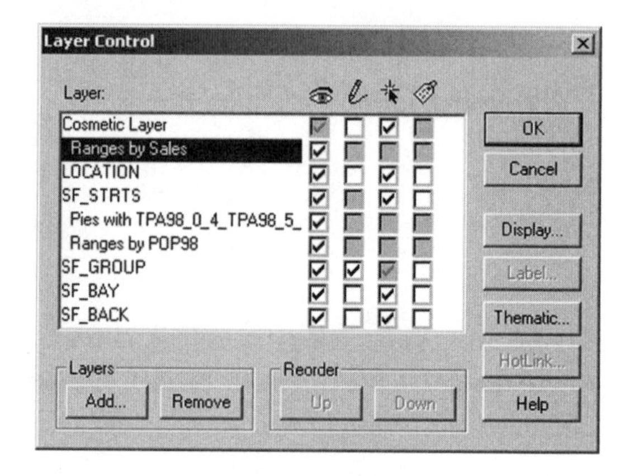

You can also access the Modify Thematic Map dialog box by double clicking on the legend of the theme you wish to modify in the Legend window, shown in the following illustration. To modify the *1998 Population Ranges* map, you can double click below the theme's title in the Legend window to access the Modify Thematic Map dialog box.

Legend window.

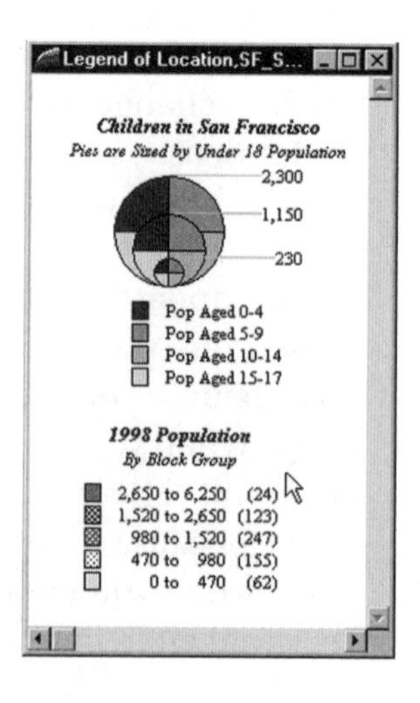

Each theme must be modified separately, even if you have created a multi-theme map. Modifying ranges/settings and styles for each type of map is reviewed in subsequent sections, followed by discussion of the Legend dialog box.

Range Themes

To modify either the range values or the number of ranges, select the Ranges button from the Create Thematic Map Step 3 of 3 dialog box or the Modify Thematic Map dialog box. Selecting Ranges in either dialog box will access the Customize Ranges dialog box, shown in the following illustration.

Five data ranges were selected instead of four, and Natural Break was chosen as the ranging method instead of Equal Count. (Four ranges and Equal Count are standard when using MapInfo Professional's default template.) After each change you make, a Recalc button will appear at the bottom of the dialog box. Select this button to put each change into effect.

The exception to selecting Recalc after every change is when you choose Custom ranges. With Custom ranges, to put the changes

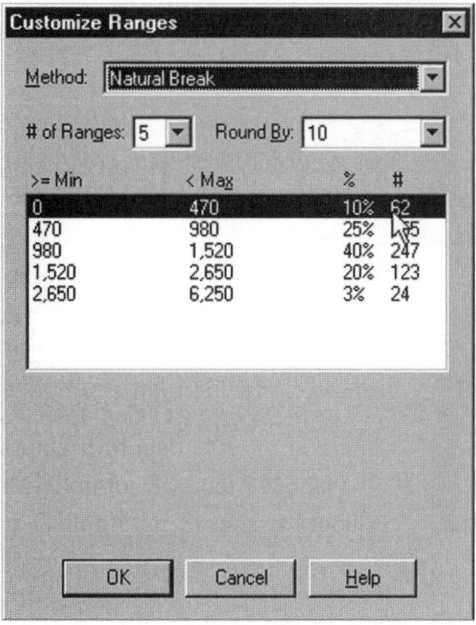

Customize Ranges dialog box.

into effect, you should enter the minimum and maximum values for all ranges prior to selecting the Recalc button. This will save you time. However, if you change the number of ranges, you will have to select Recalc before the chosen number of ranges will be available. Again, the data in each range are greater than or equal to the minimum value, and less than the maximum value. Remember to include all values, and do not create overlapping ranges.

The Round By selection allows you to force numbers to end with a zero for the predefined ranging methods (i.e., non-custom ranges). The Round By option allows you to choose from the following precision levels: .00001, .0001, .001, .01, .1, 1, 10, 100, and 1000.

To modify the map's appearance, select Styles under Customize in the Modify Thematic Map (Step 3 of 3) dialog box. The Customize Styles dialog boxes are slightly different for line, point, and region objects. The difference between styles for point and region data is discussed in material to follow.

By clicking on the top Styles region color box, you can change the region's border, fill pattern, and color in the Customize Range Styles dialog box, shown in the following illustration. If the Auto Spread option for Color is selected, MapInfo will shade the map within a single color range (e.g., from light pink to dark red).

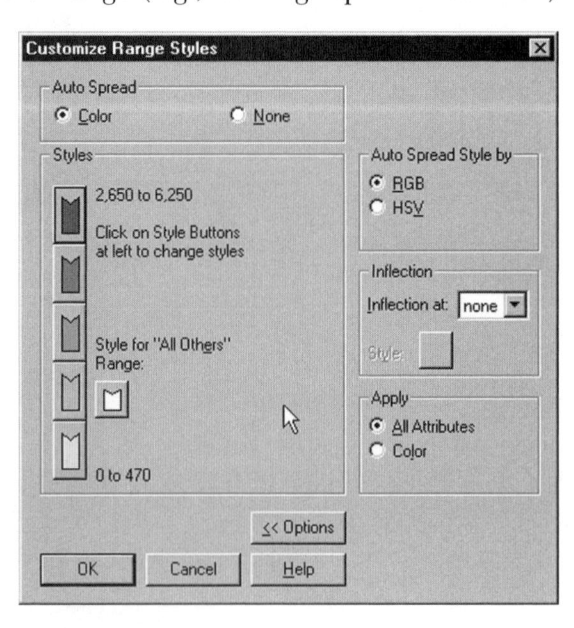

Customize Range Styles dialog box for region data.

You can also Auto Spread over two colors. To create a two-color spread for your ranges, select the top region style box, choose a color, and then select the bottom range's style box and choose another color. MapInfo will automatically spread variation of all ranges to hues between the two colors, as long as Color under Auto Spread is selected. Selecting one of the inner ranges will switch the Auto Spread option to None.

Other Options

By selecting the Options button at the bottom of the dialog box, additional selections become available to the right of the color ranges. The first option is the choice of Auto Spread by RGB or HSV. RGB indicates that the Auto Spread is using the red/green/blue color spectrum. HSV indicates that the Auto Spread is using hue saturation value. RGB works best when using a two-color spread.

Inserting an Inflection point will change the color spread of your map. If you choose a point of inflection at the center range (the third in this five-range case), the Auto Spread will converge on the inflection point from the top range's color to the inflection color, and from the bottom range's color to the inflection color. An inflection point is useful when there is a natural break in the data and you wish to have a marked difference in color above and below that point. Examples include mapping profit-and-loss data, where the inflection point would be inserted at the break-even point, and population growth, where the inflection point would be inserted at zero growth.

The use of color on maps has long been standardized in cartographic circles. Examples include mapping over red-orange-yellow, blue-green, green-red, and red-blue spectrums. The blue-green range is typically used for environmental purposes, whereas the green-red color set is used to depict dollar relationships. The red-blue spectrum is used to depict hot and cold, which could be translated to hot (and cold) oranges of your data set. For example, if you wish to identify census tracts with a large number of commuters, you could set the top range to red and the bottom range to blue to easily highlight areas of commuter concentration. MapInfo allows you to choose from many standard templates in the Step 1 dialog box.

The final option is to apply All Attributes (shading, patterns, and line thickness) or to apply Color alone. If you intend to create two thematic maps for a layer, you may wish to apply color to just one map, and then create a hatch pattern in black and white for the other. In addition, if you have changed boundary lines to a style different from the single thin line, it is easier to select the Apply Color option so that you do not have to select each range's style box and change the outline pattern.

Symbols

The Range Styles dialog box for symbols adds the choice of a Size Auto Spread, and an option to Apply all attributes, color only, or size only. The Size Auto Spread option will automatically graduate the symbols, with the smallest symbol appearing in the lowest range and the largest symbol appearing in the largest (topmost) range.

The ability to apply size or color only is very handy when you create an individual values theme or have edited the symbols and wish to preserve the unique symbols while also applying a color or size range theme to the layer. If you wish to create two themes based on a single table, you can apply color only to one theme and size only to the other. MapInfo allows you to select a different symbol for each range, and still apply Auto Spread options to the ranges. Tutorial 6-4, which follows, also uses the thematic map of population per hazardous waste site by state.

▼ TUTORIAL 6-4: CREATING A SECOND THEMATIC MAP OF POPULATION PER HAZARDOUS WASTE SITE

1 Modify the ranging method to custom ranges. Select Map | Modify Thematic Map, and select the *Pop_1990/No_Hazardous_Waste_Sites* theme. The Modify Thematic Map dialog box appears.

2 Select the Ranges button to access the Customize Ranges dialog box.

3 Select Custom as the ranging method, and 5 as the number of ranges. Select Recalc to put the changes into effect. The ranges are shown in the illustration at right.

4 Enter the ranges shown in table 6-1, which follows. To change the values, highlight a range and enter the minimum and maximum values.

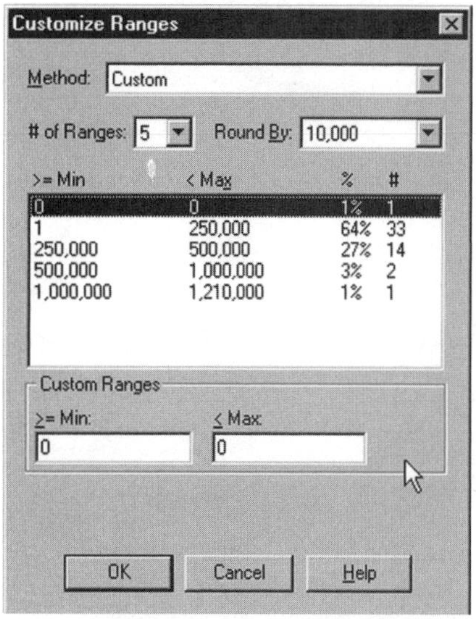

Five custom ranges.

Table 6-1: Ranges to Be Entered for Thematic Map Tutorial

Minimum	Maximum
0	0
1	250,000
250,000	500,000
500,000	1,000,000
1,000,000	1,210, 000

5 Click on OK to accept the range changes.

6 In the Modify Thematic Map dialog box, select Styles to change the colors and patterns on the map.

7 In the Customize dialog box, change the colors to reflect relative concern over the average number of people per hazardous waste site. Select the colors listed in table 6-2, which follows.

Table 6-2: Colors to Be Set in the Customize Dialog Box

Range	Color
0 to 0	Gray (This range covers Washington, DC.)
1-249,000	Red (Reflects a low number of people per waste site; that is, a high number of waste sites per person.)

Table 6-2: Colors to Be Set in the Customize Dialog Box

Range	Color
250,000-499,999	Orange
500,000-999,999	Green
1,000,000-1,210,000	Blue

8 Click on OK in the Customize Range Styles dialog box (shown below left), and on OK in the Modify Thematic Map dialog box to put the changes into effect. Your map should now resemble that shown below right.

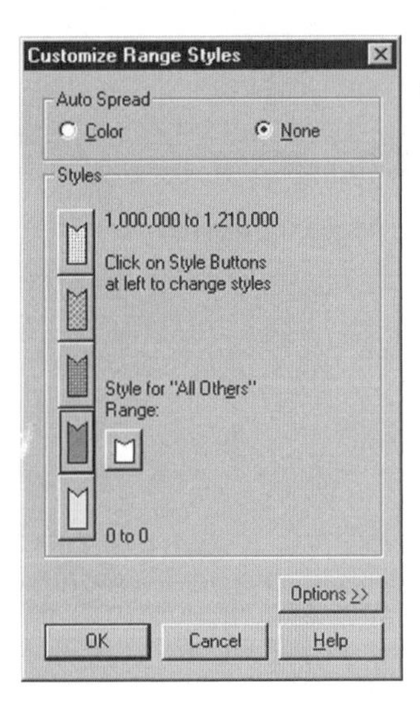

Customizing colors.

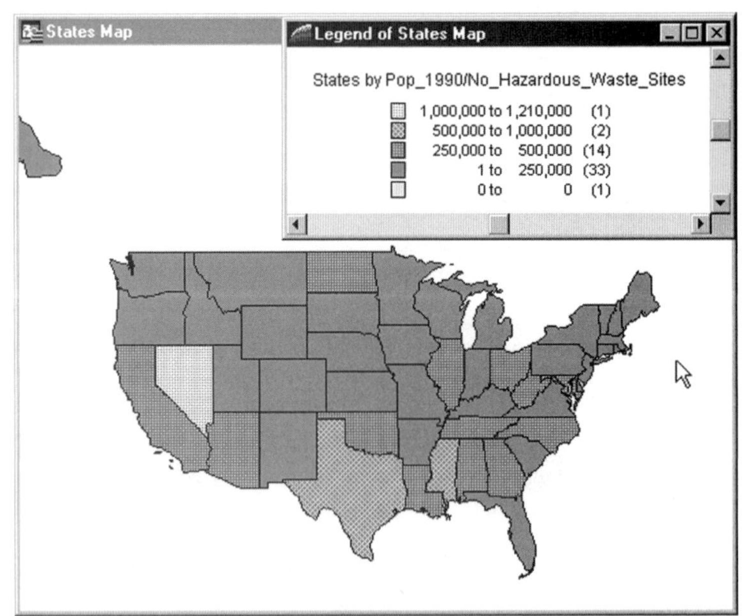

Thematic map of customized ranges.

Pie Chart Themes

Choosing to modify a pie chart thematic map will not allow you to change settings. Settings cannot be changed because the information is based on several data columns that were selected to initially create the pies, as indiated in the Modify Thematic Map dialog box shown in the following illustration.

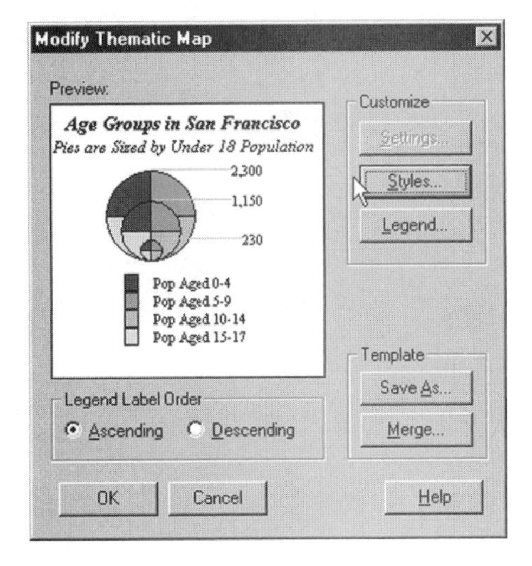

Modify Thematic Map dialog box for pie chart map.

If you wish to change the columns or expressions the theme is based on, you must recreate the pie chart map and select different data columns. Selecting Styles from the Modify Thematic Map (or Step 3 of 3) dialog box will activate the Customize Pie Styles dialog box, shown in the following illustration.

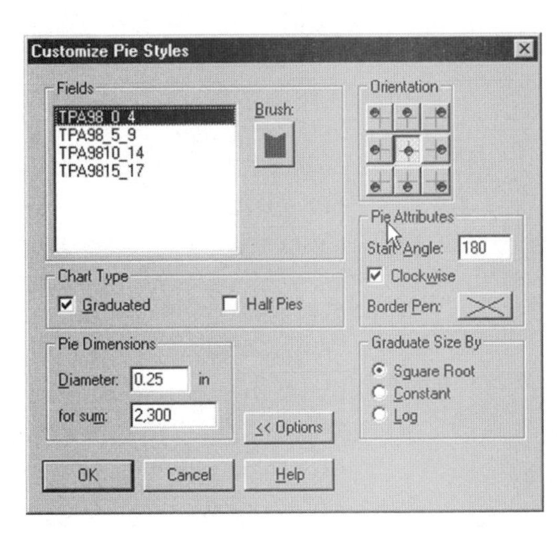

Customize Pie Styles dialog box.

In this dialog box, you can select colors and patterns for the pie slices. Highlight a field from the Fields list, and select the Brush option to change the color and pattern. Continue to select each field until you have customized the colors and patterns for all fields you wish to change. The Chart Type option allows you to choose half pies instead of full pies, as indicated in the following illustration. It also allows you to graduate the size of the pie based on the sum of the data fields. If Graduated is chosen, the pies are graduated by total diameter and sum in the Pie Dimension Option section. You may have to work with the dimension settings to make the pies appear the way you wish.

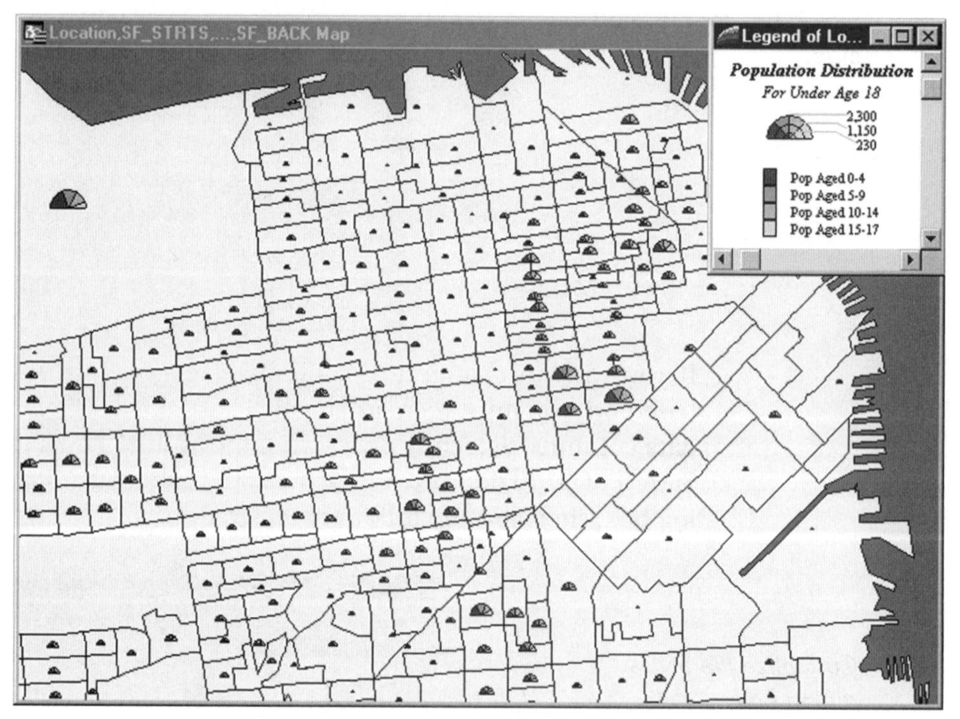

Example of pie chart theme created by selecting Half Pies option in Customize Pie Styles dialog box.

The pies are graduated by the total number of cases that fall into all slices. For example, assume you create a pie chart theme by block group based on median household income ranges of 0-0, $1-$15,000, $15,000-$30,000, $30,000-$45,000, and $45,000 and over. The pies would be graduated by the total number of households per income category in each block group.

Selecting the Options button at the bottom of the dialog box activates additional options. You can choose to graduate the pies by square root, a constant (linear graduation), or by logarithm. Other options allow you to specify where the pie appears in relation to the center of a region, the start angle for the pie, pie orientation (clockwise or not), and style for the pie outline.

Bar Chart Themes

As shown in the following illustration, the Modify Thematic Map dialog box for bar chart maps is very similar to the pie chart version. Settings cannot be changed for bar chart maps for the same reasons as pie chart maps. If you wish to change the columns or expressions the theme is based on, you must create a new thematic map based on different fields.

Modify Thematic Map dialog box for bar chart map.

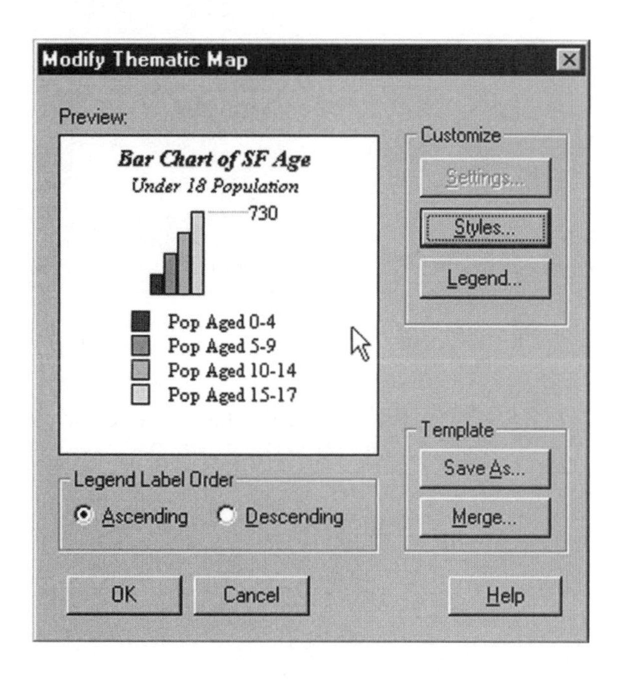

Selecting Styles will access the Customize Bar Styles dialog box. Changing the color and pattern for the bars operates the same as changing color and pattern for pies. Highlight a field from the Fields list, and select the Brush button to change the color and pattern for the field's associated bar. The Chart Dimensions option allows you to set the height and width of the bars at a specified value (the maximum value is the default).

Choosing the Options button accesses additional options similar to the Customize Bar Styles dialog box, shown in the following illustration. For example, the orientation of the bar with respect to the center of the region operates the same as in a pie chart map.

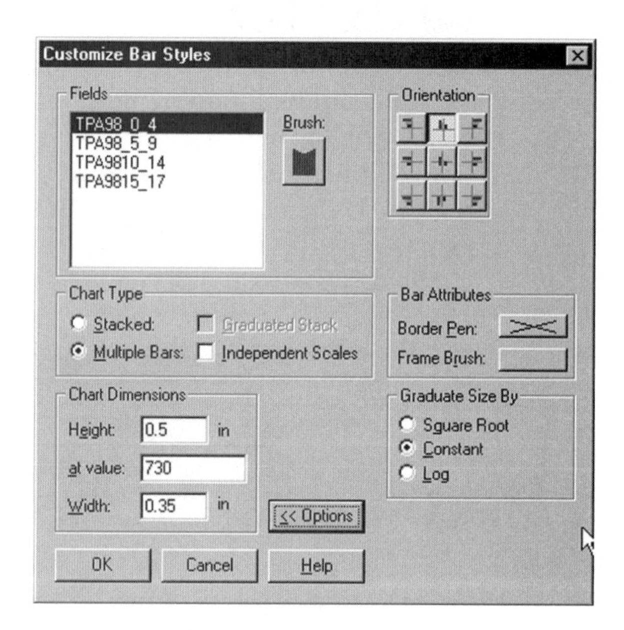

Customize Bar Styles dialog box.

The differences between bar chart and pie chart map options begin with the choice of Chart Type. Selecting Multiple Bars (the default) will place the bars side by side. Choosing Stacked will stack all data fields on top of each other to create a single bar within each region. You can choose to have a graduated or a non-graduated stacked bar. Selecting Independent Scales will treat the height and width as maximum values and will be applied to the data set's maximum value. Next, the "at value" input box will be disabled. Deselecting Independent Scales is useful when the bar charts appear too large or too small on a map.

The Bar Attributes section affects the appearance of the bar border and frame. You can alter the bar outline with the Border Pen option. In addition, you can select a pattern with the Frame Brush option to give the charts a more comparable framework. This selection allows you to more easily see the difference between the bars. A word of caution is in order here: the frame will be sized to the maximum height, and could easily fill up a map if small geographies are used, or if there are many short bars compared to the maximum height.

Graduated Themes

The Modify Thematic Map dialog box for graduated themes is shown in the following illustration. In the illustration, Styles has been disabled for graduated themes. To change the appearance (style, size, and color) of the graduated symbols, select the Settings option.

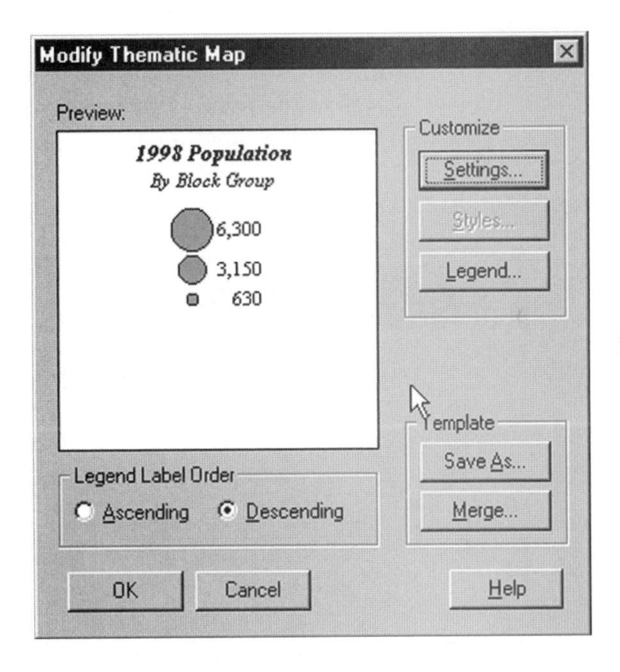

Modify Thematic Map dialog box for graduated maps showing customized settings applied to a symbol.

For graduated symbols (used for both symbol and region data), you can change the symbol, as well as symbol color and size range, in the Customize Graduated Symbols dialog box, shown in the following illustration. To change the size range, select a size for a particular data value and MapInfo automatically sizes the symbols to scale. Selecting the Options button will allow you to show graduated symbols for negative values as well. If Show Symbol is selected under Negative Values, the symbols will be graduated over the full range of data, and the values and sizes for positive values will be mirrored on the negative side in the map legend. You can select a separate color and/or symbol to depict negative values.

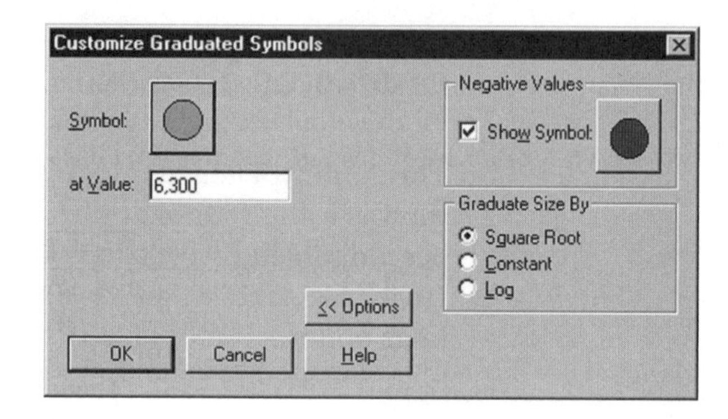

Customize Graduated Symbols dialog box.

You can also select the method by which the symbols are graduated: square root, constant, or log. The following illustrations show examples of the same data (1998 population by block group), with the symbols graduated by different methods.

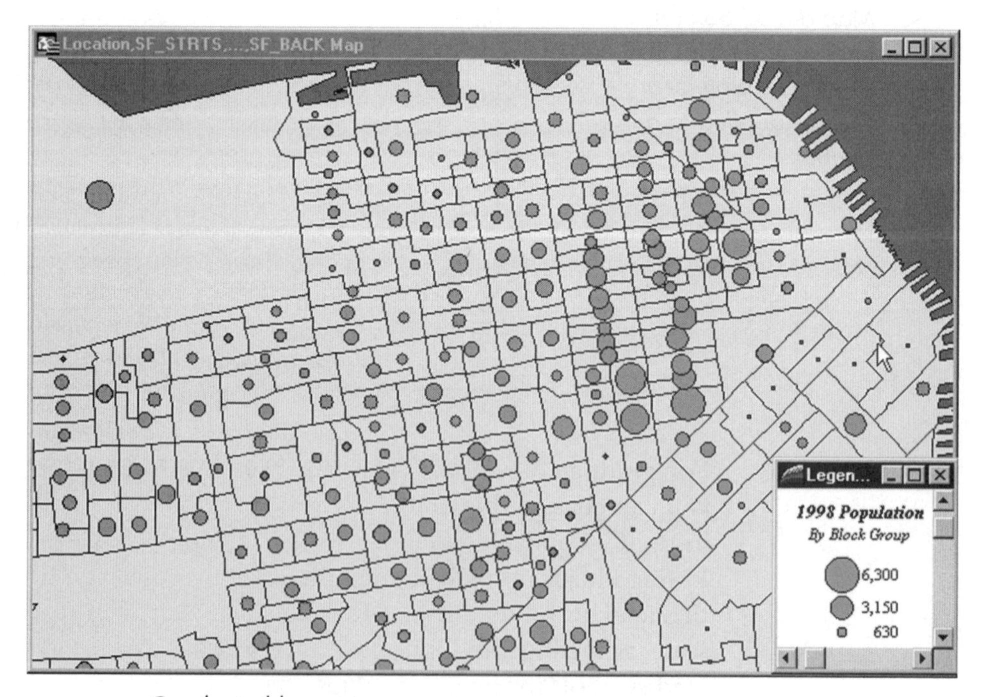

Graduated by square root.

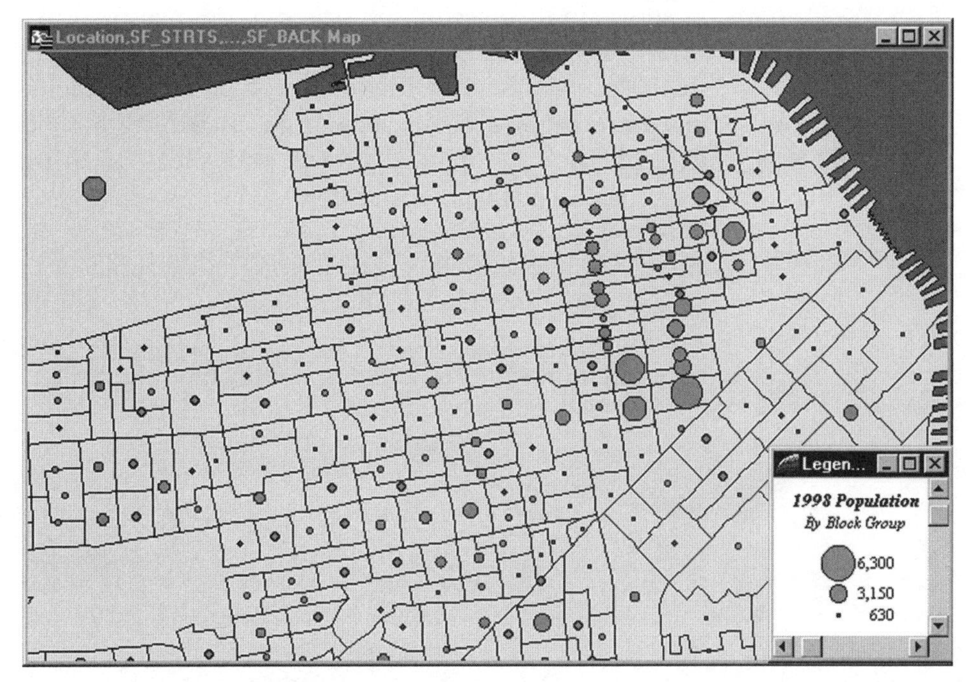

Graduated by constant.

Graduated by log.

As shown in the previous illustrations, graduating by constant provides the widest range of differentiation, especially in the case of a small spread between the range of possible values. Graduated maps are often used to show differences within a data set. For example, both population growth and sales growth typically consist of both positive and negative values. By using graduated symbols, you can easily identify areas of large growth or decline in an area. The following illustration shows an example of a population growth graduated theme; that is, population growth in an area of San Francisco between 1990 and 1998. In this map, large changes in population (1990 to 1998) are easily identified.

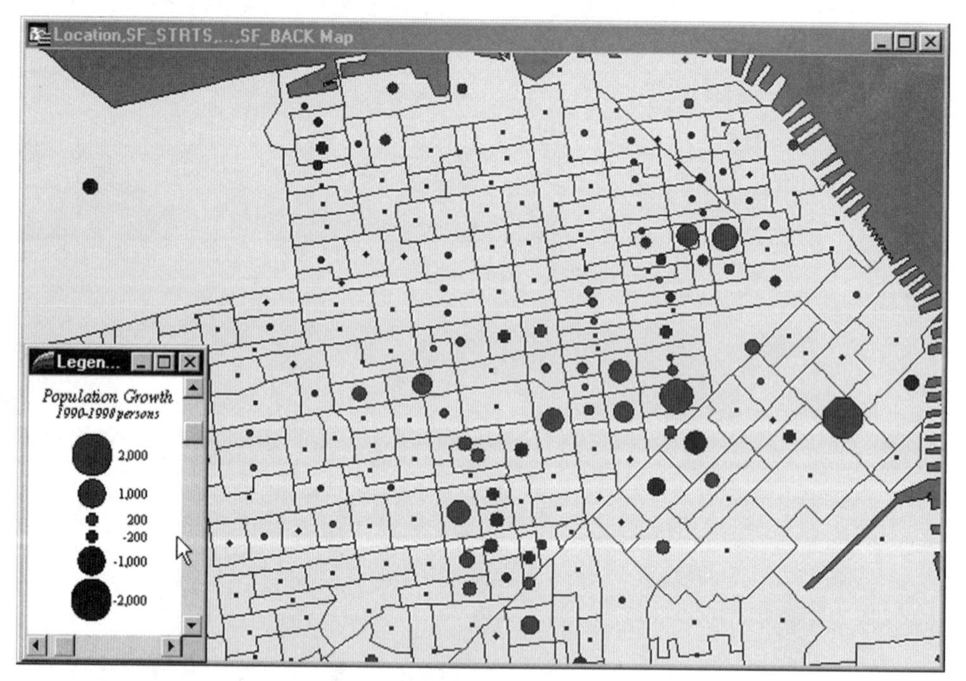

Population growth graduated theme.

Individual Values Themes

If you have chosen an individual values theme, you can customize the styles (symbol, size, and color) assigned to each value, as shown in the following illustration. To change the symbol associated with each value, highlight a value and then select the Style button. Select the color and symbol appropriate for each value.

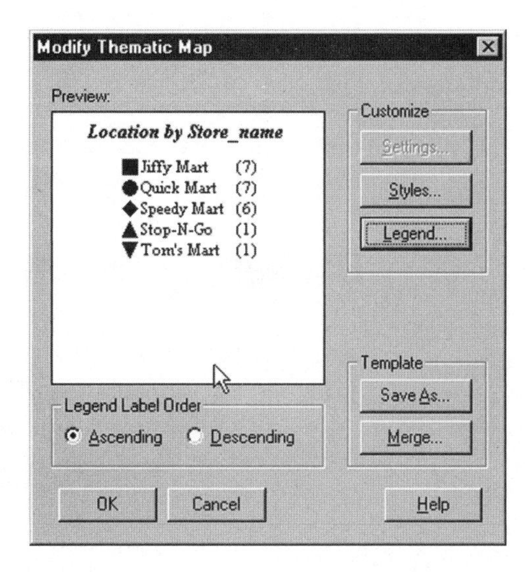

Modify Thematic Map dialog box showing customized settings applied to individual values.

If you choose a color symbol template, but wish the symbols to be different without selecting each value, deselect the Use Color Styles box at the bottom of the Customize Individual Styles dialog box, shown in the following illustration. This will change all values to different symbols, though all symbols will be black. If you wish, you can then manually change the color of the symbols representing different values.

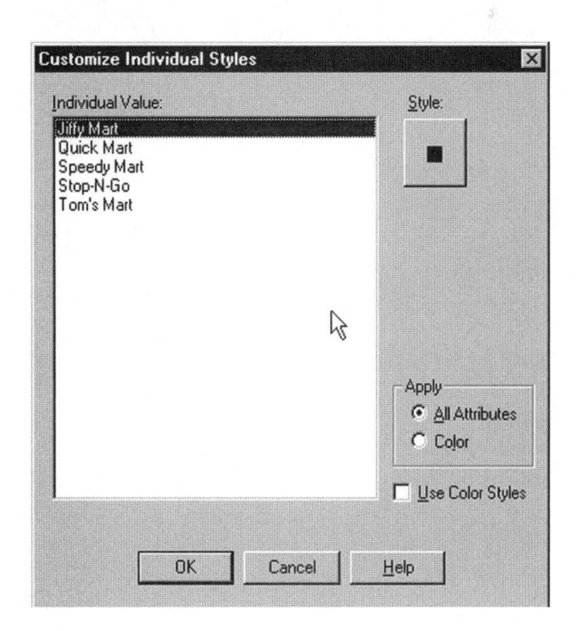

Customize Individual Styles dialog box.

If you want a single symbol to represent more than one item (as in another category), select Styles and then set the symbol and color for these items to be identical. In the legend, you can change the text associated with each group, and prevent the individual values you choose from appearing in the legend.

Dot Density Themes

You can change both the settings and legend for dot density maps, as shown in the following illustration.

Modify Thematic Map dialog box showing customized settings and legend for dot density map.

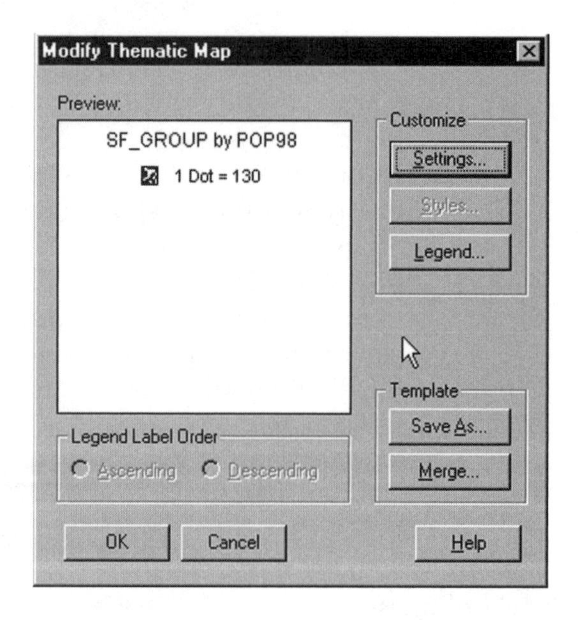

In the Settings dialog box, you can choose between large and small dots, as well as select the dot color. You can also affect the number of dots on the map by choosing how many units each dot represents. Thus, the larger the number chosen, the fewer dots on the map. An example of a dot density map is shown in the following illustration.

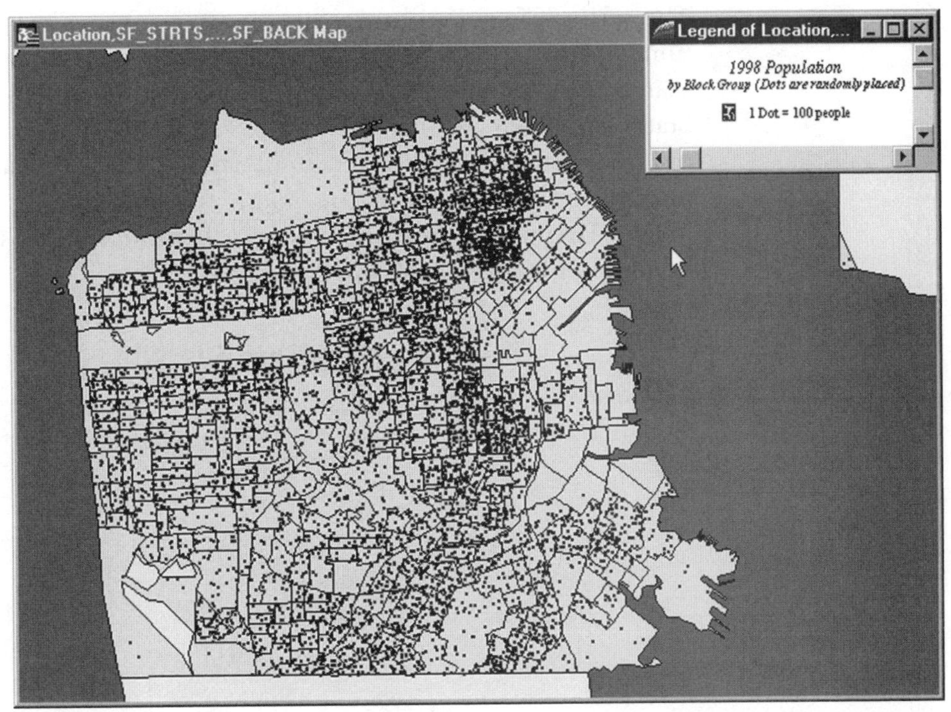

Example of dot density theme showing population concentration in San Francisco.

Grid Themes

You can change the settings, styles, and legend for grid maps. As previously discussed, grid themes are different from other theme types in that a new raster image table is created for the theme. Each change you make to the settings or style will result in MapInfo Professional recreating the image.

A raster image (i.e., a picture) is created by interpolating point data. As previously discussed, the data are broken down into a grid consisting of cells, and values are calculated and attached to each cell. The values are based on the point data within each cell, and are estimated for missing data. Unlike most themes, a grid theme is stored in a table (specified by the user), and can be opened separately. It can, however, still be modified as a theme, using Map | Modify Thematic Map.

The settings for grid themes are shown in the following illustration. MapInfo Professional's default cell size, exponent, and search radius for a temperature map of the continental United States is shown in the illustration. However, the Grid Border option was changed in order to interpolate the information missing around the edges of the boundary used to clip the raster image. In this fashion, the image will more closely follow the boundary file it is being clipped against.

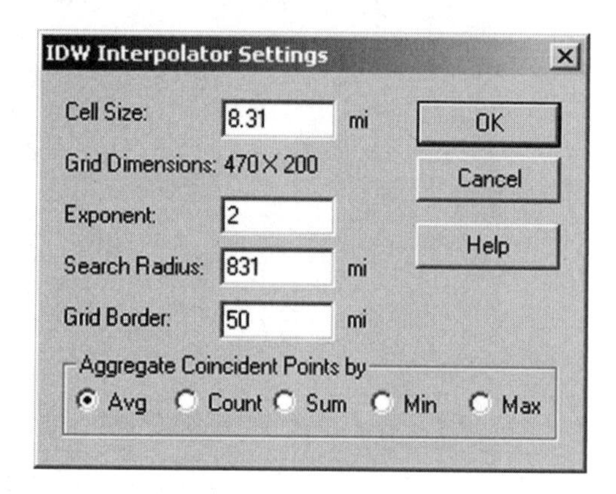

Settings for a grid theme.

Changing styles for grid themes is similar to changing styles for range themes. However, there are some pronounced differences in the appearance of the dialog box and options available. MapInfo now allows you options when defining the appearance of a grid theme. You can select the method for determining the inflection points for the colors used in the map. The available options are:

- *Equal Cell Count:* Allows approximately the same number of grids to fall into each category. This options appears as percentages.
- *Equal Value Ranges:* Breaks the data into even categories between the data minimum and maximum, and spreads the inflections accordingly. This option appears as numbers based on your data.
- *Custom Cell Count:* Allows you to determine the percentages for the inflection points (e.g., different colors).
- *Custom Value Ranges:* Allows you to determine the values for the inflection points (e.g., different colors).

MapInfo allows you the ability to change the number of inflection points (i.e., color spreads for the map) from 2 to 255. If you change the number of inflection points, MapInfo will retain the current color scheme defined and insert the additional number of specified colors at new inflection points between the two end colors. More than a few colors, however, may make the map more difficult to understand because of little variation between the colors. The five-color rainbow default is a very easy color scheme to understand: red is equal to hot (or high) value areas and blue equals cold (or low) value areas, with the other colors depicting values between the two.

The Round option allows you to choose the precision to be applied to the inflection points. However, if the spread is based on cell count, the application of this option may not be evident until the inflection values are calculated. The minimum and maximum inflection values are shown at the bottom left of the dialog box. In the Value/Percentage list (depending on the method), you can also change the values at which the "true" colors are displayed, as well as the colors.

To change a color, in the Values/Percentages list, double click on the color you wish to change, and select a color from the color palette. You can also choose grayscale or alter the contrast or brightness to adjust the colors. Changing the values for the "true" colors is easier with MapInfo 6.0. To change a value, click on a value/percentage from the value list to select it, and then type in the new percentage.

The Color Adjustment options change the appearance of the colors on the grid theme. As it implies, Contrast adjusts the contrast of the image. At 100%, only the end colors appear, whereas the middle ranges are white. At 0%, the entire spectrum appears black. You should stay close to the default of 50% to show the color variations selected. Brightness operates similarly, with the exception that at 100% the map would appear entirely white.

MapInfo added the ability to create relief maps on grid themes. Shaded relief maps add a vertical dimension to your data by using light and shadows to give the appearance of height, based on the value within each cell. To enable relief shading, check the Enabled box under Relief Shading. Once checked, the other options within this area become available. The Light Source options (Horizontal Angle and Vertical Angle) affect the orientation of the light source, making the shading appear lighter or darker.

The Vertical Scale Factor adjusts the illusion of height; the higher the number, the more the vertical distance is enhanced. You may want to adjust the Horizontal Angle, Vertical Angle, and Vertical Scale Factor values to see the difference in the maps. Processing different values for Relief Shading is very quick, so it is easy to test the effects of changes to these values to see what you prefer. MapInfo also has the capabilities to create 3D maps (discussed later in this chapter), taking this concept one step further. Style options for a grid theme are shown in the following illustration.

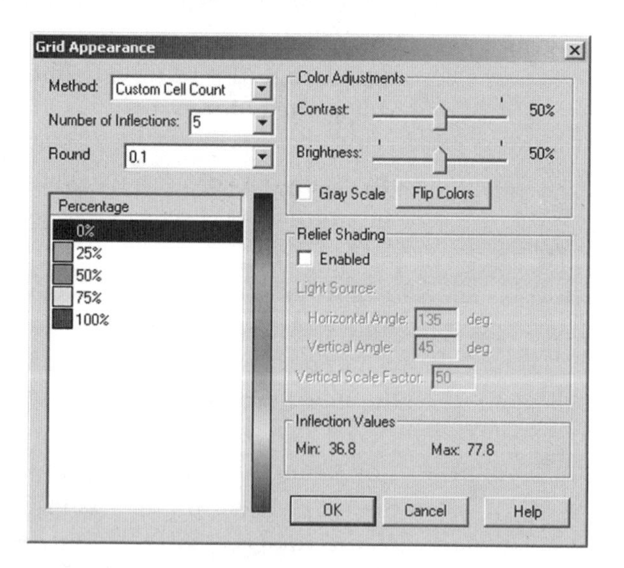

Style options for a grid theme.

When creating the grid theme, be aware that you will probably have to access the layer dialog box and move the new table up in layer order to be able view the theme. MapInfo adds the grid theme as a raster image and typically places the new table below other tables in the layer order. An example of a grid theme is shown in the following illustration.

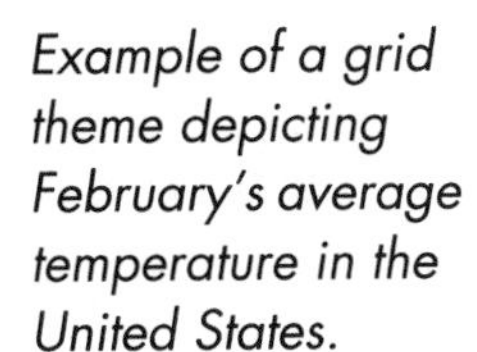

Example of a grid theme depicting February's average temperature in the United States.

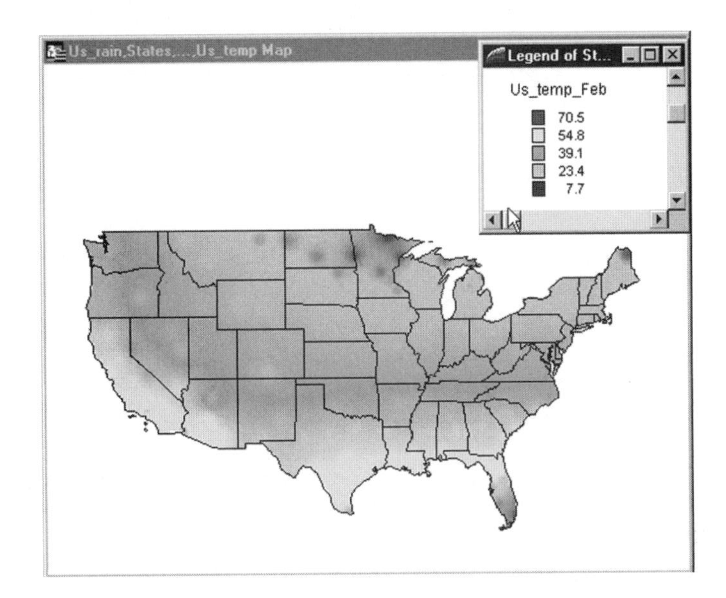

Refining the Appearance of the Thematic Legend

As previously mentioned, the map legend for all types of thematic maps operates in the same manner. You can change the legend title, which defaults to the *table_name* by *column_name*. The following illustration shows the Customize Legend dialog box for a range thematic map using block groups to depict population ranges.

Customize Legend dialog box for range thematic map.

In this dialog box, you can change the legend title and style, as well as add a subtitle to the legend. The record count for each range (the number of cases that fall into a range) can be turned on by selecting the Show Record Count option. The record count option does not apply to pies, charts, dot density, or graduated map types. The Border Style option allows you to turn the default border style on or off, or change the default border style.

If you access the Legend option on the Step 3 dialog box, you will have the choice of creating a new Legend window, placing the legend on the current Legend window, or not showing the legend. However, if you do not access Legend on the Step 3 dialog box, the legend for the theme is automatically added to the map's current legend using the default legend style.

To change a label for a range, select (highlight) the range or value in the Range Labels section, and then change the associated text in the "Edit selected range here" box. For example, you can add dollar signs to currency data (e.g., income ranges or sales figures), or change a value's text to *Other* instead of using the value itself.

In the case of a range theme, you should edit the ranges to reflect the underlying data values. When created, the ranges are continuous, which could confuse your audience as to what the ranges represent. As previously discussed, the values within each range are greater than or equal to the minimum value, and less than the maximum value. You should subtract one (1) from the maximum value of each range, and change the range labels accordingly. To ensure that your audience knows which range each value falls into, you could change the ranges in the previous illustration, for example, to the following: 2,650 to 5,280; 1,550 to 2,649; 1,060 to 1,549; 570 to 1,059; 0 to 569; and Other.

By highlighting a range and deselecting the Show this Range option at the bottom of the dialog box, the text applying to the range will not appear on the legend. The range will, however, appear on the map if any cases fall within that range. If Show this Range is deselected when you are using pie or bar chart themes, the field's color will appear in the pie/bar chart on the legend, but will have no corresponding description.

When modifying the map legend, you will typically want to change the map's title to something more descriptive, and add information in the subtitle (if appropriate). You can also change the font for the title, subtitle, and data ranges to suit the map. You will frequently turn off the Show Record Count option, as these numbers tend to confuse map audiences because they are not defined anywhere on the map.

Next, consider adding format changes (e.g., $, %, and so on) to the data ranges to make the information being displayed more clear, and modifying the range labels to match the underlying data. Once again, the thematic map of hazardous waste site/population by will be used. In tutorial 6-5, which follows, the map legend is modified.

▼ *TUTORIAL 6-5: MODIFYING A MAP LEGEND*

1 Select Map | Modify Thematic Map to access the Modify Thematic Map dialog box.

2 Select Legend to customize the legend for this theme.

3 Change the title to *The Continental United States,* using the Times New Roman font, bold, italic, 9 point.

4 Add the subtitle *1990 Pop/# Hazardous Waste Sites,* using Times New Roman, italic, 7 point.

5 As the 0 range refers only to Washington, DC, turn off the label for this layer. (Delete the x under Show this Range, with the 0 range highlighted in the Range Labels section.) Change the font to Times New Roman, 7 point. The Customize Legend dialog box should now resemble that shown in the above right illustration.

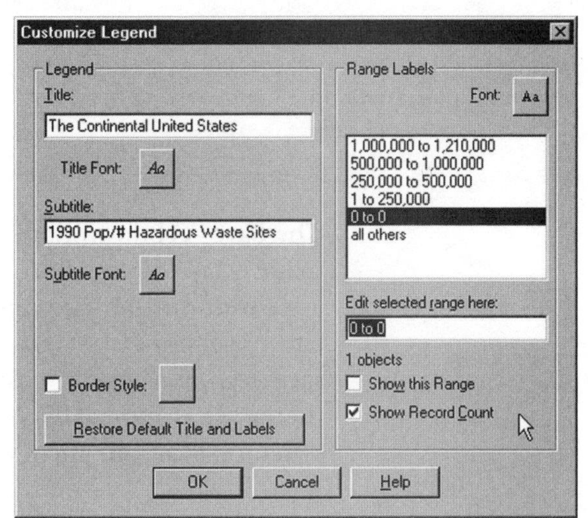

Customize Legend dialog box.

6 Click on OK to accept the changes.

7 In the Modify Thematic Map dialog box, change the Legend Label Order to Ascending.

Your map should now resemble that shown in the illustration below.

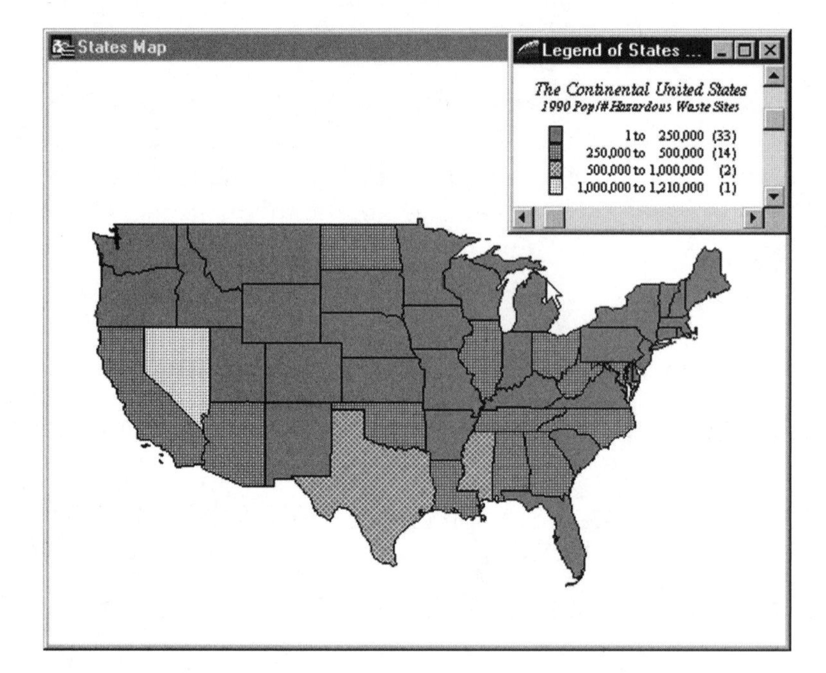

Final map.

The Save As button under Template on the Step 3 or Modify Thematic Map dialog box replaces the Save Settings button that appeared on the style dialog boxes of the various theme types in earlier versions of MapInfo Professional. This option allows you to save for later use a customized setup you have created for a theme. It is especially valuable if you use the same fonts, color schemes, symbols, and so on over and over.

Save As will prompt you for a name for the template (be sure to select a descriptive name) and ask you whether you want to save the customized legend text. Typically, you will not want to save the customized legend text. Once you save the template, it will appear in the template list as an option for the next thematic map of the same type (e.g., range, grid). However, there are some limitations to this option. For example, if you choose to save a template for a range theme where you have customized the ranges, if you choose that template at a later time, it will use the same user-defined customized ranges you created when you saved the template. You will have to manually change the ranges, or change the ranging method to reflect a data set other than the one for which it was created.

3D Mapping

New to MapInfo is the ability to create 3D Maps. This feature allows you to create depth in your maps, but is available only for a map window that contains a grid theme (discussed earlier). This feature takes relief capabilities one step further. To create a 3D map, select Map | Create 3D Map, once you have created a grid theme of your data. Make sure you only have one grid theme in the map window you are using to create a 3D map. If you have more than one grid theme on a map and select Map | Create 3D Map, MapInfo does not ask which layer you wish to base a 3D map on, and the resulting output is unusable.

As with grid maps, a raster image (i.e., picture) is created. The new map appears in its own map window. This map cannot be added to a regular map window. Correspondingly, a new menu option, 3DMap, appears on the menu bar and the Map menu option disappears. The 3D dialog box is shown in the first of the following illustrations. Options include setting the angle (Camera options), affecting the shadowing (Light options) and the appearance of the map (Appearance options). The color scheme for 3D maps based on grid themes with relief shading may appear darker than those without. An example of a 3D map of U.S. elevations using default options is shown in the second of the following illustrations.

Default options for 3D maps.

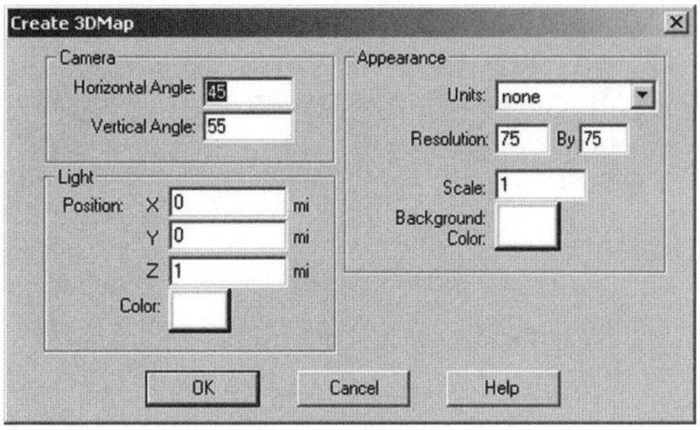

3D Map of U.S. elevations using default options.

Camera options determine the angle of the 3D map. The Horizontal Angle setting rotates the map around the center point of the grid, and options range from 0 to 360 degrees. The Vertical Angle option adjusts the elevation rotation directly over the grid. Options vary from 0 to 90 degrees. An example of applied camera angles is shown in the following illustration.

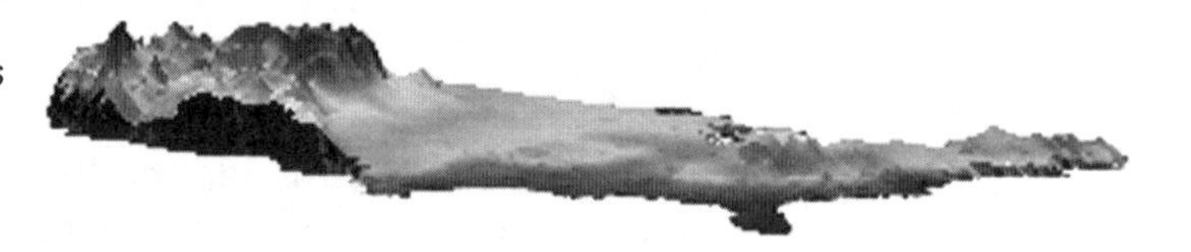

3D elevation map showing camera angles adjusted to 20° horizontal and 80° vertical.

Light Options adjust the camera's rotation around the X, Y, and Z axes. If you select a color for your map using the Color setting under Light options, you will change the appearance of the map from the color scheme of the original grid on which the 3D map is based. Color contrasts will become less pronounced. The default is white, leaving the color scheme the same as the original grid map. An example of established Light settings is shown in the following illustration.

3D elevation map showing Light options adjusted to 200 for each axis.

As the name suggests, the Appearance options affect the appearance of the 3D map. Units should only be set for grids based on some distance/height function (i.e., with relief shading). This option must be set when the 3D map is created; it cannot be changed later. Resolution determines the number of samples to take in the X and Y directions. The higher the resolution setting, the more detailed the resulting map. The resolution cannot exceed the grid dimensions (i.e., when using a 500-x-500 grid, the maximum resolution is 500 x 500). An example of this application is shown in the following illustration.

3D elevation map using feet (based on relief shaded grid).

The Scale setting affects the appearance of height (Z axis) on the 3D map. A setting of less than 1 makes the topology appear more flat, whereas a setting over 1 exaggerates (increases) the height of the topology. Finally, the Background setting allows you to select a background color for the 3D map. For example, selecting black makes the map appear to be floating in space. Remember, the background color selected will appear on your printed copy as well, so if ink is an issue, you may wish to leave the background white. An example of this application is shown in the following illustration.

3D elevation map with resolution set to 500 x 500.

Once you click on OK on the Create 3D Map dialog box, a new window will appear containing the 3D map. All features set for a 3D map can be changed, with the exception of Units, as discussed previously. To change the Horizontal or Vertical Angle for the map, either use the Arrow Select or Pan button from the Main toolbar or select 3DMap | Viewpoint Control from the Main menu (for more options). Using the menu command allows you to rotate the picture, pan across the picture, or zoom in or out. To change Light or Appearance settings, select 3DMap | Properties.

Print options available for 3D maps allow these maps to be printed to fit a page or a custom, user-defined size. To access these, make a 3D map the active window and select File | Print | Options. In exercise 6-1, which follows, you have the opportunity to practice creating a thematic map.

■ EXERCISE 6-1: CREATING A THEMATIC MAP

Creating a Ranges Map

In the first part of this exercise, you will create a ranges map for the *States* table, using the Rainfall column in the *St_info* table.

1. Select Ranges as the type of thematic map.
2. Join the tables on the *States* field.
3. Select the *Rainfall* column.
4. Change the number of ranges to 5.
5. Change the ranging method to Natural Break. Verify that Auto Spread is enabled and that Color is selected.
6. Change the colors to correspond with the rainfall ranges listed in table 6-3, which follows.

Table 6-3: Setting Colors to Correspond with Rainfall Ranges

Color	Range
Dark blue	Top range
Light blue	Second range

Table 6-3: Setting Colors to Correspond with Rainfall Ranges

Color	Range
Green	Middle range
Orange	Fourth range
Red	Bottom range

7 Change the title to *Average 1994 Rainfall*, and the font to Times New Roman, 10 pt, bold italic.

8 Change the subtitle to *By State*, and the font to Times New Roman, 8 pt, italic.

9 Change the Range Labels to add the inches symbol (") to each value (e.g., the top range would be 55.3" to 72.1"), and the font to Times New Roman, 8 pt.

Your map should resemble that shown in the following illustration.

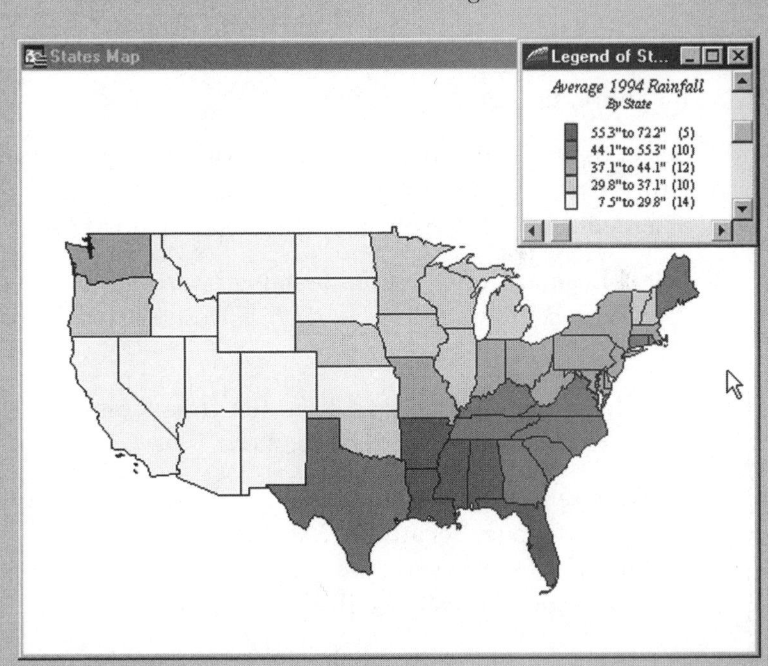

Ranges thematic map showing rainfall by state.

Now let's compare the rainfall by state to a grid theme. Continue with the following steps.

10 Open the file *us_rain*.

11 Select Map | Create Thematic map, using a grid as the type of map and Grid Default as the template.

12 Select the *us_rain* table and the *Annual* field as the basis for the map on the Step 2 dialog box, shown at right. In addition, make sure that Ignore Zeros or Blanks is checked, the output file (grid file name) is pointed to your *exercise* directory, and that you clip the image against the *States* table (as shown in the illustration at right). Click on the Next button to move to the Step 3 dialog box.

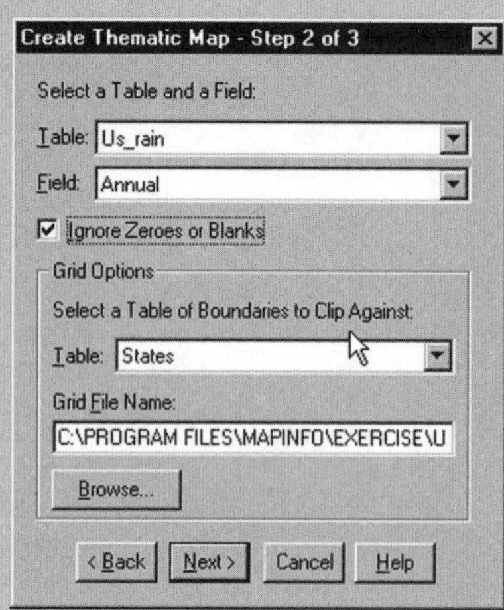

Step 2 dialog box.

13 Select Styles on the Step 3 dialog box, and change the values to the following: 3.17, 15, 30, 45, and 60.

Once the values are changed, click on OK to return to the Step 3 dialog box. Continue with the following steps.

14 Select Settings and change the grid border to 50 miles to ensure that the entire United States is covered and that missing values are accounted for. Then click on OK to return to the Step 3 dialog box.

15 Select Legend from the Step 3 dialog box, and change the title to *Average Rainfall* and the subtitle to *in the U.S.* Deselect (turn off) the Border Style option.

16 Add the symbol for inches (") to the Data Range Labels, and then click on OK to return to the Step 3 dialog box.

17 Click on OK on the Step 3 dialog box, and MapInfo will begin to create the new raster image. Because of MapInfo's hierarchical structure, you will probably not be able to see the new theme once it has been created. You will need to reorder the layers to see the new theme.

18 Access the Layer Control dialog box. Change the style for the *States* layer (access Display and select Style Override), and set the pattern to N (for none), but do not change the border style for the *States* layer. Reorder the layers so that the *States* table is on top, and the new *us_rain_annual* table is just below it. Your map should resemble that shown in the following illustration.

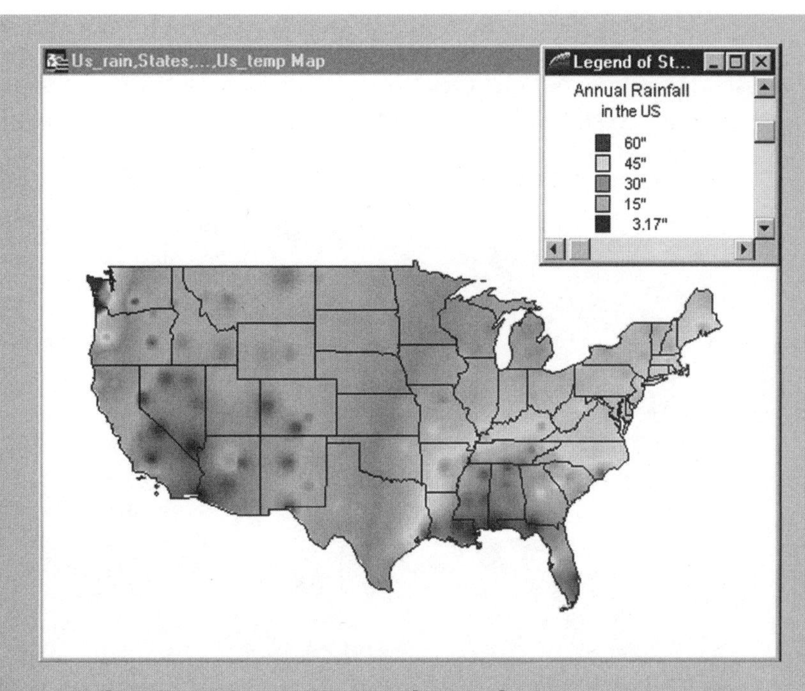

Grid thematic map showing annual rainfall in the United States.

Creating a Combination Map Based on Themes

Next, create a combination map based on the following themes: range of median household income; pie chart of age ranges 0-17, 18-24, 25-44, and 45+; and ranges for sales by location, applying size only. Your objective is to identify high-income areas where high concentrations of people aged 25 to 45 reside to determine whether this age group is influencing sales.

1 Open the *SF_GROUP, Sfgtrl00,* and *Location* tables. You may also want to open the *SF_BAY* and *SF_BACK* tables to add background and color to the map.

2 To create the range theme for median household income, you will use the *SF_GROUP* table as the base table, and join it with the *Sfgtrl00* table to obtain the income information. Create a ranges map and select the *SF_GROUP* table as the base. Select Join, and then *Sfgtrl00* as the table from which to retrieve information. Join the tables where the *Blockgroup* field in *SF_GROUP* matches the *ID* field in *Sfgtrl00.*

3 Select the *MHIN98* value as the field to be thematically mapped; then click OK.

4 From the Step 3 of 3 dialog box, select Ranges, and change the number of ranges to 5 and the ranging method to Natural Break. Click on OK once the changes are made.

5 From the Step 3 of 3 dialog, select Styles and change the color scheme to shades of green. Select the Region Styles button for the top range, and select the dark Auto Spread method. Your map should now resemble that shown in the following illustration.

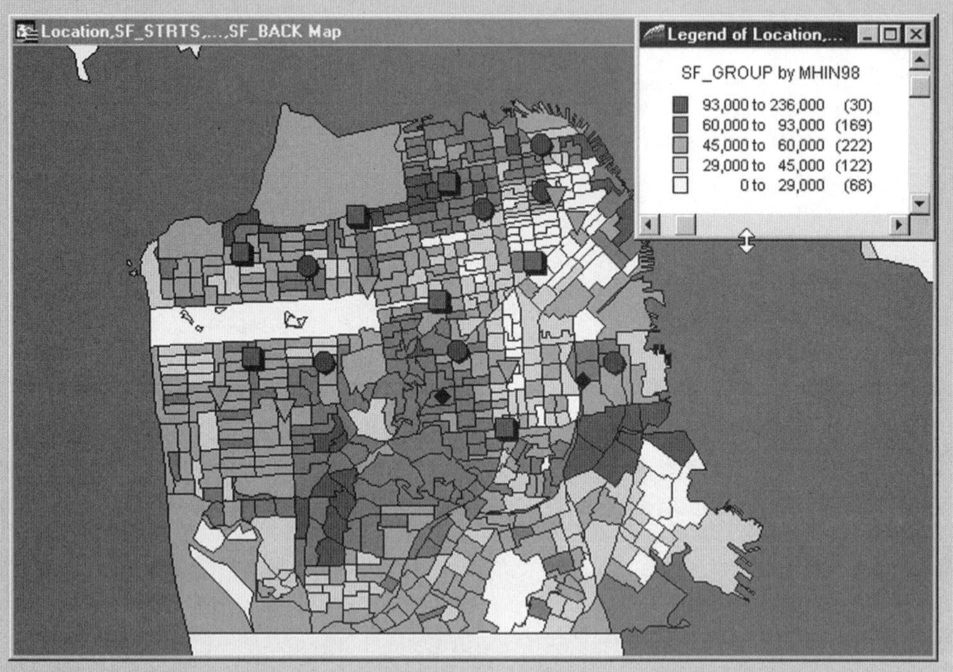

Ranges thematic map showing median household income by block group.

Zoom in to the northeast corner of the map so that the pie charts you are about to create will be easily understood. You will now create the pie chart theme based on the *SF_GROUP* table joined with the *Sfgtrl00* table. Continue with the following steps.

6 Create a pie chart map and select the *SF_GROUP* table as the base table.

7 Select Join, and then join *SF_GROUP* with the *Sfgtrl00* table where *blockgroup* from *SF_GROUP* matches *ID* in *Sfgtrl00.*

8 Select Expression (from the bottom of the "of" drop-down box), and type in the following expression: *TPA98_0_4+TPA98_5_9+TPA9810_14+ TPA9815_17.* When you click on OK, the expression should appear on the right side of the Step 2 of 3 dialog box under the Fields for Pie/Bar Chart heading.

9 Select Join again, and join the table with the *Sfgtrl00* table.

10 Select the *TPA9818_24* column, and then click on OK to return to the Step 2 of 3 dialog box.

11 Select Join again, and join the table with the *Sfgtrl00* table.

12 Select Expression, and type in the following: *TPA9825_34+TPA9835_44.*

13 In the Step 2 of 3 dialog box, select Join to create the final age range, and join the table with the *Sfgtrl00* table.

14 Select Expression, and type in the following: *TPA9845_54+TPA9855_59+TPA9860_64+TPA9865_74+TPA98_75P.*

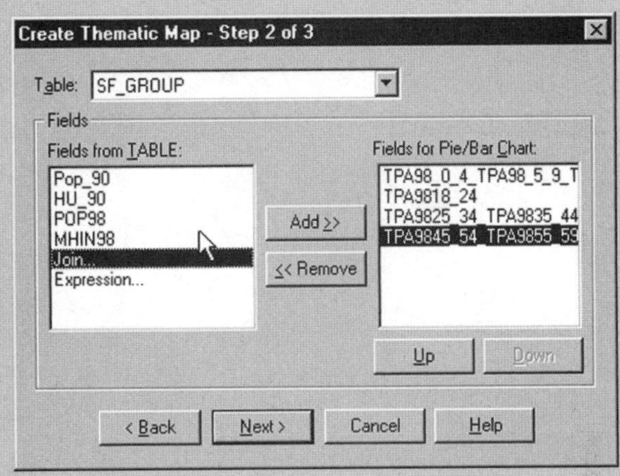

Step 2 of 3 dialog box showing expressions for building age ranges.

15 The Step 2 of 3 dialog box should contain the four expressions on the right (under Fields for Pie/Bar Chart), as shown in the above right illustration.

16 Click on OK to proceed to the Step 3 of 3 dialog box.

17 From the Styles option in the Step 3 of 3 dialog, change the color for the 25-44 age group to yellow so that it will be easily seen. Change the color for the 0-17 age group to light blue; the 18-24 group to light green; and the over 45 group to light red.

18 From the Step 3 of 3 dialog, select Legend and edit the range labels in the legend so that the age groups are clear to your audience. They should read as follows: 0-17, 18-24, 25-44, and 45+. Change the font to Times New Roman, 7 point.

19 Change the title to *Age Groups in San Francisco* using Times New Roman, bold, italic 9 point, and add the subtitle *Pie chart size represents population/block group,* using Times New Roman, italic, 7 point.

The map at this stage of the analysis appears in the following illustration. The bigger pie charts represent a larger base population than the smaller pies. The yellow pie sections show the concentration of the age group under investigation (25 to 44 age group).

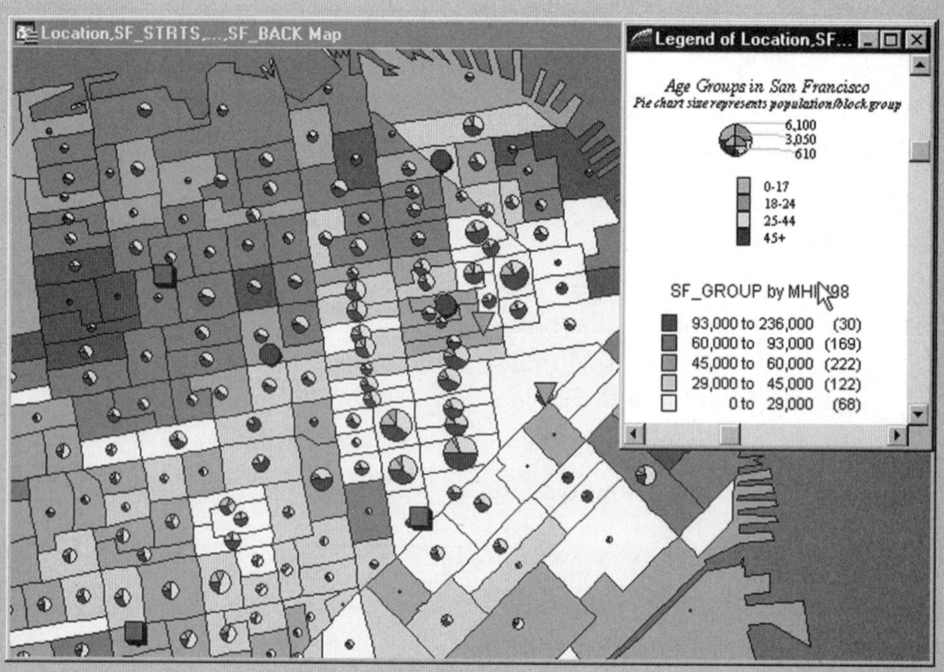

Pie chart thematic map showing age groups and income by block group.

Creating a Theme of Sales by Location

To complete the analysis, you will create the sales-by-location theme. Perform the following steps.

1. Select a ranges theme (using the point range, varying size template), with the *Location* table as the base table.
2. Select *Sales* as the column to be mapped.
3. From the Step 3 of 3 dialog box, access the Customize Ranges dialog box by selecting Ranges. Change the number of ranges to 5, and the ranging method to Natural Break.
4. From the Step 3 of 3 dialog, access the Customize Range Styles dialog box by selecting Styles. Verify that the Auto Spread option is set to Size.
5. Select the Options button at the bottom of the Customize Range Styles dialog box, and then select Size under Apply. This action will apply only the size ranges to the theme. (Color and symbol style will not be applied.) The dialog box at this stage is shown in the following illustration.

Customize Range Styles dialog box.

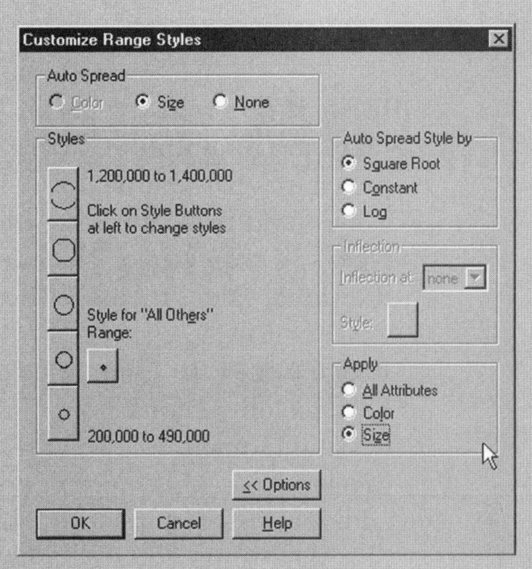

6 From the Step 3 of 3 dialog box, select Legend to add dollar signs to the range labels in the legend.

The following illustration shows the completed thematic map. Try to determine whether income level or age group is driving sales.

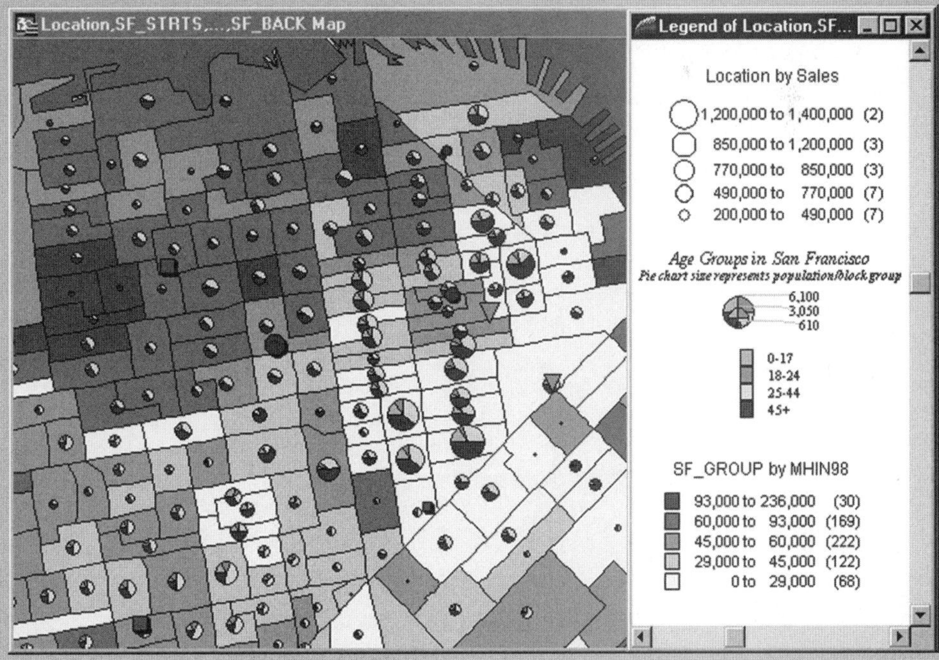

Pie chart thematic map showing age groups, income, and sales by block group.

Summary

In MapInfo Professional you can create several types of thematic maps to communicate analysis results. MapInfo supports range, pie charts, bar charts, dot density, graduated symbols, individual values, and grid themes. Although the majority of these theme types will map numeric data only, you can map text data with an individual values theme.

Range maps offer numerous methods of creating ranges for your data, including equal count, equal ranges, natural break, standard deviation, quantiles, and user-defined custom ranges. Each type of thematic map offers many options for customizing appearance. In addition, you can create and display more than one thematic map in a Map window, as well as overlay two or more themes on a single layer. Thematic mapping shows MapInfo's powerful data analysis and display capabilities.

CHAPTER 7

EDITING ATTRIBUTE AND GRAPHICAL DATA

KEEPING UP WITH CONSTANT CHANGE is an ongoing challenge, and the world of desktop mapping is no exception. For instance, the number of people living in cities and states changes through time. Such change requires that data change, and as the attributes of existing data change, you may find it desirable to add new columns of data to your data tables.

Map objects may also require modification. MapInfo's editing and drawing capabilities allow you to create and modify graphic objects on a map. Drawing tools and commands make it easy for you to build custom objects that can greatly increase your ability to analyze data. Recent releases of MapInfo provide new tools, such as object editing tools, that facilitate building custom graphic objects. This chapter explores numerous methods of data editing.

The data tables used in this chapter are provided on the companion CD-ROM. However, to edit these tables, you must copy the data files to your hard drive.

Editing Attribute Data

Chapter 4 reviewed the Browser window as a type of spreadsheet for the presentation and organization of data. This section is focused on editing the attribute data displayed with the Browser window.

Changing Data in the Browser Window

When you edit attribute data, the most common method is to use the Browser window. For example, to change the 1990 population of Alaska in the *States* table, perform tutorial 7-1, which follows.

▼ TUTORIAL 7-1: USING THE BROWSER WINDOW TO EDIT ATTRIBUTE DATA

1 Copy the *States* table (*samples* directory on the companion CD-ROM) to your hard drive. Open a Browser window of the *States* table.

2 Position the cursor with the mouse over the field you want to change; that is, the *Pop_1990* attribute for the state of Alaska. Click the left mouse button. The Browser window ready for editing is shown in the following illustration.

States Browser

State_Name	State	FIPS_Code	Pop_1980	Pop_1990	Num_Hh_80	Num_Hh_90
Alabama	AL	01	3,893,888	4,040,587	1,342,371	1,506,790
Alaska	AK	02	401,851	550,043	132,369	188,915
Arizona	AZ	04	2,718,215	3,665,228	959,554	1,368,843
Arkansas	AR	05	2,286,435	2,350,725	816,706	891,179
California	CA	06	23,667,902	29,760,021	8,644,633	10,381,206
Colorado	CO	08	2,889,964	3,294,394	1,062,879	1,282,489

Browser window of States table ready to edit Pop_1990.

3 Type in the new attribute value. The attribute data are edited.

NOTE: *The <Tab> key is handy for moving from field to field when editing a Browser window.*

Changing Data in the Info Window

Although often overlooked, the Info window can be useful for making minor changes to data. Using the Info tool can help ensure that you are editing the intended object, but this tool is only practical if the data table contains graphic objects. You can obtain the same results demonstrated in the previous tutorial with the Info window. Tutorial 7-2, which follows, takes you through this process.

▼ TUTORIAL 7-2: USING THE INFO WINDOW TO EDIT DATA

1 Open a Map window of the *States* table and position it so that you can see the state of Alaska.

2 Select the Info tool from the Main toolbar.

3 Select the state of Alaska, as shown in the following illustration. The Info window will display, showing the data attributes for the state of Alaska.

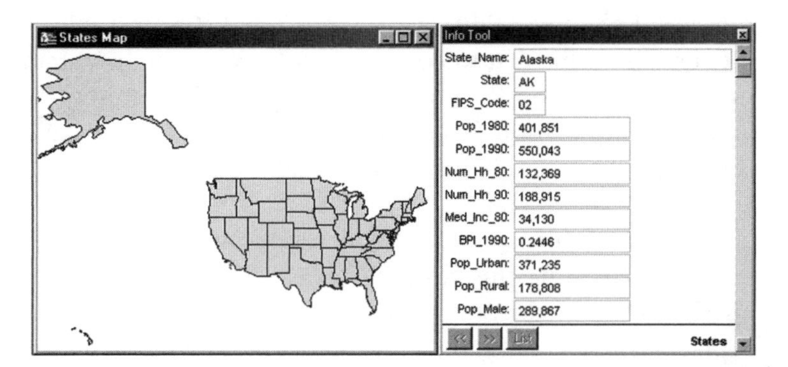

Using the Info Tool to select the state of Alaska.

4 Position the cursor with the mouse over the *Pop_1990* attribute in the Info window, and click the left mouse button. The Info window ready for editing is shown at right.

5 Type in the new attribute value. The attribute data are edited.

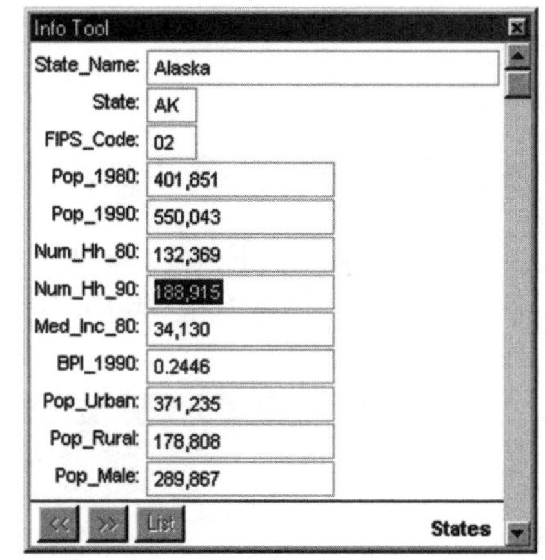

Info window for Alaska ready to edit Pop_1990.

NOTE: *The <Tab> key is also handy for moving from field to field when editing an Info window.*

Saving Changes

Once you have made changes to the attributes in a MapInfo Professional table, you will want to save the modifications. Use the

File | Save Table command to save the active document and its current name, location, and file format. If you want to change the name, location, or format of an existing document before you save it, select the File | Save Copy As command.

Undoing Changes

MapInfo Professional provides two functions to help you reverse unwanted changes: Edit | Undo and File | Revert Table. If possible, use the Edit | Undo command to reverse the last change you made. The name of the command changes, depending on the last action, such as Undo Paste or Undo Edit. Once you have undone something, the command name changes to Redo.

The Revert Table command will reverse all user changes made since the last time the table was saved. This command will undo all changes you do not want to make permanent. Revert Table replaces the selected table currently in memory with the last saved version. For instance, to revert or undo the changes made to the *States* table, perform tutorial 7-3, which follows.

▼ *TUTORIAL 7-3: USING THE REVERT TABLE COMMAND*

1 Select File | Revert Table. The Revert Table dialog, shown in the following illustration, is displayed, in which you can select the *States* table to be reverted.

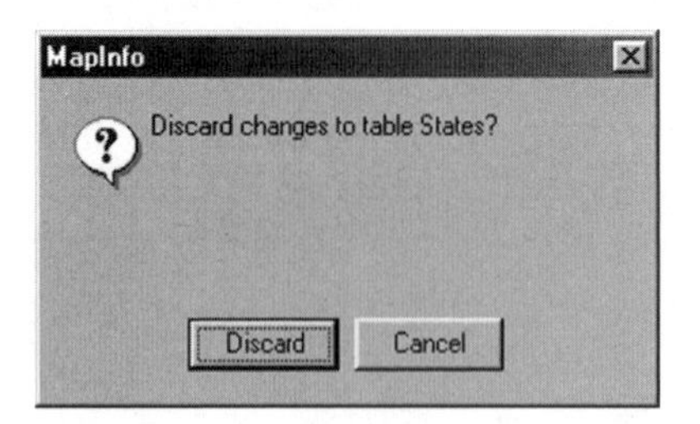

Revert Table question dialog.

2 Once you have chosen a table to revert, MapInfo prompts you on whether you wish to discard the changes you have made in the table.

3 Select Discard. MapInfo discards the changes you have made to the table.

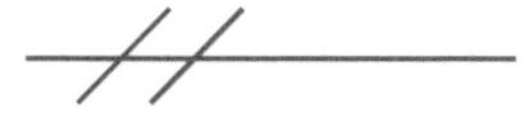

Adding and Deleting Records

As you build and maintain data tables pertaining to your special needs, you will often want to add or delete rows of data. For instance, to maintain a list of businesses in a particular city, you would need to add a row to the table for each new business, and delete a row for each business that closes.

To delete rows from a Browser window, simply select the row you wish to delete and either press the <Del> key on the keyboard or select the Edit | Cut option from the Main menu bar. The Browser window will show the deleted row as grayed out and containing no data, as indicated in the following illustration.

Browser table showing a deleted row.

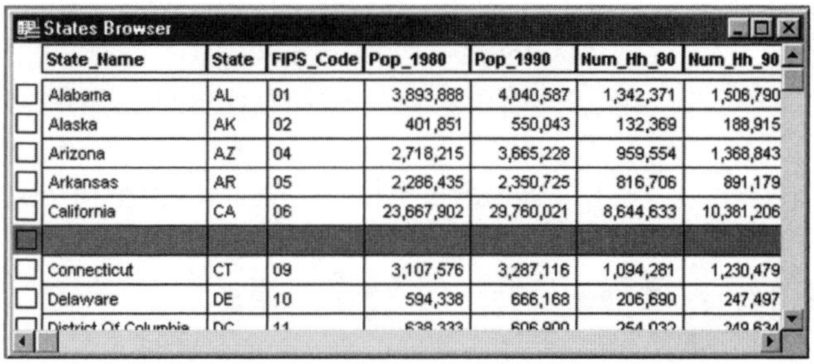

States Browser

State_Name	State	FIPS_Code	Pop_1980	Pop_1990	Num_Hh_80	Num_Hh_90
Alabama	AL	01	3,893,888	4,040,587	1,342,371	1,506,790
Alaska	AK	02	401,851	550,043	132,369	188,915
Arizona	AZ	04	2,718,215	3,665,228	959,554	1,368,843
Arkansas	AR	05	2,286,435	2,350,725	816,706	891,179
California	CA	06	23,667,902	29,760,021	8,644,633	10,381,206
Connecticut	CT	09	3,107,576	3,287,116	1,094,281	1,230,479
Delaware	DE	10	594,338	666,168	206,690	247,497
District Of Columbia	DC	11	638,333	606,900	254,032	249,634

Empty or deleted rows can be removed from the table using the Table | Maintenance | Pack Table function on the Main menu bar. To add rows to a Browser window table, select the Edit | New Row option from the Main menu bar. The new row will be added to the end of the table and can be seen as a row of blank values at the bottom of the Browser window, as shown in the following illustration.

Browser table showing an added row.

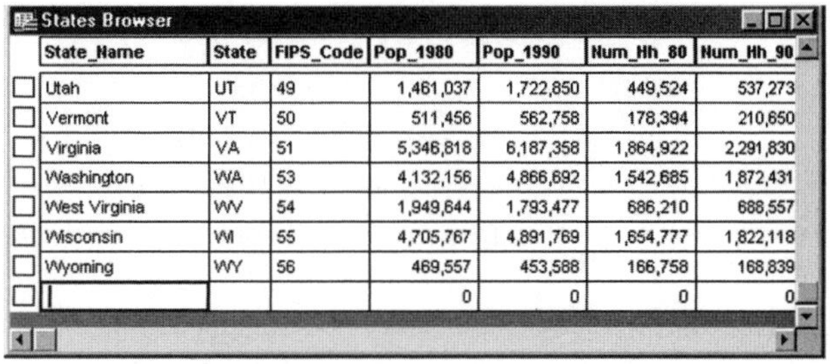

States Browser

State_Name	State	FIPS_Code	Pop_1980	Pop_1990	Num_Hh_80	Num_Hh_90
Utah	UT	49	1,461,037	1,722,850	449,524	537,273
Vermont	VT	50	511,456	562,758	178,394	210,650
Virginia	VA	51	5,346,818	6,187,358	1,864,922	2,291,830
Washington	WA	53	4,132,156	4,866,692	1,542,685	1,872,431
West Virginia	WV	54	1,949,644	1,793,477	686,210	688,557
Wisconsin	WI	55	4,705,767	4,891,769	1,654,777	1,822,118
Wyoming	WY	56	469,557	453,588	166,758	168,839
			0	0	0	0

If you are uneasy about editing an existing data table, use the File | New option from the Main menu bar to create a new table and add rows to the new table. When the new data has been entered and checked, use the Table | Append Rows to Table option to add the rows from the new table to your existing data table. The dialog for performing this action is shown in the following illustration.

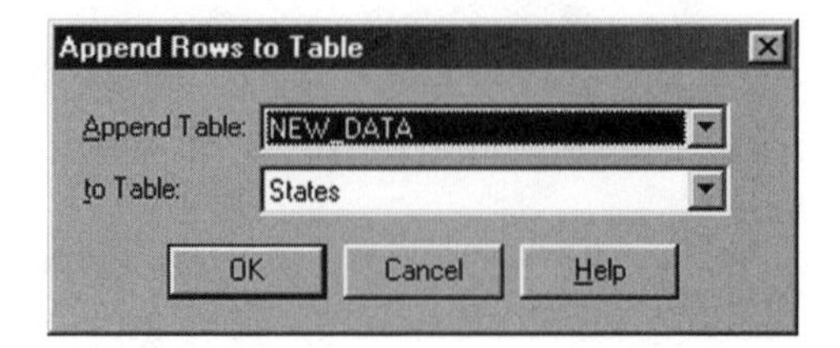

Dialog for appending rows to a table.

The preceding discussion included adding and deleting rows from tables using the Browser window. Although this is the only method available for data-only tables, for a table containing associated graphic objects you can use the Map window to add and delete objects. The next section discusses various editing tools for adding records to the Map window, or adding a blank row to the end of the table. Likewise, deleting objects from the map will remove the row of data attributes.

Editing Graphic Objects

MapInfo Professional provides two modes of editing graphic objects: creating new graphic objects, and modifying existing graphic tables. Before editing of graphic objects can occur, the layer must be made editable by using the Layer Control dialog box. The toolbars for drawing and editing will remain grayed out and unavailable to you until a map layer is editable.

Tools for Creating New Graphic Objects

There are several methods for creating tables of new graphic objects. If you create a detailed street network, you would probably elect to digitize the street network from existing hardcopy maps.

If you have a raster image (a digital picture) of the desired graphic objects, you can register the image, and trace or draw on top of the raster image. (Digitizing and image registration are discussed in Chapter 10.) If you want to create a quick table for analysis purposes, you can use the drawing tools to draw simple points, lines, or shapes. MapInfo provides a set of nine drawing tools on the Drawing toolbar for freehand drawing of graphic objects.

Use the Arc tool, shown at left, to draw arcs. You can change the start and end point of the arc as well as the angle. To draw a circular arc, hold down the <Shift> key while using the Arc tool.

Use the Ellipse tool, shown at left, to create ellipses and circles. To draw a circle, hold down the <Shift> key while using the Ellipse tool.

Use the Line tool, shown at left, to draw straight lines. To draw a horizontal, vertical, or 45-degree line, hold down the <Shift> key while using the Line tool.

Use the Polygon tool, shown at left, to create regions.

Use the Polyline tool, shown at left, to draw non-straight lines.

Use the Rectangle tool, shown at left, to draw rectangular shapes. To draw a square, hold down the <Shift> key while using the Rectangle tool.

Use the Rounded Rectangle tool, shown at left, to draw rectangular shapes with rounded corners. To draw a rounded square, hold down the <Shift> key while using the Rounded Rectangle tool.

Use the Symbol tool, shown at left, to create points.

Use the Text tool, shown at left, to add text.

These drawing tools work much the same as similar tools in popular drawing software packages. MapInfo also supports the standard cut, copy, and paste commands found in most drawing software. A few drawing tools will be employed in examples later in this chapter to demonstrate some of the more complicated drawing operations possible in MapInfo.

TIP: *MapInfo Professional is equipped with a snap function that is not commonly found in drawing software. Snap mode helps you to get the points of various objects to line up exactly. Use the <S> key to toggle the map editor to and from Snap mode. You can determine the current map mode by looking for the SNAP keyword in the lower status bar of the MapInfo main window. With Snap mode on, you will see a large cross hair appear when your cursor location is "locked" on a node. Any new point will snap to the nearest node within the snap tolerance.*

Tools for Modifying Existing Graphic Objects

MapInfo Professional provides an additional set of six buttons on the Drawing toolbar to help you edit graphic objects. These tools allow quick button access to tools for changing the shape or style of graphic objects.

The Reshape tool, shown at left, allows you to add, delete, or remove nodes in an editable layer.

The Add Node tool, shown at left, allows you to place additional nodes in a polyline or polygon object.

The Line Style tool, shown at left, permits access to the Line Style dialog used to change the line color and style.

The Region Style tool, at left, permits access to the Region Style dialog used to change the region color, fill pattern, and border type.

The Symbol Style tool, shown at left, accesses the Symbol Style dialog used to change the symbol type, color, and size.

The Text Style tool, shown at left, accesses the Text Style dialog used to change the font, size, color, and background of the text object.

TIP: *When you double click the left mouse button with the Select tool on a graphic object, an object description dialog displays. The dialog describes the location of the object. If the layer of the selected object is editable, you can make certain object edits in the dialog box. This is especially useful for changing the text in text objects as you move around a map.*

Editing Nodes

The sections that follow explore the Reshape and Autotrace functions. Tutorials are provided for practice in using these functions.

Reshape

The Reshape option allows you to add or remove nodes in a polyline or region. This option can be used to modify boundaries, such as city limits, county boundaries, or voting districts. The Reshape option can also be used to modify lines, such as in extending a street, reshaping a river bank, or moving a storm water drainage pipe.

Each region or polyline consists of a series of nodes connected by lines to form the object. When using the Reshape option, each node is marked with a small black square and can be moved. For example, let's experiment with the command on the state of Texas in the *States* table. Tutorial 7-4, which follows, takes you through this process.

▼ *TUTORIAL 7-4: USING THE RESHAPE OPTION*

1 Open a Map window of the *States* table and position it so that you can see the state of Texas.

2 Confirm that the *States* table layer is editable in the Layer Control dialog box.

3 Select the state of Texas.

4 Select the Reshape tool, which became active upon selecting Texas, as shown in the illustration at right.

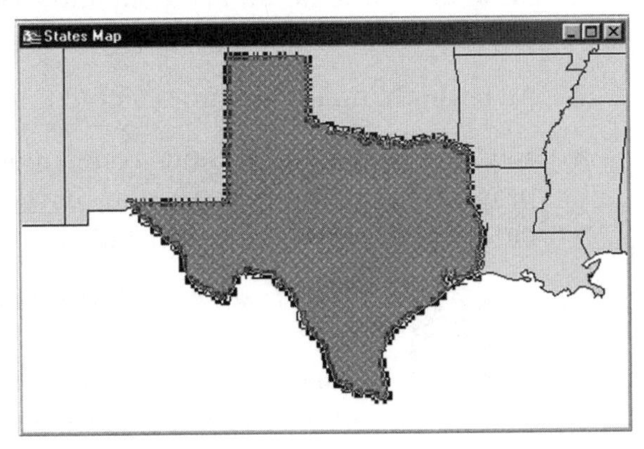

Nodes displayed for the Texas region.

5 Select and hold down the mouse button on the node at the far south tip of Texas, and drag the node to the east. The shape of Texas has been changed, as shown in the following illustration.

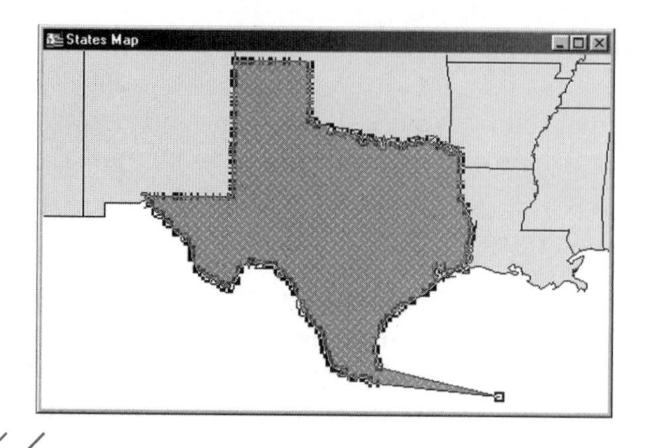

State of Texas with a node moved to the southeast.

Any single node of a region or a polyline may be moved in the manner previously described. Sometimes it may be desirable to delete, copy and paste, or move multiple nodes belonging to the object in a single edit. For example, to move a set of nodes in the Texas example, perform tutorial 7-5, which follows.

▼ TUTORIAL 7-5: MOVING MULTIPLE NODES

1 Select a node on the boundary of Texas.

2 Move along the boundary a bit, and then hold down the <Shift> key as you select a second node.

3 All nodes between the two selected nodes will be highlighted.

4 You can now apply any editing command you wish. The entire section of nodes will be modified. In this example, the cut (delete) command has been chosen to remove the selected nodes, the result of which is shown in the following illustration.

TIP: *To select the shortest range between two nodes, click on a node to select it, and then <Shift>+click on another node. To select the longest range between two nodes, click on a node to select it, and then <Ctrl>+click on another node. To select all nodes in an object, click on a node and then <Ctrl>+click on the same node.*

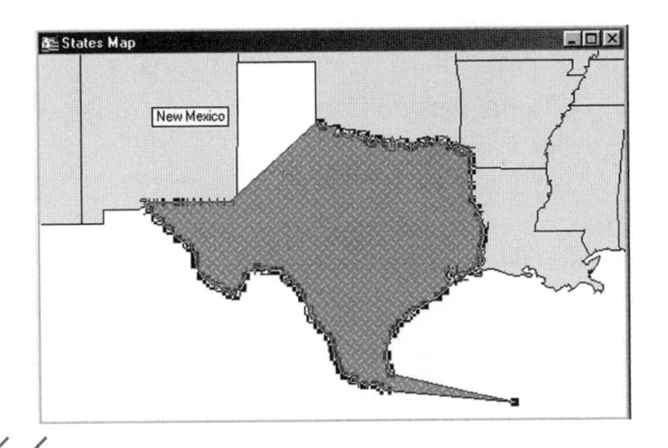

Multiple nodes deleted from the state of Texas.

Autotrace

The Autotrace function makes it easier for you to create new lines or polygons to align with existing graphic objects. In tutorial 7-6, which follows, you will experiment with the Autotrace function by tracing a portion of the Virginia State boundary.

▼ *TUTORIAL 7-6: USING THE AUTOTRACE FUNCTION*

1. Verify that the *States* table layer is editable (Layer Control dialog box).
2. Press the <S> key to activate Snap mode.
3. Select the Polyline drawing tool.
4. Click on a node of the Virginia State boundary (polyline/polygon).

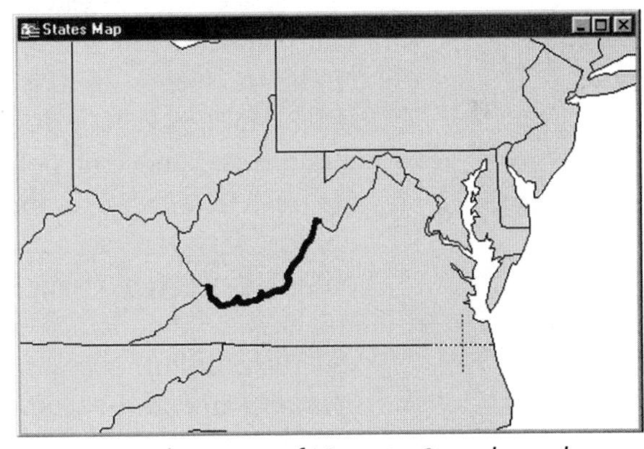

Autotraced portion of Virginia State boundary.

5. Move the mouse to another node of the same object. To draw a polyline, hold down the <Shift> key, and click. (To draw polygon objects, you would hold down the <Ctrl> key, and click.) The resulting polyline has been traced over the Virginia State boundary, as shown in the above right illustration.

When pressing the <Shift> or <Ctrl> keys, the trace path will show you a highlighted path of what will be traced. Click to automatically trace the segments between the nodes, and add them to the polyline/polygon you are drawing.

NOTE: *Autotrace works on only one object at a time. Clicking nodes in different objects will produce a straight line between the two nodes.*

Moving the Boundary Between Two Polygons

In the reshape example, the surrounding state boundaries stayed in place when the nodes were moved along the border of Texas. This behavior is a configurable option; you may want the surrounding boundaries to move when editing boundary files.

To set the method of movement of surrounding regions, use the Map Window Preferences dialog box (shown at right), accessed from Options | Preferences on the Main menu bar. The "Move Duplicate Nodes in" section of Map Window Preferences controls the way the surrounding polygons are treated. If "None of the Layers" is set, the surrounding polygons are not changed. If you select "the Same Layer," the surrounding polygons will change shape as the one being edited changes shape.

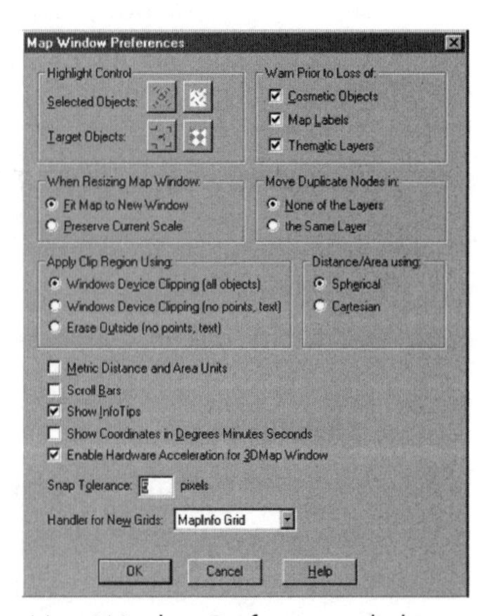

Map Window Preferences dialog box.

Add Node Tool

Occasionally, you may wish to add more nodes to a line or a region. The Add Node tool makes this process easy. For instance, assume you want to add new nodes to the Florida coastline. In tutorial 7-7, which follows, you will do just that.

▼ TUTORIAL 7-7: USING THE ADD NODE TOOL

1. Open a Map window of the *States* table and position it so that you can see the state of Florida.
2. Verify that the *States* table layer is editable in the Layer Control dialog box.
3. Display all nodes along the Florida border as described when reshaping objects.
4. Click on the Add Node tool in the Drawing toolbar. The Add Node tool displays as a cross hair when positioned in the Map window.
5. Move the cross hair to the point along the coast where you want to add the new node.
6. Click the left mouse button to add the node. The result is shown in the following illustration.

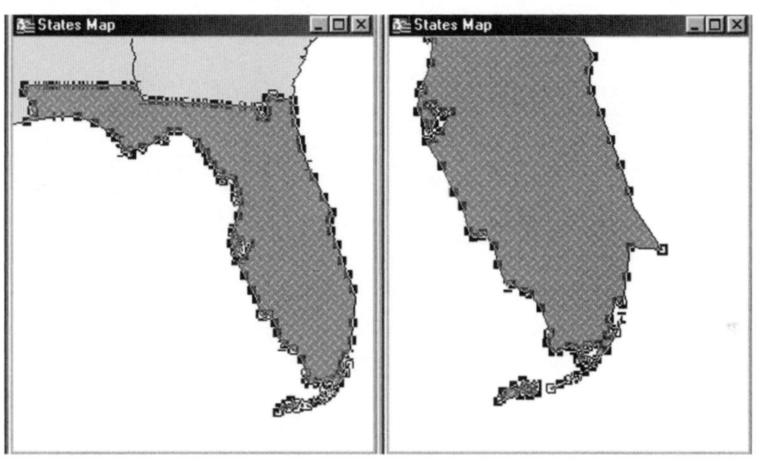

Adding a node to the Florida coastline.

Changing Object Styles

You will frequently want to change the appearance of an object in a Map window to highlight specific features or enhance its appearance. MapInfo offers several style dialogs to assist in changing the color and style of graphic objects.

Line Style

With the Line Style dialog box, shown in the first of the following illustrations, you can change line style, color, and width. Additionally, MapInfo Professional will allow intersecting lines to be interleaved. For example, a street intersection may be interleaved instead of one street line lying on top of the other. The illustrations that follow show examples of options for changing line style, width, and color, as well as an example of interleaving.

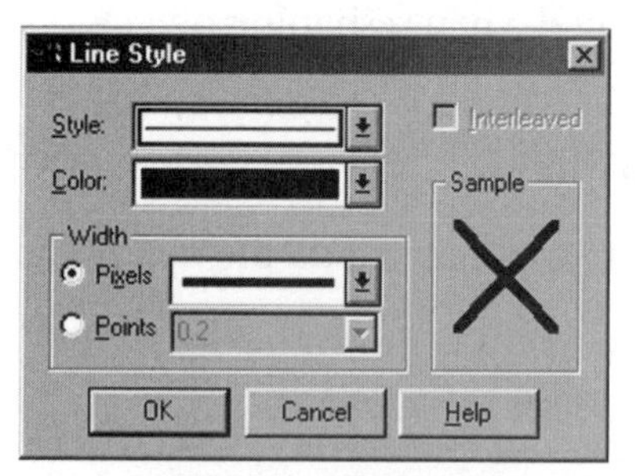

Line Style dialog box.

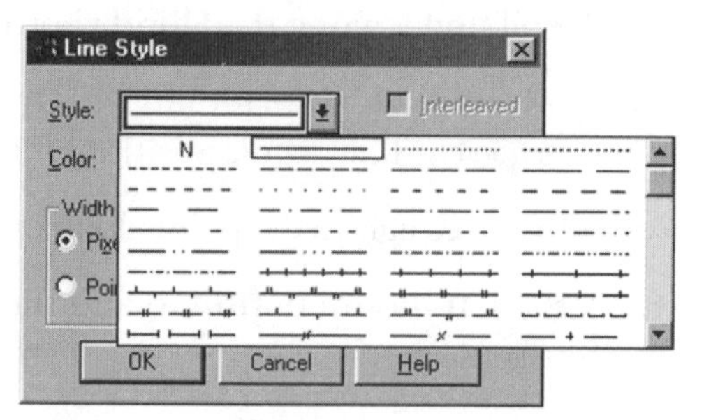

Options for changing line style.

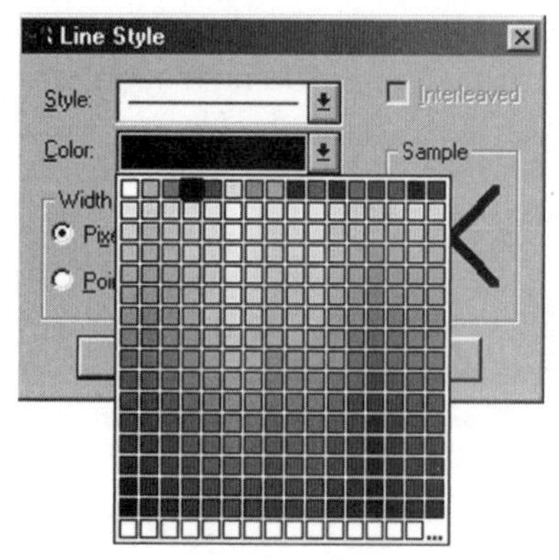

Options for changing line color.

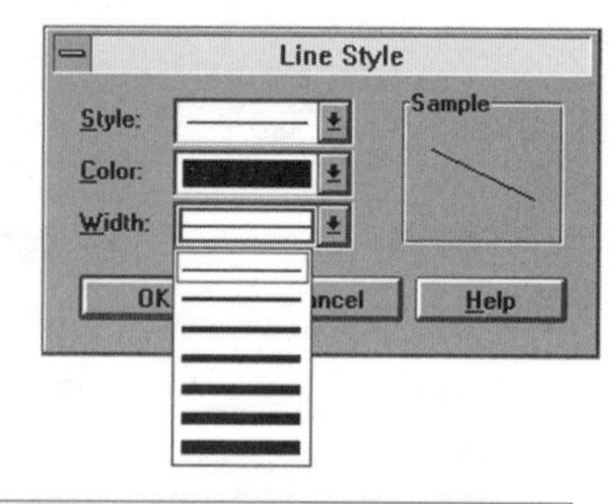

Options for changing line width.

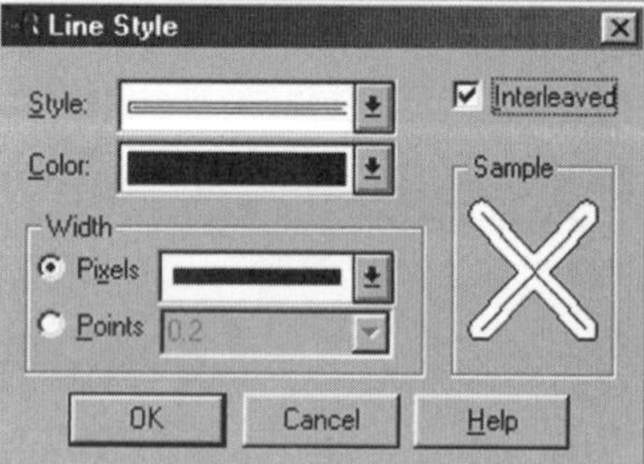

Example of interleaved line intersections.

Region Style

With the Region Style dialog box, shown in the first of the following illustrations, you can change the region and border style. The border style is controlled in the same way as the line style. The fill style has an additional option allowing for a pattern fill, which can have both foreground and background colors, as indicated in the second of the following illustrations.

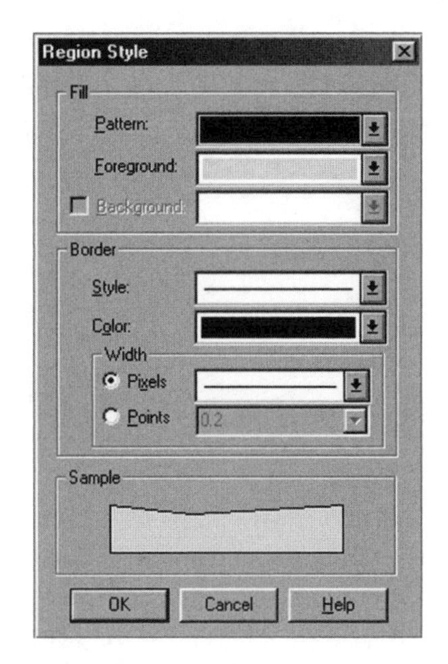

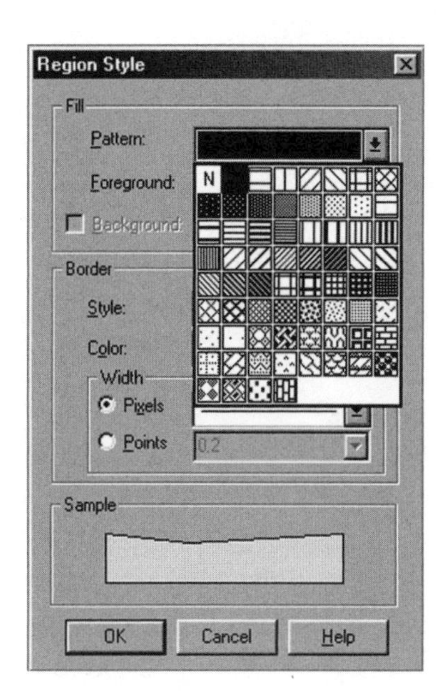

Left: Region Style dialog.

Right: Options for changing the region fill pattern.

Symbol Style

With the Symbol (points) Style dialog, you can change the font file source, symbol, color, and size. You can also select various background and effects settings. (See also "MapInfo Symbol Sets" in Chapter 5 for information on symbol set font and bitmap files, and creating and using customized symbols.) The illustrations that follow show options for changing symbol sets, the Symbol Style dialog box, and options for changing a symbol.

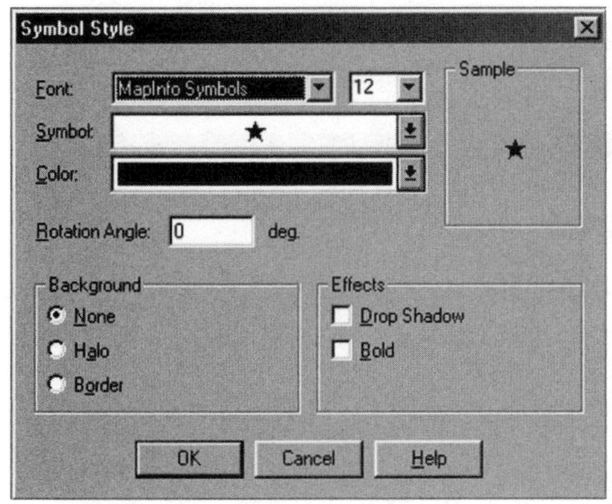

Options for changing symbol sets.

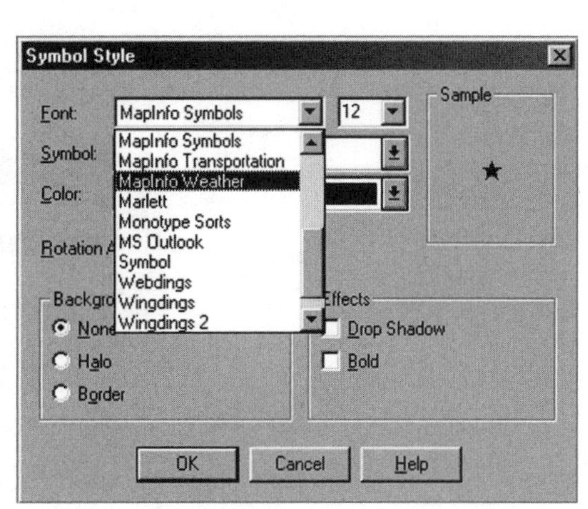

Symbol Style dialog.

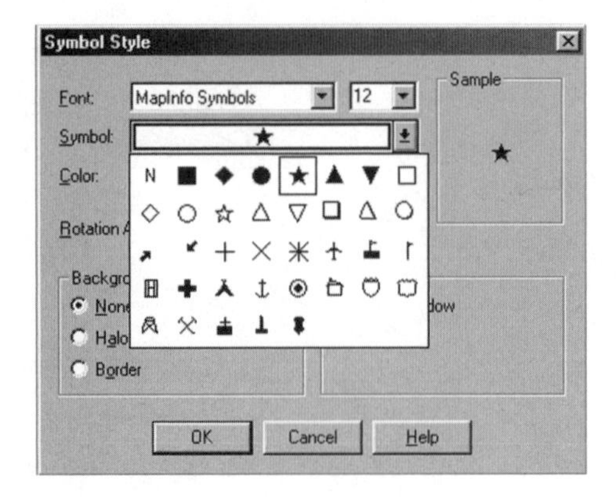

Options for changing the symbol.

Text Style

With the Text Style dialog box, shown in the first of the following illustrations, you can change the font, color, size, and style. The fonts listed in the font list will vary, based on the fonts loaded on your system, as indicated in the second of the following illustrations.

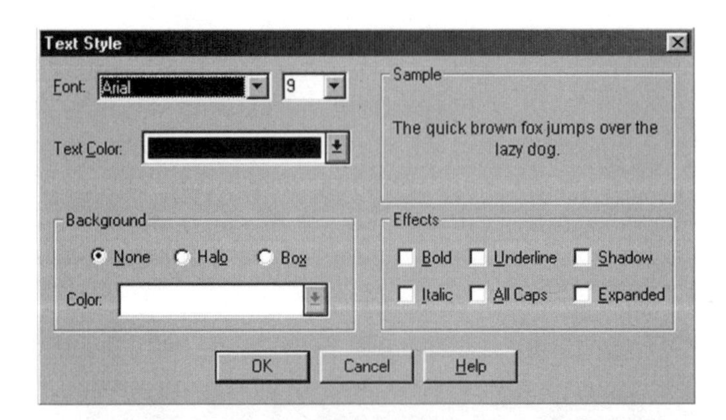

Text Style dialog box.

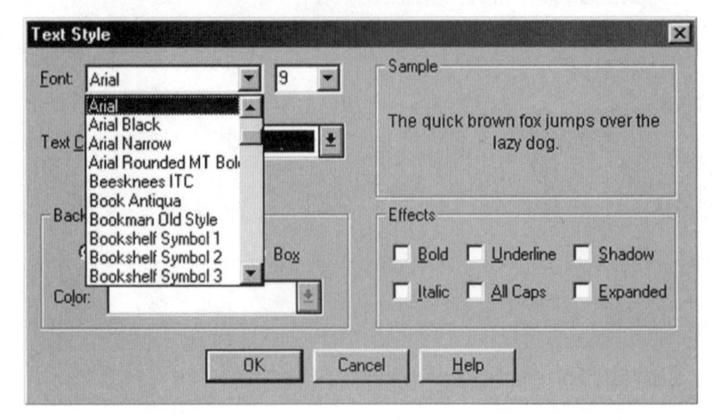

Options for changing the font.

Example of Changing Object Styles

All style dialog boxes work in the same manner. Typically, these dialog boxes are used to change the current system style setting before objects are drawn, or to alter the style of a selected set of objects that have already been drawn.

The first use is fairly self-explanatory. When MapInfo Professional is started, each of the object styles is set to a respective default, and all objects drawn in a Map window will be set to the default style. If you would prefer fat red lines instead of the default thin, black lines, make the change in the Line Style dialog box. From that point forward, every line you draw in the Map window will be thick and red.

Changing the style of existing objects is a more common operation. Tutorial 7-8, which follows, demonstrates this feature. In this tutorial, assume a hurricane has affected the eastern United States, and you wish to change the region style of the states most directly impacted by the hurricane.

▼ TUTORIAL 7-8: CHANGING OBJECT STYLES

1 Open a Map window of the *States* table and position it so that you can see the states along the eastern coastline.

2 Verify that the *States* table layer is editable in the Layer Control dialog box.

3 Select the southeastern states of Florida, Georgia, South Carolina, North Carolina, and Virginia, as shown in the illustration at right.

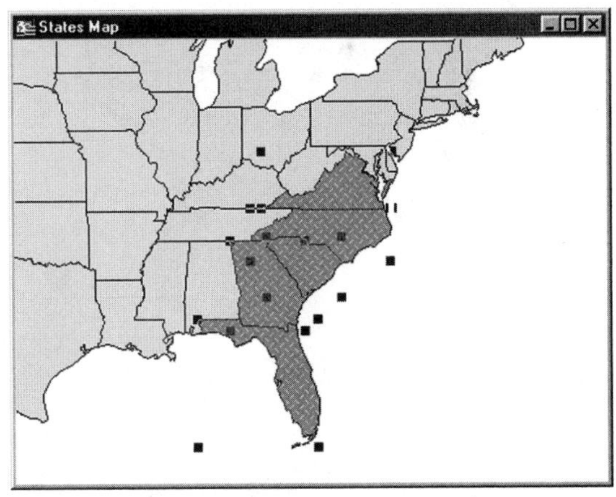

Map of selected southeastern coastal states.

4 Select the Region Style dialog, shown in the following illustration, and choose a fill style, dark foreground color, and thick border style.

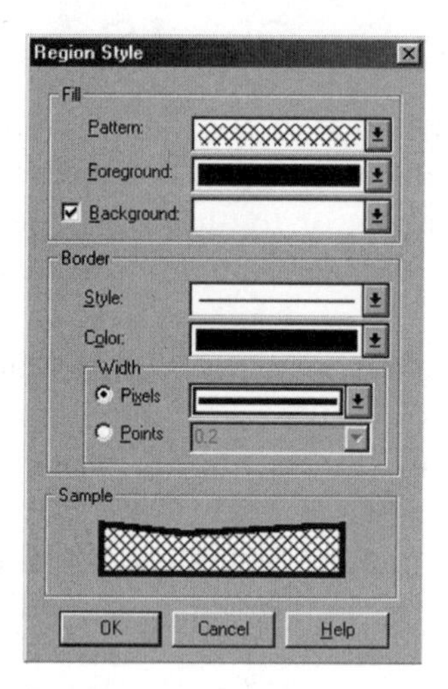

Region Style dialog box with changes.

5 When you click on the OK button, the selected regions are modified, as shown in the following illustration.

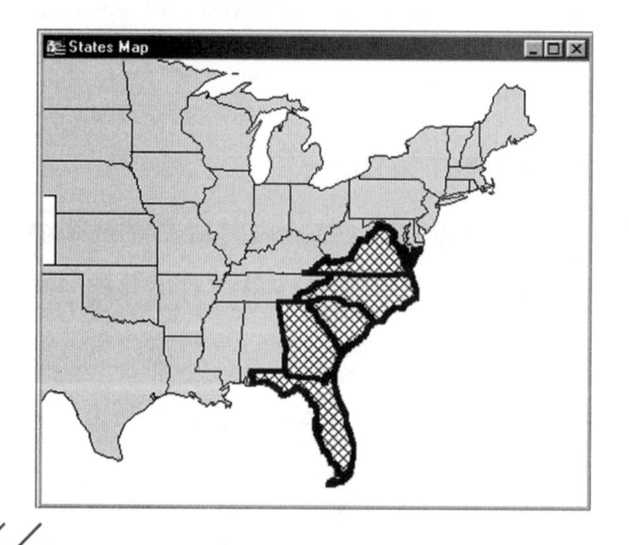

Resulting map with edited region style.

Altering Object Tools

The sections that follow explore the Convert to Regions and Convert to Polylines functions, and the Smooth and Unsmooth functions. A tutorial is provided for practice in using the Smooth function.

Convert to Regions/Convert to Polylines

The functions for converting to regions and polylines are located under the Objects option on the Main menu bar. These functions allow you to change regions to behave as polylines, and to change polylines to behave as regions.

These functions may not sound very useful to readers who have never experienced the frustration of not having the correct type of object. If you are trying to build a Select statement to find all objects that surround or touch a given shape, a polyline select will intersect only shapes that touch the line, whereas a region will intersect all shapes that touch the line as well as the interior or the region.

When working with multiple region layers, the selecting, layering, and fill types can sometimes become a nuisance. The nuisance factor can be minimized by converting some of the layers to polyline layers.

You will also find that certain editing functions will work on polylines, but not on regions. This characteristic is explored in the next section.

Smooth/Unsmooth

The Smooth and Unsmooth functions are located under the Objects option on the Main menu bar. When you use the Smooth option on a jagged or angled line, the line will be changed to a curved line. The Unsmooth function allows you to undo the results of using the Smooth command.

The Smooth function may be very useful for smoothing a jagged shoreline that has been traced or digitized. For example, to smooth the Louisiana coastline, you would switch the Louisiana region to a polyline and then smooth it. In tutorial 7-9, which follows, you will do just that.

▼ *TUTORIAL 7-9: USING THE SMOOTH FUNCTION*

1 Open a Map window of the *States* table, and position it so that you can see the state of Louisiana.

2 Verify that the *States* table layer is editable in the Layer Control dialog box.

3 Select the state of Louisiana.

4 The object representing the state of Louisiana is currently a region. Because the Smooth option will not work on regions, you must convert the region to a polyline, as shown in the illustration at right. Select Objects | Convert to Polylines from the Main menu bar.

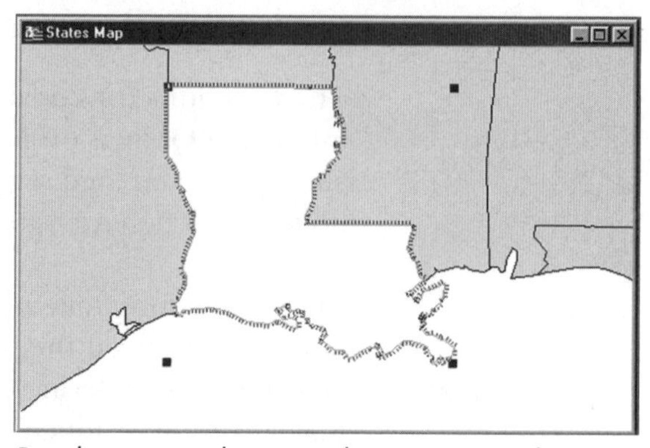

Resulting map showing object converted to a polyline.

5 Now you are ready to smooth the polyline representing the state of Louisiana. Select Objects | Smooth from the Main menu bar. In the map shown in the following illustration, the width of the Louisiana state line has been expanded to more easily perceive the smoothing that has occurred. Zoom in on the Gulf shore to examine the results of the smoothing.

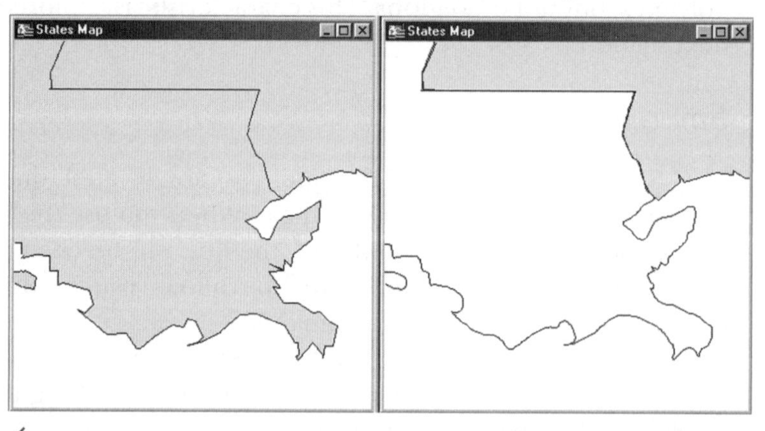

Resulting map showing smoothed Louisiana shoreline.

You may have noted that the entire state boundary for Louisiana was smoothed. If you want to smooth only the shoreline, you would first split the polyline representing the entire Louisiana state line into two separate lines. One of the lines is the shore, and the other represents the rest of the boundary. Next, you would smooth only the shoreline section, and then join the two line sections. The Split and Join functions are explored in the next section.

Advanced Editing Tools

Beginning with release 3.0, MapInfo Professional's advanced editing functionality allows you to combine, split, and erase map objects, and to overlay nodes using a "set target - apply action" editing model. This model allows you to use objects from either the same table or another table to create new objects. Data aggregation and disaggregation methods permit you to calculate new data values appropriate for the newly created objects.

Combining Objects

The Combine object function allows you to combine two or more objects into one object and then aggregate the associated data. You may want to use this function to join multiple street sections into a single section or to join multiple shapes into a larger region for defining sales districts.

TIP: *You cannot use a single Combine function to combine polylines and regions. If you want to combine regions and polylines, you must first use the Convert To Polylines command on the regions or the Convert To Regions command on the polylines. Do not use Combine with points or text objects.*

The Combine option not only allows you to join the object shapes as a single object but to aggregate the data associated with the objects as you combine them. Table 7-1, which follows, describes aggregation method options.

Table 7-1: Aggregation Options

Option	Effect
No Change	Leaves the selected column unchanged.
Blank	Stores blank values in the selected column. To store blank values in all columns, use the No Data item.
Value	Stores a specific value, or the value displayed in the edit field in the row. You can enter the appropriate value in the field.

Table 7-1: Aggregation Options

Option	Effect
Sum	Calculates a sum based on the column values from all objects to be combined. The sum option is only available for numeric columns.
Average	Calculates the average of the column values for all objects to be combined.
Weight By	This option is available when you select the Average option. Allows you to choose a column by which to perform weighted averaging.
No Data	Leaves all column values blank.

Using the Data Aggregation dialog is best described and shown with an example session. An example of joining North Carolina and South Carolina into a single region is reviewed in tutorial 7-10, which follows.

▼ TUTORIAL 7-10: USING THE DATA AGGREGATION DIALOG

1 Open a Map window of the *States* table and position it so that you can see the states of North and South Carolina.

2 Verify that the *States* table layer is editable in the Layer Control dialog box.

3 Select the states of North Carolina and South Carolina, as shown in the illustration at right.

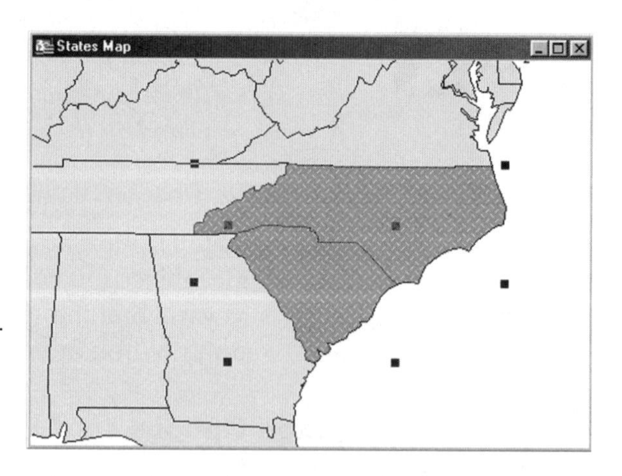

Map window with North Carolina and South Carolina selected.

4 Select Objects | Combine from the Main menu bar. The Data Aggregation dialog box displays on the screen, as shown in the following illustration.

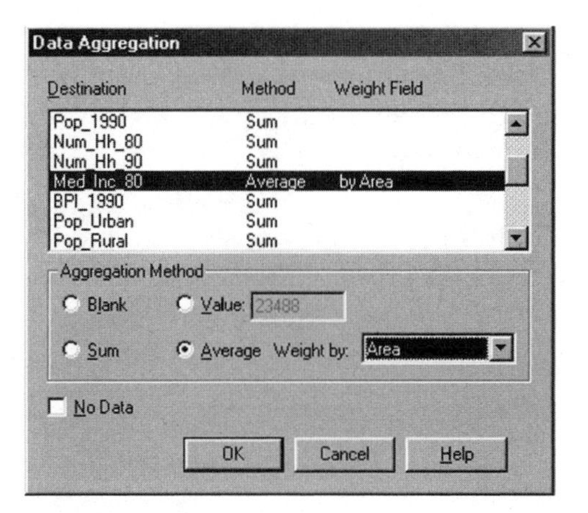

Data Aggregation dialog box.

5 MapInfo makes a guess at the type of data aggregation you wish for each of the attributes in the *States* table. Assume you want to change the type of aggregation for the *Med_Inc_80* to an average weighted by the square miles of the objects, as shown in the previous illustration.

6 Click on the OK button. The resulting map displays the combined object, as shown in the illustration at right.

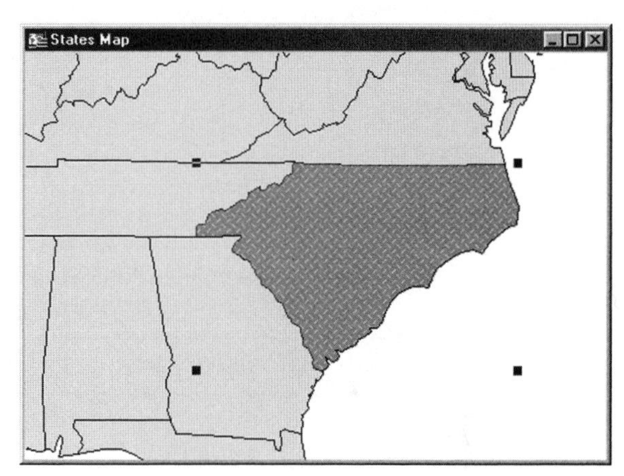

Combined area of North and South Carolina.

7 Upon investigation of the Browser window for the *States* table, you will see that the two rows that were formerly North Carolina and South Carolina have been deleted. A new row has been added at the bottom of the table for the new, combined, region.

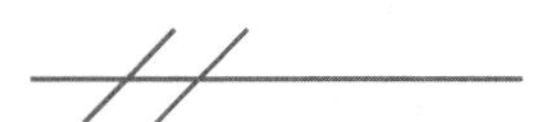

Splitting Objects

The Split function is used to break map objects into smaller parts, using the currently selected objects as the cutter. To split an object from a set of objects, the target set of objects to be edited is selected, and then the object or set of objects to split by. This splitting function is used to divide boundaries, such as voter precincts and census tracts, into subregions.

Much like the Combine option previously described, when splitting objects MapInfo Professional provides a data disaggregation function that splits the data associated with a map object into smaller parts to match the new map objects. Table 7-2, which follows, describes disaggregation method options.

Table 7-2: Disaggregation Options

Option	Effect
Blank	The value contained in the data field of the target object is deleted in the new object. To store blank values in all columns, use the No Data item.
Value	Stores a specific value, or the value displayed in the edit field in the row. You can enter the appropriate value in the field.
Area Proportion	Numeric values of the target object are proportioned for each new object, based on the area of the new objects.
No Data	Data will not be carried over into the new objects.

To demonstrate how the Split function works, in tutorial 7-11, a circle will be drawn over the four-corner area of Utah, Arizona, New Mexico, and Colorado, and the state boundaries will be split based on the circle.

▼ TUTORIAL 7-11: USING THE SPLIT FUNCTION

1 Open a Map window of the *States* table and position it so that you can see the states of Utah, Arizona, New Mexico, and Colorado.

2 Verify that the *States* table layer is editable in the Layer Control dialog.

3 Select File | New from the Main menu bar to create a new file that will be added to the current Map window.

4 Enter the attribute name of *poly_region* as a "character 10" field, and save the table to your local hard drive as *cir.tab.*

5 Select the Ellipse button on the Drawing toolbar and position the cursor over the four-corner area. Draw a circle in the newly created table with a radius of approximately 100 miles, as shown in the following illustration.

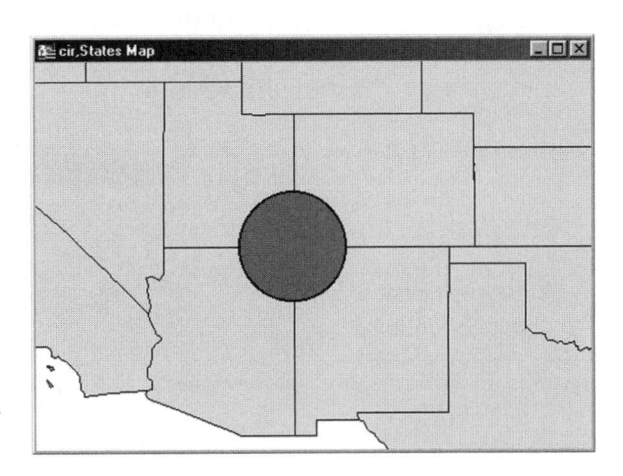

100-mile circle containing the four-corner area of Utah, Arizona, New Mexico, and Colorado.

TIP: *Pressing the <Shift> key while dragging the ellipse will create a circle.*

6 Access the Layer Control dialog, and change the *States* table to the editable layer.

7 Select Utah, Arizona, New Mexico, and Colorado.

8 Select Objects | Set Target from the Main menu bar.

9 Select the newly created circle object from the new CIR table.

10 Select Objects | Split from the Main menu bar. The Data Disaggregation dialog displays on the screen, as shown at right. The population and household counts should be set to disaggregate the data in a proportional manner, as shown in the dialog at right.

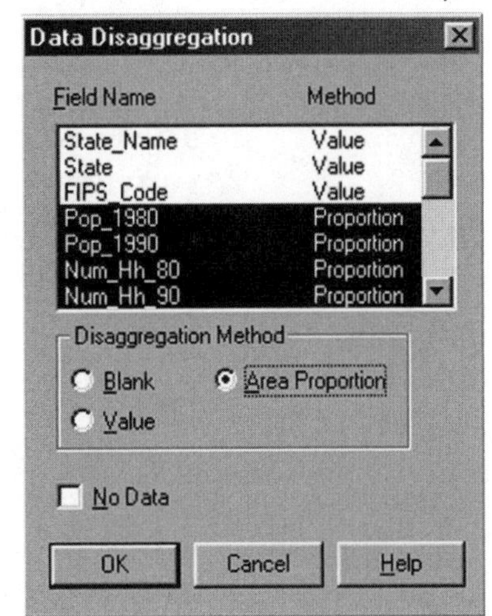

Data Disaggregation dialog.

11 Click on the OK button. MapInfo will now split the state boundaries using the circle region. If you remove the *cir* layer from the Map window, you can select each of the various shapes and see how the split worked.

Upon examining the Browser window, note that four new rows were added to the bottom of the *States* table, as shown in the following illustration.

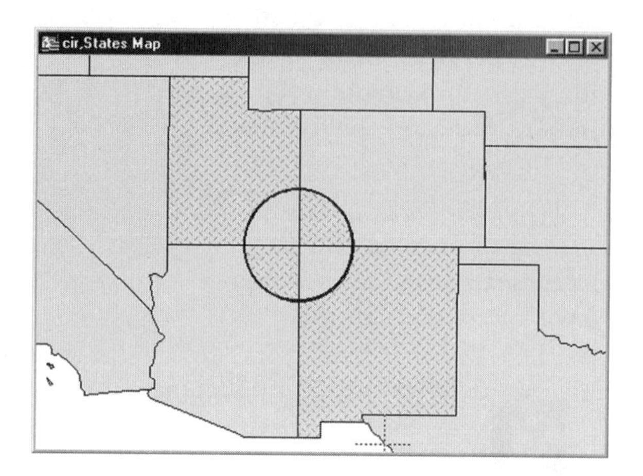

Map window of split state boundaries.

Erase/Erase Outside

The Erase command removes a portion of a map object using the currently selected objects as the eraser. The portion of the target object covered by the erasing object is removed. The Erase Outside command does just the opposite: the portion of the target object *not* covered by the erasing object is removed.

Both of the erase commands use the data disaggregation method (described previously) to proportionally remove data during the erase process. The steps for using the Erase or Erase Outside command are the much the same as described for the Split command. Instead of walking through a step-by-step example, results will be reviewed. If you draw an irregular polygon in the middle of the state of Montana, and set Montana as the target area to be edited, the Erase and Erase Outside commands will provide the object editing results seen in the following two illustrations.

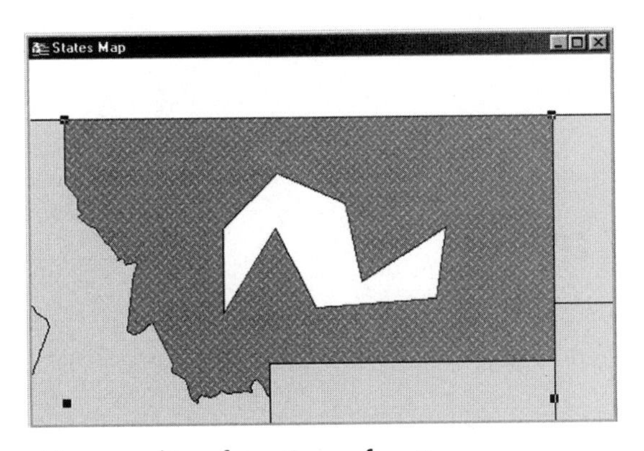

Map resulting from Erase function.

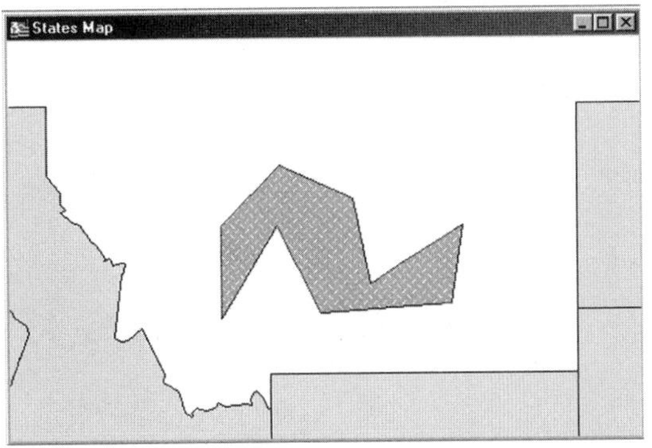

Map resulting from Erase Outside function.

Overlay Nodes

Use the Overlay Nodes command to add nodes to the target objects at all points where the target objects intersect the currently selected objects. An example of using the Overlay Nodes command is adding a street that crosses an already existing street. Two streets usually cross at an intersection, and you would want to be sure that there is a node on each street where the two streets intersect. Once the two street objects share a common node, MapInfo Professional's Find command can locate the intersection of the two streets using the two street names separated by a double ampersand (*First Street && Lincoln Avenue*).

Tutorial 7-12, which follows, takes you through an example of adding a new street, as described previously. Use the *Sf_strts* file (*samples* directory on the companion CD-ROM), and add a new street to the table.

▼ TUTORIAL 7-12: USING THE OVERLAY NODES COMMAND

1 Open a Map window of the *Sf_strts* table, and zoom in to 1/2 mile or less.

2 Verify that the *Sf_strts* table layer is editable in the Layer Control dialog.

3 Select the Line tool on the Drawing toolbar and draw a new street line crossing existing streets, as shown in the following illustration.

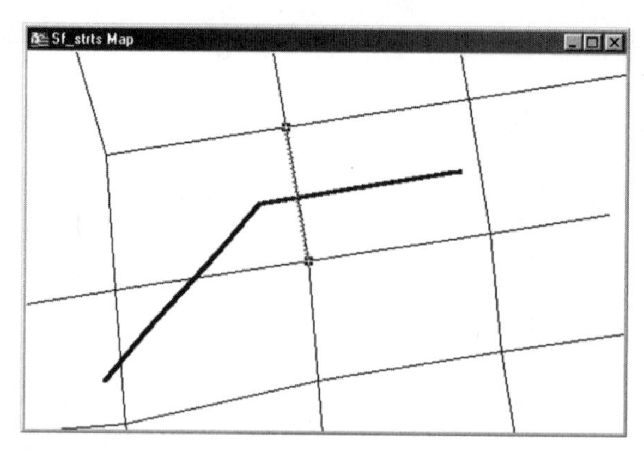

New street and nodes of existing streets.

4 Select the cross streets from the *SF_STRTS* file, and select Objects | Set Target from the Main menu bar.

5 Select the new street you drew.

6 Select Objects | Overlay Nodes from the Main menu bar.

7 Upon selecting one of the existing cross streets and viewing the nodes of that street, you will see that a new node has been added to the street at the exact location where the new street crossed it, as shown in the following illustration.

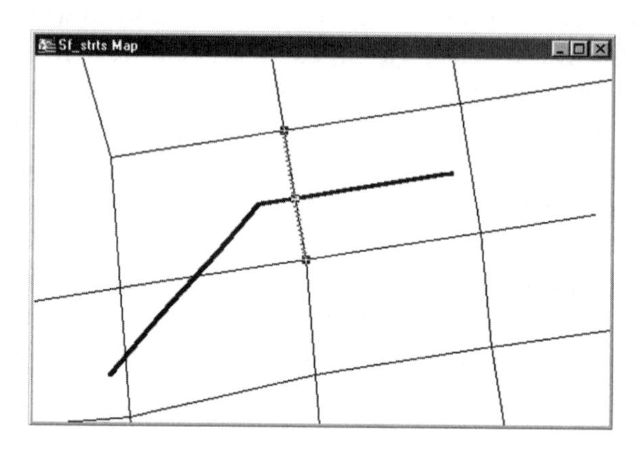

New streets and nodes of existing street after Overlay Nodes command.

In this tutorial, the new street was drawn too long to ensure that the added node was successfully created. Normally, before you finish with this new street, you would edit to remove the end nodes of the street so that it terminated exactly at the intersecting point of the streets.

Creating Special Polygons

Do you have points, but need polygons to do analysis? You may have a single point, such as a school. You want to find out all of the students that live within 5 miles of the school. You know the point, but you need to create a polygon (or 5-mile circle) around the school point. MapInfo's buffering tool is perfect for creating these types of polygons.

Do you have a lot of points you need to work as one polygon? The Convex Hull tool allows you to create a single polygon around a grouping of points. These tools and buffering concepts are explored in the sections that follow.

Buffering

Buffering is the process of creating a region around one or more map objects, including points, lines, and polygons. The following illustration shows buffers around each of these types of objects (the dotted line represents the buffer).

Buffers around points, a line, and a region.

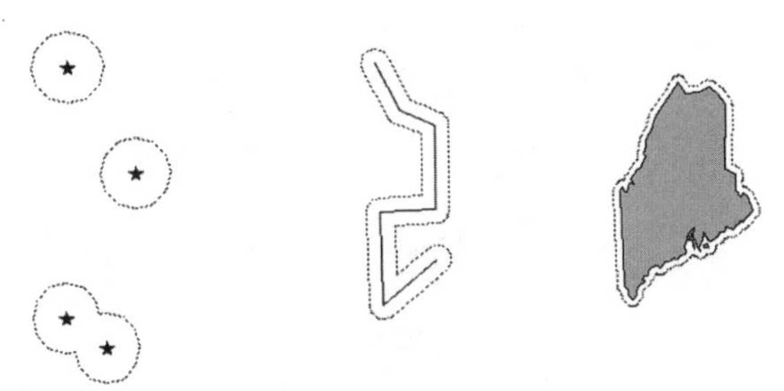

Buffering is a powerful tool to help analyze objects within a specified distance of other objects. Because buffers are simply region objects, you can search for objects inside them.

Creating Buffers

The most difficult aspect of creating buffers is determining the distance to obtain the best region for analysis. In tutorial 7-13, which follows, you will analyze coverage areas for hospitals in the San Francisco area, and create a half-mile buffer around each hospital.

▼ TUTORIAL 7-13: CREATING A BUFFER

1. Open the *Sf_landm* table in a Map window. Using the Layer Control dialog box, make the *Sf_landm* table editable.
2. Select all hospitals from the *Sf_landm* table. Use the SQL Select dialog box to select all objects from the *Sf_landm* table where *class* = *hospital.*
3. Select Object | Buffer from the Main menu bar.
4. Set the Buffer Objects dialog box, shown in the illustration at right, to create a buffer of .5 miles around each of the selected hospitals.

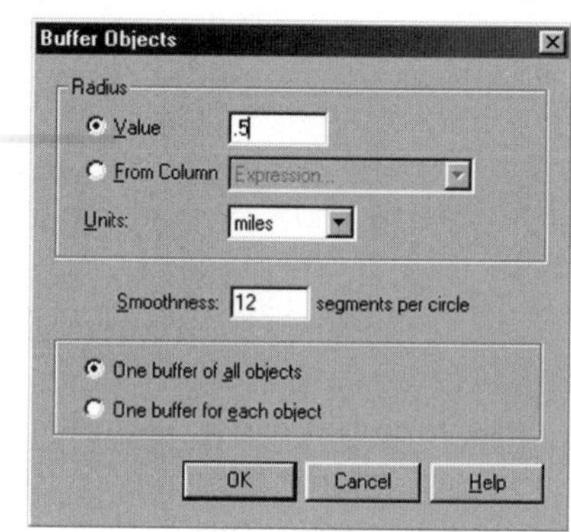

Buffer Objects dialog box.

5. Click on the OK button. MapInfo Professional will create the buffer regions established in the *Sf_landm* table, as shown in the following illustration.

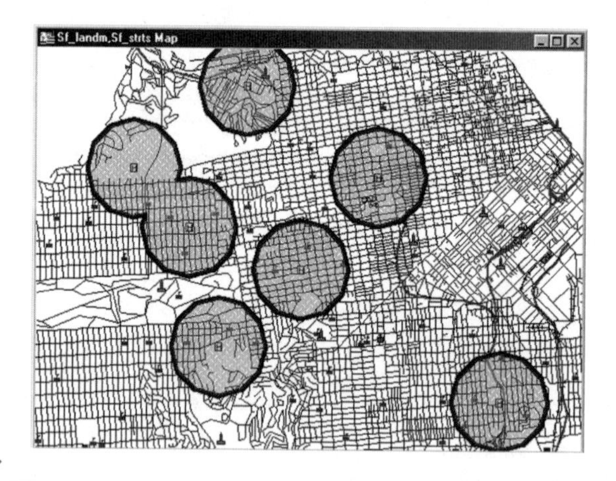

Half-mile buffers around hospitals in the San Francisco area.

The proximity of the hospitals and respective trade areas is now fairly easy to perceive. This type of analysis can help you determine which hospital is closest for patients and which areas of the city lack adequate hospital accessibility. Although the example uses hospitals, the same method would be appropriate for simple service coverage, site analysis, and telecommunications coverage, among other applications.

The Buffer Objects dialog box provides a field for specifying a buffer smoothness parameter. The higher the number used for this parameter, the smoother or greater the number of segments MapInfo will use to draw the buffer. The example shown in the following illustration compares a buffer smoothness of 6 versus 30.

Buffer smoothness comparison.

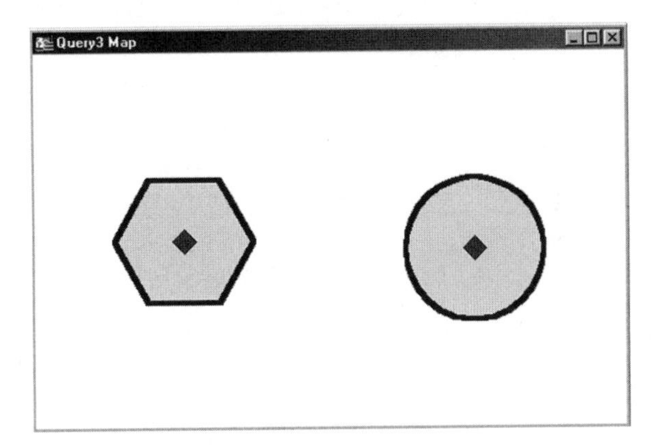

Upon viewing this comparison you might jump to the conclusion that the higher the smoothness number the better. When choosing a buffer smoothness parameter, remember that the higher the number, the greater the processing time to create the buffer. You should also consider the node number limit for region objects. The higher the smoothness number, the more nodes the region buffer object will have. Thus, if you are working with a complex object, MapInfo may exceed the node limit in trying to create a buffer for the object. In general, choose the smallest number for smoothness that provides an adequately smooth shape for the analysis you need to perform.

TIP: *MapInfo will sometimes notify you of errors while creating buffers. When this happens it is generally due to a combination of the shape of a particular region or polyline and the smoothness factor you have chosen. When a buffer creation error occurs, retry the buffer function with a smoothness factor one or two numbers higher or lower.*

Advanced Uses of Buffering

Advanced uses of buffering include data-driven buffers, negative buffers, and buffer zones. These advanced uses are briefly explored in the following sections.

Data-driven Buffers

MapInfo Professional has the ability to create buffers using values in a table or values derived from the table values through an expression. For instance, assume that each transceiver tower owned by a cellular telephone company is represented by a point on a map of the company service area. Certain associated attributes further describe the tower. These attributes can provide a more detailed understanding of telephone service coverage.

In tutorial 7-14, which follows, three telephone towers are plotted on a map of San Francisco. Signal strength and height of each tower are represented in two columns of data in an associated table. Although you could use either column of data values to draw the buffers, in this tutorial an expression created from both columns is used instead.

▼ TUTORIAL 7-14: CREATING A DATA-DRIVEN BUFFER

1 Open the *tower.tab* file (*samples* directory on the companion CD-ROM).

2 Verify that the tower point layer is the editable layer in the Layer Control dialog.

3 Select all tower points.

4 Select Objects | Buffer from the Main menu bar.

5 Set the buffer radius based on an expression of tower *Height * Power * 5*, as shown in the dialog box at right.

6 When you click on the OK button, the map shown in the following illustration is

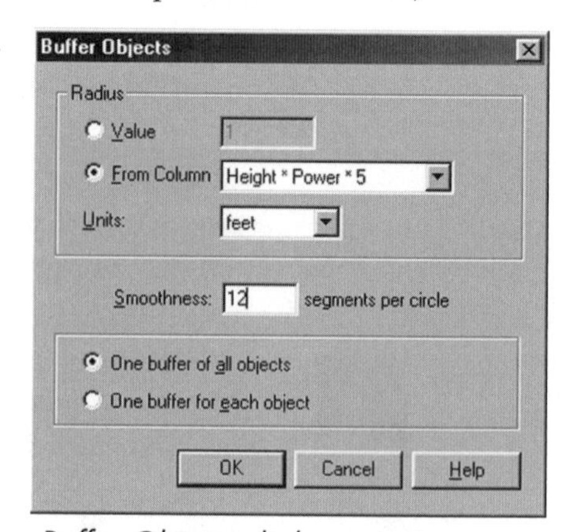

Buffer Objects dialog using an expression.

drawn. The buffers represent the tower strength for cellular telephone service in the sample area.

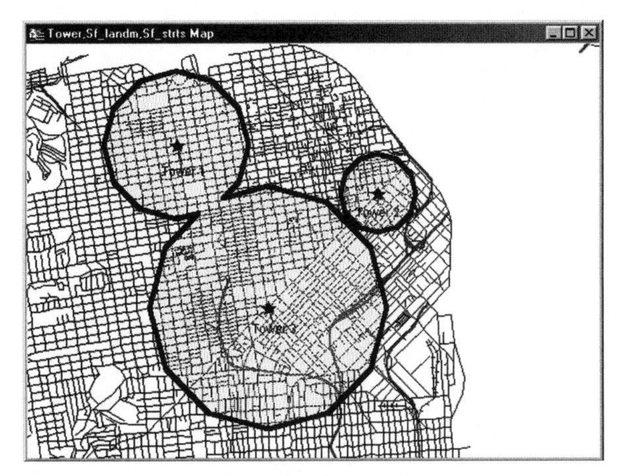

Buffers with distance based on expression.

Negative Buffers

At first glance, negative buffers appear to be a useless function. However, if you want to locate all data points that fall within a specific distance of a particular entity's border, a negative buffer is just the function you need. To see negative buffers in action, consider the problem of an insurance adjuster who wants to locate all insurance policyholders within 20 miles of the California border. Tutorial 7-15, which follows, takes you through this process.

▼ TUTORIAL 7-15: CREATING A NEGATIVE BUFFER

1 Open the *States* table in a Map window. Size and position the map so that California is easy to see.

2 Verify that the *States* table is editable.

3 Select the state of California.

4 Select Objects | Buffer from the Main menu bar. The Buffer Objects dialog box settings for a negative buffer are shown in the illustration at right.

Buffer Objects dialog box settings for a negative buffer.

5 Set the buffer radius with a value of -20 miles.

6 Click on the OK button. The buffer object drawn is 20 miles inside the border of California, as shown at right.

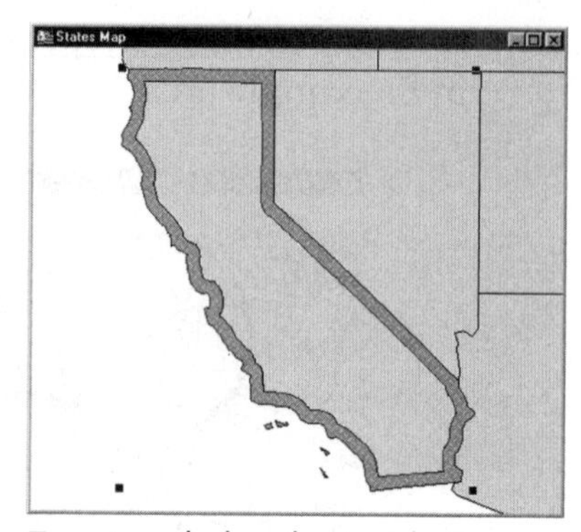

Twenty-mile border inside the state of California.

To continue this analysis, assume you want a shape representing the 20-mile area inside the California border. To accomplish this task, you would make use of the techniques reviewed in the chapter on object editing. Tutorial 7-16, which follows, takes you through this process.

▼ TUTORIAL 7-16: PERFORMING ANALYSIS WITH A NEGATIVE BUFFER

1 Select the shape that represents the entire state of California.

2 Select Objects | Set Target from the Main menu bar.

3 Select the shape that is 20 miles inside the state of California.

4 Select Objects | Erase from the Main menu bar.

5 The resulting shape is the 20-mile area that lies between the California border and the interior of the state, as shown in the illustration at right.

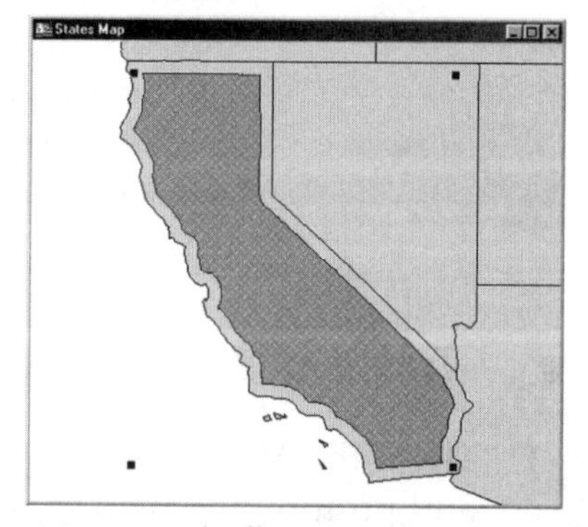

Negative buffer 20 miles short of the California border.

You can now open the table of policyholders or other point data with the newly created map shape, and locate all policyholders or points located within 20 miles of the California border.

Zones

When performing trade area analysis of retail sites, analysis of customer characteristics in various zones surrounding the sites is often desirable. For a simple analysis, a two-mile radius buffer might be sufficient. In a more complex analysis, you might wish to examine the characteristics of customers who live within one mile of the site and compare these characteristics to those of customers who live between one and two miles from the site. This section reviews how to perform the latter type of zone analysis.

For the following review, buffer zones are built around Green Hospital in the San Francisco sample data. A zone is then created to focus on the number of people and homes within a half-mile radius of the hospital. Assume you wish to analyze the number of people and houses located in the half-mile versus one-mile radii surrounding the hospital.

To set up for this analysis, open the *Sf_landm* and *Sf_group* tables in a Map window. Next, create a table named *buff* for the buffer zones you will create. Continue working with zones in tutorial 7-17, which follows.

▼ TUTORIAL 7-17: WORKING WITH ZONES

1 Verify that the *buff* table is editable.

2 Locate and select the point object representing Green Hospital.

3 Select Object | Buffer from the Main menu bar.

4 Draw a buffer with a one-mile radius from the hospital point.

5 Repeat and draw another buffer with a half-mile radius from the hospital point.

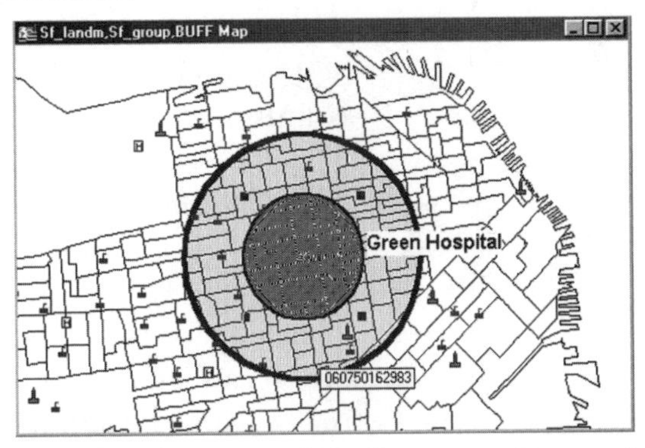

One-half and one-mile buffers around hospital point.

6 At this point, you will see the map shown in the above right illustration, with the two buffer circles displayed.

At this point the center hole will be cut out of the larger circle so that you have a ring or donut shape for a region between a half-mile and one mile of the hospital. Continue with this example by performing tutorial 7-18, which follows.

▼ TUTORIAL 7-18: MODIFYING THE ZONE

1 Verify that the *buff* table is editable.

2 Select the larger buffer circle.

3 Select Object | Set Target from the Main menu bar.

4 Select the small buffer circle.

5 Select Object | Erase from the Main menu bar.

6 You now have a half-mile radius circle and a half-mile to one-mile donut shape, as shown in the illustration at right.

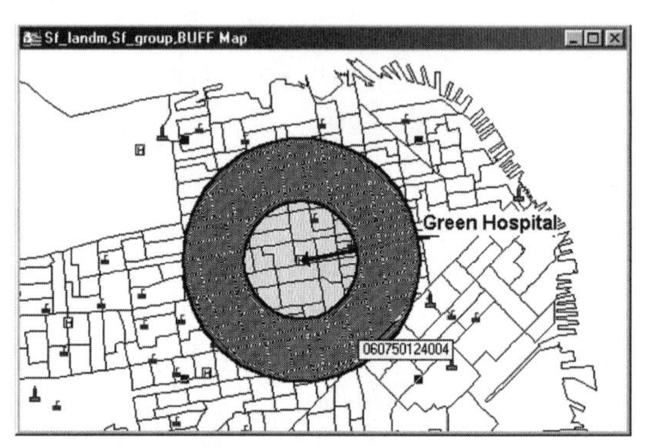

Half-mile to one-mile ring buffer around hospital point.

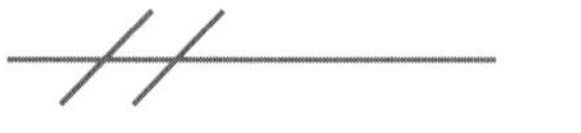

Now you can use these two objects with the polygon tool or the SQL Select dialog to analyze the population and household counts in the census block groups of the *Sf_group* table.

TIP: *The MapBasic program named* r_rings.mbx, *located in the* tools *subdirectory, will create concentric ring buffers.*

The concept of creating several buffers and using one to edit the other can greatly enhance your ability to obtain the data you really want to examine. Not only does this concept apply to rings or donuts, as seen previously, but also to overlapping buffers. If buffers of two site locations overlap, you can use editing capabilities to remove the overlap or assign the overlap area to one region or the other. In the map shown in the following illustration, the overlapping area between Green Hospital and Letterman General Hospital has been removed from the Letterman buffer region, and retained only in the Green Hospital buffer region.

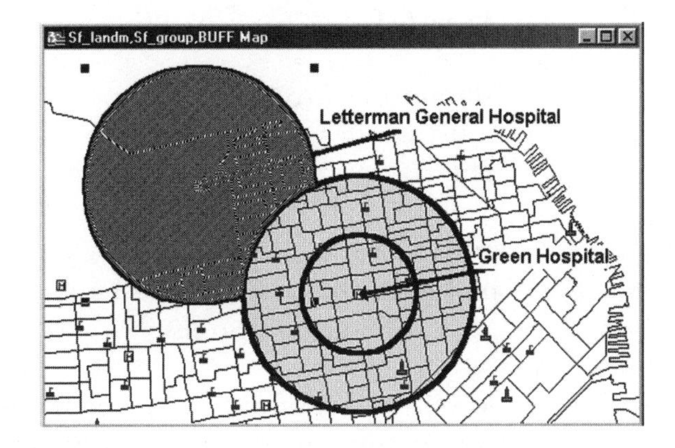

Region overlap removal.

Convex Hull

When you have a group of points you want to process as a single polygon, the Convex Hull tool may help. This tool was added with version 6.0 of MapInfo. This section will first describe what a convex hull polygon is, and then explain how to use this MapInfo tool.

What Is a Convex Hull?

The following is a technical definition of a convex hull polygon. Given a set of points, a convex hull is an enclosing boundary line about the points using the positions of the extremal points as the inflection points for the line.

A more simple way of grasping the nature of a convex hull polygon is to visualize a set of points; for example, nails on a board. The convex hull polygon can be represented by a rubber band placed around the outside nails on the board. The following illustration shows a convex hull polygon around a small collection of points.

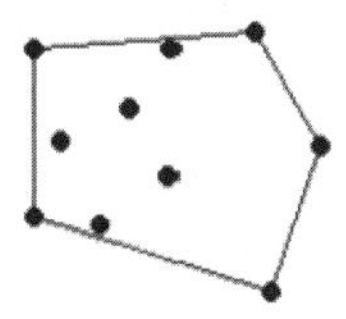

Convex hull polygon.

MapInfo's Convex Hull

MapInfo's convex hull returns a region object that represents the convex hull polygon based on the nodes from the input object. The convex hull polygon can be thought of as an operator that places a rubber band around all of the points. It will consist of the minimal set of points such that all other points lie on or inside the polygon. The polygon will be convex; no interior angle can be greater than 180 degrees.

As stated previously, the most difficult aspect of creating buffers is determining the distance to obtain the best region for analysis. In tutorial 7-19, which follows, you will analyze coverage areas for hospitals in the San Francisco area, and create a half-mile buffer around each hospital.

▼ TUTORIAL 7-19: CREATING A CONVEX HULL

1 Open the *Sf_landm* table in a Map window. Using the Layer Control dialog box, make the *Sf_landm* table editable.

2 Select Query | Select All from *Sf_landm* from the MapInfo Main menu bar.

3 Select Objects | Convex Hull from the Main menu bar. This accesses the Create Convex Hull dialog box, shown in the above right illustration.

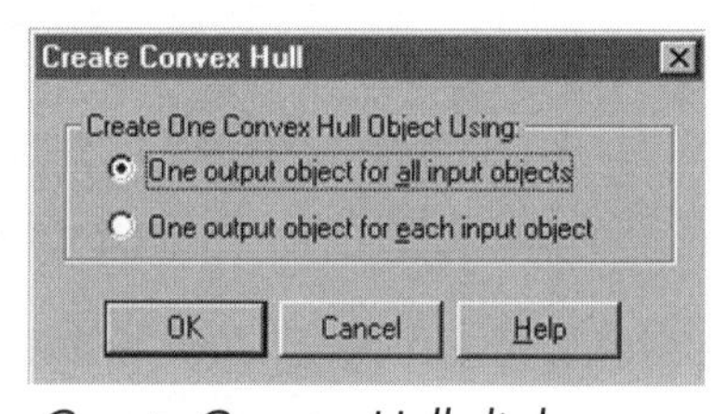

Create Convex Hull dialog box.

4 You can choose to create one convex hull object from all of the input objects, or one convex hull object for each input object. Accept the default of creating one polygon.

5 Click on the OK button. MapInfo Professional will create the convex hull region in the *Sf_landm* table, as shown in the illustration at right.

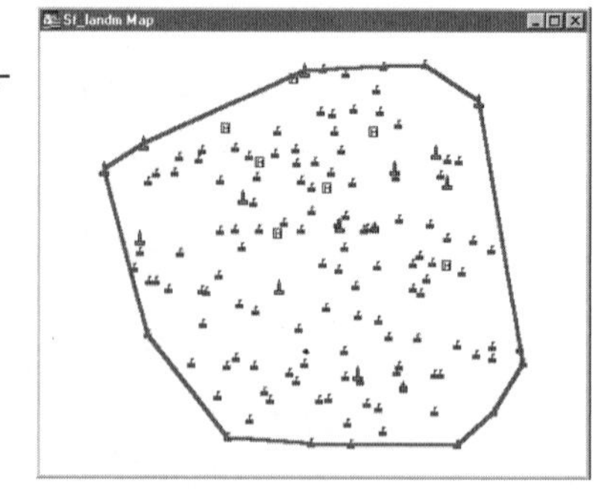

Resulting convex hull polygon for the Sf_landm *table.*

NOTE: *A convex hull requires three input nodes to produce results.*

Enclose Function

The MapInfo Enclose function has been added with version 6.0. This function will create regions from polygonal areas enclosed by polylines. The Objects Enclose feature can produce more than one object. For example, use this feature to create regions from linear tables, such as road networks. These regions represent the parcels between the roads. In tutorial 7-20, which follows, you will create regions from polygonal areas enclosed by polylines.

▼ TUTORIAL 7-20: USING THE ENCLOSE FUNCTION

1 Select polylines in an active Map window with an editable layer.

2 Select Objects > Enclose.

3 Open a Map window of the *Sf_strts* table and position it so that you can see the street lines.

4 Select File | New from the Main menu bar to create a new file that will be added to the current Map window.

5 Enter the attribute name of *poly_region* as a "character 10" field, and save the table to your local hard drive as *region.tab*.

6 Check the Layer Control dialog box to ensure this new region table is editable. This is the table in which the neighborhood boundaries will be built.

7 Select a subset of the streets in the *Sf_strts* table.

8 Select Objects | Enclose from the MapInfo Main menu bar. This will create the boundaries in the *region* table, as shown at right.

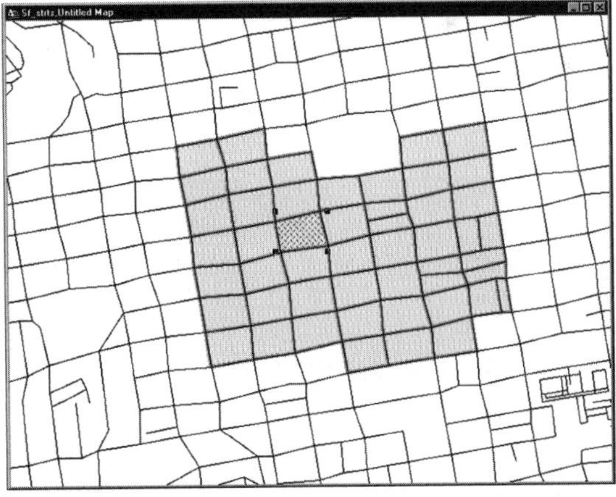

Enclosed Streets become Neighborhood polygons.

9 Use the Layer Control dialog box to move the *region* table to the bottom of the map display. Set the region display color, fill, and line patterns so that you can see and understand what this function has done for you.

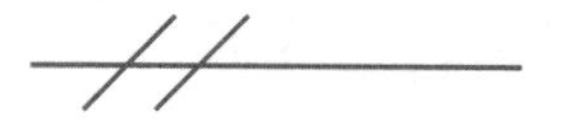

Checking Regions/Tables for Incorrect Data

When creating your own data files, finding data errors is always a problem. Some types of regions or polygons are prone to causing errors in other processes and really should be corrected. With MapInfo 6.0, there is a new function to help you determine these potentially troublesome polygons. The polygon problem areas checked include the following:

- Line segments within a region that cross each other (self-intersecting polygons)
- Nodes within a single polygon of a region where the polygon touches itself
- Overlapping polygons (if desired)

In tutorial 7-21, which follows, you will check a file for incorrect data.

▼ TUTORIAL 7-21: CHECKING FOR INCORRECT DATA

1 Select File | New from the Main menu bar to create a new file that will be opened in a new Map window.

2 Enter the attribute name of *poly_region* as a "character 10" field, and save the table to your local hard drive as *region.tab.*

3 Check the Layer Control dialog box to ensure this new region table is editable. This is the table in which you will create some erroneous polygons.

4 Select the polygon drawing tool and draw one polygon in the shape of a bow tie. Draw a second polygon that intersects or lies over the first one. Your map should look something like that shown in the following illustration.

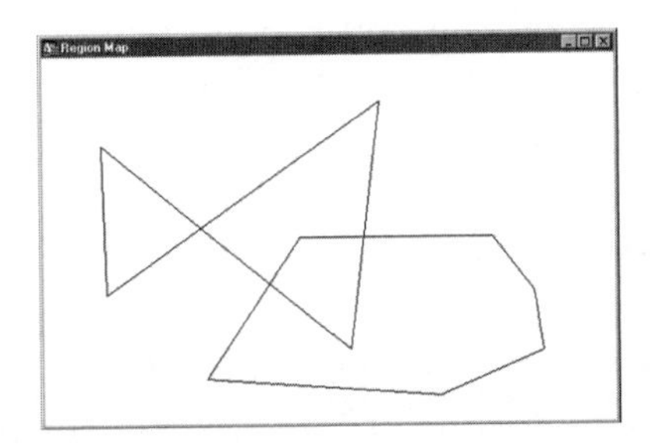

Polygon figures to be error checked.

5 Select the two polygon objects using the selection tool.

6 Select Object | Check Regions from the Main menu bar. The Check Regions dialog box (shown in the following illustration) will be displayed.

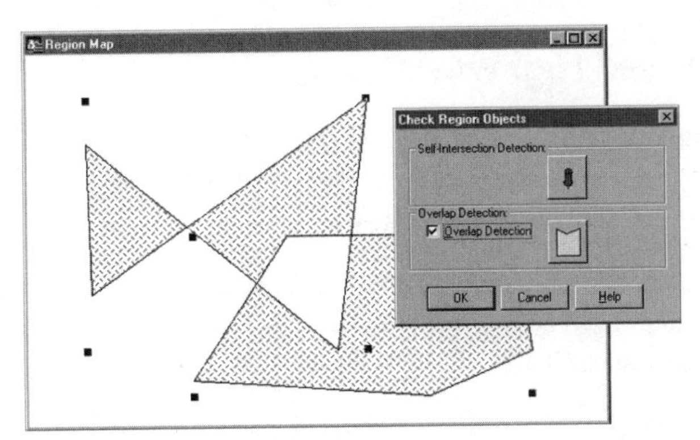

Check Regions dialog box.

7 Select to check for overlap and self-intersection and click on the OK button. The resulting map is shown in the following illustration.

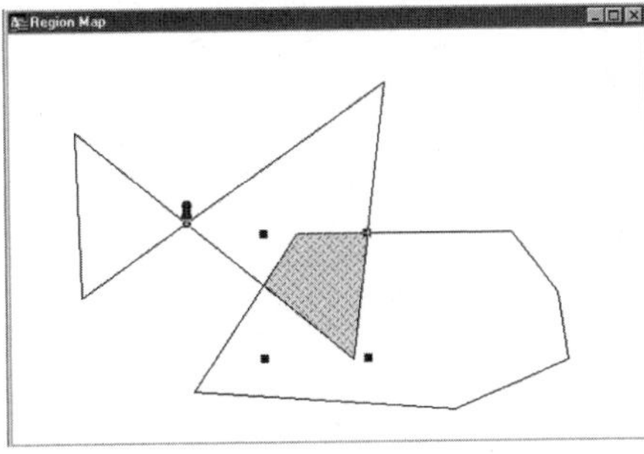

Polygon Check Regions resulting map.

If there are no data problems, you will see a dialog with the message "Check Region did not find any data problems." If there are data problems, a point object is created and placed into the output table, as in the map shown previously. If you select to check for overlapping objects, the overlapped regions will be placed in the output table, as in the previous map. In exercise 7-1, which follows, you have the opportunity to practice editing graphic objects.

■ EXERCISE 7-1: EDITING GRAPHIC OBJECTS

The following exercise is focused on map applications typically required of city planning offices. City planners monitor growth and decline areas to make recommendations for road improvement projects. In the following, a map is prepared showing where a proposed new highway will pass through northern San Francisco, as well as roads that will have been repaved or widened in 1999.

Drawing the Proposed Highway

First, the proposed highway will be drawn. Because proposed roads are typically shown as dashed lines on a road map and highways as red lines, the new road will be depicted as a red dashed line.

1 Select the File | Open Table option from the Main menu bar, and open the *SF_STRTS.tab* table (*samples* directory on the companion CD-ROM). This is shown in the following illustration.

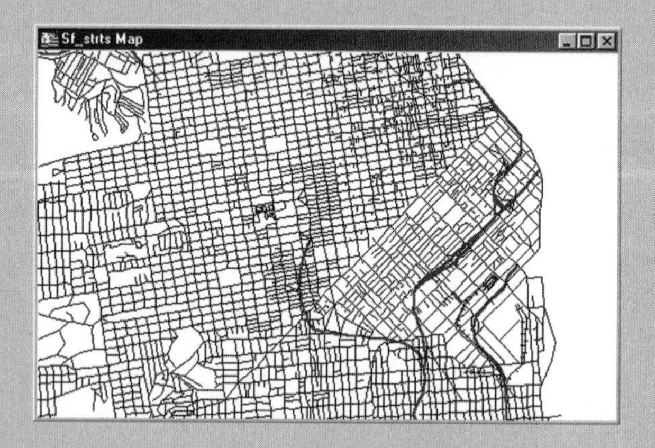

Map window of the SF_STRTS *table.*

Note the section of highway in the northeastern part of San Francisco, and another section toward the central western part of the city. The road project in question proposes to connect these two highway sections, and you need to map the route the highway will take. A new table for the proposed road will be created because the planning office is in the planning stages and may have to change the plan many times before it is approved.

By using a new table, the integrity of the original *SF_STRTS* file is maintained. Continue with the following steps.

2 Select the File | New option from the Main menu bar. This accesses the Create New Table dialog box, shown in the illustration at right.

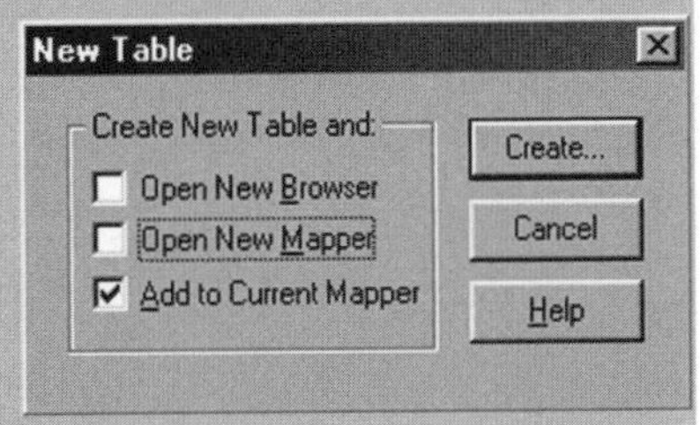

Create New Table dialog box.

3 Select to add the new table to the current mapper, as shown in the previous illustration, and click on the Create button.

4 Using the New Table Structure dialog box, and add the *name* attribute column to the table. The new column is a 40-character text field, as shown in the following illustration of the New Table Structure dialog box.

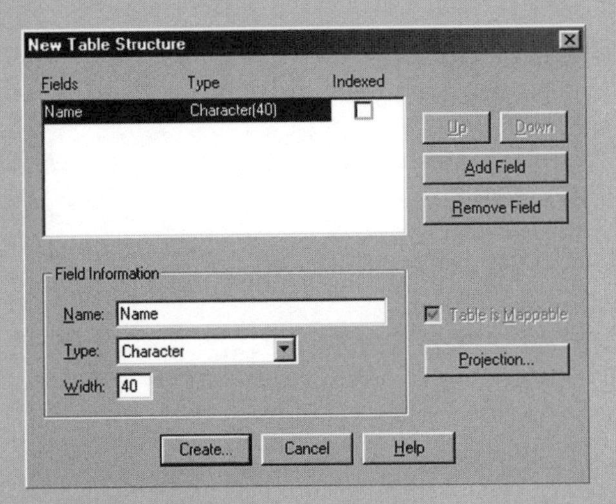

New Table Structure dialog box.

5 Click on the Create button to activate the File dialog box. Set the dialog to save the table, with the name *NEW_ROAD*, to your local hard drive. Click on the OK button. The table will be built and added as a new top layer to the Map window.

You are now prepared to draw the proposed road. At this point, you must set the line symbology for the drawing, and use the Polyline drawing tool to draw the proposed road. Continue with the following steps.

6 Zoom in on the Map window so that the highway area is easy to see and work with in the Map window.

7 Select Options | Line Style from the Main menu bar.

8 Set the line Color to red, the line Width wider, and the line Style to a dashed line such that what is shown in the dialog box matches the illustration at right.

9 Click on the OK button. Press the <S> key to toggle on the Snap mode.

10 Select the Polyline tool from the Drawing toolbar.

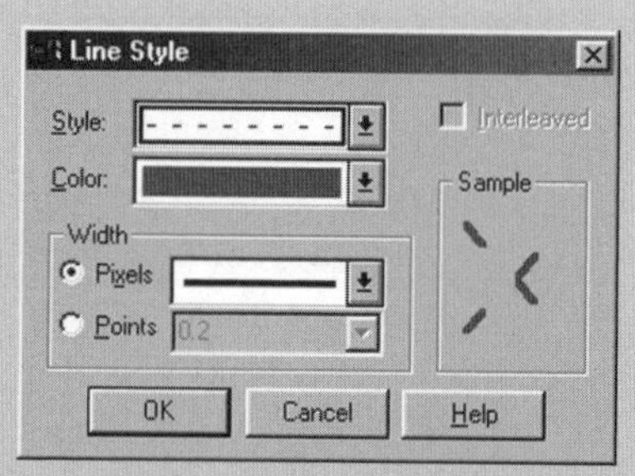

Line Style dialog box for proposed highway symbology.

11 Click with the mouse at an end of the old highway and trace a path through the neighborhood streets to connect to the other section of highway. Your Map window should now display the proposed road, as shown in the following illustration.

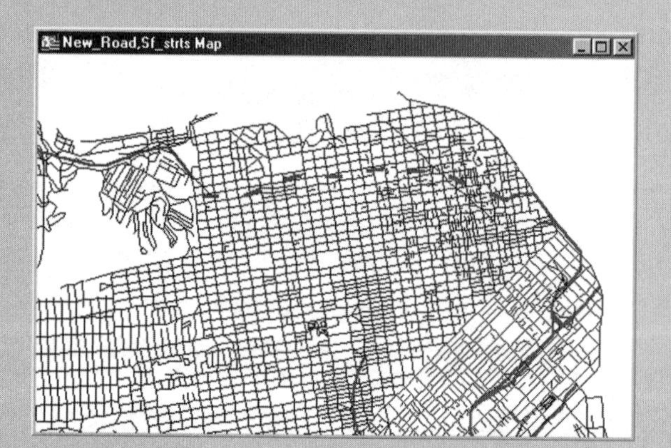

Map window showing the proposed road.

Next, you want to show which roads will undergo construction as a part of the proposed 1999 street improvement plan. Heavy black lines will be used to indicate the roads included in the plan, and these streets will be added to the new *NEW_ROAD* table.

12 Select a few streets from the *SF_STRTS* table.

13 Select Edit | Copy from the Main menu bar.

14 Select Edit | Paste from the Main menu bar. The selected streets will be copied into the *NEW_ROAD* table.

15 Set the symbology of these streets to a heavy black line. Leave the copied streets selected in the *NEW_ROAD* table and select Options | Line Style from the Main menu bar.

16 Set the line Color to black, and the line Width wider. Click on the OK button. The resulting Map window is shown in the following illustration.

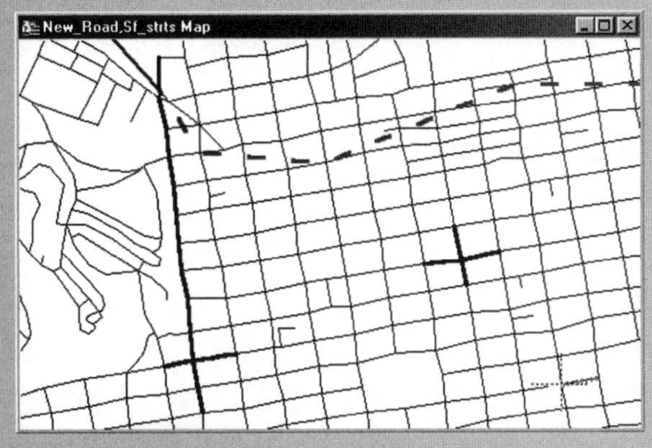

Map window of San Francisco roads to be improved.

At this point, the focus will be switched to another community problem: maintaining voter precincts. As neighborhoods grow and shrink, voter precincts are often split or combined to more easily manage the voting process. Assume that the *SF_GROUP* table polygons represent voter precinct boundaries for the city of San Francisco.

First, select the File | Open Table option from the Main menu bar. Open the *SF_GROUP.tab* table, shown in the following illustration, and zoom in on the northeastern part of San Francisco. Continue with the exercise in the section that follows.

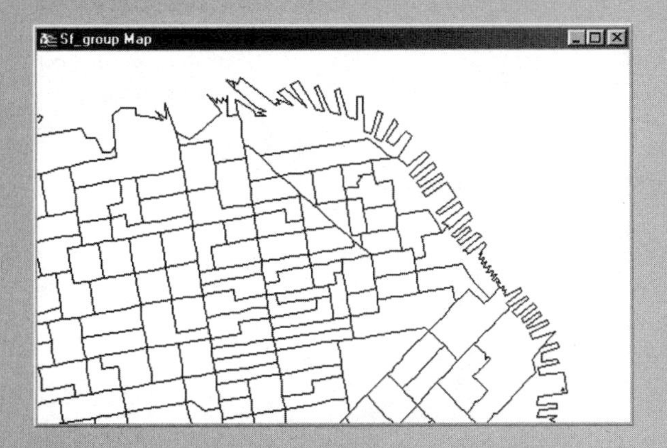

Map window of the SF_GROUP *table.*

Joining Regions

Now you are ready to join certain regions in the *SF_GROUP* table into a single region to enlarge a voting precinct.

1 Select Map | Layer Control from the Main menu bar.

2 Check the editable box to make the *SF_GROUP* table editable.

3 Select three or four of the regions in the central part of the map, as shown in the following illustration.

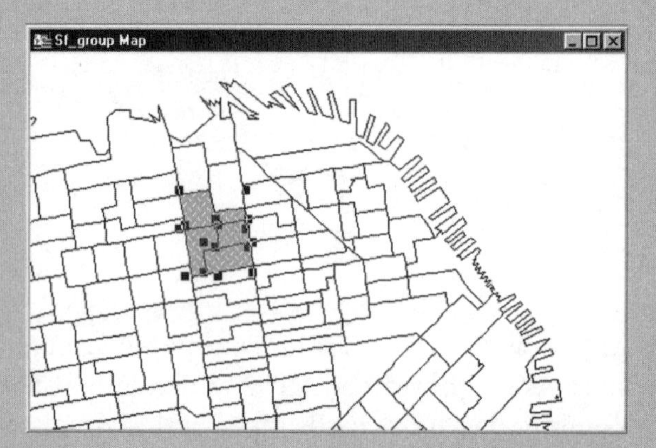

Selected regions in San Francisco.

4 Select Objects | Combine from the Main menu bar.

5 Set the Data Aggregation dialog box to leave the first several columns of the new object blank. The only data columns you really want to maintain are the total population and number of households in the new region. Set the Data Aggregation dialog box as shown in the following illustration.

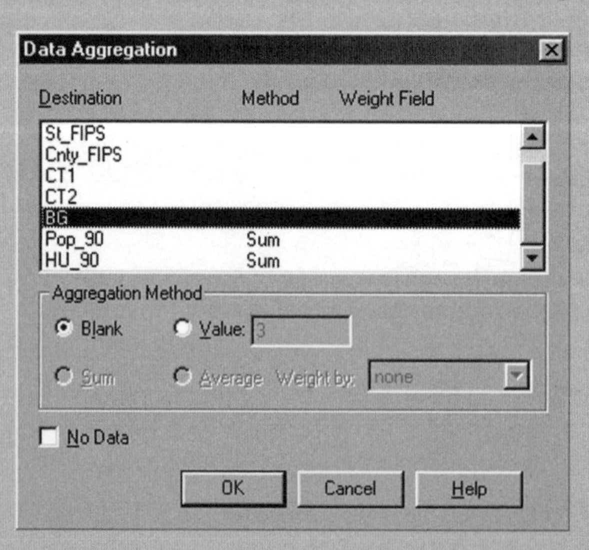

Data Aggregation dialog box.

6 Click on the OK button. The resulting Map window will show a single combined region that replaces the three or four smaller regions. A Browser window

of the *SF_GROUP* table will show the total population and number of households for this new combined region, as shown in the following illustration.

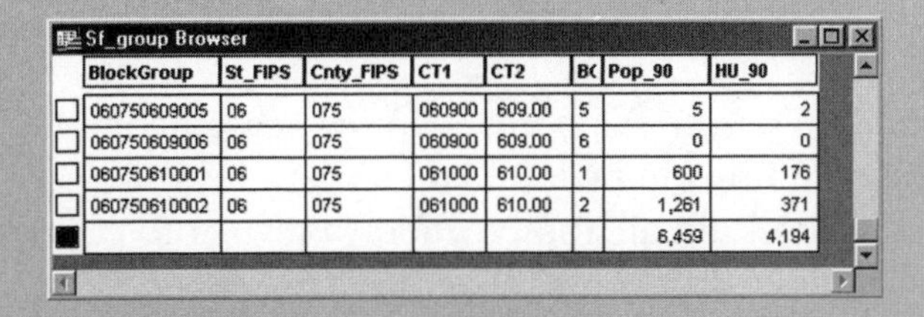
Sf_group Browser

BlockGroup	St_FIPS	Cnty_FIPS	CT1	CT2	B(	Pop_90	HU_90
060750609005	06	075	060900	609.00	5	5	2
060750609006	06	075	060900	609.00	6	0	0
060750610001	06	075	061000	610.00	1	600	176
060750610002	06	075	061000	610.00	2	1,261	371
						6,459	4,194

Resulting Browser for combined voter precinct map.

Splitting polygons is a little more complicated in that you have to decide where you want to split the region, and then draw a shape that will allow a split at the desired location. Continue with the following steps.

7 Zoom in on a single voter precinct you intend to split, and select one of the larger regions in the precinct, as shown in the following illustration.

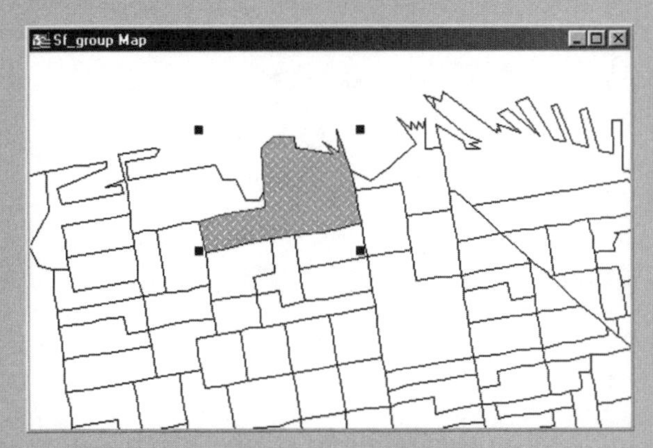

Target voter precinct for splitting.

8 Select the Region tool from the Drawing toolbar and draw a polygon around the right half of the selected region, as shown in the following illustration.

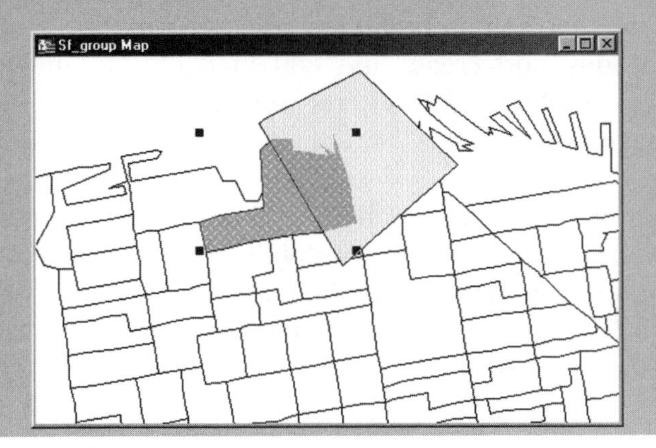

Region to use for splitting voter precinct.

9 Select the region to be split.

10 Select Objects | Set Target from the Main menu bar.

11 Select the newly drawn region to be split.

The data disaggregation will be set the same way as in the region combine example. All data attribute columns will be left blank except for the population and the number of household attributes where you want to calculate the area proportional values. After you click on the OK button, MapInfo will split the region in two pieces and calculate the requested data attributes. The resulting split shape is shown in the following illustration.

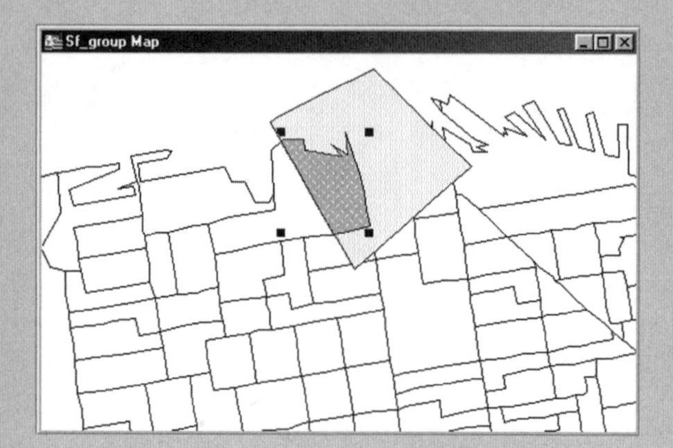

Split voter precinct.

Summary

This chapter focused on creating and editing map objects. These abilities can provide you with valuable information that normally would be very difficult or even impossible to obtain. The Drawing toolbar contains tools and commands used to switch to Reshape mode, add nodes, and create and edit map objects, including drawing lines, polylines, ellipses, regions, arcs, rectangles, and manipulating text.

MapInfo's advanced editing capabilities allow you to build custom objects that can greatly increase your ability to analyze data. The object creation commands Combine, Split, Erase, Overlay, Enclose, and Check Regions greatly enhance your ability to create accurate custom objects.

CHAPTER 8

CREATING GRAPHS

GRAPHS REPRESENT DATA IN INTERESTING and attractive ways that enhance the viewer's ability to understand and evaluate the data portrayed. Graphs can also help you analyze and compare data. When you create a graph based on attributes in a MapInfo table, the values from the table are represented by bars, lines, columns, slices, dots, and other shapes. Groups of data values originating from a single row or column are grouped in data series. Each data series is then distinguished by a unique color or pattern, or both.

By using MapInfo's graph function, you can show trends in data or relationships between different types of data. For example, you can create a graph that shows trends over time in a city's population growth or ethnic composition.

A graph by itself allows you to visualize the statistical relationships in a data set. In MapInfo, the simultaneous viewing of data in Map, Browser, and Graph windows facilitates data analysis. For example, you may wish to compare the population of the three largest cities in the United States. You can use the Map window to select and visualize city locations, and the Graph window to compare respective populations.

Graphs allow you to compare relationships of numerical data. Therefore, the Graph window function is enabled only if at least one table with at least one numeric column is open.

Types of Graphs

Certain issues should be considered when selecting the type of graph to use for displaying data. What relationships are you trying to show? Will the graph display a single set of data, or will it compare multiple data sets? Do you want to illustrate the relationship between parts of a whole? Are you showing growth or change over time? MapInfo supports five main types of graphs, discussed in the following sections: area, bar, line, pie, and scatter.

Area Graphs

Area graphs represent quantities as the size of an area on the graph. This type of graph is similar to a line graph, with the area between the line and the label axis filled. Each filled area on the graph represents a series and is identified by a different color or pattern. Values are plotted on the vertical (Y) axis, and categories are plotted on the horizontal (X) axis. An example of an area graph is shown in the following illustration.

Area graph comparing population of four states.

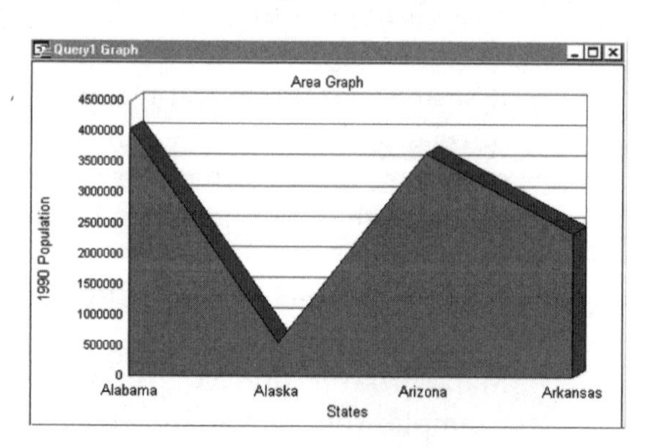

When you select the Stacked option, MapInfo stacks one data series on top of another so that the shaded areas are proportional to data values. A stacked area graph, an example of which is shown in the following illustration, displays a series of data sets successively added together. As you add new data to the area chart, the line representing each set is added to the value of the previous set. Use area graphs to emphasize the relative importance of values over a period of time. An area graph displays the magnitude of change rather than the rate of change.

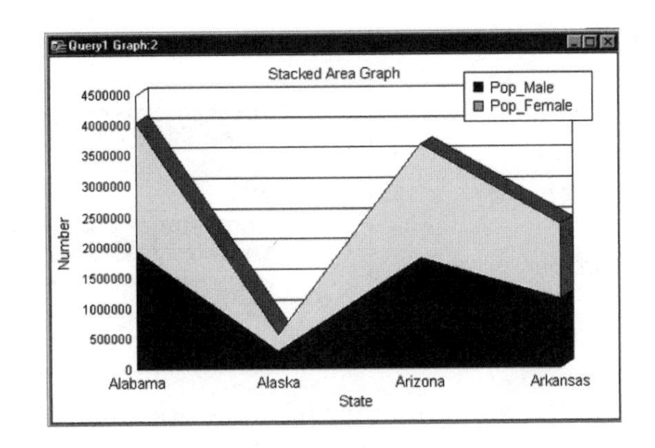

Stacked area graph comparing the male and female populations in four states.

Bar Graphs

Bar graphs, also known as bar charts, are used to compare and contrast a relatively small number of discrete items. Quantities are represented by the length of the bars placed side by side. MapInfo allows you to graph up to any number of variables in a single bar graph. Columns for the variables are positioned side by side unless you choose the Stacked option. An example of a bar graph is shown in the following illustration.

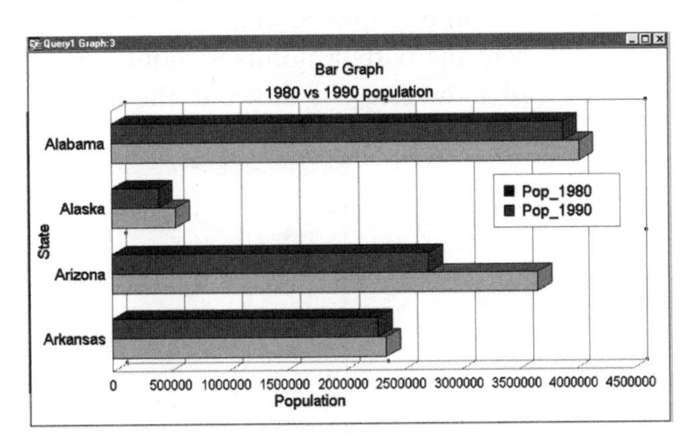

Bar graph comparing 1980 and 1990 population totals in four states.

When you select the Stacked option, MapInfo stacks the bars for each variable. The stacked bar graph is ideal for showing the relative contribution of components to an overall result, such as what proportion of total sales by district are derived, respectively, from service and support, reorders, and new customer purchases. An example of a stacked bar graph is shown in the following illustration. Use bar graphs to compare one item to another, or to com-

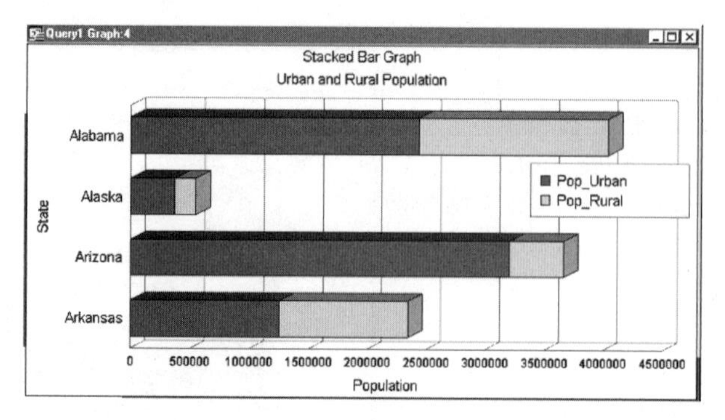

Stacked bar graph comparing urban and rural populations in four states.

pare a number of items over time. These graphs are particularly effective at showing big changes from one category to another.

Line Graphs

Line graphs plot data points at equal intervals along the label axis, and the points are connected with a line. MapInfo allows you to simultaneously plot up to four different variables. Different colored points are automatically used for each variable. To set the colors, use the Series option on the Graph menu.

Use line graphs to depict trends over time. A line chart can represent many data points without cluttering the graph. An example of a line graph is shown in the following illustration.

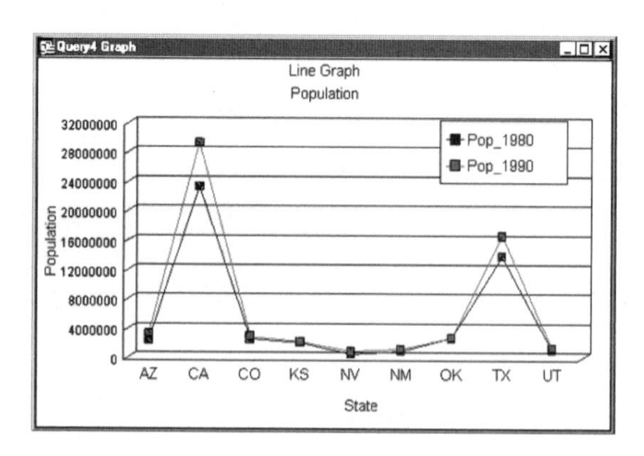

Line graph depicting median income in nine states.

Pie Graphs

Pie graphs, also known as pie charts, show the relative proportions of items, with quantities represented by the area of the pie wedge.

The pie represents a category, and each slice of the pie represents a value in the category. Showing small value differences in a pie graph is difficult. Consequently, you should try to limit the number of wedges in the pie. Negative values cannot be displayed in a pie graph.

In general, you should only use a pie graph if data values add up to the total of a given quantity. For example, if you are graphing sales among territories, you should use a pie graph only if you are graphing sales for all territories. In this way, the size of a territory's wedge accurately represents its contribution to the total sales effort. When you are graphing sales of a portion of territories, say five of 13, you should use a bar chart. A bar chart allows you to compare the sales of selected territories. In contrast to a pie chart, a bar chart would not imply that the selected territories represent the entire sales effort. An example of a pie graph is shown in the following illustration.

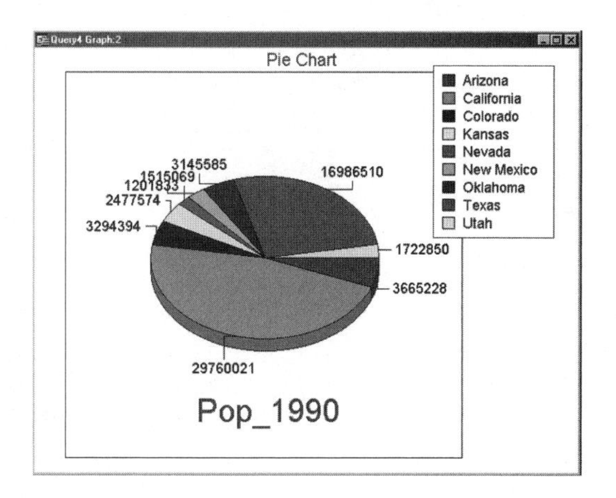

Pie graph showing population by state.

Scatter Graphs

Scatter graphs are also known as scatter plots. These graphs plot points according to X and Y coordinates without connecting lines. Use a scatter graph when you want to examine the correlation between variables. When you specify the columns to be graphed (Window | New Graph Window), the first column is graphed along the X (label) axis, and the second along the Y (value) axis. Use scatter graphs to plot two groups of numbers as a single series of XY coordinates. An example of a scatter graph is shown in the following illustration.

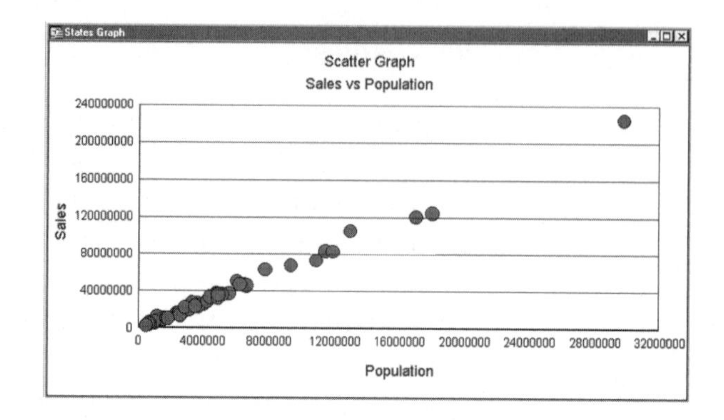

Scatter graph showing sales compared to population by state.

Surface Graphs

Surface graphs plot points according to X, Y, and Z coordinates. A bubble graph will plot the "bubble" at the X and Y location and size the point by the Z value. A surface graph will create a 3D view of the data. An example of a surface graph is shown in the following illustration.

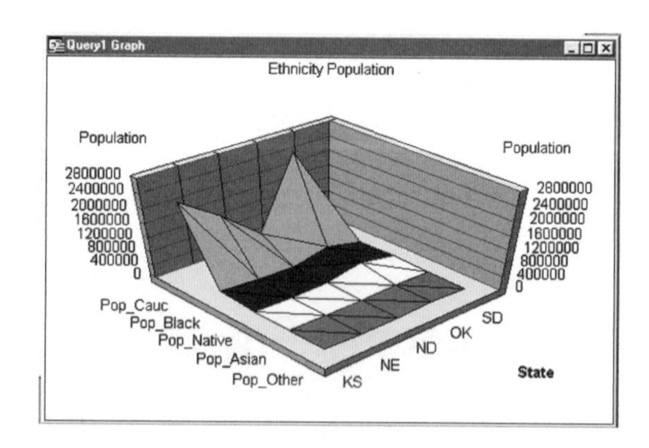

Surface graph showing ethnicity by state.

The following table provides a quick review of all graph types in MapInfo.

Table 8-1: MapInfo Graph Types

Graph	Type	Best Use
3D bar-column histogram	Bar graph	Compares one item to another or a number of items over time.
Area	Area graph	Emphasizes the relative importance of values over a period of time.
Line	Line graph	Depicts trends over time.
Pie	Pie graph	Use if values add up to the total of some quantity. Shows proportion of the whole.
Scatter	Scatter graph	Plots correlation between two numeric variables or data sets.
Bubble surface	Surface graph	Plots correlation between three numeric variables or data sets.

Creating Simple Graphs

Creating graphs with MapInfo is quick and simple. In tutorial 8-1, which follows, the *STATES* table is used to create a graph comparing the 1980 and 1990 populations of Alabama, Florida, Georgia, Mississippi, and South Carolina.

▼ *TUTORIAL 8-1: CREATING A SIMPLE GRAPH*

1 Select AL, FL, GA, MS, and SC, as shown in the following illustration.

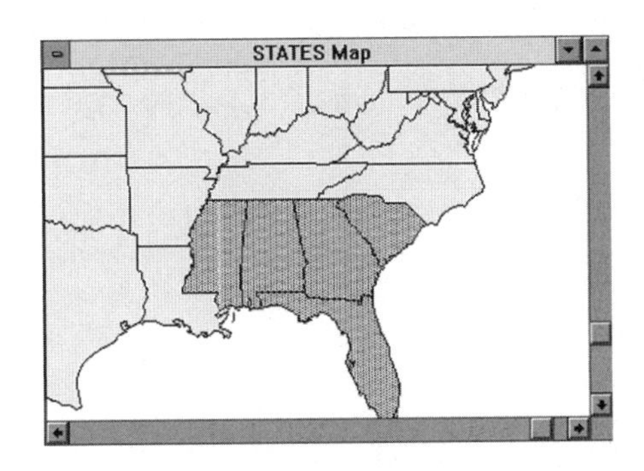

STATES map with AL, FL, GA, MS, and SC selected.

2 Select Window | New Graph Window from the Main menu bar.

3 From the Graph Wizard dialog box, shown in the illustration at right, select to create a bar graph with a clustered template.

4 Set the New Graph Window dialog box to graph the *Pop_1980* and *Pop_1990* columns of the table selection. Select to create a column graph and to label with the *State* column, as shown in the following illustration.

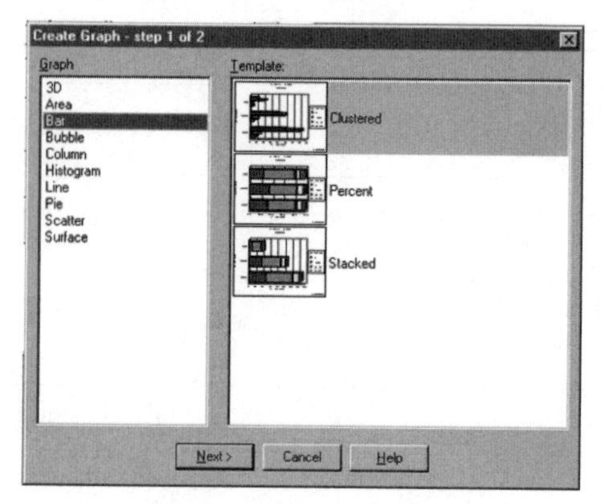

New Graph Window dialog box.

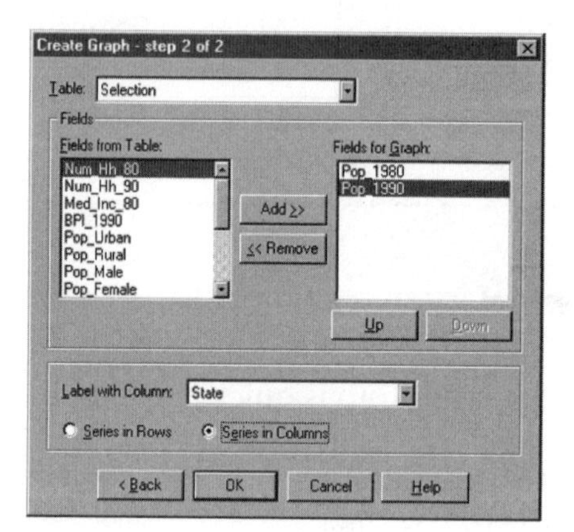

Settings established in the New Graph Window dialog box.

5 Click on the OK button. The bar graph window shown in the following illustration will display.

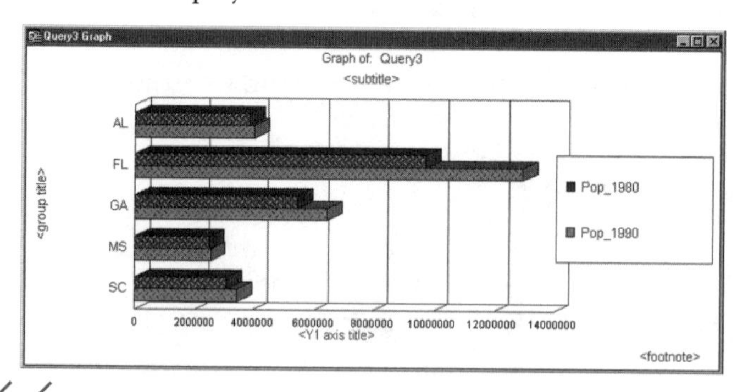

Resulting bar graph comparing 1980 and 1990 populations.

TIP: *Note that in the preceding example a graph was created based on only five states. When building a graph, a large number of rows is difficult to display. Carefully select your graph type and your select statements*

to build meaningful graphs. To demonstrate this problem, the following illustration shows how MapInfo handles graphing a population comparison of all 50 states.

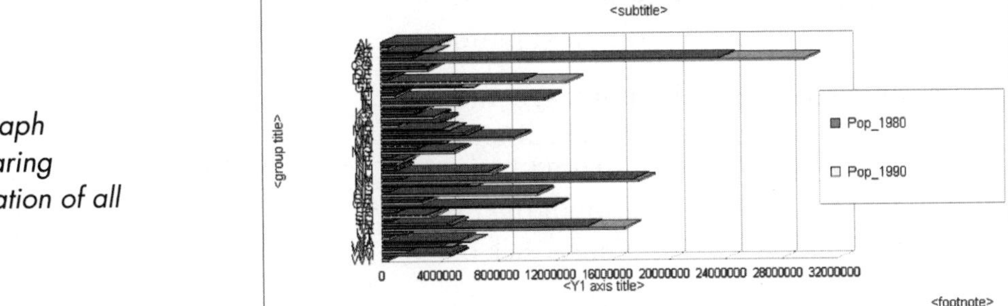

Bar graph comparing population of all states.

Using the Graph window and some of the SQL Select techniques reviewed in Chapter 4, you can graph more than just the attributes of selected data. Using the SQL Select dialog box, shown in the following illustration, you can identify all western U.S. states with a population greater than 2 million. In tutorial 8-2, which follows, you will compare sales per square mile in the western states.

SQL Select
Select Columns: State, State_Name, Sales_1990 / Area(obj, "sq mi")
from Tables: States
where Condition: CentroidX(obj) < -95.0 And Pop_1990 > 2000000
Group by Columns:
Order by Columns:
into Table Named: Selection
Browse Results
Tables | Columns | Operators | Aggregates | Functions
Save Template | Load Template
OK | Cancel | Clear | Verify | Help

SQL Select dialog for selecting western states with populations greater than 2 million.

▼ TUTORIAL 8-2: USING SQL SELECT AND GRAPHS

1 Build the SQL Select statement to select the data you wish to graph. Set the dialog to select the *state* and *state_name* columns, as well as to derive a "sales per square mile" column. The *Where* clause is used to select the western states with a population in excess of 2 million. The dialog should be constructed as seen in

the previous illustration. The results of the SQL selection criteria should be eight states.

2 Select Window | New Graph Window. Set the Graph dialog to build a clustered bar graph, as shown in the illustration at right. Click on the Next button.

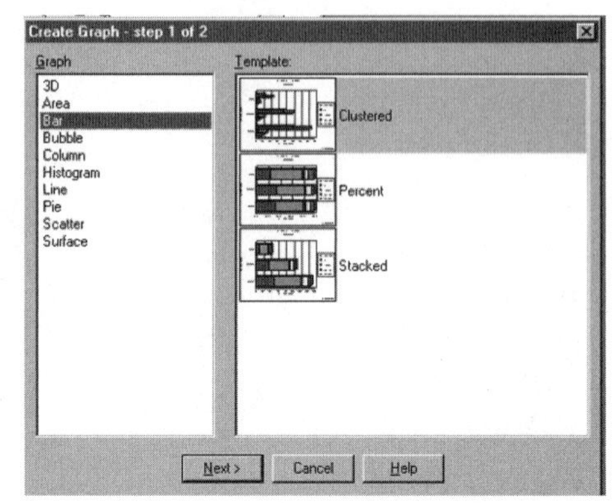

Graph wizard to build bar graph from selection.

3 Set the second graph wizard screen with the SQL result table and the column of data derived in steps 1 and 2. The dialog should look like that shown in the first of the following illustrations. Click on the OK button. The resulting graph is shown in the second of the following illustrations.

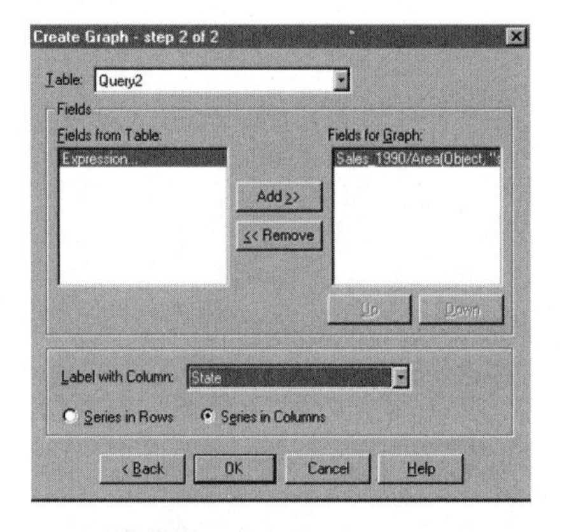

Graph wizard table and column selection dialog.

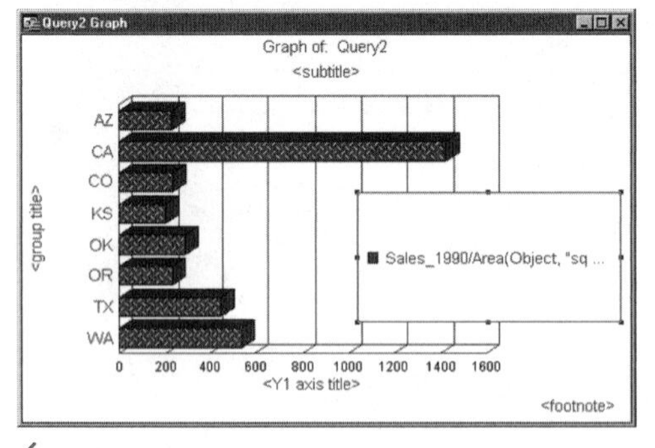

Graph of sales per square mile in western states.

***TIP:** The composition of effective graphs requires that you perform many SQL Group By commands to consolidate data into fewer rows of data to*

be graphed. For instance, to analyze the total population of the major cities in a state, you would issue the following SQL Select statement: Select Sum(Tot_pop) from City_1k Group By State. *This Select statement reduces the rows in the* City_1k *table from 1,000 to 51. Because 51 rows still exceed the number of rows effectively graphed with MapInfo, you should further consolidate the data before constructing a graph.*

Customizing Graphs

Once a graph has been generated, you will want to enhance its appearance. When you select the data to display in a Graph window, MapInfo will automatically place the data in a bar graph with default display settings. You should first consider whether the type of graph is the correct one for the type of data to be displayed.

The means of customizing a graph to add emphasis and clarity are numerous. MapInfo allows you to modify titles and borders, select fonts and shading, and adjust scaling and data gaps through use of the Graph menu, shown in the following illustration. Methods of graph customization are discussed in the following sections.

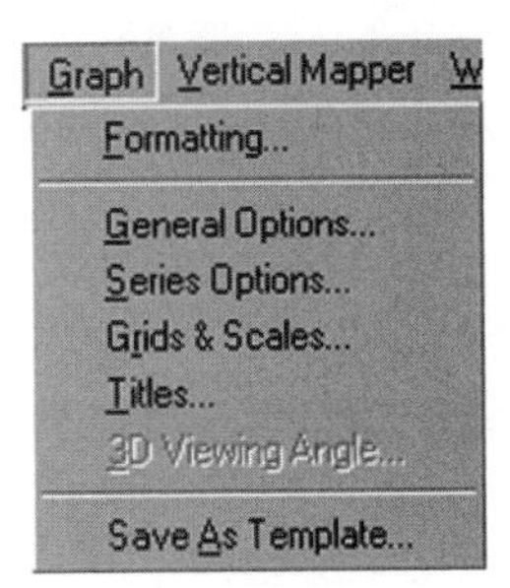

Graph menu.

Formatting

Create a simple graph with data from the *states* table, as you did earlier in this chapter. Once this graph has been created and displayed on the screen, the Graph window will be available on the MapInfo Main menu bar.

The first option on the Graph menu allows the user to control formatting of the various graph objects. Selecting the Graph | Format menu option will display the Graph Formatting dialog box, shown in the following illustration.

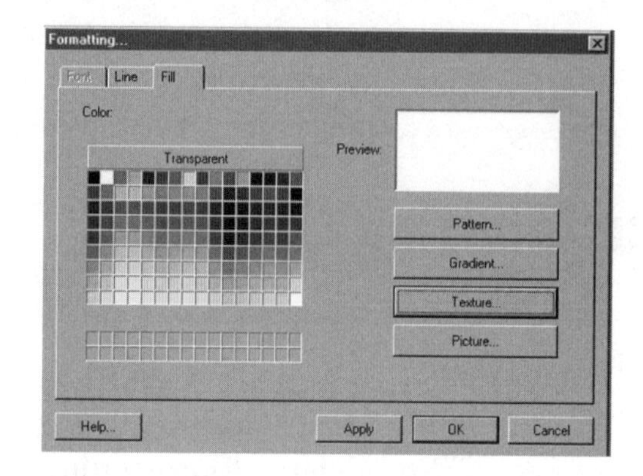

Graph Formatting dialog box.

This formatting dialog allows the user to alter the Font, Line, or Fill style of any of the graph items. For example, select the bars on the graph, and you can change the color of their fill or the weight of their lines. The graph shown in the following illustration has had some of the color fills, line weights, and fonts altered to add emphasis.

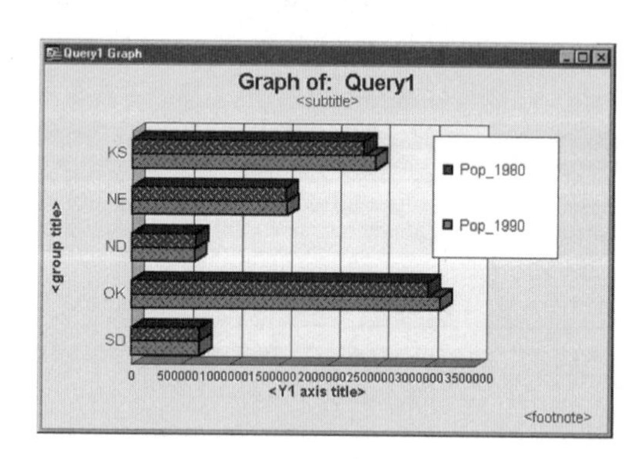

Graph window with some altered format items.

General Options

The next Graph menu item allows you to change the general options. After a graph has been built, selecting Graph | General Options from the MapInfo Main menu bar will display the General Options dialog box, shown in the following illustration.

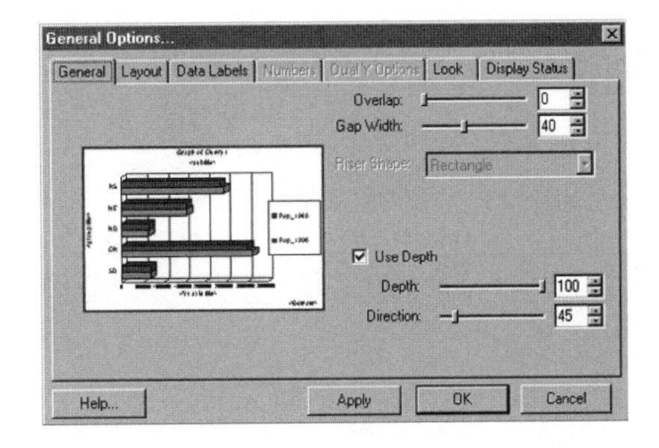

General Options dialog box.

The general options allow you to adjust the basic appearance of the graph. The layout and display of labels may be adjusted. The actual items you will be able to adjust will vary, depending on what type of graph you are displaying. The graph shown in the following illustration has an adjusted gap width, labels added to the graph bars, and legend displayed as a box with beveled edges.

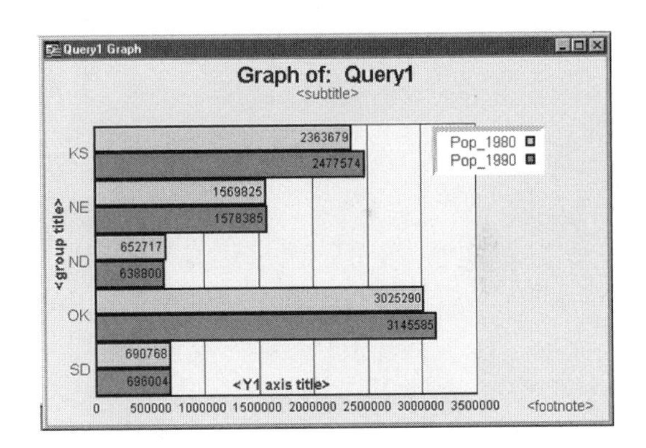

Graph with some adjusted general options.

Series Options

The Series Options dialog box is used to apply formatting options to an individual series in a graph. The Series Option in the Graph menu will only be available for selection if you previously selected a series in your graph. If a series is not selected, a message box displays this message: *Please select the series that changes should be applied to.*

To display the General Options dialog box, select a riser (area, bar, line, marker, pie slice, and so on) or a legend marker, and select Graph | Series Options from the MapInfo Main menu bar. The General tab shows general formatting options for a series for each Graph type. The Series dialog box is shown in the following illustration.

- The Data Labels tab shows formatting and display options for data labels for the currently selected series.
- The Number tab shows formatting options for data label numbers. This tab is only activated when data labels are selected in the Data Labels section of Graph Options.
- The Trendlines tab shows formatting and display options for trend lines. This selection is not available in 3D and pie graphs.

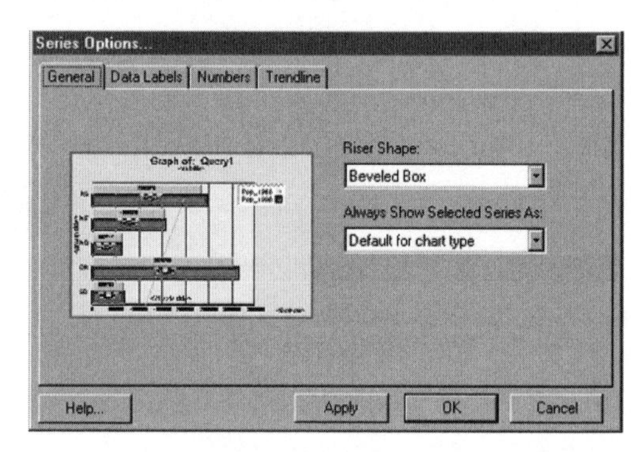

Series dialog box.

The graph shown in the following illustration has a style of the series adjusted to be a beveled edge. The series also has the number centered in the bar.

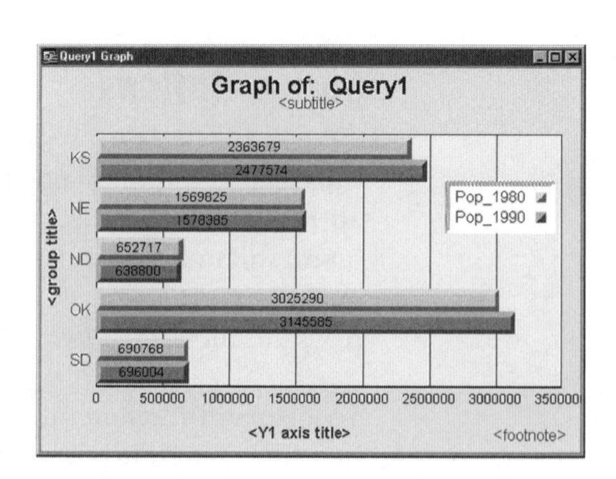

Graph with some adjusted series options.

Grids and Scales

In the Grids and Scales dialog box, shown in the following illustration, you can format all of the axes in your graph. The tabs on the left of the dialog show the available axes in the graph: Category Axis, Y1 Axis, Y2 Axis (for dual-axes graphs), X Axis (for bubble and scatter graphs), and Series Axis (for 3D graphs). When you select an axis tab, the tabs at the top of the dialog will also change to represent the formatting options for each axis.

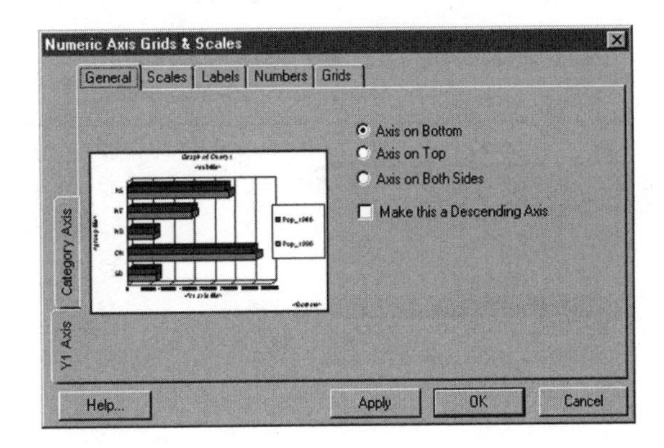

Grids and Scales dialog box.

The Grids and Scales dialog box will contain different options, depending on which axis and formatting tab is selected. The graph in the following illustration shows some adjustments to the grid lines and grid titles.

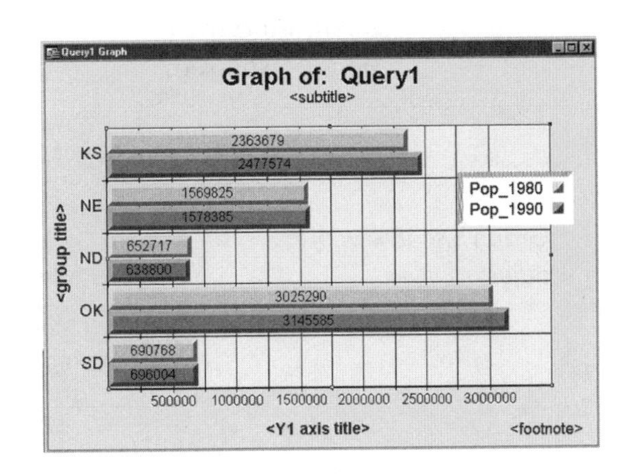

Graph with adjustments made to grid lines and titles.

Titles

Let's go change those default titles and really make this graph window make sense. All graphs can include a graph title, subtitle, and footnote. Most graphs can also include a Category Axis title and a Numeric Y1-Axis title. Other axis titles may be available, depending on the graph type. The Titles and Labels dialog box, shown in the following illustration, can be used to add, change, or delete a Title, Subtitle, Footnote, Category Title, X-Axis title, Value Title (Y1 and Y2), and a Series Axis title in your MapInfo graph.

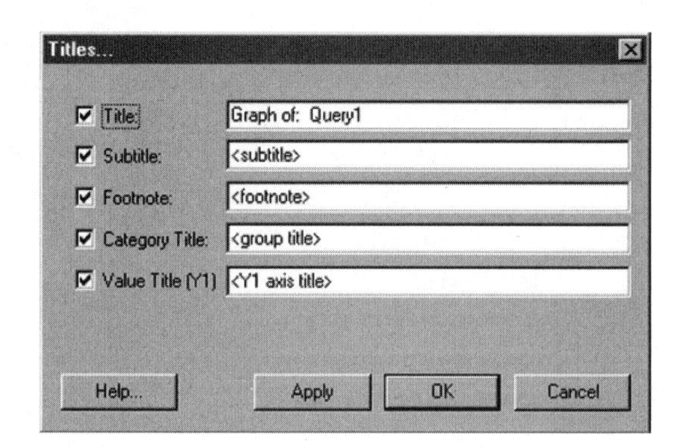

Titles and Labels dialog box.

The graph window shown in the following illustration displays a more appropriate title and subtitle for the graph you have been working with in this section. The axis titles also now represent what is being displayed on each axis. In addition, the footnote has been altered to give credit to the originating source of the data being displayed.

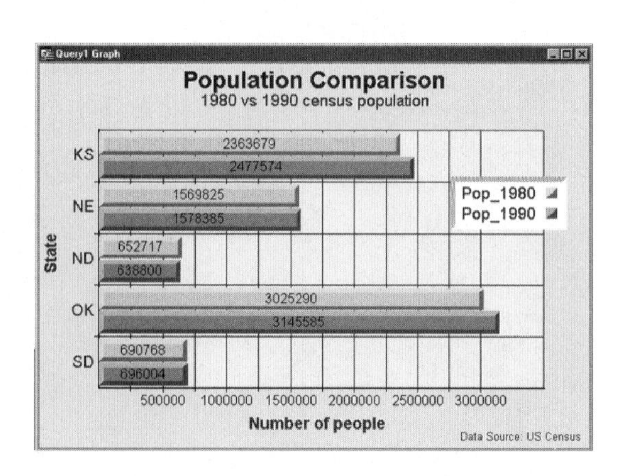

Graph window with adjusted titles.

3D Viewing Angles

The 3D Viewing Angle dialog box is accessed by selecting Graph | 3D Viewing Angle from the Main menu bar. This function allows you to rotate, move, or resize the frame of a 3D graph.

The graph you have been working with throughout this section is not a 3D graph, so this menu item is grayed out. The following illustrations shows exactly the same graph created in the 3D bar graph template.

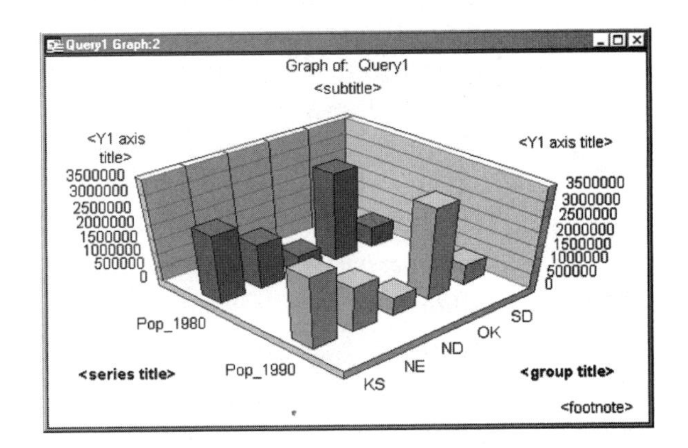

3D bar chart graph.

Once you have created a 3D graph, the 3D viewing angle menu item is enabled. The 3D Viewing Angle dialog box, shown in the following illustration, allows you to adjust your 3D graph, with control over the twisting and turning of the graph.

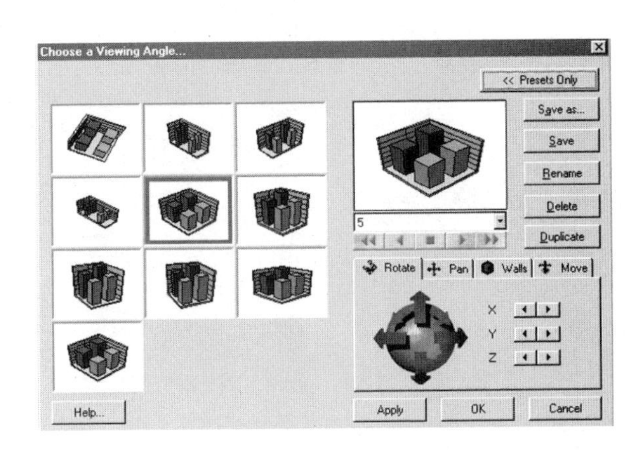

3D Viewing Angle dialog box.

Save As Template

Do you have a graph you create often or a specific color scheme you always want to use for your graphs? If so, the Save As template is just what you are looking for. Using this template, you can create your graph and make all the alterations to the style and appearance of the graph.

Now you are ready to create your own template file. With this graph in the active window, select the Graph | Save As Template menu item. The Save Graph Template dialog box will be initialized to the template folder that corresponds to the type of graph being saved. The next time you create a graph window, the saved template file will be available. In exercise 8-1, which follows, you will create and customize a graph.

TIP 1: *If you are using another software package for visualizing data, you can export a MapInfo file to that program. The data can be exported as an ASCII file in comma-delimited format or as a DBF file.*

TIP 2: *To remove titles from the graph, delete the text in the Graph window. If you intend to print the graph, you may wish to remove the titles and annotate the graph with the greater flexibility allowed in the Layout window. The Layout window is discussed in Chapter 9.*

■ EXERCISE 8-1: GRAPH CREATION AND CUSTOMIZATION

In this exercise, the San Francisco area is analyzed to create a mailing list campaign for the San Francisco Opera House. Use the San Francisco area tables and the *SF_CUSTD* table in the *samples* directory on the companion CD-ROM. The Opera House is located in the *SF_LANDM* table, and the *SF_CUSTD* table contains a list of companies that support the Opera House.

Setting Up the Map Window

1. Select File | Open from the Main menu bar.
2. Select *SF_STRTS.tab*, *SF_ZIPS.tab*, *SF_CUSTD.tab*, and *SF_LANDM.tab* in the File | Open Table dialog box, shown in the following illustration.

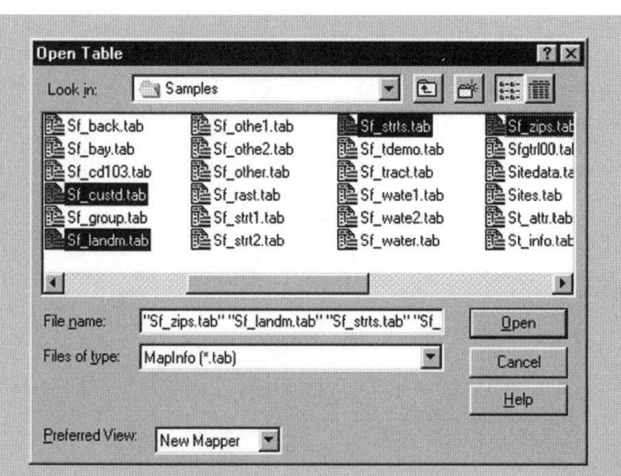

Open Table dialog box for SF_STRTS, SF_ZIPS, SF_LANDM, *and* SF_CUSTD *tables.*

3 The Map window in the illustration at right shows the tables to be analyzed. To locate the Opera House, select Query | Find from the Main menu bar.

4 Set the Find dialog box to find a name in the *SF_LANDM* table, as shown in the following illustration.

5 Click on the OK button. Type in *San Francisco Opera House* as the name to be found.

6 The resulting Map window will show the location of the Opera House in relation to the objects in the *SF_CUSTD* table, as shown in the following illustration.

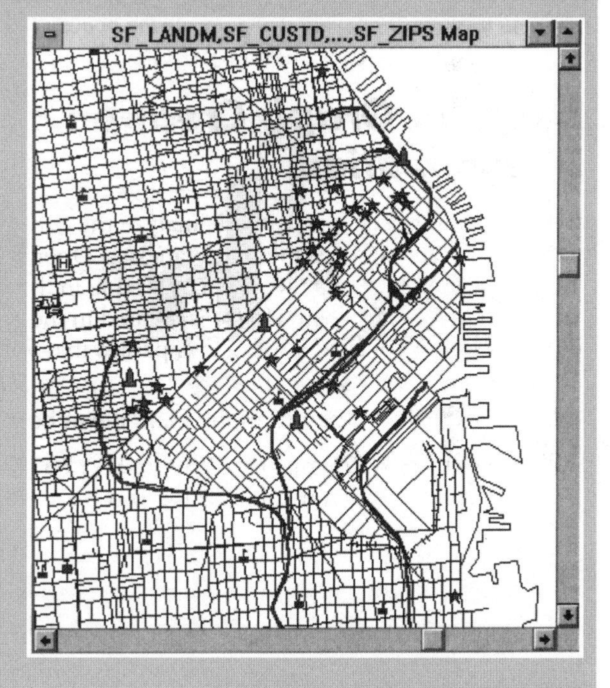

Map window of the four tables.

Find the San Francisco Opera House.

Map window showing location of the San Francisco Opera House.

To begin analyzing the companies that support the Opera House, they will be aggregated by zip code. With only 50 points or companies in the *SF_CUSTD* table, this step may not seem significant. However, if you were to examine thousands of points, analysis would quickly become cumbersome. Gathering the data into geographic regions makes trends or profiles easier to detect.

Aggregating Data by Zip Code

Through a Browser window of the *SF_CUSTD* table, note that most of the zip codes are five digits, and that a few rows contain a nine-digit zip code, as shown in the illustration at right. To successfully aggregate the data by zip code, all zip codes in the table must be standardized to five digits.

zip_5 Browser

Company	zip_code
Cantelope Software	94102
Ranstrom Financial Services	94102
Tohn Engineering	94102
WHKG Radio	94102
Touster Medical Supply	94102
Pacific Union Resdential Broker	94102
Hidden Grove Vinyards	94102
San Francisco Parking Authority	94102
Peterson Entertainment	94102
Quick Change Oil	94102
Walker Publishing	94103
Sunset Tours	94103
S.F. Tribune	94103

Browser window of selected five-digit zip codes.

1 Select Query | SQL Select from the Main menu bar.

2 Build an SQL Select statement to select the company name and the leftmost five digits of the zip code column. The five digits are used to create a temporary working table named

ZIP_5. The SQL Select dialog box containing the statement is shown in the following illustration.

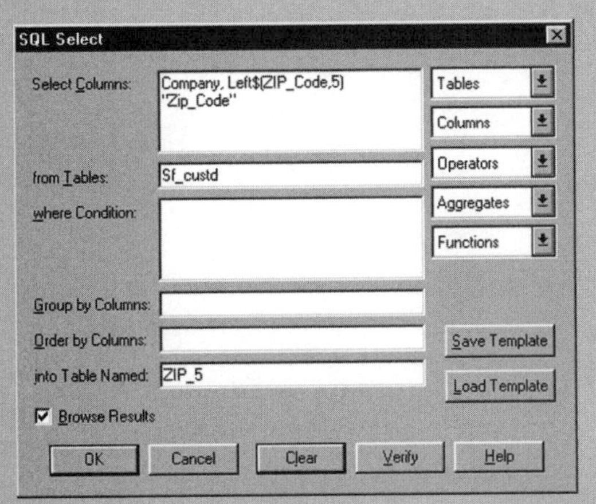

SQL Select dialog box for standardizing all zip codes to five digits.

3 At this point, assume you wish to know the number of companies within each zip code in the *zip_5* table. Select Query | SQL Select from the Main menu bar. Select *Zip*_Code and the *Count(*)* aggregate function from the *zip_5* table. Select Group By on the *Zip_Code* data column. The query statement is shown in the illustration at right.

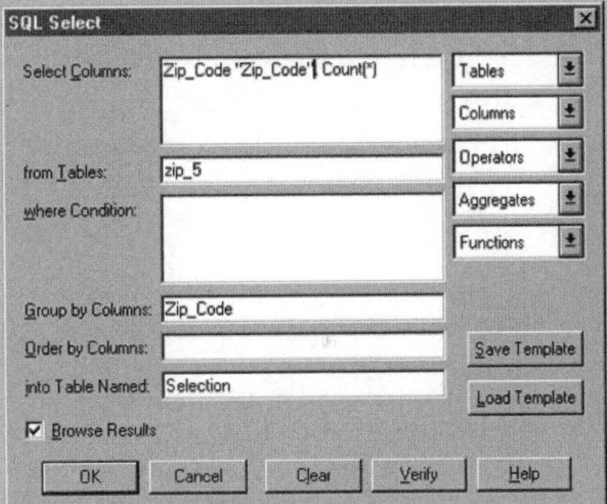

SQL Select dialog box for aggregating companies by zip code.

The resulting table shows the six zip codes in the sample data, along with a count of the supporting companies located in each zip code area. The data preparation phase that precedes graph creation is almost always relatively time consuming. You need to understand the data before you build the graph.

Building the Graph

At this point, you know that several companies supporting the Opera House are located in each zip code area. You are ready to build a graph to visualize the results of companies supporting the Opera House.

1 Select Window | New Graph Window from the Main menu bar.

2 Set the Graph Window dialog box to create a column graph with the Clustered template, as shown in the following illustration. Then click on the Next button.

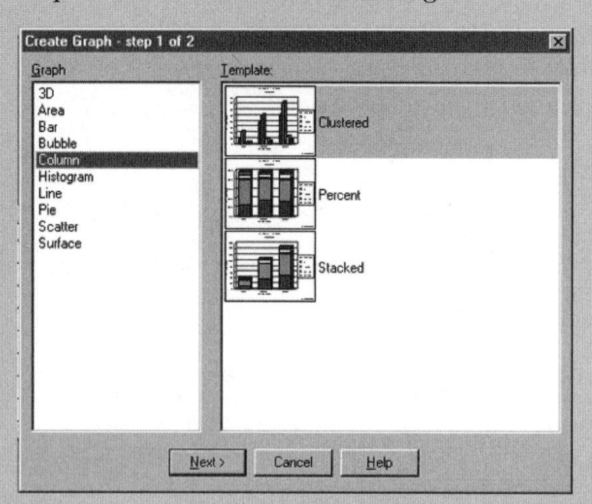

First dialog box for creating new graph window.

3 In the step 2 dialog, shown in the illustration at right, set the table to be graphed as the one you created in the previous sequence of queries. You will be graphing the number of companies in each zip code.

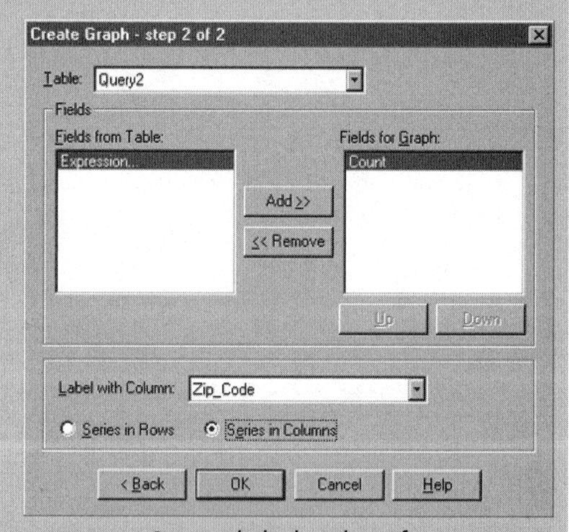

Second dialog box for creating new graph window.

4 Click on the OK button. The resulting graph, shown in the illustration below, displays the number of companies supporting the Opera House in the surrounding zip codes.

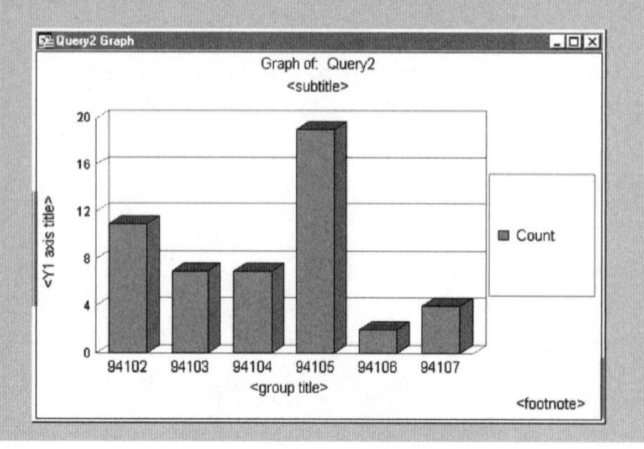

Graph showing the count of companies in each zip code.

5 Give the graph a meaningful title. Select Graph | Titles from the Main menu bar. This will access the Graph Type dialog box, shown in the illustration at right. Next, type in *Opera House Support* as the new title. Finally, set new axis labels.

6 Click on the OK button. The resulting graph is shown in the following illustration.

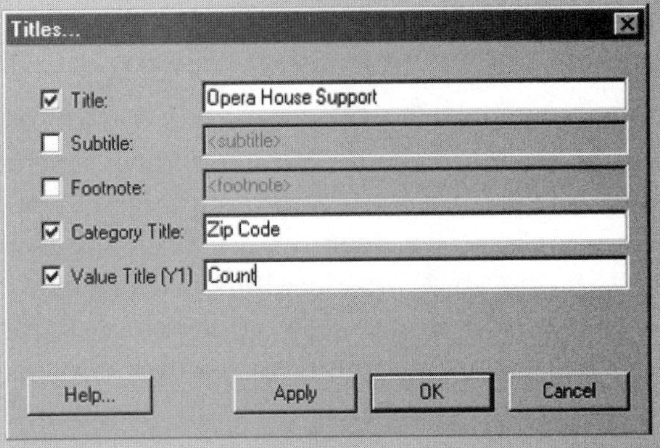

Graph Type dialog box.

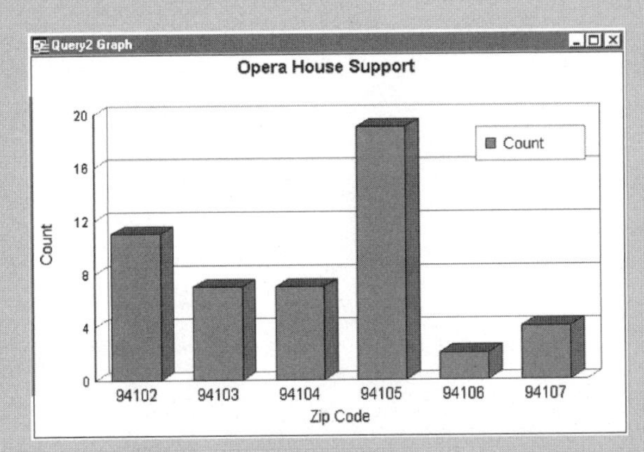

Resulting customized graph.

If you were to continue working with the Opera House mailing list task, you could examine company profiles and the zip code demographic profiles of the people who are current supporters of the Opera House. For instance, you could compare the dollar amount of support in each zip code area in relation to the number of companies that contributed. By analyzing this type of information, you could determine the profile of companies most likely to respond to a fund-raiser mailing campaign.

In contrast, you could also compose a profile of the companies that do not respond to a standard mailing. Companies who fit the latter profile could then become targets for an alternative mailing campaign. Once you have drawn conclusions about the type of mailing campaign to recommend, you would use a combination of Map, Browser, and Graph windows to create a final presentation for the San Francisco Opera House.

Summary

You can expand your data analysis capabilities by using graphs alongside the browse, selection, and mapping features covered in previous chapters. The following illustration shows how the four features are used in concert. Assume you wish to compare the populations of Kentucky's major cities. The Map window is used to select all cities in the state of Kentucky. Next, the Browser window is used to select major cities, along with respective population counts. Finally, the Graph window is employed to compare the populations of the major cities.

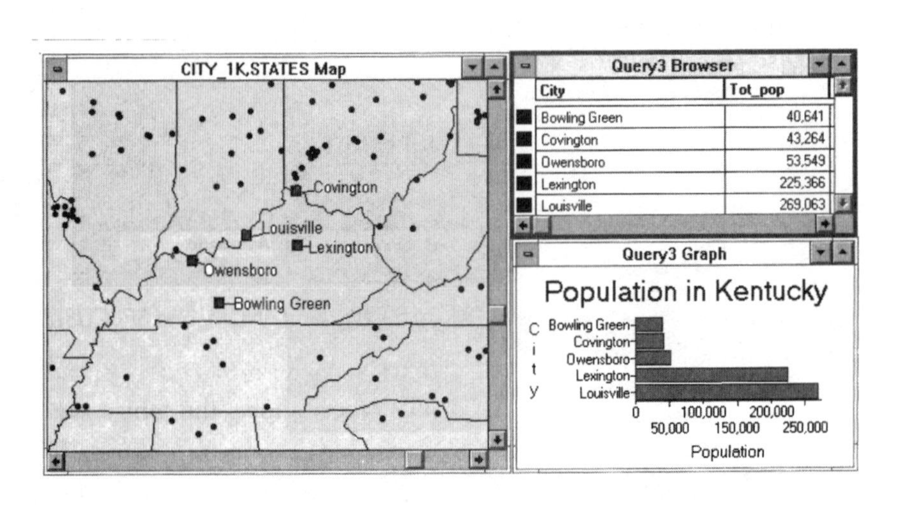

Selecting, browsing, mapping, and graphing the populations of Kentucky's major cities.

CHAPTER 9

PREPARING HARDCOPY OUTPUT

NO MATTER HOW WONDERFUL THE SOFTCOPY DISPLAY of maps and data, you will always need printed material. The combination of MapInfo Professional and native operating system print capabilities makes it easy to create hardcopy output. However, creating attractive hardcopy output requires some time and effort. This chapter reviews options for preparing printed maps.

Print Setup

Literally hundreds of printers appropriate for printing MapInfo Professional data are available. MapInfo is device independent and should print with any printer make and model as long as it has a print driver. To specify the device to which you want output directed, select File | Page Setup from the Main menu bar. The Page Setup dialog box, shown in the following illustration, shows the page layout for your current or default printer. You can select an alternate printer by pressing the printer button at the bottom of the dialog.

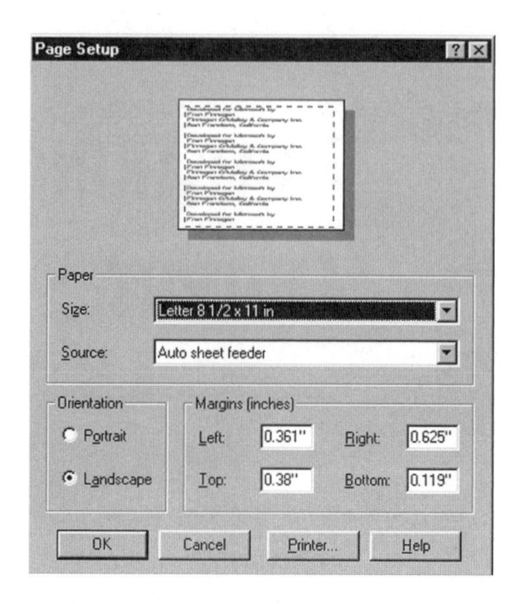

Page Setup dialog box.

The Printer Page Setup dialog allows you to select another printer, as shown in the following illustration. The dialog also allows the user access to the various printer properties as provided by the printer's device drivers.

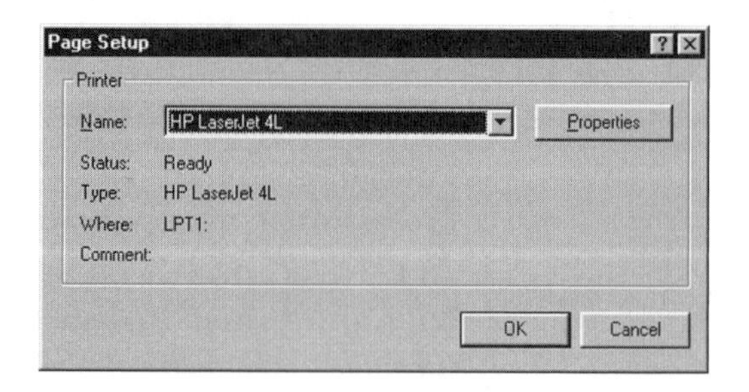

Printer section of Page Setup dialog box.

Printing a Window

All MapInfo Professional windows (Map, Browser, or Graph) are easy to print. To prepare a window for printing, size it according to how you want it to appear on the printed page. For example, landscape orientation is usually appropriate for a Map window that is wider than it is tall.

Once the window is ready to print, select File | Print from the Main menu bar. The Print dialog, shown in the following illustration, will display, describing the hard copy you are about to print.

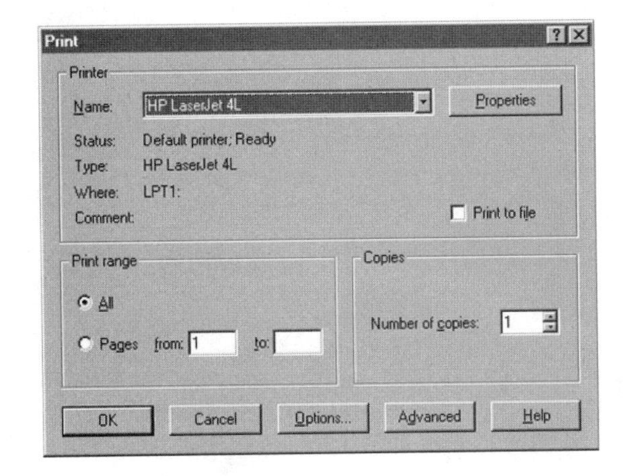

Print dialog box.

The Print dialog allows you to specify the number of copies and the pages to be printed. The Options button accesses the Map Print Options dialog, shown in the following illustration. This dialog is set with initial values, but you can customize the map content, size, and scale.

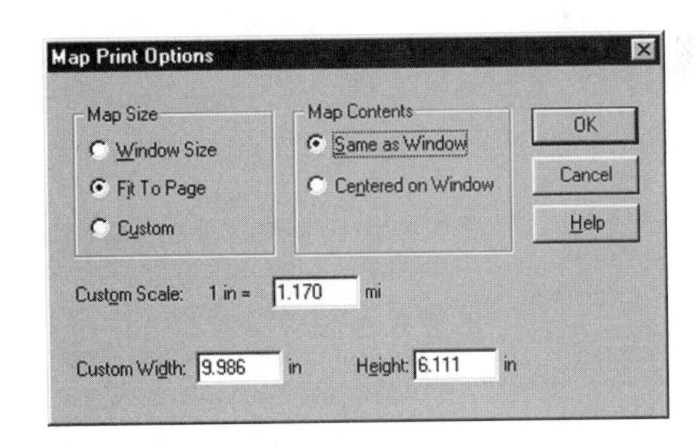

Map Print Options dialog box.

The Print dialog also has an advanced button across the bottom of the dialog. One option in this dialog, shown in the following illustration, allows you to print using MapInfo's enhanced metafile handling. Other options allow you to specify special handling of the transparency printing issues.

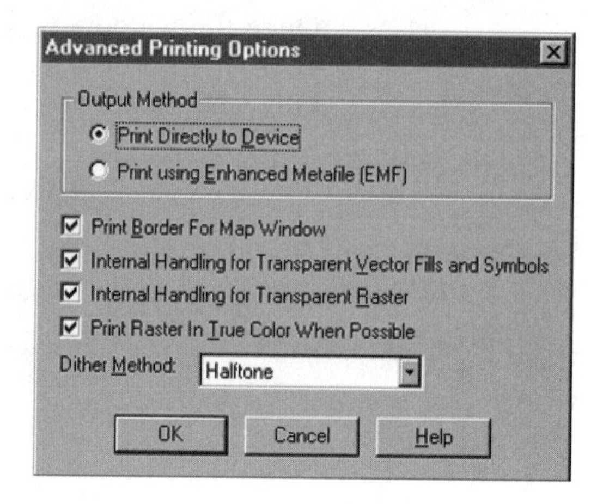

Map Print Advanced Printing Options dialog.

Printing draft copies of a Map window is typical during project analysis. For attractive maps and data presentation, however, you will want to use the Layout window.

TIP: *To create electronic output, use File | Save Window As. With this command, you can save the content of a window in a graphics format appropriate for the platform you are using. These electronic window files can then be imported into presentation managers such as PowerPoint or Microsoft Word for output preparation.*

Layout Window

The Layout window allows you to arrange multiple MapInfo Professional windows for preparing a single hardcopy output. Map, Browser, and Graph windows (along with additional elements such as titles, legends, scale bars, and north arrows) can be arranged in the Layout window. This window allows you to design a map and preview its appearance before you send it to a specific printer or plotter.

In the Layout window, you can assemble window elements through the use of frames. A frame is a container object that is sized and positioned on a Layout window and then filled with a Map, Browser, Graph, Thematic Legend, or other window.

Layout with a Single Map Window

In tutorial 9-1, which follows, you will work with the Layout window by placing the *States* table into a Layout window and preparing it for printing.

▼ TUTORIAL 9-1: MANIPULATING A TABLE IN THE LAYOUT WINDOW

1 Open the *States* table in a Map window using the File | Open Table option on the Main menu bar.

2 Select Window | New Layout Window from the Main menu bar. The New Layout Window dialog box, shown in the illustration at right, will appear.

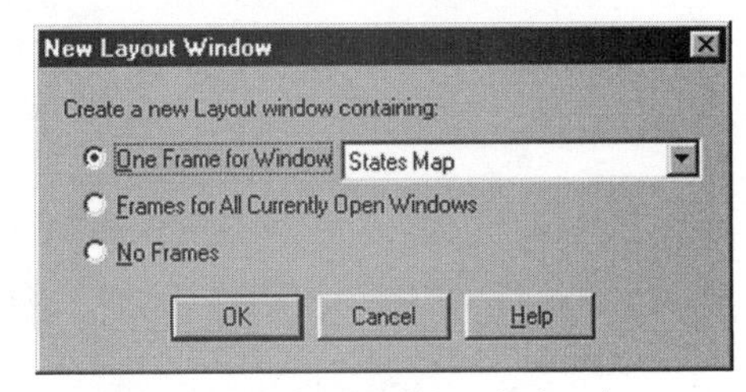

New Layout Window dialog box.

3 Because only one window is open, selection of the one Map window or all currently open windows options does not matter. Press the OK button and MapInfo will build the default Layout window. Remember that the size and shape of the Layout window are based on the print setup, discussed previously.

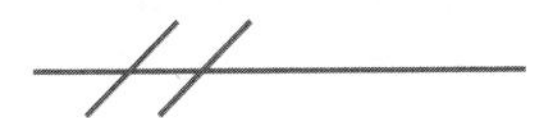

A review of the available Layout menu options and tools on the toolbars will indicate that you can zoom in and out on the Layout window. Using the View Entire Layer command is recommended for a preview of the printed output. The Layout window for the *States* table is shown in the following illustration.

TIP: *Use the Layout | View Actual Size function to ensure that the font sizes in the Layout window are appropriate. Some fonts (particularly in the Legend window) typically overrun the window borders, yet will not appear to overrun unless viewed at full size.*

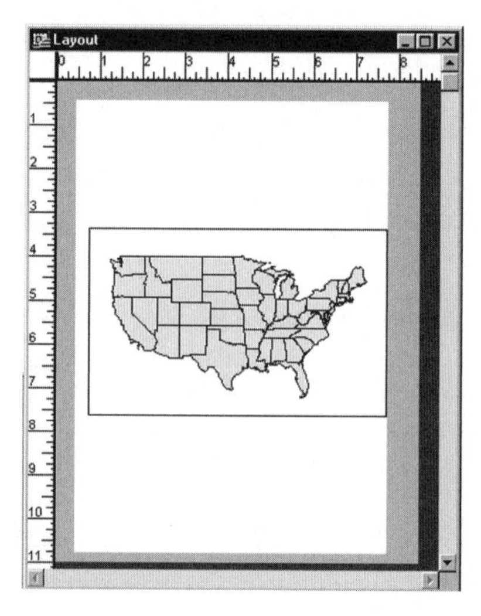

Layout window of the States *table.*

In tutorial 9-2, which follows, you will make a printed copy of the *States* table that is twice as large as the one created in the previous steps. In this tutorial, you will be using the Layout Options panel.

▼ Tutorial 9-2: Using the Layout Options Panel

1 Select Layout | Options from the Main menu bar.

2 In the Layout Display Options dialog box, shown in the illustration at right, set the width at 2 pages.

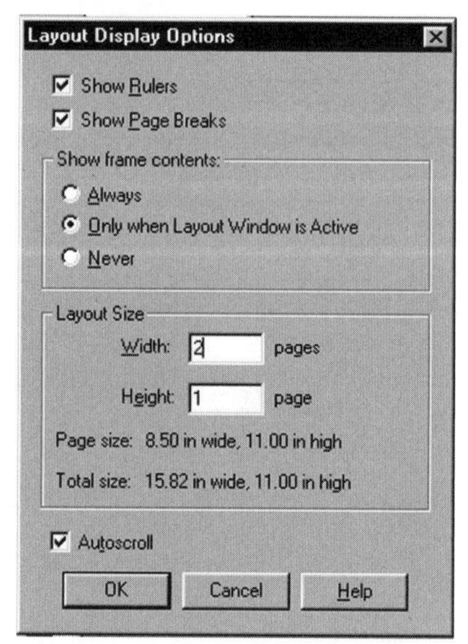

Layout Display Options dialog box.

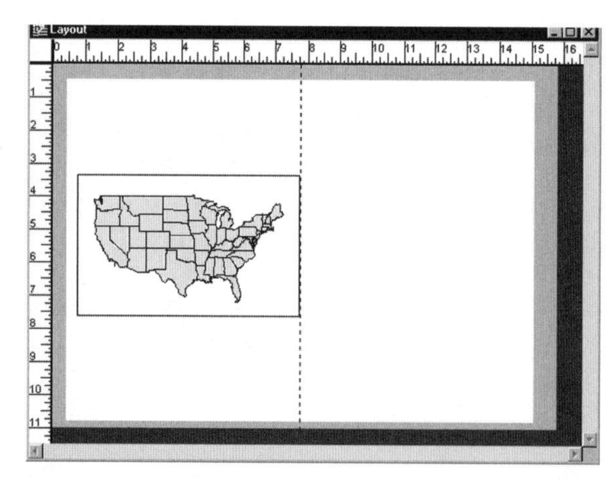

Two-page layout width.

3. Click on the OK button. The Layout window will be two pages wide, as shown in the previous illustration.

4. When you select the frame that contains the *States* map object, you will see a large rectangle with small black squares at each corner of the frame, as shown in the illustration at right. To make the map wider, grab the lower right black square with the left mouse button and drag it across to the lower right edge of the second page.

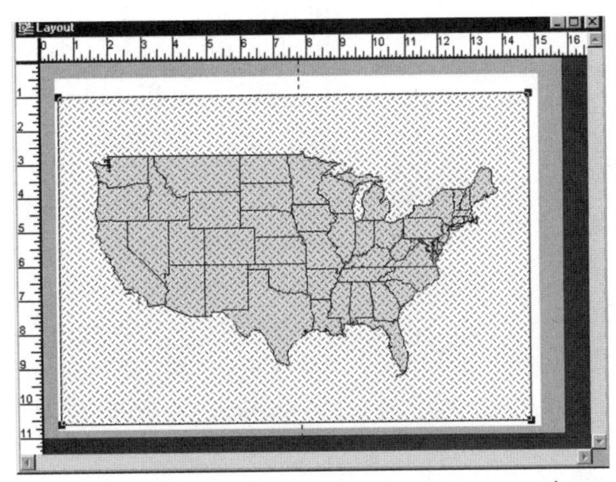

Two-page width of States *map in Layout window.*

For new users, MapInfo Professional's process for sizing windows within layout frames may be confusing. MapInfo provides two methods for positioning the map window within the frame on the Layout window. The Fill Frame with Contents option controls these methods. This option is accessed by double clicking the left mouse button over the map frame in the Layout window. The Frame Object options window is shown in the following illustration.

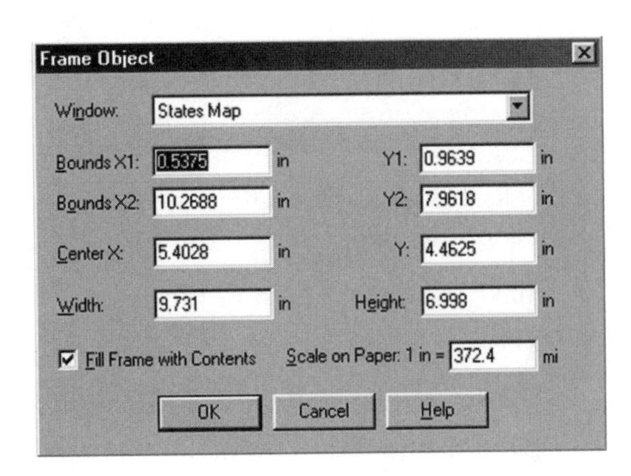

Frame Object options window.

With the Fill Frame option turned off, the map window will appear with the same positioning and shape as the source Map window. As the frame is made larger or smaller, MapInfo will resize the *States* map as large as possible within the frame. When the Fill Frame option is turned off, the Map window will grow to the full size of the frame in the layout window. With this option, the Layout window may show more of the map than is shown in the source Map window.

TIP: *In the Layout Display Options dialog, you can specify when to show frame content. Redrawing windows when working with large map tables can be time consuming. These dialog options (Always, Only when Layout Window is Active, or Never) allow you to determine when the content within the frames in a Layout window should be redrawn. If you are experiencing performance problems while using the Layout window, changing the option to Only when Layout Window is Active is recommended.*

Frame Tool

When you begin assembling and arranging a Layout window for printing, you will often wish to add an additional frame to the layout. Using the Frame tool button on the Drawing toolbar, you can easily add frames to the Layout window. In tutorial 9-3, which follows, instead of showing the plain *States* map, a thematic view of the map showing 1990 population will be created, and a legend frame will be added to the Layout window.

▼ *TUTORIAL 9-3: USING THE FRAME TOOL*

1. Make the *States* Map window the active window.
2. Select Map | Create Thematic Map. Set the Thematic dialog for a range shading of the *States* table on the *Pop_1990* data column to build the thematic map.
3. Make the Layout window the active window.
4. At this point, the objective is to add the population range legend to the Layout window. Select the Frame tool from the Drawing toolbar.
5. Position and drag a frame in the lower right corner of the Layout window.
6. Select the Legend window as the content of the new frame. The result is shown in the following illustration.

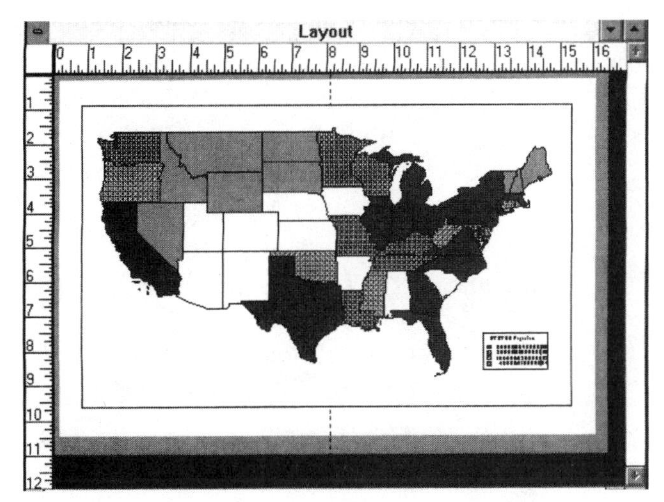

Added legend frame to States *map layout.*

TIP: *To change the content of a frame, double click on the window and choose a new window from the drop-down list of available windows.*

Adding Text and Altering Frame Borders

In tutorial 9-4, which follows, you will add a title to the map and remove the extra black border line drawn around the states.

▼ *TUTORIAL 9-4: WORKING WITH TEXT AND FRAMES*

1 To add a title in a large bold font, use the Options | Text Style command. Select the Text tool from the Drawing toolbar and type the title text into the Layout window. The text will be added with the style changes you specified in step 1.

2 To further alter the text style, select the title after it is typed, select the Options | Text Style dialog, and change the text style. The titled map is shown in the illustration at right.

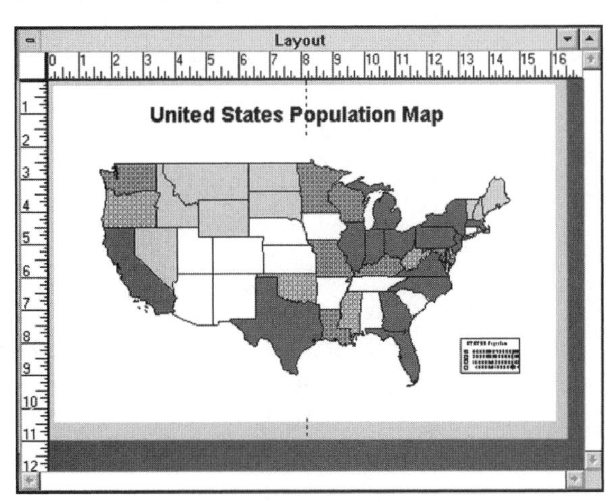

Titled States *map.*

For best results when using pen plotters, be sure to use fonts dsigned specifically for plotters. To identify these fonts, use the following command sequence: Main Program Group | Control Panel | Fonts Icon. In the Fonts dialog box, plotter type fonts are flagged by the word in parentheses, or *(Plotter)*. Generally, you will encounter Modern, Roman, and Script fonts.

TIP 1: *A thin, black outline is placed around a MapInfo object whenever you place one in the Layout window. This line is easily changed using the Options | Region Style command. To remove the outline border, verify that the frame is selected and use the Region Style dialog to set the border style to None.*

TIP 2: *When preparing a map for final printing you will often wish to include a scale bar. MapInfo Professional ships with a small MapBasic application named* scale.mbx *that will draw a scale bar on any Map window.*

Multiple Windows in a Layout Window

Once you have prepared each of the MapInfo elements on the screen as they are to appear on the printed page, you are ready to create a Layout window. For example, assume you are working on a population study of the United States. Map, Browser, and Graph windows could be used to represent the states with the largest populations. In tutorial 9-5, which follows, you will create a Layout window.

▼ *TUTORIAL 9-5: CREATING A LAYOUT WINDOW*

1 Open the *States* table in a Map window.

2 Select Query | SQL Select from the Main menu bar and build the following query. Table 9-1, which follows, summarizes this query.

```
Select * from states where Pop_1990 > 8000000
```

Table 9-1: Summary of Step 2 Query

Action	Query Expression
Select Columns	*
from Tables	STATES
where Condition	Pop_1990 > 8000000

3 Display the results in a Browser window.

4 Select Window | New Graph Window and graph the results of the query previously shown.

5 You now have Map, Browser, and Graph windows prepared and are ready to create a Layout window. Select Window | New Layout Window from the Main menu bar, and select the option to display all currently open windows in the new Layout window. The Layout window contains the three windows displayed with initial sizes and positions, as shown in the illustration at right.

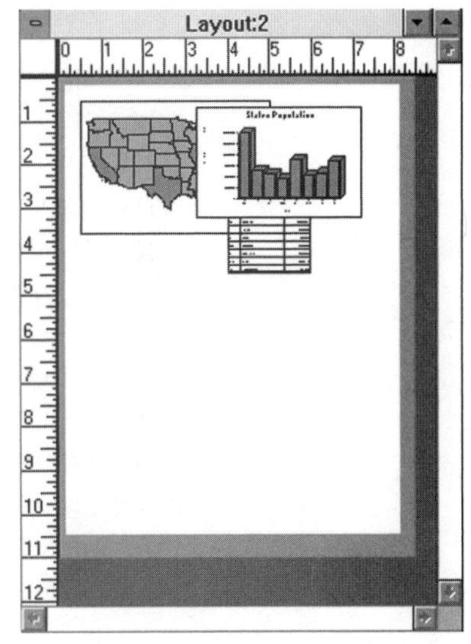

Initial Layout window.

6 Move and resize each of the windows to organize an attractive presentation in the Layout window.

7 Using the Text tool, add a title at the top of the page. The completed Layout window is shown in the following illustration.

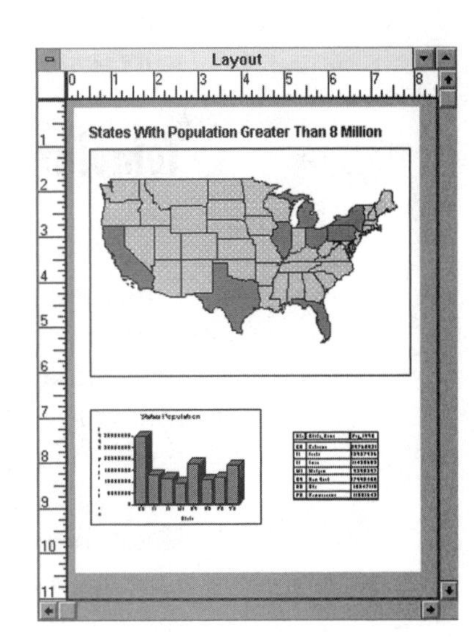

Adding a title at the top of the page.

Workspaces

Creating effective output can require a large time investment. To save your work in the Layout window, select File | Save Workspace. If you want to use this presentation again, simply select File | Open Workspace. The workspace will open up the files used during this session and recreate the windows, including the Layout window with all of its settings.

Remember that workspaces are created based on information in permanent data tables only. In the preceding example, the Layout window contains a Graph window based on a query. The query results are stored in a temporary table. If this table were saved to a workspace, the graph would not be recreated when the workspace was opened.

Tips and Techniques

The sections that follow explore templates, creating drop shadows, aligning objects, and output preferences. In tutorial 9-6, you will create a template, using these various features.

Templates

A Layout window can also be designed to serve as a template for future hardcopy production. This is done by creating a Layout window containing empty frames, or frames that have not yet been linked to any MapInfo window. Templates can be stored as workspaces for future use, at which time the links to windows may be established interactively.

If you need to create the same document every week, month, or quarter, a template layout can save you time. In a template layout there is little or no variation of the placement of objects on the page. Only the content of the actual map and/or title changes.

Let's build a simple template Layout window to better understand how you can simplify a repetitive output process. Assume that upon creating population comparison charts for each state in the

United States, you want a map of the state, a graph showing the comparison, and a browser showing the raw data. In tutorial 9-6, which follows, you will create these elements.

▼ TUTORIAL 9-6: CREATING A TEMPLATE LAYOUT WINDOW

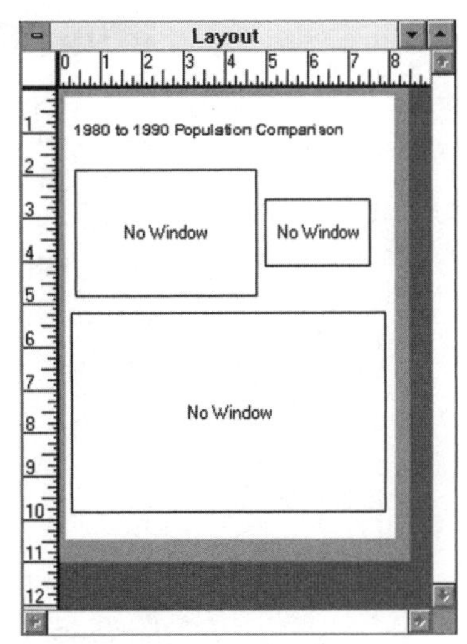

Template layout window.

1. To build the template Layout window, select Window | New Layout Window from the Main menu bar.
2. Using the Frame tool, draw three frames for a map, browser, and graph. Assign None as the window linked to each frame.
3. Add a title at the top of the page.
4. You now have a template Layout window, shown in the illustration at right. Save the template as a workspace and it will be ready to receive windows when you need it.

To create output using a template Layout window, create all of the windows you wish to include in the layout. Double click on each of the frame objects in the layout to associate a window with the frame, as shown in the following illustration.

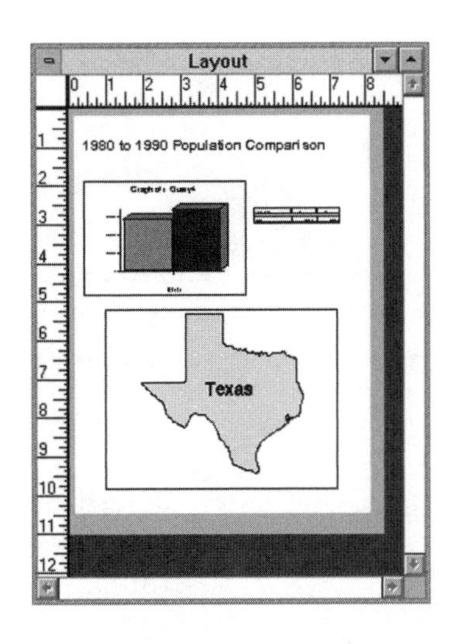

Template layout with windows assigned.

TIP: *Create MapInfo Professional tables or raster images (see Chapter 10) of your preferred north arrow representation and company logo for use in Layout windows. Creating such templates prevents unnecessary duplication, and establishes a company standard.*

Drop Shadows

You can add drop shadows to window objects in the Layout window. Drop shadows can be used to give a simple layout a little extra design punch. This option is available under Layout | Create Drop Shadow on the Main menu bar. The following illustration shows the preceding example template layout after adding drop shadows to the Layout window.

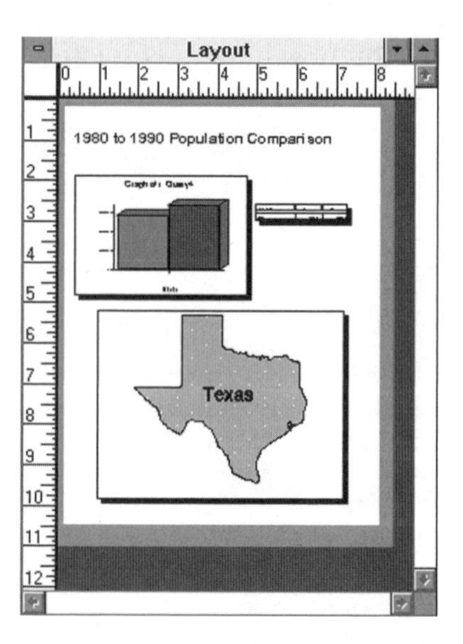

Layout with drop shadows.

TIP: *When you move or resize the layout object, the drop shadow does not resize or move. Do not create a drop shadow until you are certain of the layout object position and size. If you have already created the drop shadows, you can use the <Shift>+select method or the Marquis Select tools from the Main toolbar to select both the object and its shadow. The object and shadow can then be moved or resized together.*

The drop shadow appears behind the selected object. Because a drop shadow is a rectangle object, you can delete it or change its fill pattern and line style.

Align Objects

The Align Objects function allows you to align objects in the window in both horizontal and vertical directions. Although often overlooked, in a Layout window you can draw or create graphic objects with the drawing tools the same as you can with an editable Map window. For example, you may wish to draw a legend in a Layout window to describe the point symbol usage on your map. The Align Objects function is a very handy tool for improving presentation appearance.

▼ Tutorial 9-7: Using the Align Objects Function

1 Open a new Layout window by selecting the Window | New Layout Window from the Main menu bar.

2 Using the Drawing tools, select Symbol Styles and place points in the Layout window to represent hospitals, airports, and churches, as shown in the illustration at right. Place a legend title and point type text field next to each point.

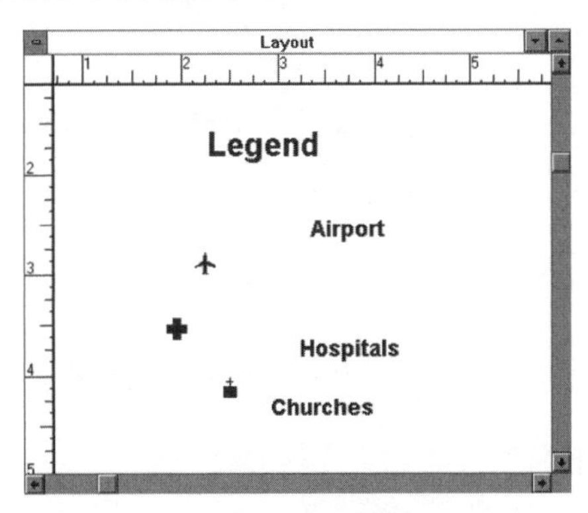

Building a legend in the Layout window.

3 Select each of the symbol objects using the Select tool.

4 Select the Legend | Align Objects function from the Main menu bar. This will access the Align Objects dialog box, shown in the illustration at right.

Align Objects dialog.

5 Use the Legend | Align Objects function to align and distribute the points. Next, align and position the text objects until all objects in the legend are lined up. The aligned legend is shown in the following illustration.

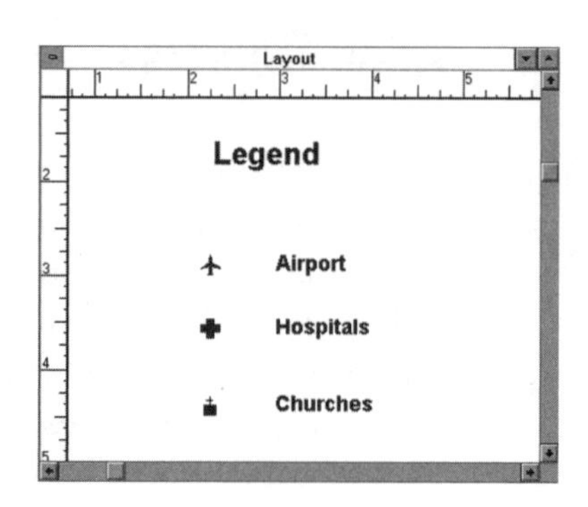

Aligned legend in Layout window.

NOTE 1: *Your alignment settings are saved from one use to the next within a session. Be sure to check both the vertical and horizontal settings before you click on OK.*

NOTE 2: *A printed map or layout may appear slightly different from the screen representation. This is most noticeable in the case of text objects because they may appear larger or smaller in the hardcopy print than on the screen.*

Output Preferences

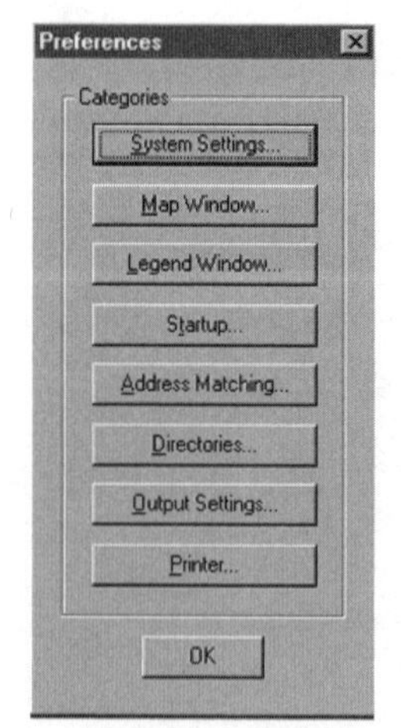

Preferences dialog box.

If you do a lot of printing of specific types of maps, you may find setting output preferences useful. For example, it may be appropriate to have your computer default set to print documents to a laser printer. However, you always want your maps to be sent to a color printer. The Output Preferences dialog allows you to set map printing preferences so that each time MapInfo starts, these are the defaults. Select Options | Preferences from the MapInfo Main menu bar to view the Preferences dialog box, shown at left.

The next to last button on the Preferences dialog box is the Output settings button. Selecting this button displays the Output Preferences dialog box, shown in the following illustration.

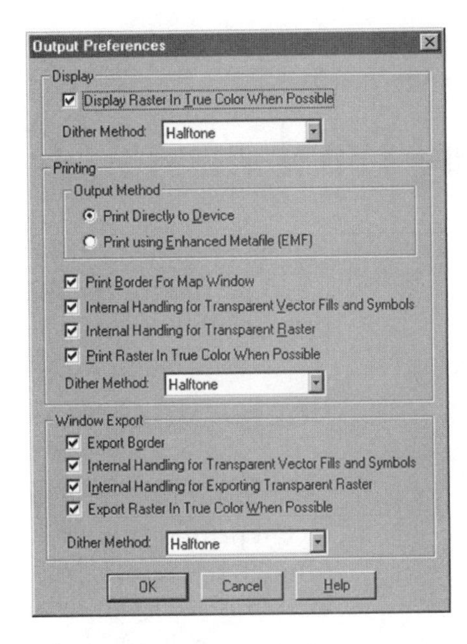

Output Preferences dialog box.

In the Output Preferences dialog, you can specify display, printing, and window export preferences. Many of the options relate to the handling of transparent fills and raster display. Also found on the Preferences dialog box (the last button) is the Printer button. Selecting this button displays the Printer Preferences dialog box, shown in the following illustration.

Users will find this dialog very useful for specifying the printer to use for printing MapInfo maps. Until this box was added, MapInfo always used the default printer until you specified a different

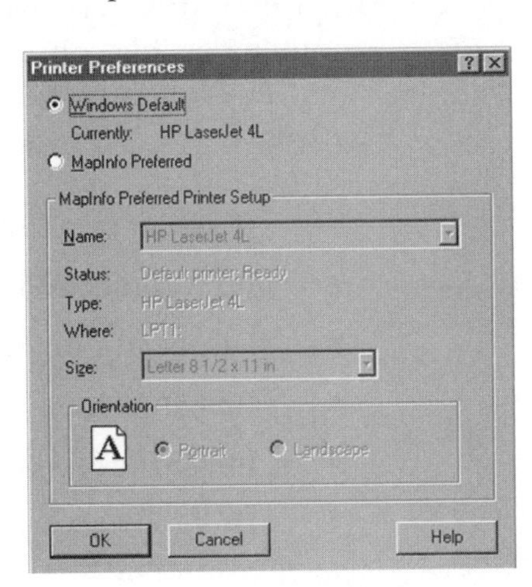

Printer Preferences dialog Box.

one each time you started MapInfo. With this dialog, the user can specify an alternate printer to always be used for map printing.

Printing Problems

Printing problems have numerous sources. Because MapInfo Professional runs in the Microsoft Windows environment, it uses the printers and drivers set up and provided for by Windows.

Printer Drivers

Some printing problems may be resolved by obtaining the most recent print driver available for your printer. In the Windows environment, MapInfo Professional uses the windows graphical device interface (GDI) to draw on the screen and to the printer. If the hardcopy printout does not look the same as your display, it may be because the printer driver is not interpreting the GDI calls properly. A more recent printer driver may correct this problem. To check printer driver versions, contact the printer manufacturer. The manufacturer will either send you the most updated driver or inform you about where you can download the driver from an electronic service.

Printing Software

If printer driver replacement does not resolve your problem, another possible solution is to invest in third-party printing software. The following applications offer sophisticated printing alternatives.

- JETPRO by ACRS
- PowerPress Color Printing by Colossal Graphics
- DiceNet Color Server by Dice America
- Image Alchemy PS by Handmade Software
- Imagez Professional System by Onyx Graphics
- DoPlot by Pacesetter Labs

- Wasatch Poster Maker by Wasatch Computer
- SuperPrint by Zenographics

NOTE: *SuperPrint has been tested by MapInfo Corporation.*

A Note About Printing Hardware

There are literally hundreds of printers on the market that are appropriate for printing MapInfo Professional data. If you are planning to purchase a new printer for your system to print MapInfo data, there are several questions that should be addressed. The first big question is color versus black and white. Although color printing technology is becoming more affordable every day, it is still much more expensive than black and white. Ink and paper for color printing are also more expensive than for black and white.

The second question involves quality. What resolution is required? Are you printing draft copies, or do you need a printer for final client deliverables? Printer specifications include a dots per inch (dpi) rating to help judge print quality. These dpi ratings vary enormously among printers and manufacturers.

Another question that must be addressed is speed. Do you print a few pages a week or many large reports per day? Printer specifications usually list the number of pages per minute you can expect to produce on a given printer.

Next is the question of paper size. If you need to print wall-size maps, your printing requirements are different than those of a typical business. Several available plotters used with CAD programs are capable of handling wide rolls of paper (36 inches to 48 inches wide). Printing for customers abroad may also require the handling of a wide variety of paper sizes.

Finally, you need to consider printer technology. When reviewing printer technologies you will see dot matrix, plotter, InkJet, Laser, and PostScript. Each of these technologies has characteristic strengths, weaknesses, and price tags.

If you are selecting new printing hardware for the sole purpose of printing maps from MapInfo Professional, it may help to test the hardware first. This is a relatively easy process that can save you from making a purchase you will regret. Tutorial 9-8, which follows, steps you through the process.

▼ TUTORIAL 9-8: TESTING MAP PRINTING HARDWARE FOR USE WITH MAPINFO

1. Obtain a copy of the printer driver from the printer vendor or manufacturer and install it on your machine.
2. Build a complex map in MapInfo that represents all of the items you want to print. You want to see how the layers, colors, lines, points, and fonts appear on a printed sample.
3. Use the Print Setup command and select to send the map to that printer driver with the output actually going to a file.
4. Use the Print command to create the printer file for the map.
5. Take the file on disk to the printer vendor and request that the file be printed. Then examine the result in regard to your map requirements.

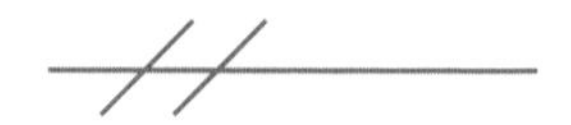

Cartographic Design

Cartography is defined by the International Cartographic Association as the "art, science, and technology of making maps." In GIS, the science and technology of making maps often consumes the bulk of project development time, and the creative process of design (the art of making maps) is often ignored or poorly planned.

The old adage "A picture is worth a thousand words" was never more true than in the art of cartography. People respond to visuals, and given that maps are a natural choice in visually presenting data relationships, the importance of carefully designed maps to effectively communicate your intent and purpose is evident.

Confusion and clutter on a map is a design failure. You must find design strategies that reveal the details and complexity of the data rather than fault the data for excessive complexity, or blame users for lack of understanding.

Five Questions for Effective Communication

Maps are a highly specialized form of communication. A useful way to begin the cartographic design process is to answer the five basic questions in communication planning, also known as the five Ws. These questions are explored in the sections that follow.

- Who is your target audience?
- What is the message for the target audience?
- Why are you presenting this message to the target audience?
- Where would this message best be given to the audience?
- When is the best time to present this message?

Who is your audience?

You are making an effort to communicate to a certain group of people. Who are they? What do you know about them as a group? What are the demographics of the people and/or the businesses represented? Understanding your audience will help you understand how to reach them at their level.

What message do you have for the target audience?

Perhaps you are presenting the results of an analysis, or the conclusions to important research. Whatever the message may be, it must be clearly focused in your own mind before you can expect to effectively present it to others.

Why are you presenting this message to this audience?

What is the purpose behind the map presentation? For a navigation map, for example, the map must be easy to read, with clear representation of the roads and road names. For an environmental impact study map, the colors and psychological impact of the map are much more important.

Where will you present this message to the audience?

The conditions under which your map is presented will influence map design. For instance, maps prepared for oral presentation versus print publication are different. For an oral presentation, consider issues such as room size, lighting, and projection systems. Oral presentation slides will generally contain minimal detail to facilitate a short viewing time. For written publications, you will often want to reduce the use of color or even prepare the maps for black-and-white printing.

When is the best time to present this message?

You rarely have full control over the time available to design and create the map, much less for the presentation of the map. It is crucial to use the control you have over the process to your best advantage.

General and Specific Points to Consider in Design

When you are ready to design your map product, consider the following general guidelines.

- *Be observant of other maps and graphic design styles, as well as industry-specific standards (e.g., weather maps have established symbology).* Which colors work best? Which text styles are easiest to read? Take note of what works, and what you like and dislike. This will help you develop your own style.
- *Develop an organizational style or standard to consistently follow.* This can be a very simple and effective marketing tool if all maps from your company have a consistent look.
- *Once you have developed a style, create layout templates that can be used when you have to produce maps in a hurry.* Keep the templates simple, and allow for flexibility within the design.

Specific points to consider in designing maps and pages follow.

- *Take advantage of the power of white or negative space.* These are areas that are not part of your map. White space is a very powerful design tool.
- *Build contrast between the geographic figure (the area of focus to your map purpose), and the background.*
- *Prioritize the geographic layers of your map based on the map purpose.* Which elements are most important? The importance of elements may be based on one of several criteria, including chronological order, alphabetical order, or largest to smallest. These elements should visually stand out more than other elements of lesser importance. Create a worksheet establishing the "cartographic order."
- *Use one or more visual variables, such as color, shape, texture, and size, to create levels of contrast between the map elements.* The point here is to create the visual illusion of a cartographic order.
- *Ensure that your map consistently presents related data.* Verify that features of the same type of map element are presented similarly. For example, for

annotation, stick to one style per feature type. Avoid labeling similar objects with different fonts, sizes, or colors.

- *Consider the entire page.* Where does the map fit best considering its shape? Where will the title and legend fit best? Think in terms of the map's "flow." What does your eye focus on first, and where does it naturally move from there? Use this natural flow to guide the viewer throughout the logical sequence of your map "story."
- *As you near completion, work through a mental checklist of the basic map elements.* Have you placed a prominent title? Did you include a scale bar? Did you include an adequate legend with your name and/or company name, date, and north arrow? Have you listed both map and data sources? Have you considered a "locator" map? Each of these items has the potential to contribute to and clarify your presentation.

Another set of considerations involves color. The maps you design must make a statement about the geography, rather than about color. In general, you should use lighter colors for most of the map, saving the darker or more prominent color for the points of purpose in the map.

Rules of Color Composition

The Swiss cartographer Eduard Imhof composed the following rules of color composition. [From Eduard Imhof, *Cartographic Relief Presentation* (Berlin, 1982), edited and translated by H. J. Steward.] You may find yourself bending or breaking some of these rules as you pick colors for your GIS map output, but they are good rules of thumb.

- Pure, bright, or very strong colors have loud, unbearable effects when they stand unrelieved over large areas adjacent to each other, but extraordinary effects can be achieved when they are used sparingly on or between dull background tones.
- The placing of light, bright colors mixed with white next to each other usually produces unpleasant results, especially if the colors are used for large areas.
- Large area background or base colors should do their work most quietly, allowing the smaller, bright areas to stand out most vividly, if the former are muted, grayish, or neutral.
- If a picture is composed of two or more large, enclosed areas in different colors, then the picture falls apart. Unity will be maintained, however, if the colors of one are repeatedly intermingled in the other.

In exercise 9-1, which follows, you have the opportunity to try your hand at the various hardcopy map output techniques previously discussed. The exercise uses the *States* map, and incorporates a section of a map adjacent to or within a base map. This arrangement is typical in the case of including the state of Hawaii in a base map of the United States. An individual object, the state of Alaska, will also be copied and pasted from a Map window directly onto the Layout window. This procedure offers another method of calling attention to a particular map feature.

■ EXERCISE 9-1: TECHNIQUES IN HARDCOPY OUTPUT PREPARATION

First, to create a map inset of the state of Hawaii you must prepare the two Map windows before you can build a Layout window.

Preparing Map Windows

1 Select the File | Open Table option from the Main menu bar. Open the *States.tab* table, shown in the illustration at right.

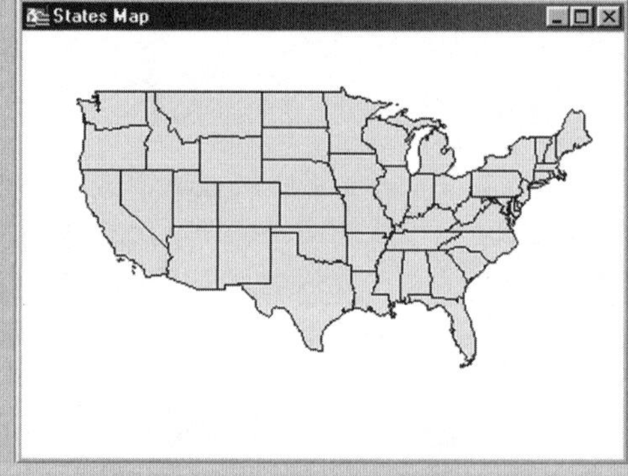

Map window of the *States* table.

2 Select Window | New Map Window. An additional *States* Map window will be opened.

3 Select Map | View Entire Layer from the Main menu bar to resize the map so that you can see the state of Hawaii.

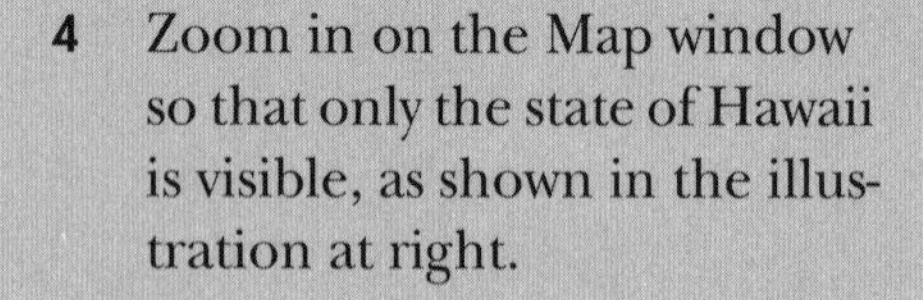

4 Zoom in on the Map window so that only the state of Hawaii is visible, as shown in the illustration at right.

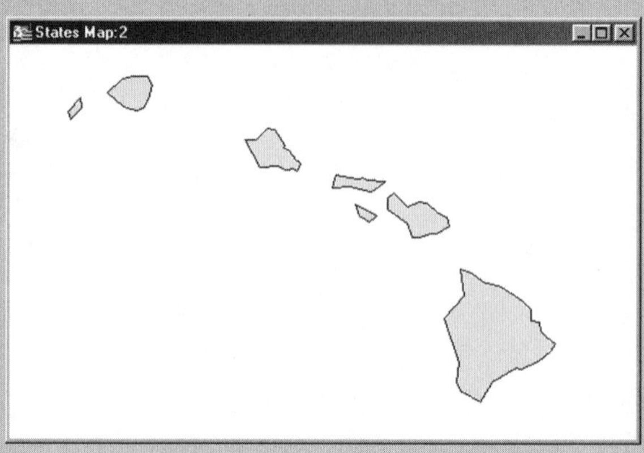

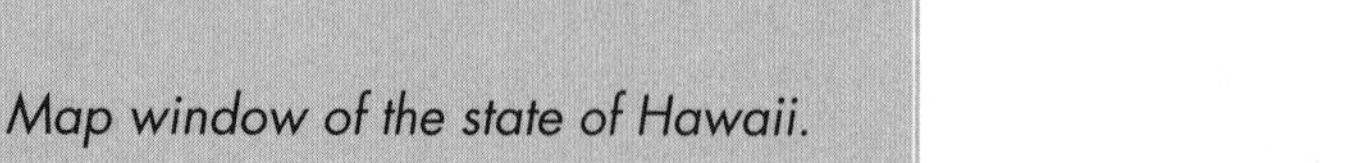

Map window of the state of Hawaii.

Creating the Layout Window

Now that both the continental United States and Hawaii maps are prepared, you can create the Layout window.

1 Select Window | New Layout Window from the Main menu bar.

2 The New Layout Window dialog box, shown in the illustration at right, will be displayed. Select the option to include frames for all currently open windows in the new layout.

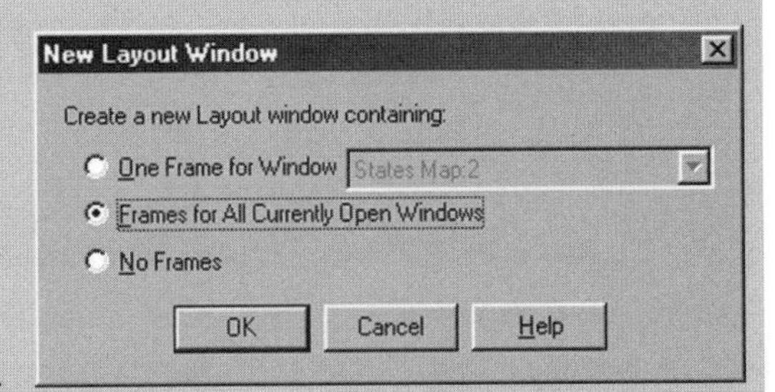

New Layout Window dialog box.

3 Click on the OK button and the Initial Layout window, shown in the illustration below left, will be displayed.

4 Resize the *States* map to fill the upper half of the layout page.

5 Move and resize the *Hawaii* map to cover the lower left portion of the *States* map. The illustration below right shows the Layout window at this stage.

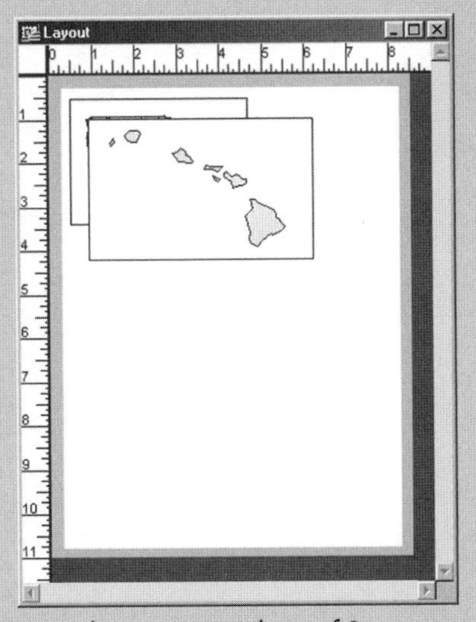

Initial Layout window of States *and* Hawaii *maps.*

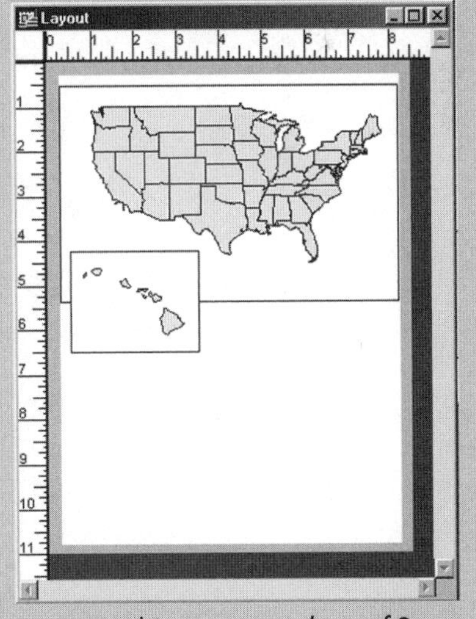

Arranged Layout window of States *and* Hawaii *maps.*

Before you move on to working with the state of Alaska, map scales deserve mention. In the previous layout, the scale at which you show the *States* and

Hawaii maps are much different. For some presentations this is acceptable; however, for others the maps should be at relatively the same scale. If you double click the left mouse button on the Hawaii frame in the Layout window, you are presented with the Frame Object dialog box, shown in the following illustration.

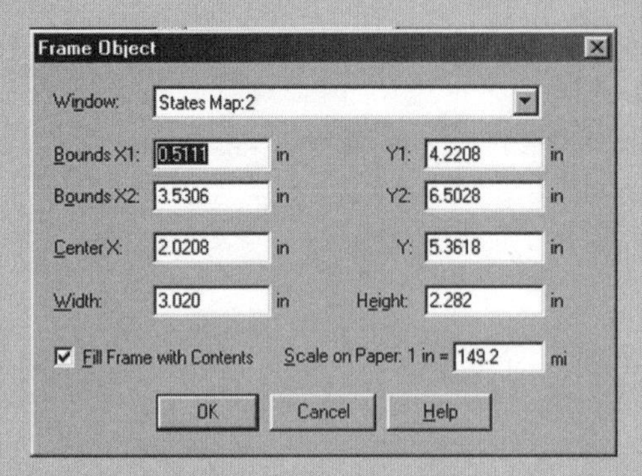

Frame Object dialog box for Hawaii.

At the bottom of the dialog an entry shows Scale on Paper: 1 in = 112.6 mi. In this dialog box, the number of miles can be adjusted for any frame such that the scales of the maps in the Layout windows are the same.

Creating the Inset

To review another method for creating an inset, you will copy and paste the state of Alaska item into the Layout window.

1 Select Window | New Map Window from the Main menu bar to create yet another Map window of the *States* table.

2 Select Map | View Entire Layer from the Main menu bar to resize the map so that you can see the state of Alaska.

3 Zoom in on the Map window so that only the state of Alaska is visible, as shown in the illustration at right.

4 Select the state of Alaska with the Select tool.

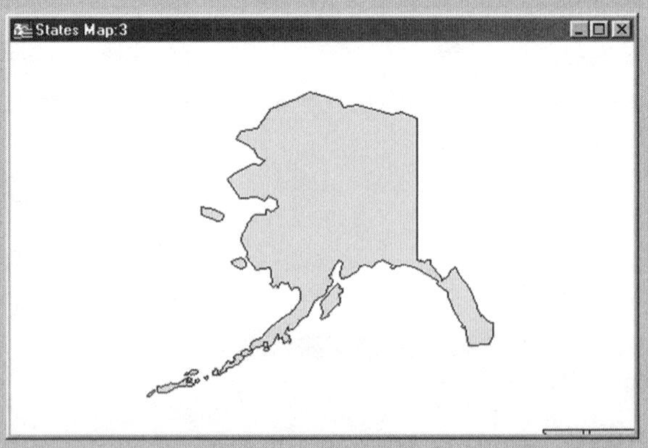

Map window of the state of Alaska.

5 Select Edit | Copy from the Main menu bar. This action will copy the Alaska map object to the Clipboard.

6 Make the Layout window the active window.

7 Select Edit | Paste from the Main menu bar. The Alaska map object will now be pasted into the Layout window. Move and resize the Alaska map object to position it in the lower portion of the Layout window, as shown in the illustration at right.

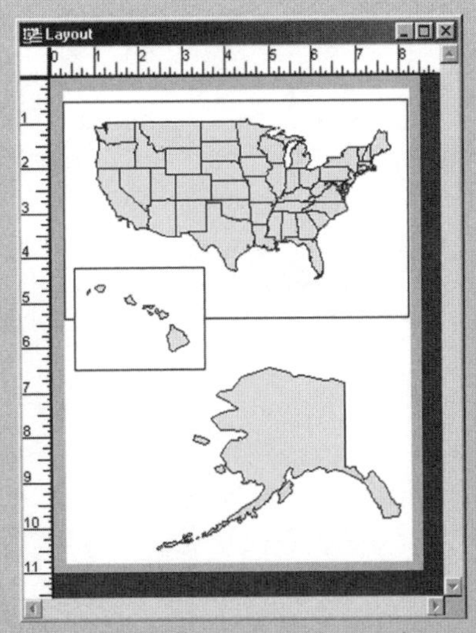

Arranged Layout window of States, Hawaii, and Alaska.

Creating a Drop Shadow

One of the advantages of pasting an object into a Layout window is that it becomes an object that can be manipulated in the layout. Therefore, it is possible to create drop shadows for the object, resize it, or change its style. The remaining steps are focused on creating a drop shadow for the state of Alaska.

1 Select Alaska in the Layout window.

2 Select Layout | Create Drop Shadow from the Main menu bar. This will access the Create Drop shadows dialog box, shown in the illustration at right.

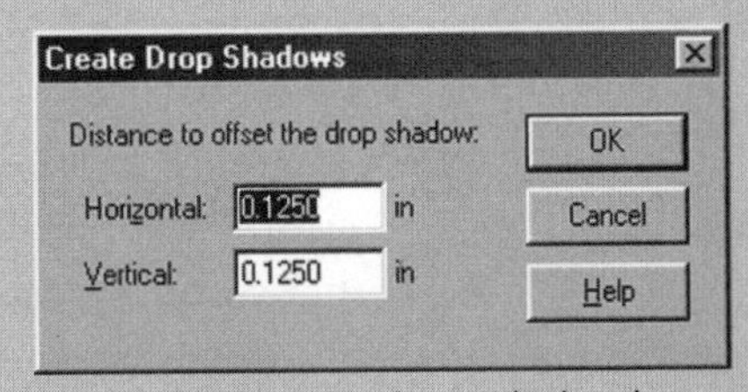

Create Drop Shadows dialog box.

3 Click on the OK button and a drop shadow will be created around the state of Alaska, as shown in the following illustration.

You can now reshape and alter the styles of both Alaska and its drop shadow, as shown in the following illustration.

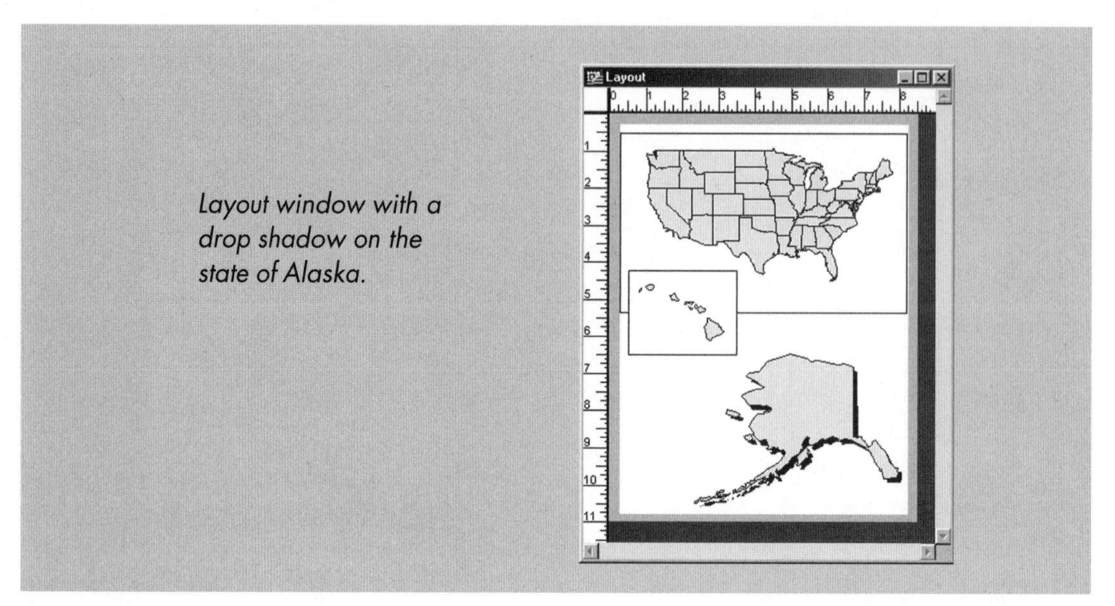

Layout window with a drop shadow on the state of Alaska.

Summary

MapInfo Professional has provided many options to help you prepare output for GIS data analysis results. Although each MapInfo window may be quickly printed by itself, using the Layout window provides you with many options for combining multiple windows, graphics, and text to make an impressive data display.

Printer hardware often causes problems. When choosing a printer for your output, be sure to clearly define your printing requirements for quality and quantity. The first step in troubleshooting printer problems is to check with the printer vendor to ensure that you have the most current version of the printer driver.

CHAPTER 10

OTHER KEY TOOLS

CHAPTERS 1 THROUGH 9 FOCUSED ON MapInfo Professional's core functionality. Chapters 10 through 12 address advanced tools, data manipulation options, and customization. This chapter covers hot links, geocoding, redistricting, real-time map updates, statistics, and Crystal Reports (report writing tool).

Hot Links

MapInfo Hot Links maybe better known in the Internet world as hypertext links. These links are the signature characteristic of the World Wide Web. The ability to jump from one page to another through hypertext links makes the Web exciting and attractive to viewers. MapInfo now allows you to create MapInfo tables that have links to web pages.

Displaying Hot Links from a Browser Window

HotLink button.

The HotLink button, shown at left, on the MapInfo Main tool bar is enabled when the table being browsed has HotLink options stored in its metadata.

If one of the browser field's expressions matches the HotLink Filename Expression, the text in that browser field is underlined and the HotLink button is enabled, as shown in the following illustration.

HotStates Browser

Sales_1990	Total_Area	HTML_Link
26,373,205	51,832.5	http://www.state.AL.us
4,668,675	652,868.2	http://www.state.AK.us
26,137,116	114,016.3	http://www.state.AZ.us
15,386,039	53,058.3	http://www.state.AR.us
225,065,880	158,508.5	http://www.state.CA.us
24,383,148	104,001.5	http://www.state.CO.us
27,729,253	5,021.6	http://www.state.CT.us
6,041,091	2,049.2	http://www.state.DE.us
3,815,320	70.6	http://www.state.DC.us
105,303,987	58,907.2	http://www.state.FL.us
46,748,222	58,958.5	http://www.state.GA.us
11,203,513	6,257.4	http://www.state.HI.us
6,004,215	83,313.0	http://www.state.ID.us
83,478,926	56,276.0	http://www.state.IL.us
37,574,006	36,091.7	http://www.state.IN.us
18,817,884	56,203.8	http://www.state.IA.us
16,655,815	82,246.6	http://www.state.KS.us
23,860,802	40,437.6	http://www.state.KY.us

Enabled HotLink tool.

Clicking with the HotLink tool on the underlined text will launch the URL or file specified by the text. In tutorial 10-1, which follows, you will use the HotLink tool in a Browser window.

▼ TUTORIAL 10-1: USING HOTLINK IN A BROWSER WINDOW

1. Using the File | Open Table option on the Main menu bar, open the *HotStates* table in a Browser window.
2. Scroll to the far right of the *HotStates* table. The far right column, named *HTML_Link,* contains web links to each state's main Internet home page.
3. Select the HotLink button from the Main tool bar. Select the Hotlink or Hyperlink option for one of the states. Control will now jump to that state's main web page.

Displaying Hot Links from a Map Window

The HotLink tool allows you to launch the URL or file associated with an active object by clicking on the object or its label in a MapInfo Map window. In tutorial 10-2, which follows, you will launch a URL or file name using the HotLink tool in a Map window.

▼ TUTORIAL 10-2: USING HOTLINK AND THE MAP WINDOW

1 Open a Map window of the *HotStates* table, used in the previous tutorial. Click on the HotLink button (lightning bolt) in the Main tool bar. The cursor appears as a pointing hand.

2 When the cursor is positioned over one of the states' map objects, the normal arrow cursor (shown in the following illustration prior to cursor change) becomes a pointing hand holding a lightning bolt. The link associated with the active object displays in the status bar.

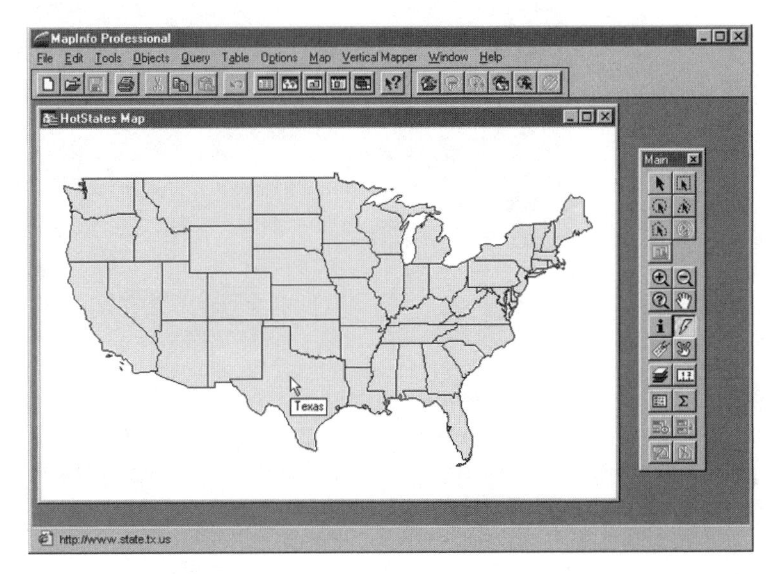

HotLink cursor selected on Main menu.

3 Click on the object or label to launch the associated URL or file.

NOTE: *The HotLink tool is available for Map windows containing at least one active layer. A layer is active if it is selectable or editable and its HotLink options have been set (in particular, the Filename Expression option).*

Creating Hot Links in a MapInfo Table

The previous sections have shown how to use a MapInfo table that has hot links. This section briefly describes what steps must occur to create your own links in your own MapInfo tables.

To create link objects to a URL or file in a MapInfo table, add a character field to the MapInfo table using the Modify Table Structure dialog, shown in the following illustration. Make sure the field is wide enough to accommodate the longest URL or file name you will want to add.

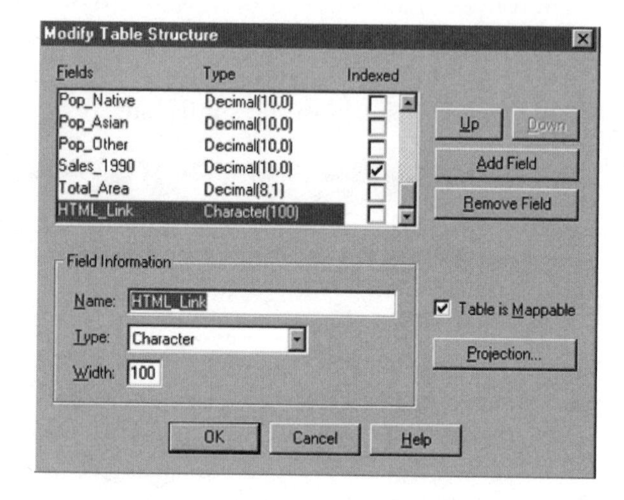

Creating a text field.

Now you are ready to open a Browser window and enter the file names or URLs you want to associate with each row of your table. The final step in creating hot links is to make MapInfo understand that these are URL or file names.

Open your table as a MapInfo Mapper window and access the Layer Control dialog box, shown in the following illustration. This dialog box has a HotLink button on the lower right side.

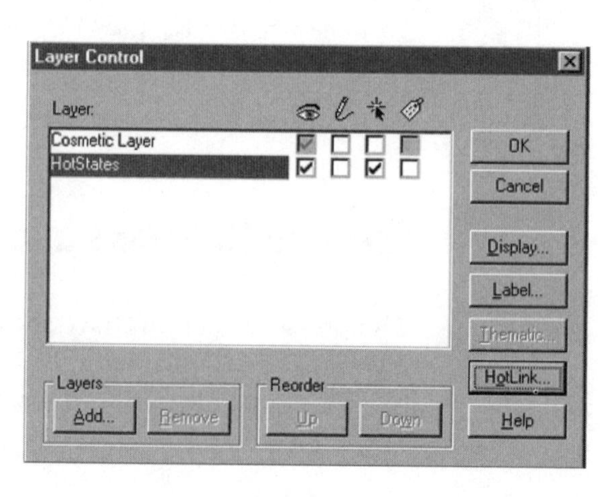

Layer Control dialog with HotLink button.

On the HotLink dialog box, select the Filename Expression column to be the column you have created, as shown in the following illustration. Select the options for the conditions under which the hot link will be active. In addition, check the "Save options to table metadata" box if you want these Hotlink options automatically restored the next time you use the same table in a Map or Browser window.

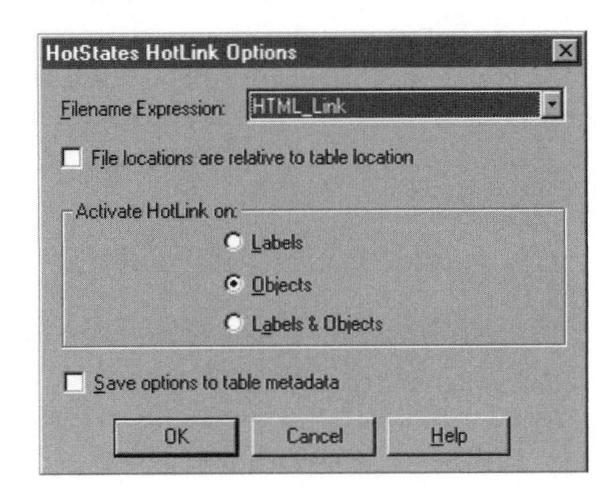

Selecting the Filename Expression column.

Once these steps are complete, the HotLink tool bar button will be available when this table is active in a Map or Browser window.

Geocoding

Most organizations in the private and public sectors work with data containing geographic references, such as address, zip code, county, and/or state. Geocoding is the process of assigning X and Y coordinates to records in a table based on location information in a mappable database. Many database records contain some type of geographic information. By matching this information to locations on a map, you can display data in a map. With this ability, you open doors to geographic and spatial analyses that were not possible when viewing data in tabular form alone. Geocoding or address matching consists of the following components.

- A data file containing street addresses or any other mappable data (e.g., zip codes, cities, and countries).
- The streets database to be used as the geographic base.

- Software, or MapInfo Professional's geocoding function, for bringing the addresses and streets together to derive X and Y coordinates that can be used for mapping and analysis.

The address field in many databases can be used to locate customer addresses in the *streets* layer. The zip codes file can be used to refine the search. Because several streets of the same name (e.g., Main Street) may appear in a given county, you need a refining boundary such as zip codes to ensure that your records geocode as accurately as possible. In this way, the address will be placed on the correct street because both the street address and the zip code must match.

Geocoding Methods

To geocode a table with street addresses, MapInfo Professional will use the street names and address ranges in your street files. With this type of geocoding, the data records will be assigned X and Y coordinates such that the record will be spotted at the record's street address, and even on the proper side of the road.

Another popular method of geocoding is to a boundary level. Some data can be matched to county, city, or zip code levels. With this type of geocoding, the data records will be assigned X and Y coordinates located at the centroid of the appropriate region.

Occasionally you will find data files that already contain X and Y coordinate fields as part of the attribute data. In this case, you would use the Table | Create Points function on the Main menu bar. By setting the Create Points dialog box, shown in the following illustration, all points will automatically be created at the locations specified in the coordinate columns.

Create Points dialog box.

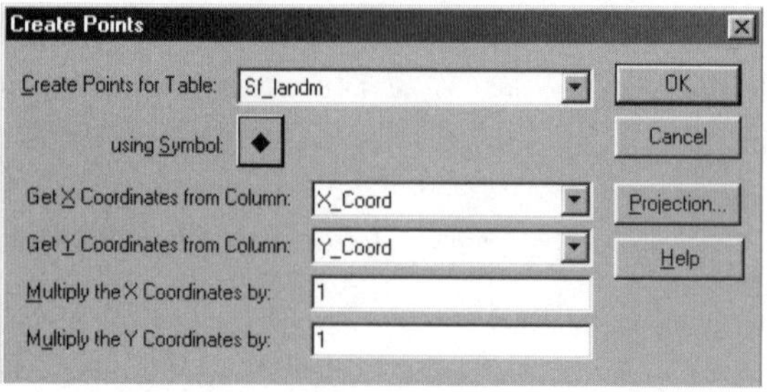

The Science of Geocoding

Street addresses are the most common form of geographic data. Nearly all of us work with addresses every day. Address geocoding in MapInfo Professional allows you to create points associated with each row of a data file based on an address. A 100-percent match of street addresses against the geocoded layer occurs only when both the address table and the geocoded street file are 100-percent properly coded. A 100-percent match during geocoding rarely occurs.

There are ways, however, to improve your chances of obtaining a match between the address and a point on the map. First, you can ensure that address data are coded accurately, and that the address prefixes and suffixes are coded in a consistent fashion with the format used in the street network. For example, the address 1400 N. 16th Ave. may result in a match, whereas the address 1400 N. 16th Ae may not match.

Third-party software packages are available that can enhance your geocoding success rate. These software packages will "scrub" addresses by cleaning the street names of the addresses in your file to be consistent with U.S. Postal Service standards for street names.

Extensive quality control on address entry will not give you a match if the corresponding street in the street network is missing or incorrectly coded. In order to ensure a high percentage of matches, equal attention must be devoted to guaranteeing that your geocoded street network is accurate and up to date.

The raw TIGER street files from the U.S. Census Bureau are likely to contain errors and omissions. The errors are not sufficient to render TIGER files unusable, but certainly enough to reduce your geocoding hit rate. It is possible to edit the TIGER files to improve accuracy, but depending on your specific requirements you may wish to purchase a revised street network from a commercial data provider.

Certain companies specialize in providing address matching services and software. An example is MapInfo Corporation's MapMarker. MapMarker assigns latitude/longitude at the street, zip code, zip + 2, and zip + 4 levels in a single pass of the data and will locate virtually any address or place name anywhere in the United States. MapMarker geocodes large files quickly (with high, accurate hit rates) and has a very friendly user interface. Embedded within MapMarker are sophisticated matching rules to find and match addresses containing misspellings, omissions, and even incorrect data. Data used for matching with MapMarker are derived from the most current Census Bureau and U.S. Postal Service data and are updated every six months.

Most of the current discussion has centered on U.S. data files and addresses. International data files introduce different addressing issues that must be considered.

First, the MapInfo abbreviation file must take into account different language strings and abbreviations. Some nations (e.g., many European countries) place the street name first, followed by the street number. MapInfo began providing for this reversed address order with release 3.0.

Japan poses an entirely different geocoding problem. Japanese addresses are based on a named and/or numbered region scheme in which regions get progressively smaller until the address unit is identified. In Japan, the MapInfo provider has created a separate geocoding module that specifically deals with the Japanese addressing system.

How Does Street Address Geocoding Work?

To understand how geocoding works, this section focuses on a street file and associated data. The *Sf_strts* file directory on the companion CD-ROM is examined here. In the street file, shown in the following illustration, find Gough Street near the center of the San Francisco area.

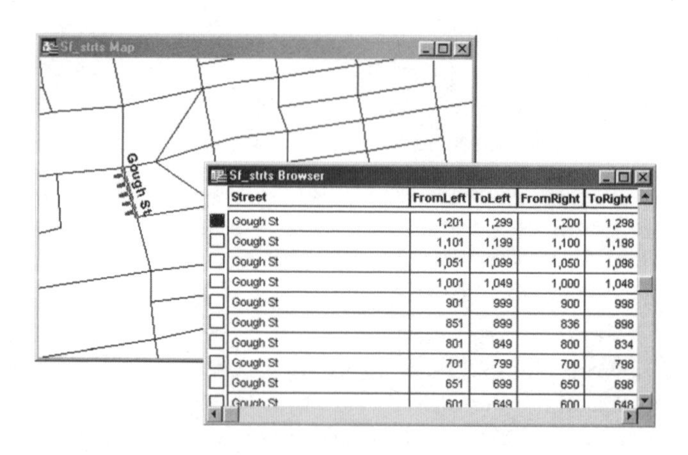

San Francisco street file.

Upon examining Gough Street, you can see four columns of data in the *Sf_strts* table: *FromLeft, ToLeft, FromRight,* and *ToRight.* Based on this information, you know that odd numbered addresses from 1201 to 1299 are located on the left side of the road, whereas even numbers from 1200 to 1298 are on the right side. To geocode a point, MapInfo Professional first locates the street segment with the correct name and range of addresses. It then determines whether the street address number is an odd or even number, and thus, location on the left or right side of the road. The number will then be proportionately placed along the street segment on that side of the road.

For example, 1280 Gough Street is on the right side of the road, and is located much nearer the "to" end of the street. The map shown in the following illustration displays the geocoded location of 1280 Gough St.

NOTE: *The data table must be editable to perform either geocoding or ungeocoding operations.*

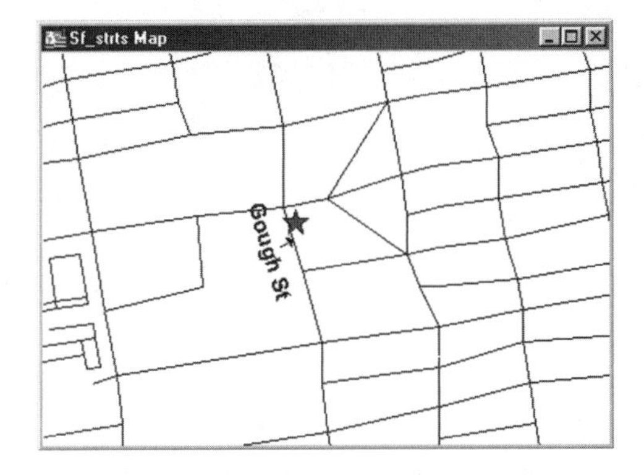

1280 Gough St. in San Francisco.

Ungeocoding Records

Before examining how to geocode records, a quick look at ungeocoding data records or removing the existing graphic point object from the table is in order. Because the *Sf_custd* table is used for the geocoding review in subsequent sections, all current point locations from the table will be removed here. To do this, perform tutorial 10-3, which follows.

▼ *TUTORIAL 10-3: REMOVING POINT LOCATIONS*

1 Open the *Sf_custd* table (*samples* directory on the companion CD-ROM).

2 Select Table | Maintenance | Table Structure on the Main menu bar. This accesses the Modify Table Structure dialog box, shown in the following illustration.

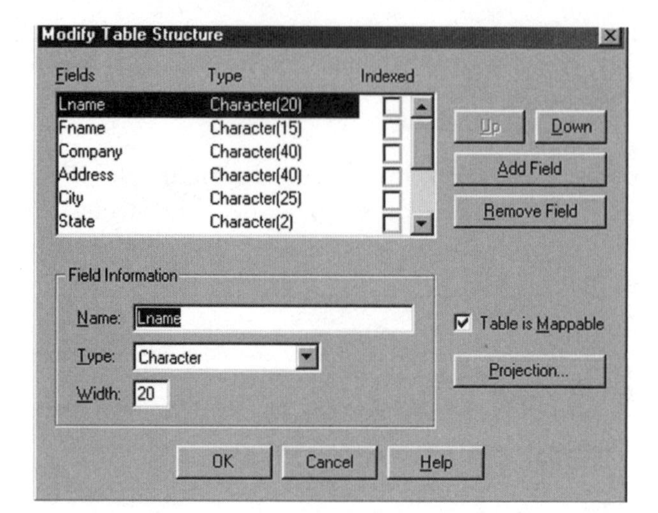

Modify Table Structure dialog box.

3 Deselect the Table is Mappable check box option and press the OK button.

4 The ungeocoding operation cannot be undone and MapInfo will display the warning dialog box shown in the following illustration. After pressing the OK button in the dialog, MapInfo will discard the graphic objects for the *Sf_custd* table.

5 You are now unable to open a new Map window of the *Sf_custd* table because mappable graphic objects are no longer associated with the table.

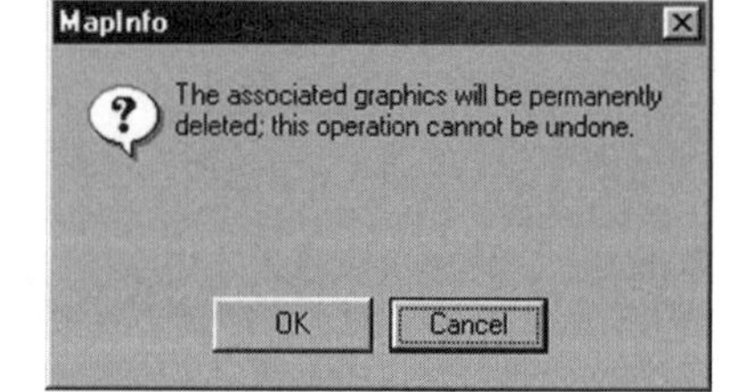

Ungeocoding warning message.

Automatic Geocoding

During the automatic geocoding process, only exact matches are geocoded. When data fail to match exactly (often the result of typographical errors), you need to geocode in interactive mode to match the near misses by hand. It is generally best to execute two passes on a table; the first pass would be set on Automatic and the second on Interactive. To geocode the *Sf_custd* table in the automatic mode, perform tutorial 10-4, which follows.

▼ Tutorial 10-4: Geocoding a Table

1. Open the *Sf_custd* and *Sf_strts* tables.
2. Select Table | Geocode from the Main menu bar and the Geocode dialog box, shown in the following illustration, will activate.

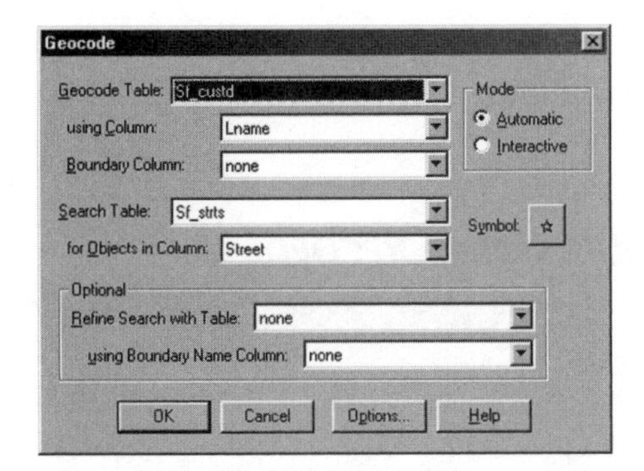

Geocode dialog box.

3. Set the dialog to geocode the *Sf_custd* table using the Address column of the data. Next, set the dialog to use *Sf_strts* as the table to geocode against.
4. As this is the first pass through geocoding, set the Mode to Automatic. After you press the OK button, MapInfo Professional will proceed to automatically geocode all records in the *Sf_custd* table. The geocoding progress dialog box, shown at top right, displays so that you can monitor geocoding progress.

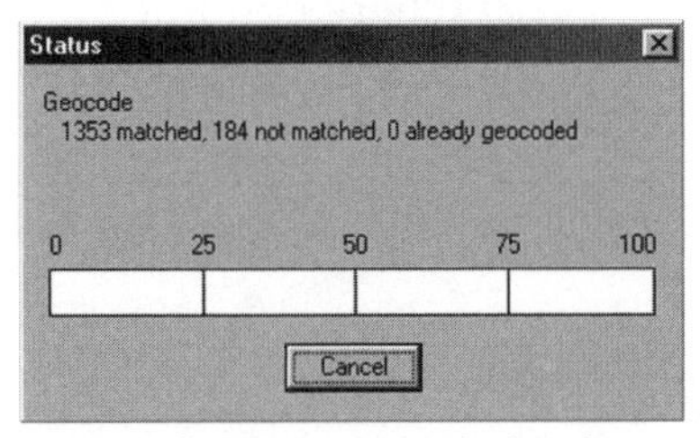

Geocoding progress dialog box.

5. The geocoding progress dialog shows the current address being matched and statistics about how many records have been matched and missed. At the end of geocoding, MapInfo displays the completion notice dialog box, shown at bottom right. This dialog shows that MapInfo was successfully able to geocode 44 of the *Sf_custd* records, and six records were not geocoded.

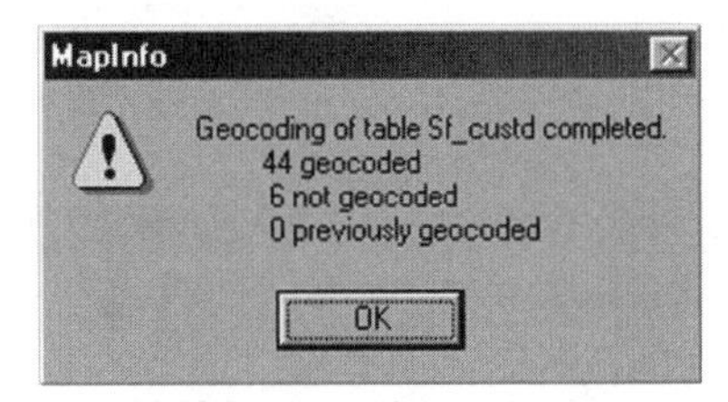

Geocoding completion notice.

When the automatic geocoding process is complete, a few records that were not geocoded will (almost invariably) remain. These records can be geocoded during the interactive geocoding process.

Finding Nongeocoded Records

Before moving on to the interactive geocoding phase, in tutorial 10-5, which follows, you will locate all nongeocoded records in the *Sf_custd* table.

▼ TUTORIAL 10-5: LOCATING NONGEOCODED RECORDS

1 Select Query | SQL Select on the Main menu bar.

2 Set the dialog to select all records from the *Sf_custd* table using the *not obj* Where condition shown in the illustration at right.

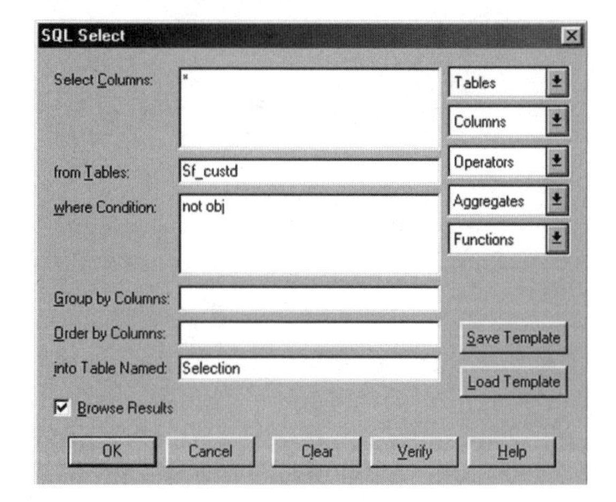

SQL Select for nongeocoded records.

After clicking on the OK button, you will see the following Browser window containing the six nongeocoded records in the *Sf_custd* table.

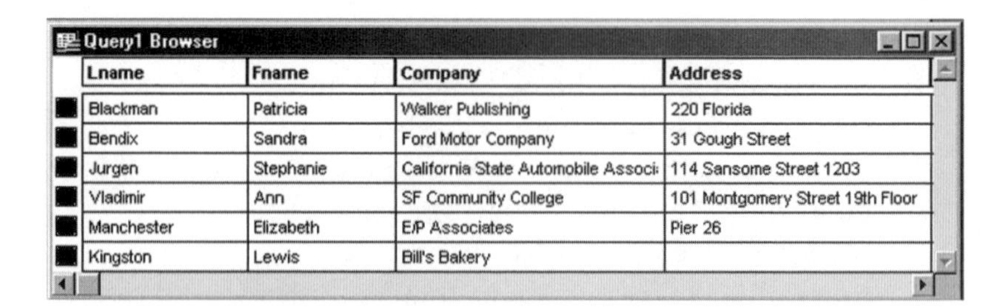
Query1 Browser

Lname	Fname	Company	Address
Blackman	Patricia	Walker Publishing	220 Florida
Bendix	Sandra	Ford Motor Company	31 Gough Street
Jurgen	Stephanie	California State Automobile Associ	114 Sansome Street 1203
Vladimir	Ann	SF Community College	101 Montgomery Street 19th Floor
Manchester	Elizabeth	E/P Associates	Pier 26
Kingston	Lewis	Bill's Bakery	

Browser containing nongeocoded records in the Sf_custd *table.*

Interactive Geocoding

In the interactive mode you are given various street name options that could be potential matches. Continuing with the example of geocoding the *Sf_custd* table, in tutorial 10-6, which follows, you will make an interactive geocoding second pass through this table.

▼ TUTORIAL 10-6: INTERACTIVE GEOCODING

1. Select Table | Geocode on the Main menu bar.
2. Select the Interactive option in the Geocode dialog box and click on the OK button.
3. The first nongeocoded record you see is 220 Florida. The Geocode dialog box shows that Florida Street may be the street name match you are looking for, as shown at right. Click on the OK button and this record will be geocoded to the Florida St. street object.

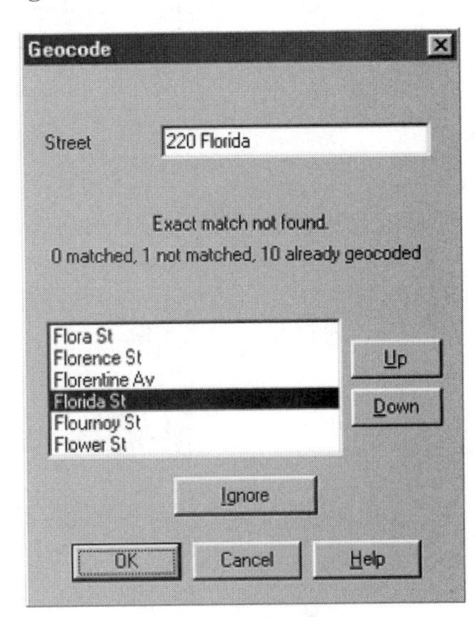

Interactive geocoding street name match.

4. In the manner previously described, continue matching records in the *Sf_custd* table through three more records. When you arrive at the fifth nongeocoded record in the *Sf_custd* table, you will see no possible match for this address, as shown in the following illustration. Select the Ignore button.

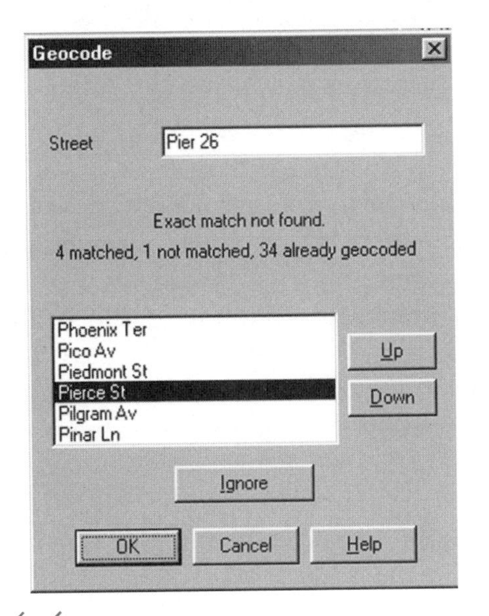

Interactive geocoding showing no street name match.

You cannot geocode the sixth record in *Sf_custd* because the address field has been left blank. At the end of the interactive geocoding process, you will have matched an additional four records, and two records will remain unmatched, as shown in the following illustration.

Additional records geocoded during interactive geocoding.

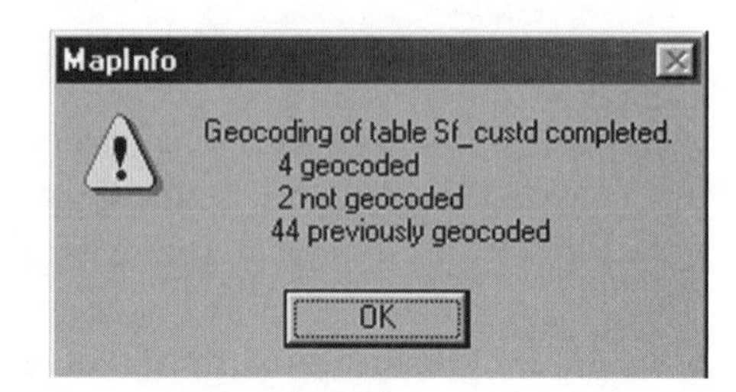

To complete the geocoding of the *Sf_custd* table you would find either the address error or the street file error of the Pier 26 street address, and enter an address for the Bill's Bakery record.

Modifying the Abbreviation File

If many records were not geocoded during the automatic geocoding procedure, you may wish to examine the unrecognized records to identify common problems. One common problem in geocoding involves street extensions and abbreviations. When MapInfo Professional geocodes, it looks for an exact match between the records in your data file and the street file. If the extensions do not exactly match, MapInfo will not find a match.

MapInfo uses an abbreviation file to help resolve extension differences that can help you achieve a higher address-matching rate. This file contains common names that may be used in the address file to be matched that may not appear in the street file. Examples are Drive versus DR., Rd. versus Rd., and Street versus St. The abbreviation file may also contain other types of common abbreviations. A sample portion of the abbreviation file is shown in the following illustration.

Each line in the file consists of a pair of items. The first item, a character string, is found in the address data to be matched. If MapInfo Professional finds this string, the string is translated as the second item in the pair when seeking an exact match.

Portion of abbreviation file.

FIRST	1ST
THIRD	3RD
SOUTH	S
WEST	W
AVENUE	AV
BOULEVARD	BLVD
CIRCLE	CIR
DRIVE	DR
HIGHWAY	HWY

SECOND	2ND
NORTH	N
EAST	E
ALLEY	AL
AVE	AV
BRIDGE	BR
COURT	CT
EXTENSION	EXT
INTERSTATE	I

The abbreviation file, *mapinfow.abb*, is located in the MapInfo program directory. To achieve a higher address matching success rate, you can add abbreviations to this file using any text editor.

Using Result Codes

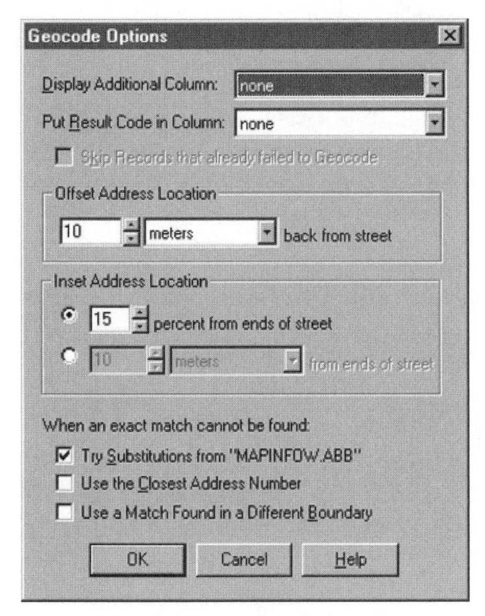

Geocode Options dialog box.

During the geocoding process, MapInfo Professional can generate result codes indicating whether a record was geocoded or not. To review the result codes in the previous geocoding example, you would need to add an integer column to the *Sf_custd* table. Moreover, you would set the Geocode Options dialog, shown in the illustration at left, to place the result codes in this new column that you have named *geo_results*. Upon using the result codes in the *geo_results* column during the previous geocoding example, you would obtain the values shown in the following illustrations.

Browser window of result codes.

Sf_custd Browser

State	ZIP_Code	Telephone	Sales	No_Employ	Sales_Rep	geo_results
CA	94103	408 730 6713	120	5	Kadison	5
CA	94103		130	7	Exler	2
CA	94103	415 553 8517	140	50	Armstrong	2
CA	94103	415 239 3751	6,850	100	Welch	602
CA	94103	415 978 2349	160	175	Caris	2
CA	94103	415 477 1506	6,830	40	Parker	2
CA	94103	415 995 2604	11,820	105	Armstrong	2
CA	94104	415 397 3278	8,810	22	Caris	2
CA	94104	415 391 1333	6,800	432	Kadison	2
CA	94104	415 788 3100	210	59	Exler	2
CA	94104	415 989 7223	220	17	Parker	5

What do these result codes mean? The result codes appear in a one- to three-digit format. A negative number means an inexact match. Numbers in the ones place of the result code refer to how MapInfo deals with the street name. Table 10-1, which follows, lists street name codes and their meanings.

Table 10-1: Street Name Codes

Code	Meaning
1	Exact match found
2	Abbreviation file applied
3	Exact match not found
4	No street specified
5	User picked a name from the list

Numbers in the tens place of the result code refer to how MapInfo deals with the address range. Table 10-2, which follows, lists address range codes and their meanings.

Table 10-2: Address Range Codes

Code	Meaning
00	Exact address range and side of street found
10	Address range found, but could not determine side of street
20	Address range not found, but within minimum and maximum ranges

Table 10-2: Address Range Codes

Code	Meaning
30	Address range not found, but beyond minimum and maximum ranges
40	Address range not specified, but matched to the minimum range
50	Streets do not intersect
70	User picked an address from the list

Numbers in the hundreds place of the result code refer to how MapInfo deals with the refining boundary. Table 10-3, which follows, lists refining boundary codes and their meanings.

Table 10-3: Refining Boundary Codes

Code	Meaning
100	Address range found in only one boundary other than specified boundary
200	Address range found in more than one boundary other than specified boundary
300	No boundary specified, but found in only one
400	No boundary specified, but found in more than one
500	Exact street address found more than once in the specified boundary
600	User picked a boundary from the list
1000000	User typed in something new

The codes work in a cumulative manner. Thus, in the case of a –43 code, the negative sign means that the record did not geocode, the 40 means that the address range was not specified but matched to the minimum range, and the 3 means that an exact match for the street name was not found. By examining result codes you can better determine where to spend additional effort to improve geocoding success rates.

Manual Geocoding

The remaining records that were not geocoded will be manually located. You can manually geocode the Bill's Bakery record with the use of the Pushpin drawing tool.

To prepare for manual geocoding, a Map window containing the *Sf_strts* and *Sf_custd* tables must be open. You also need a Browser table of the ungeocoded records. In tutorial 10-7, which follows, you will place the Map window and Browser table side by side.

▼ TUTORIAL 10-7: OPENING A MAP WINDOW AND A BROWSER

1 Verify that the *ST_CUSTD* layer is editable.

2 Select the Browser table row record for Bill's Bakery.

3 Make the map the active window and position the map for a location to place Bill's Bakery.

4 Select the Pushpin drawing tool.

5 Click the spot on the map to locate Bill's Bakery. The result is shown in the illustration at right.

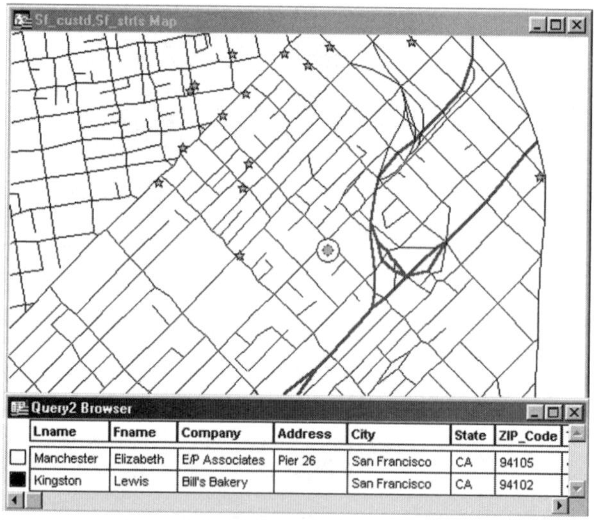

Manually geocoding a record.

For each record you are unable to automatically or interactively geocode, you will have to use manual geocoding. As seen in this review, each level of geocoding becomes more labor intensive. If you have many data records to geocode, it is worth spending the time to ensure the best geocoding success rates at each level.

Redistricting

Redistricting is the process of assigning map objects to a group or district. A district is an aggregation of both map objects and underlying data from a given table. For example, a selection of states and associated data such as sales totals can be grouped to create a sales district. In the Map window, thematic shading reveals the map objects that belong to particular districts, and in the Browser each record contains summary information about each district.

Use redistricting techniques when data are subject to frequent changes, and there is a need to experiment with different realignment scenarios. Redistricting techniques are appropriate for a wide variety of applications, such as sales territories, school and voter districts, emergency service coverage areas, delivery routes, natural resource management areas, and telecommunication coverage areas.

MapInfo Professional provides redistricting and load-balancing capabilities through the Redistrict window, a special type of Browser window. With this feature you can assign map objects to groups that have a common field. MapInfo calculates totals for each group and displays the groups in a Redistrict Browser window.

Redistricting is indispensable for creating and managing territories when you need to continually adjust them based on the changing distribution of data sets. The load-balancing capabilities of the Redistricting function calculate totals for each group, with the groups then displayed in a Redistrict Browser window. This load balancing is executed as you perform what-if scenarios. Once you are happy with the district realignments, you can make the districts permanent. If you need to change the districts, realigning them is easy. The Redistricting window is accessed by selecting Window | New Redistrict Window on the Main menu bar.

Creating Districts

In the following, you will create four sales districts using the *Sf_group* table (*samples* directory on the companion CD-ROM). Once you have created the groups, you will review the functionality for realigning the districts. In tutorial 10-8, which follows, you will prepare the *Sf_group* table for creating districts.

▼ TUTORIAL 10-8: PREPARING A TABLE FOR CREATING DISTRICTS

1 Open the *Sf_group* table.

2 Select Table | Maintenance | Table Structure from the Main menu bar.

3 In the Modify Table Structure dialog box, add a new integer data field named District to the *Sf_group* table, as shown in the following illustration. Click on the OK button.

4 Select Window | New Map Window from the Main menu bar to open a map of the *Sf_group* table.

Now that the table is prepared, *Sf_group* polygons are divided into four approximate initial territories, and a unique territory indicator is assigned in the new District column. If the data you are working with already contain territory indicators, the following steps would not be necessary.

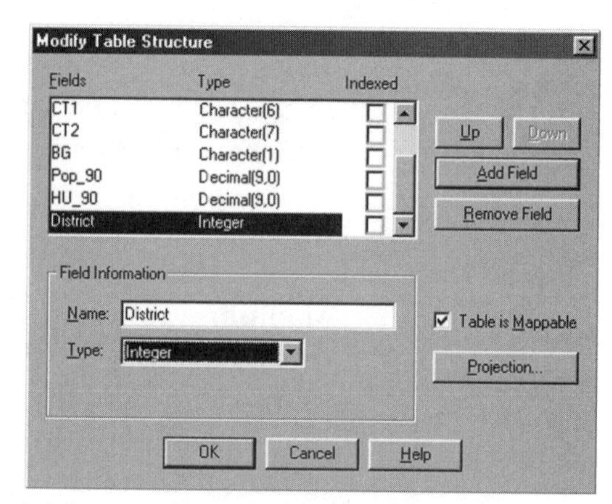

Adding District column to the Sf_group table.

5 Select the northwest quarter of the polygons in the *Sf_group* table, as shown in the illustration at right.

6 Select Table | Update Column from the Main menu bar.

7 Set the Update Column dialog box to update the District column in the selection set to the Value of 1, as shown in the following illustration. Click on the OK button.

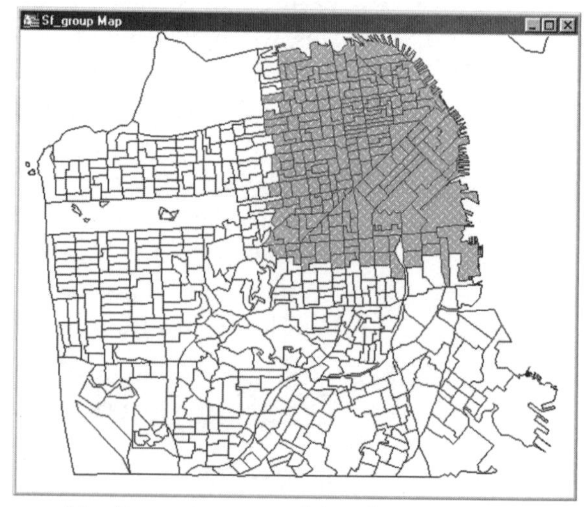

Northwest territory of the Sf_group map.

8 Repeat the preceding steps for the remaining three-quarters of the *Sf_group* table, and assign values of 2, 3, and 4 to respective quarters.

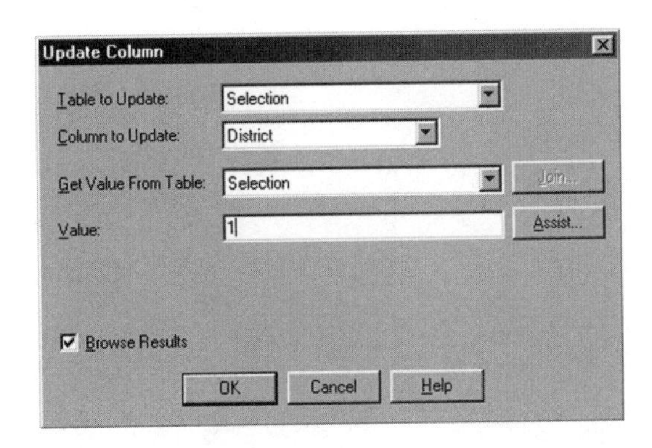

Updating District column with Value of 1 to represent first territory.

Four districts were created. However, assume you have no idea whether they are divided equally for the four sales territories. The objective is to obtain approximately the same number of households per territory in order to maintain equal opportunity for the sales representatives. This is where the Redistricting function really starts to work for you. In tutorial 10-9, which follows, you will look at the alignment of the four territories you created.

▼ Tutorial 10-9: Examining the Alignment of Districts

1 Select Window | New Redistrict Window from the Main menu bar.

2 Set the New Redistrict Window dialog box, shown in the following illustration, to use the *district* field in the *Sf_group* table. Because you also want to view the total number of households in the district, you need to add the *SUM(HU_90)* field to the Redistrict Browser.

3 When you click on the OK button, the redistricting map and Browser window will appear on the screen, as shown in the following illustration.

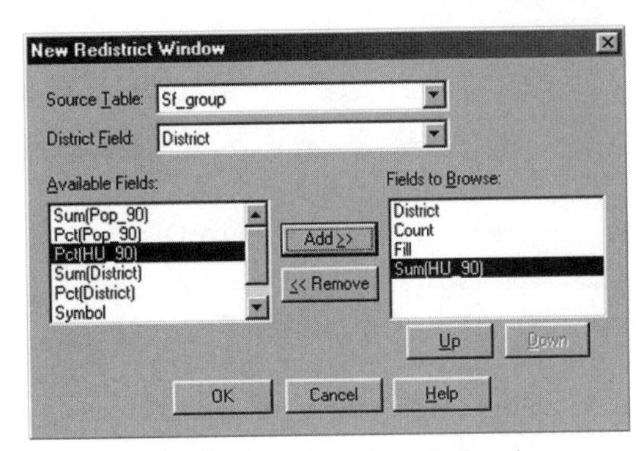

New Redistrict Window dialog box.

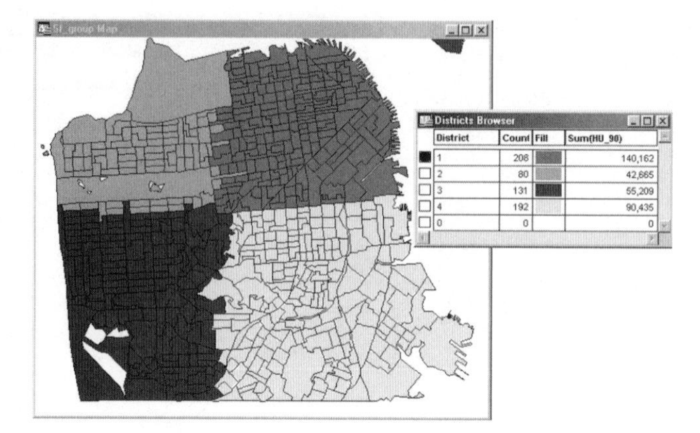

Four initial districts of Sf_group.

Note that District 1 contains far more households than Districts 2 and 3. In the next section, Districts 1 and 2 are realigned so that household numbers in the two districts are approximately equal.

Realigning Districts

In this section, the load-balancing capabilities are investigated in order to remove area and households from District 1 and add them to District 4. The first thing you must do to realign a district is set a target district. All regions selected from the other districts will be added to the target district. Setting the target district can be accomplished using one of the following methods.

Set Target tool.

- Using the Set Target tool (shown at left).
- Select the Redistrict | Set Target From Map option on the Main menu bar.
- Select a row in the Redistrict Browser window.

To realign Districts 1 and 2, perform tutorial 10-10, which follows.

▼ *TUTORIAL 10-10: REALIGNING DISTRICTS*

1 Select a polygon contained in Region 2.

2 Select Redistrict | Set Target From Map from the Main menu bar. Region 2 is now the target district.

3 Select some of the polygons contained in Region 1. As you select polygons in Region 1, note that the *Count* and *Sum(HU_90)* fields adjust, as shown in the illustration at right.

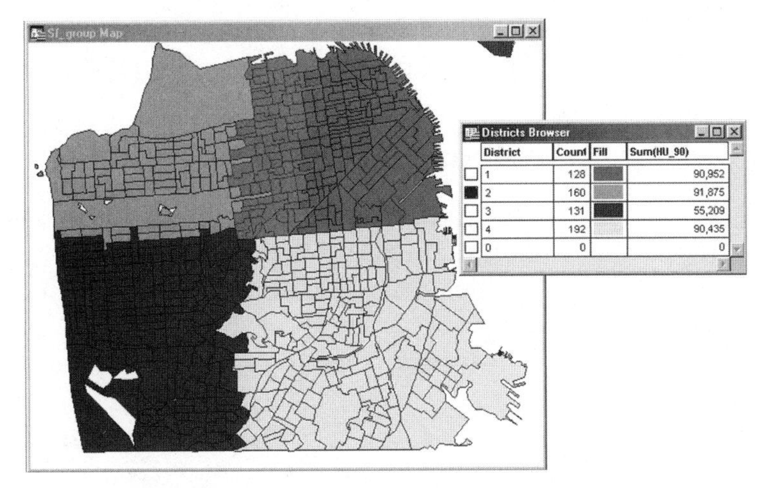

Redistricting alignment scenario.

You would continue selecting polygons until the distribution of data meets your requirements. Up to this point, all region selections are temporary.

When you are satisfied with the distribution, you can make the assignment permanent with either the Redistrict | Assign Selected Objects command or the Assign tool. The realigned districts are shown in the following illustration.

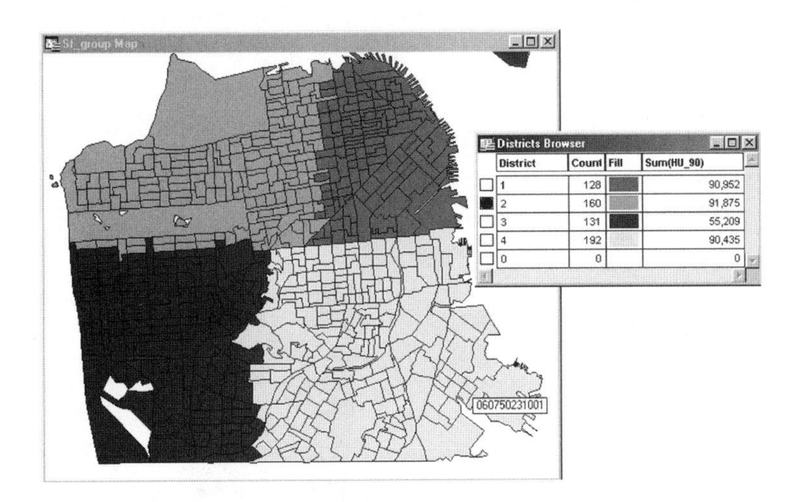

Realigned Districts 1 and 2.

To further align this example, you would need to assign more households to District 3. You can continue the process of setting targets, testing scenarios, and making permanent reassignments until all districts contain a roughly equal number of households.

Adding or Removing Districts

Selecting the Redistrict | Delete Target District option will remove a district. Upon deleting District 1, as in the previous tutorial, the regions from that district move to the bottom unassigned group of regions, as shown in the following illustration. These unassigned regions can then be realigned into the other existing districts.

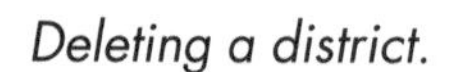

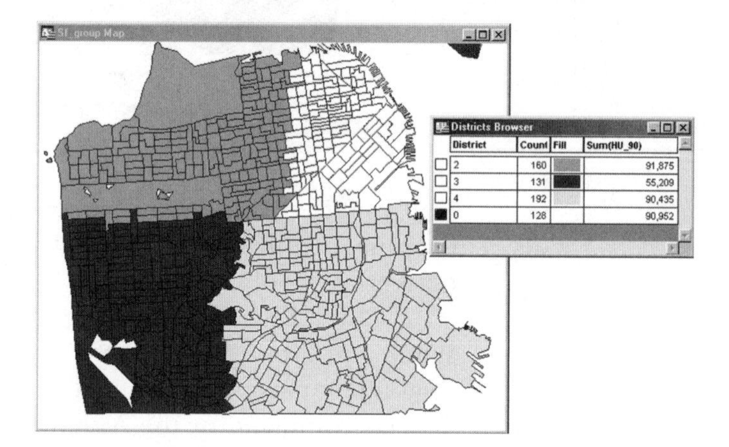

Deleting a district.

Selecting the Redistrict | Add District option creates a new record for a new district. When you add a district, as shown in the following illustration, MapInfo Professional will add a row to the District Browser table and create a new identifier and new symbology for the district. The new region is now ready for realignment.

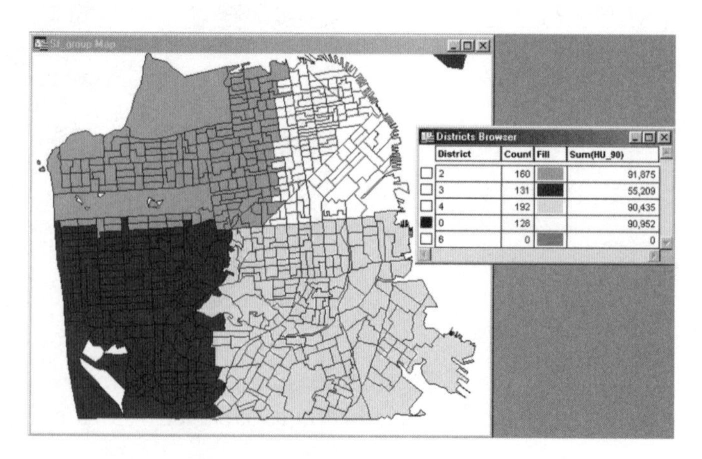

Adding a district.

MapInfo also provides configurable redistricting options from the Redistricter | Options command on the Main menu bar. In the Redistricter Options dialog box, shown in the following illustration, you can customize the order and appearance of items in the Redistricter Browser window.

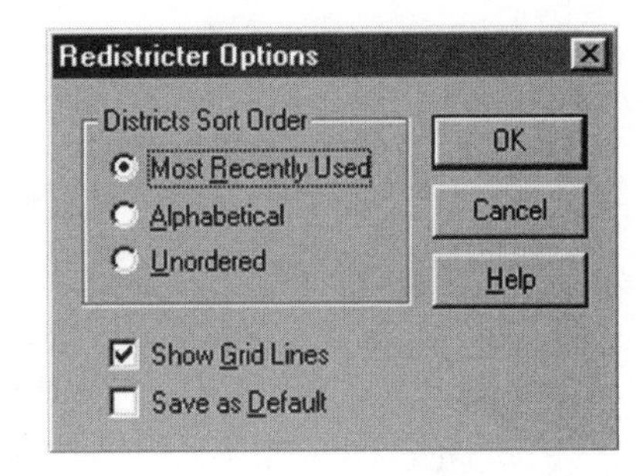

Redistricter Options dialog box.

More Redistricting

Thus far you have reviewed a redistricting example using polygons. However, the same redistricting functions can be applied to points or lines. With a set of point data, you may wish to use the redistricting function to group points into categories. The following illustration shows the results of redistricting points rather than polygons.

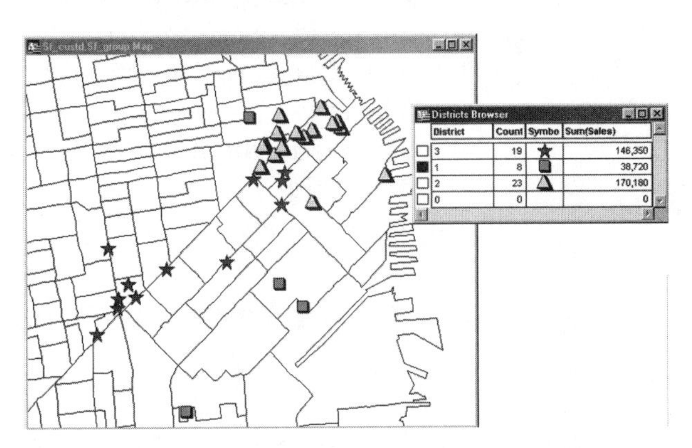

Redistricting Map window of point data.

Redistricting with a data file of lines could be useful in the analysis of routing problems. The following illustration shows three bus routes for the Pacific Heights School. Not only are the paths easy to see, but mileage for each route can be displayed in the District Browser window.

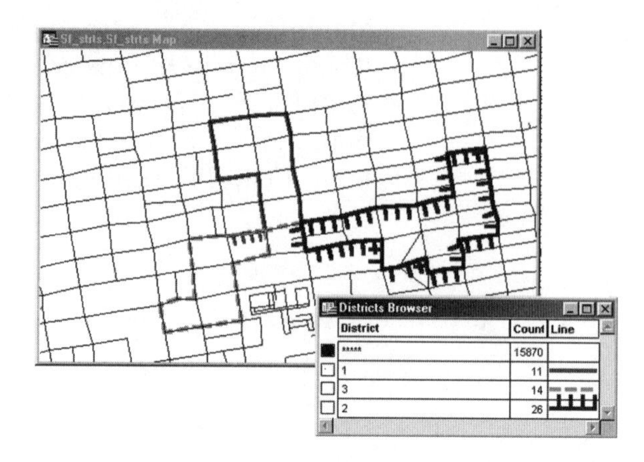

Redistricting using lines for bus routes.

Statistics

Statistics can be generated for any numeric field in a MapInfo Professional table. MapInfo provides two mechanisms for generating statistics. First, the Statistics window will display statistics for all numeric attributes in a table. Second, the Calculate Statistics command provides statistics for the values in a specified attribute and table.

The floating Statistics window contains the sum and average of all numeric fields for currently selected objects or records. The number of records chosen is also displayed in the title section of the Statistics window. As the selection changes, the data are recalculated, and the floating window is automatically updated. In tutorial 10-11, which follows, using the *States* table, you will work with the Statistics window.

▼ *TUTORIAL 10-11: WORKING WITH THE STATISTICS WINDOW*

1 Open the *States* table in a Map window.

2 Select Options | Show Statistics Window from the Main menu bar. This function is also provided on the Main tool bar via the Statistics tool.

3 Select the state of Florida and the Statistics window shown in the following illustration (at right) will be displayed.

4 Hold down the <Shift> key and select more southeastern states. Note that the Statistics window automatically recalculates the statistics, as shown in the illustration below.

Statistics

Table: States
Records Selected: 1

Field	Sum	Average
Pop_1980	3,025,290	3,025,290
Pop_1990	3,145,585	3,145,585
Num_Hh_80	1,118,191	1,118,191
Num_Hh_90	1,206,135	1,206,135
Med_Inc_80	20,869	20,869
Med_Inc_90	23,577	23,577
BPI_1990	1.0765	1.0765
Pop_Urban	2,130,139	2,130,139
Pop_Rural	1,015,446	1,015,446
Pop_Male	1,530,819	1,530,819
Pop_Female	1,614,766	1,614,766

Statistics window for the state of Florida.

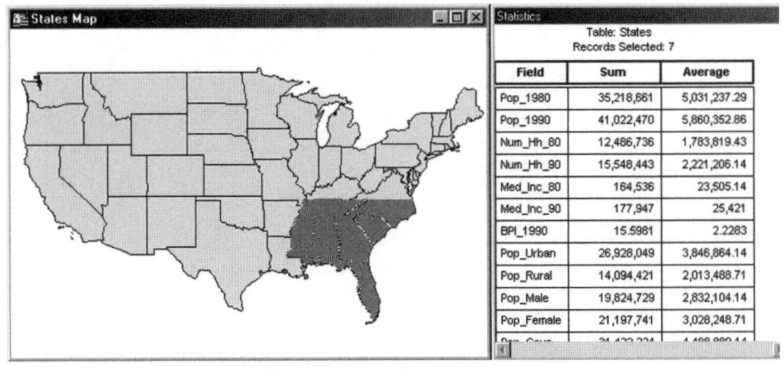

Statistics

Table: States
Records Selected: 7

Field	Sum	Average
Pop_1980	35,218,661	5,031,237.29
Pop_1990	41,022,470	5,860,352.86
Num_Hh_80	12,486,736	1,783,819.43
Num_Hh_90	15,548,443	2,221,206.14
Med_Inc_80	164,536	23,505.14
Med_Inc_90	177,947	25,421
BPI_1990	15.5981	2.2283
Pop_Urban	26,928,049	3,846,864.14
Pop_Rural	14,094,421	2,013,488.71
Pop_Male	19,824,729	2,832,104.14
Pop_Female	21,197,741	3,028,248.71

Statistics window for the southeastern states.

The previous method of reviewing statistics shows the average and sum of the numeric data fields, or very high-level summary type statistics. Using the Query | Calculate Statistics option provides more statistics, but limits the calculations to a single attribute. Included here are the column's count, minimum, maximum, range, sum, mean, variance, and standard deviation. In tutorial 10-12, which follows, you will review the Calculate Statistics options for the entire United States.

▼ *TUTORIAL 10-12: USING THE CALCULATE STATISTICS OPTION*

1 Open the *States* table.

2 Select Query | Calculate Statistics from the Main menu bar. This accesses the Calculate Column Statistics dialog box, shown in the illustration at right.

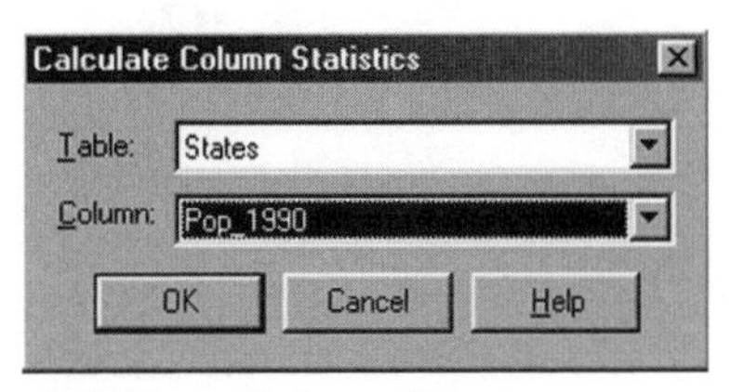

Calculate Column Statistics dialog box.

3 Set the dialog box to calculate statistics for the *Pop_1990* attribute of the *States* table. When you click on the OK button, the statistics are displayed, as shown in the following illustration.

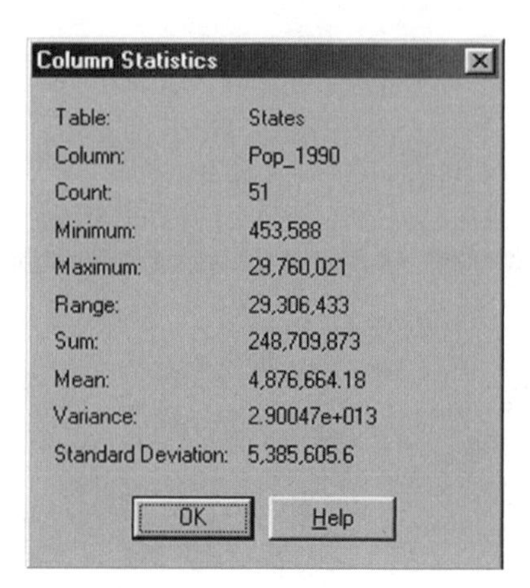
Column Statistics

Table:	States
Column:	Pop_1990
Count:	51
Minimum:	453,588
Maximum:	29,760,021
Range:	29,306,433
Sum:	248,709,873
Mean:	4,876,664.18
Variance:	2.90047e+013
Standard Deviation:	5,385,605.6

1990 population statistics for the United States.

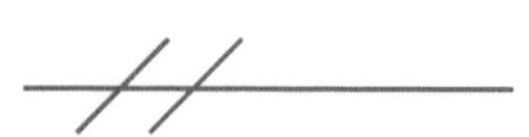

Although the preceding example was calculated based on an entire table, you could just as easily have used any subselection in the table.

NOTE: *Statistics can also be generated using the SQL Query function. The aggregate functions of* Avg, Count, Min, Max, Sum, *and* WtAvg *offer the same statistics provided in the previously described functions.*

Crystal Reports

MapInfo Professional 5.0 and above include the Seagate Crystal report writing functionality. Now you have the option of outputting tabular data in a Browser window or in the more formatted view of Crystal Reports.

Building a Crystal Report

In this section you will step through building a very simple ethnicity comparison report of U.S. population. Begin building the report in tutorial 10-13, which follows.

▼ TUTORIAL 10-13: BUILDING A CRYSTAL REPORT

1 Open the *States* table in a Map window.

2 Select Table | SQL Select from the MapInfo Main menu bar.

3 Select the *State_name, Pop_1990, Pop_Cauc, Pop_Black, Pop_Native, Pop_Asian,* and *Pop_Other* columns from the *States* table.

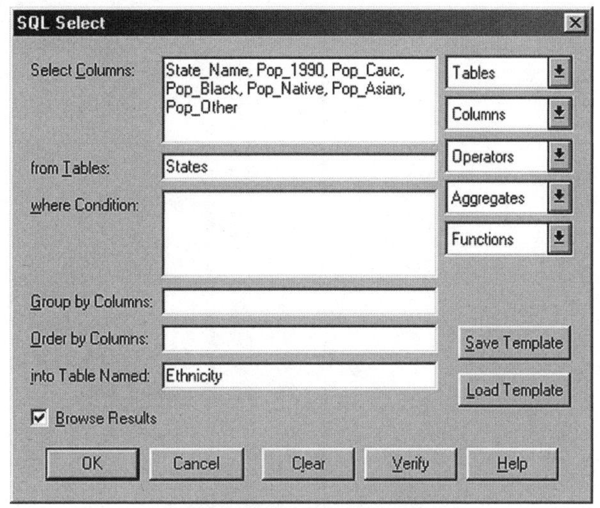

SQL Select dialog box.

4 Select this data into a table named *Ethnicity*.

5 The SQL Select dialog box should appear, as shown in the illustration at right.

6 The values you wish to include in the ethnicity report are now contained in a temporary selection table named *Ethnicity*.

7 Select Tools | Crystal Reports | New Report from the Main menu bar. The Crystal Reports menu is shown in the following illustration (at left). The New Report dialog box will display, as shown in the following illustration (at right).

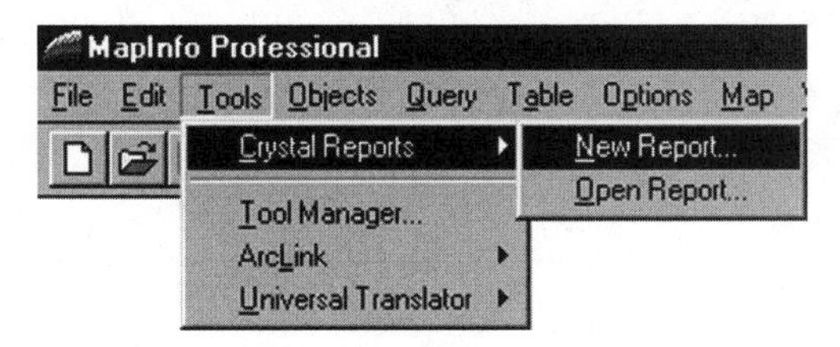

Crystal Reports menu.

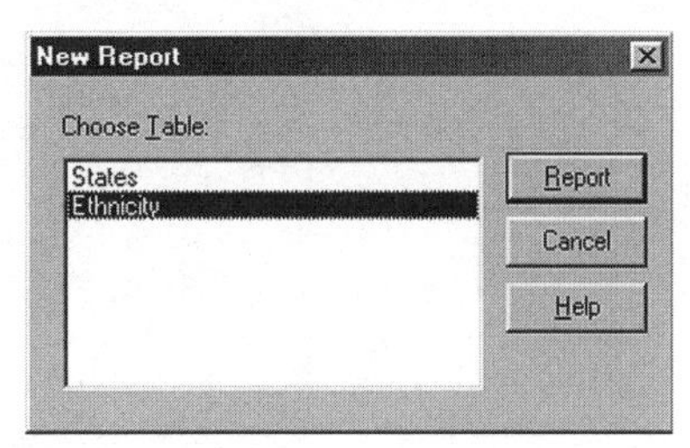

New Report dialog box.

8 Select the *Ethnicity* table from the Choose Table list and click on Report.

The Crystal Report window displays an automatically created report, and the report contains the data table you selected, as shown in the first of the following illustrations.

To improve the report's appearance, consider adding headers, footers, and titles. Click on the Design tab in the Crystal Reports window to view the report layout, shown in the second of the following illustrations. Now you are ready to make the report more descriptive and informative.

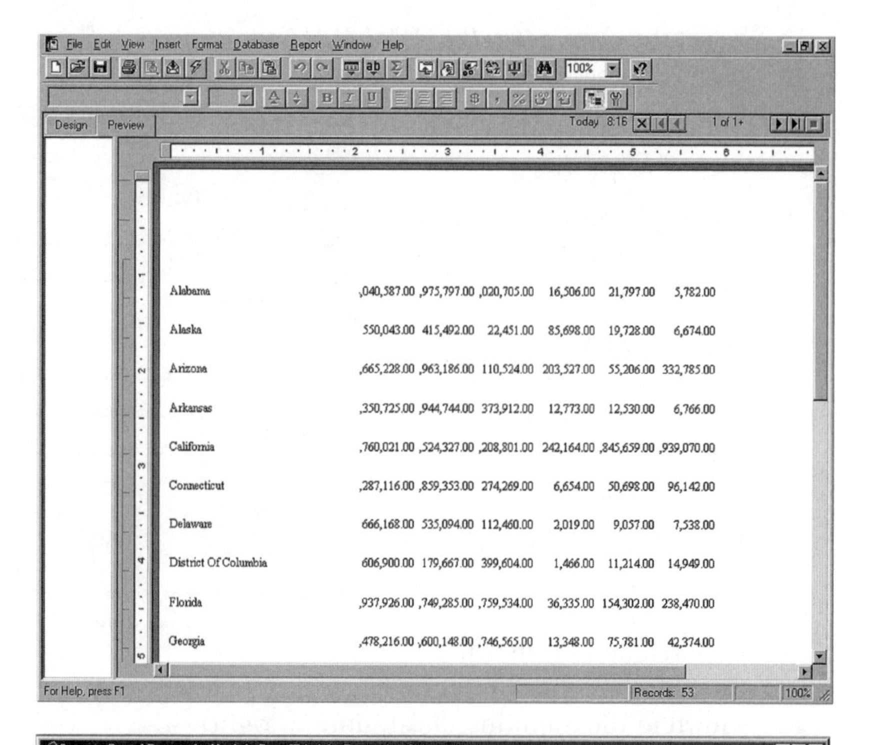

Initial ethnicity report.

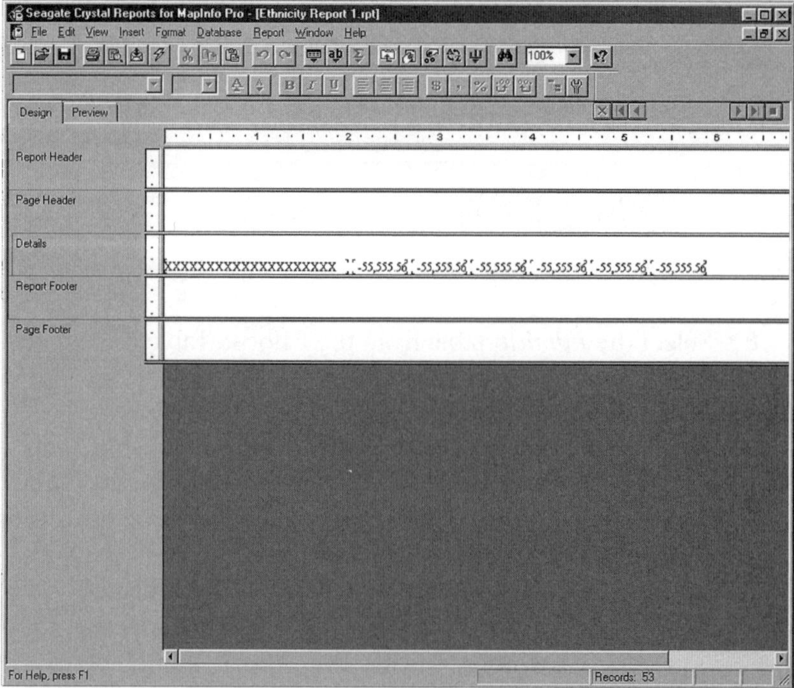

Design view of the ethnicity report.

Modifying a Crystal Report

Creating the body of a report from the MapInfo Professional interface is reasonably straightforward. Appearance enhancement is the next step. This section highlights selected Crystal Reports functions to add common components to the U.S. ethnicity report previously created.

Adding Titles

In tutorial 10-14, which follows, you will add a title and column labels to the report previously created.

▼ *TUTORIAL 10-14: ADDING A TITLE AND COLUMN LABELS*

1. Select Insert | Text Object from the Crystal Reports Main menu bar.
2. When you drag the mouse into the body of the report, it will display a bounding box. Move the cursor to the Report Header portion of the report and place the bounding text box.
3. Type in *U.S. Ethnicity* for the title of the report, as shown in the illustration at right.

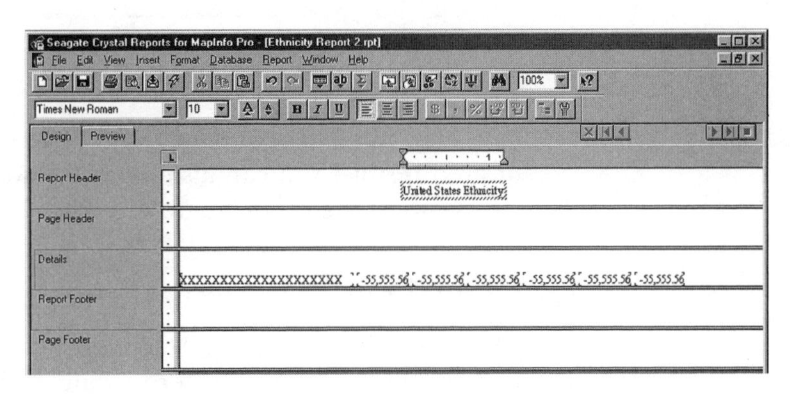

Adding report title.

4. Note that the report title has defaulted to a Times New Roman font, size 10. Select the title text and use the various font format controls on the screen to increase font size, as shown in the following illustration.

5 Repeat steps 1 through 4, using the Insert | Text Object menu item to add column headings to the page header portion of the report, as shown in the following illustration.

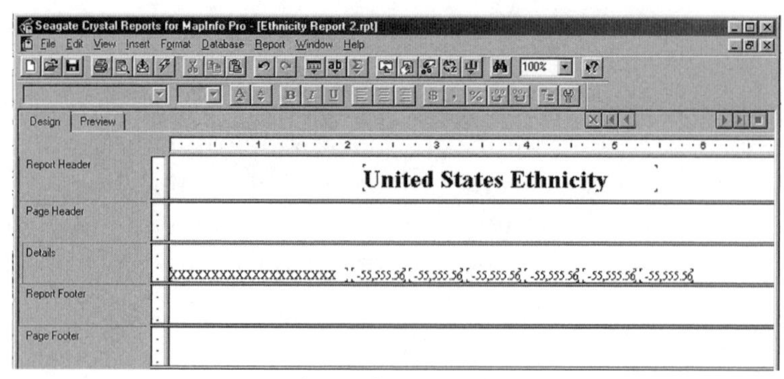

Enlarged report title.

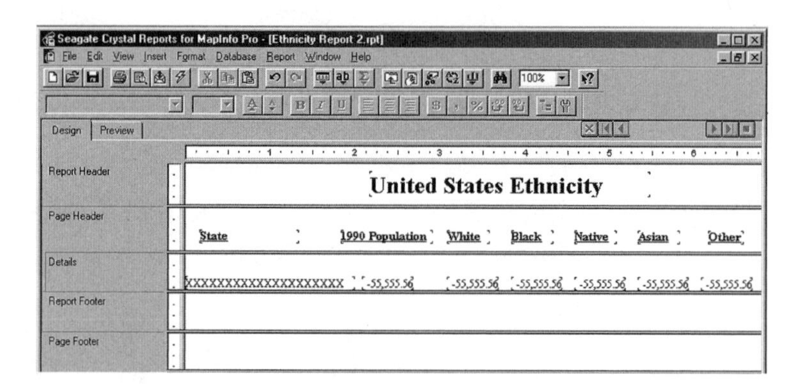

Report with column titles.

You may need to rearrange some of the fields to obtain the desired results. You can also toggle between the Design and Preview views of the report to observe the progress of report design.

Adding Footers

The report header is now starting to take shape, but what about the footer? Items to consider placing at the bottom of the report include page numbers, date, company logo, and confidentiality statements. In tutorial 10-15, which follows, you will add a page number and date to the footer of the report.

▼ TUTORIAL 10-15: ADDING FOOTERS

1 Select Insert | Special Field | Page Number Field.

2 Move the mouse and position the field in the middle of the page footer section of the report.

3 Select the Insert | Special Field | Print Date Field.

4 Move the mouse and position the field on the right side of the page footer section. The result is shown in the illustration at right.

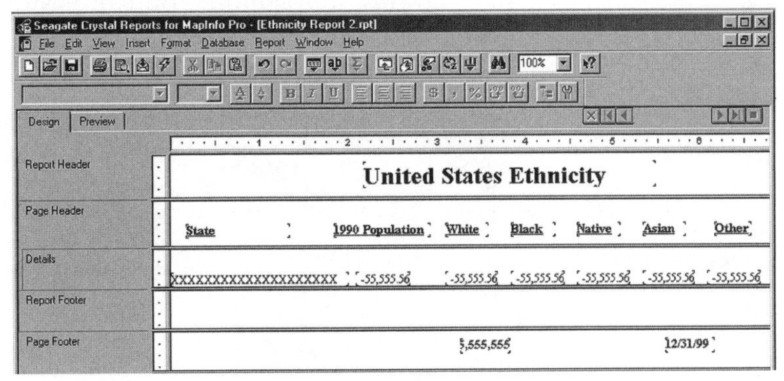

Report with page number and date fields.

Formatting Fields

In tutorial 10-16, which follows, you will format the fields of the report created previously.

▼ Tutorial 10-16: Formatting Fields

1 Select the numeric data fields in the detail section of the report; that is, all fields except the state name.

2 Select Format | Format Objects from the Crystal Reports Main menu bar.

3 Click on the Number tab of the Format Editor dialog box. Change the Decimals number in the dialog to 1. The dialog should appear as shown in the illustration at right.

4 Click on the OK button. The numeric fields no longer have numbers to the right of the decimal place. Additional rounding, as well as many other types of special formatting, could be accomplished with this dialog.

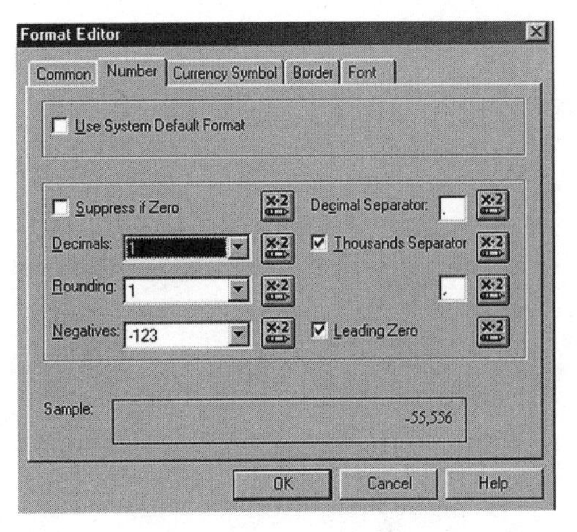

Format Editor dialog box.

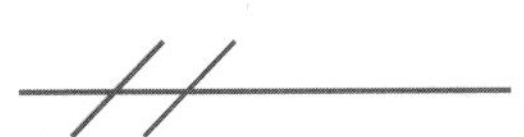

Now when you look at the preview version of the report, shown in the following illustration, you can easily understand the information presented.

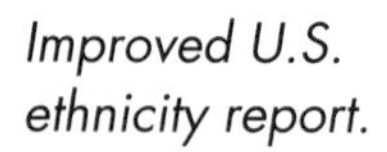

Improved U.S. ethnicity report.

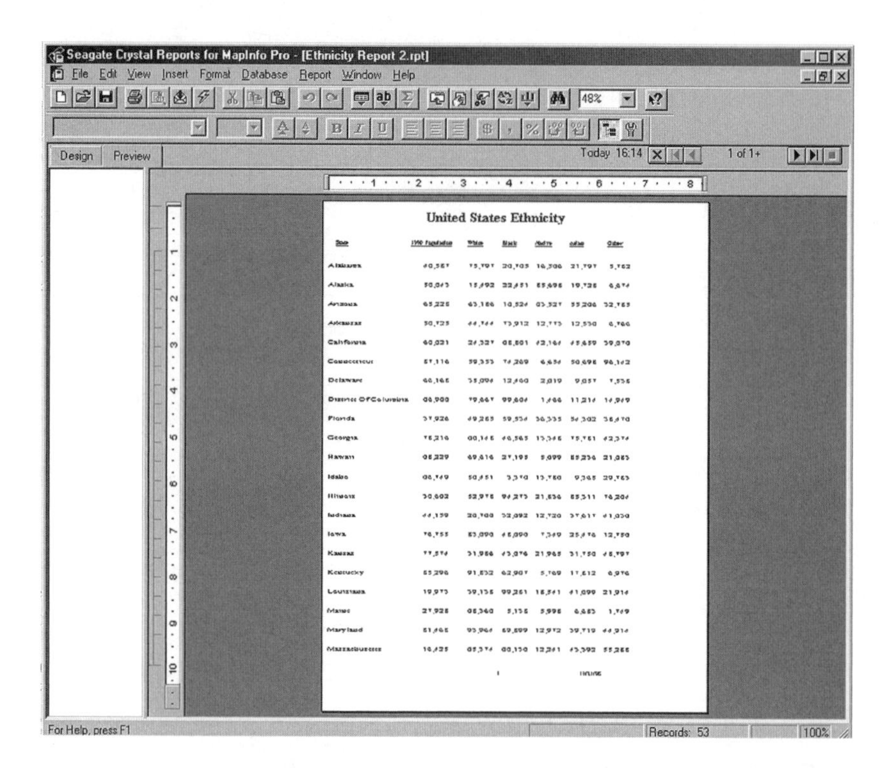

Within Crystal Reports, you can save this report with the File | Save menu options. The next time you need the report, you can use the MapInfo Tools | Crystal Report | Open Report Main menu item to reopen the report. In exercise 10-1, which follows, you will dive into a site location analysis.

■ Exercise 10-1: Site Location Analysis

Several topics in this chapter are combined in this exercise, aimed at selecting the best location for a new retail establishment. Assume that after working with a real estate agent, you have narrowed the search to two available properties in the San Francisco area. GIS techniques will be used to evaluate which of the two sites is best for your purposes.

Geocoding Potential Sites

The first step is to locate or geocode the two potential sites: 665 Green Street and 1274 Balb Street. One method is to type each of the addresses into a new table and use the geocoding function, as discussed earlier in the chapter. However, because you are interested in only two locations, using the Find command is easier.

1 Open the *Sf_strts* table.

2 Use Map | Change View to size the Map window.

3 Select Query | Find from the Main menu bar.

4 Set the Find dialog box, shown in the illustration below right, to search the *Sf_strts* table for objects in the street column.

5 Click on the OK button. MapInfo will display the next Find dialog box, shown in the illustration below.

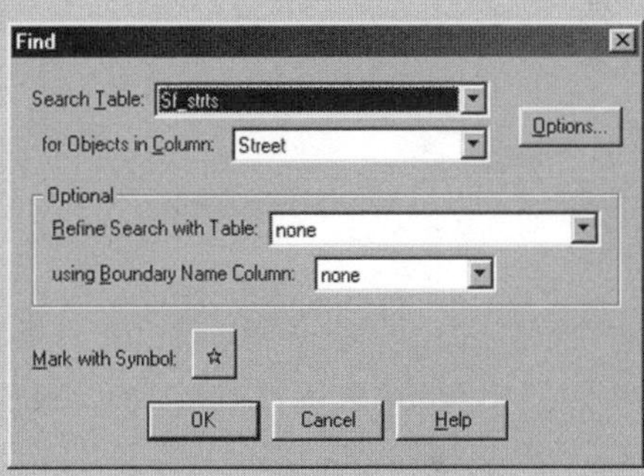

Find dialog to locate 665 Green St.

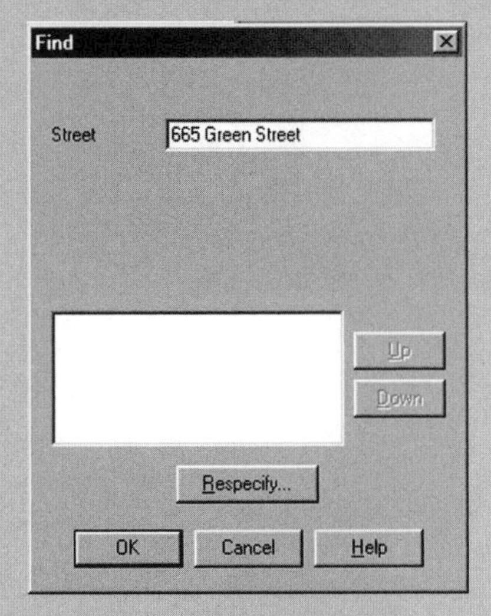

Find dialog to locate an address.

6 Enter *665 Green St* as the street to be found.

7 Click on the OK button. MapInfo will locate 665 Green Street, place a star symbol on the address, and center the map on the location, as shown in the following illustration.

Now you will locate the second property. Continue with the following steps.

8 Select Query | Find from the Main menu bar.

9 Type in *1274 Balb St*, as shown below.

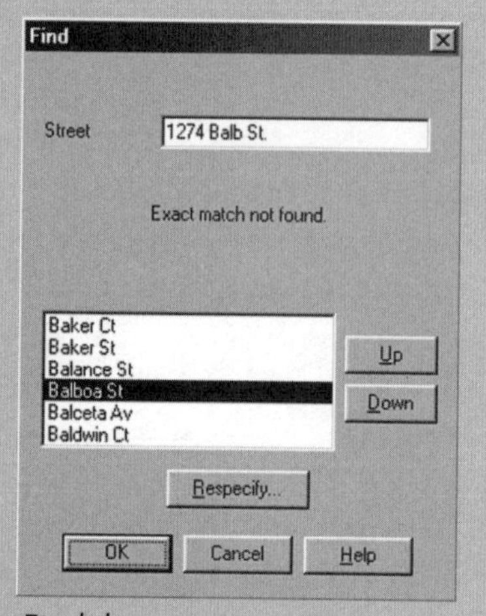

Find the correct street name.

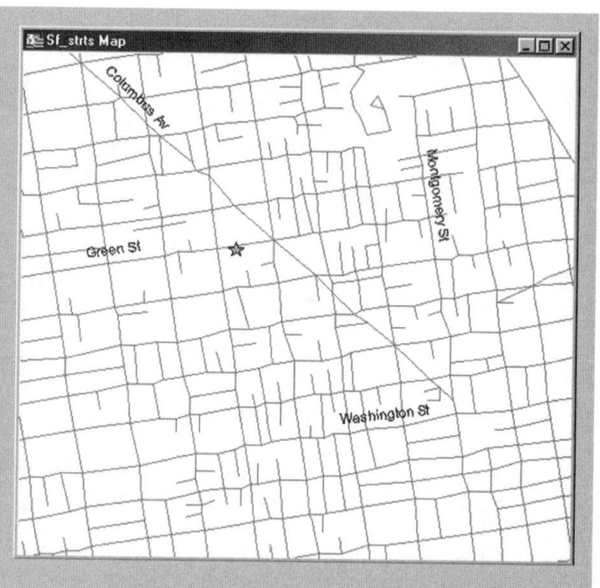

665 Green Street property location.

10 Click on the OK button. MapInfo does not find Balb St and presents you with a list of possible street matches. Click on the OK button to indicate that you want *Balboa* St.

11 MapInfo will locate 1274 Balboa St, place a star symbol on the address, and center the map on the location, as shown in the illustration at right.

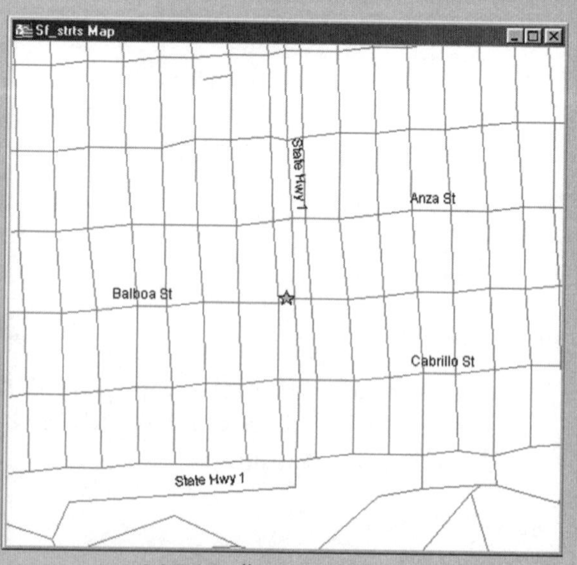

1274 Balboa St location.

Creating Buffers for Analysis

With both properties located, you are ready for the next step. Let's evaluate the number of people residing within a 1-mile radius of each prospective site. In the following, 1-mile buffers are created around each of the properties located on the map.

1 Create a new table for the buffer regions by selecting File | New from the Main menu bar.

2 Select to Add the new table to the current Map window.

3 Create an attribute as a 10-character string named *buffer_name.*

4 Click on the OK button, and save the table with the name *buffer.* The new table will be added to the Map window, and to the map as the editable layer.

5 Size and position the Map window so that you can see both sites. A zoom scale between 6 and 6.5 miles will provide a good view of the city areas.

6 Select Map | Save Cosmetic Objects from the Main menu bar.

7 Select to save the objects to the newly created buffer table.

8 Select the two points geocoded at the beginning of this exercise. These two points are now located in the *buffer* table.

9 Select Objects | Buffer from the Main menu bar. This accesses the Buffer Objects dialog box, shown in the illustration at right.

10 Set the Buffer Objects dialog box to buffer with a value of 1 mile, as shown at right, and a smoothness of 12 segments per circle. Set the dialog for One buffer for each object.

11 Click on the OK button. MapInfo will draw two circle regions on the map, around each of the retail properties you are evaluating, as shown in the following illustration.

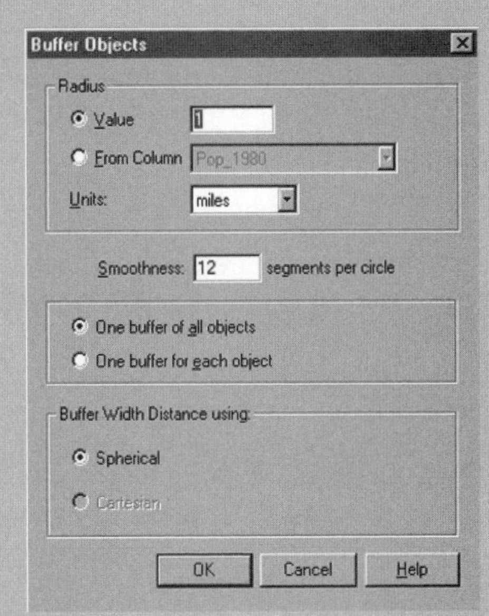

Buffer Objects dialog set for 1-mile-radius buffers.

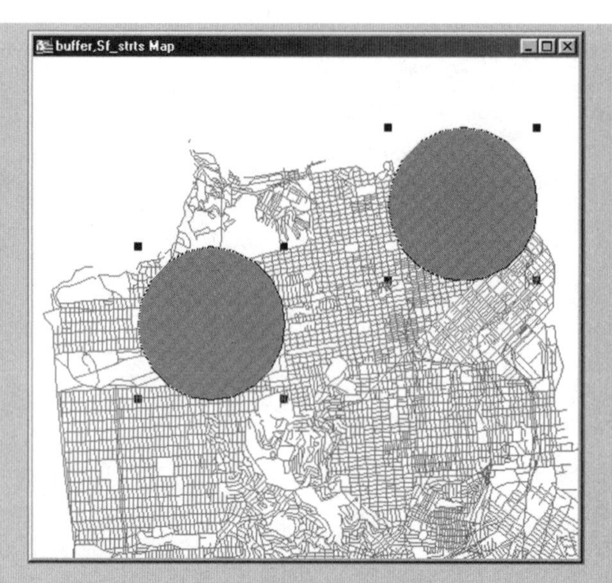

Buffers drawn on the map.

Using Statistics Functions for Analysis

Now you have all map data items prepared for analysis. You will use the *Sf_group* table to review information about people residing in the areas surrounding each of the proposed sites. The statistics functions are useful tools for quick initial analysis of this data.

1 Open the *Sf_group* table.

2 Set the layers of the Map window to contain the *buffer* table and the *Sf_group* table. Set the Cosmetic Layer as nonselectable.

3 Select the Region Select tool from the Main toolbar.

4 Select the buffer region for the area near the 1274 Balboa St location. All *Sf_group* regions will be selected.

5 Select the Statistics Summary tool from the Main toolbar. In the Statistics window, you can see that in 1990 the population count was 55,557, and the

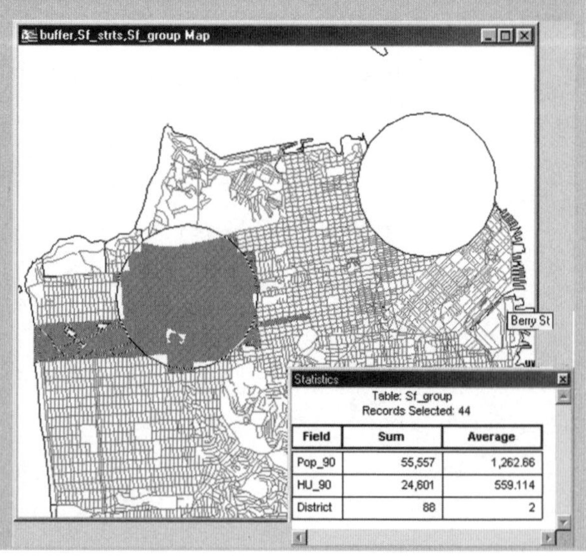

Map and statistics for the 1274 Balboa location.

household count 24,601, as shown in the previous illustration.

6 Select the Region Select tool from the MapInfo Main toolbar.

7 Select the buffer region for the area near the 665 Green St location. All *Sf_group* regions will be selected.

8 Select the Statistics Summary tool from the Main toolbar. As seen in the Statistics window shown in the following illustration, in 1990 this area had a population count of 92,968 and a household count of 53,594.

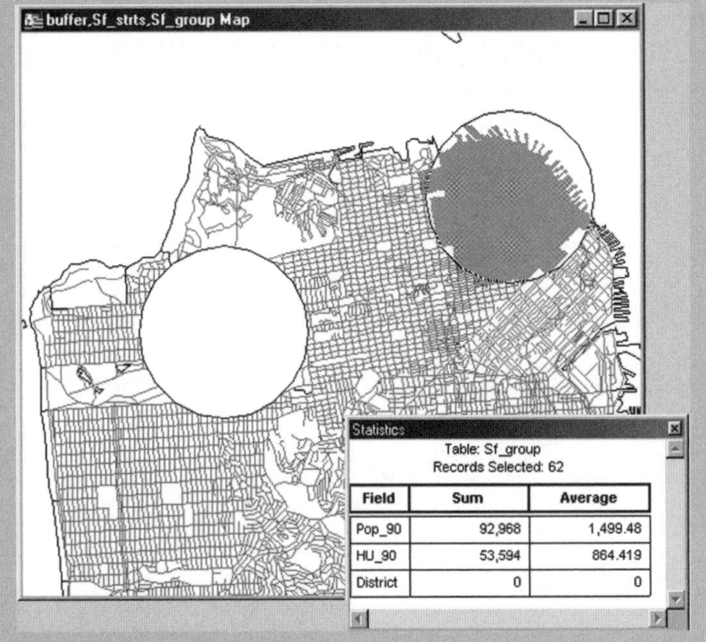

Map and statistics for the 665 Green St location.

Given these count statistics, you would most likely choose to build the new site at the 665 Green Street location because a significantly larger number of people and households are found in the area. Although in this instance the summary statistics revealed all you needed to know, you could also use the Statistics window to examine the demographic information in greater detail.

Summary

This chapter reviewed five special options provided by MapInfo Professional: the HotLink tool, geocoding, redistricting, statistics, and Crystal Reports. The HotLink tool allows you to specify World Wide Web type hyperlinks into your MapInfo Maps and Browser tables. Click on a map object or table row and jump to the associated information. Geocoding helps you to attach geographic references to data that have some type of geographic reference, such as an address, zip code, county, and/or state.

For users who maintain districts, redistricting is an easy means of redistributing objects within districts. The "scenario" approach to redistricting allows you to try several assignment scenarios and examine their effects before making the assignment permanent. Crystal Reports functionality enables the creation of sophisticated reports based on map data attributes.

CHAPTER 11

ADVANCED FEATURES

THIS CHAPTER IS FOCUSED ON ADVANCED DATA MANIPULATION features, including raster imagery, the animation layer, ODBC (open database connectivity), generating seamless map layers, using selected data translators, and digitizing.

Metadata

Metadata describes the details of a data set. This information is stored within MapInfo tables. MapInfo has built a Metadata Browser that allows you to query and review data that may meet your needs.

What Is Metadata?

In the MapInfo Professional environment, metadata is data stored in a table's **.tab* file. The metadata is not displayed within the MapInfo environment when a user opens a table, but is visible when the user opens the **.tab* file in a text editor. Viewable metadata includes the following:

- Creator(s) of a particular data set
- Sources used to create the data set
- When and by whom the last edits were performed
- Copyright notices

Why Is Metadata Desirable?

Metadata helps solve one of the biggest problems encountered in terms of data once created and distributed. Metadata stores such documentation with the data set, and is never separated from it. The following is a very simple example of metadata; in this case, included in a **.tab* file.

```
Author(Creator):
Name:
Pat Generic
E-mail:
pgeneric@somesite.com
Organization:
Generic Data Sets R Us
```

The following is the *DLG00.met* (metadata) file for the U.S. Geological Survey's 1:100,000 scale DLG data.

```
Identification_Information:
Citation:
Citation_Information:
Originator:
U.S. Geological Survey or another mapping agency in cooperation with
  USGS.
Publication_Date:
The date the DLG was entered into the National Digital Cartographic
  Data Base (NDCDB).
Title:
1:100,000-scale digital line graphs (DLGs)
Publication_Information:
Publication_Place:
Reston, Virginia
Publisher:
U.S. Geological Survey
Description:
Abstract:
Digital line graph (DLG) data are digital representations of
  cartographic information. DLGs of map features are converted to
  digital form from maps and related sources. Intermediate-scale DLG
```

data are derived from USGS 1:100,000-scale 30- by 60-minute quadrangle maps. If these maps are not available, Bureau of Land Management planimetric maps at a scale of 1:100,000 are used. Intermediate-scale DLGs are sold in five categories: (1) Public Land Survey System; (2) boundaries; (3) transportation; (4) hydrography; and (5) hypsography. All DLG data distributed by the USGS are DLG - Level 3 (DLG-3), which means the data contain a full range of attribute codes, have full topological structuring, and have passed certain quality-control checks.

Purpose:

DLGs depict information about geographic features on or near the surface of the Earth, terrain, and political and administrative units. These data were collected as part of the National Mapping Program.

Time_Period_of_Content:

Time_Period_Information:

Range_of_Dates/Times:

Beginning_Date:

19870619

Ending_Date:

present

Currentness_Reference:

publication date

Status:

Progress:

In work

Maintenance_and_Update_Frequency:

Irregular

Spatial_Domain:

Bounding_Coordinates:

West_Bounding_Coordinate:

-124.7333

East_Bounding_Coordinate:

-067.9500

North_Bounding_Coordinate:

49.3833

South_Bounding_Coordinate:

24.5333

Keywords:

Theme:

Theme_Keyword_Thesaurus:

```
None.
Theme_Keyword:
digital line graph
Theme_Keyword:
DLG
Theme_Keyword:
hydrography
Theme_Keyword:
transportation
Theme_Keyword:
boundaries
Theme_Keyword:
U.S. Public Land Survey System
Theme_Keyword:
hypsography
Place:
Place_Keyword_Thesaurus:
U.S. Department of Commerce, 1977, Countries, dependencies, areas of
  special sovereignty, and their principal administrative divisions
  (Federal Information Processing Standard 10-3):Washington, D.C.,
  National Institute of Standards and Technology.
Place_Keyword:
US
Place_Keyword:
CA
Place_Keyword:
MX
Place_Keyword_Thesaurus:
U.S. Department of Commerce, 1987, Codes for the identification of the
  States, the District of Columbia and the outlying areas of The
  United States, and associated areas (Federal Information Processing
  Standard 5-2):Washington, D.C., National Institute of Standards and
  Technology.
Place_Keyword:
FIPS codes for states covered
Access_Constraints:
None
Use_Constraints:
None. Acknowledgement of the U.S. Geological Survey would be
  appreciated in products derived from these data.
```

In brief, metadata provides a very useful place to look for documentation about the data file you are working with. Moreover, if you are creating a data file, consider including the appropriate metadata for your users.

MapInfo's Metadata Browser

MapInfo's Metadata Browser (MDB) is an easy-to-use application for building and launching geospatial metadata. It also helps you analyze the metadata you receive.

Metadata Browser button.

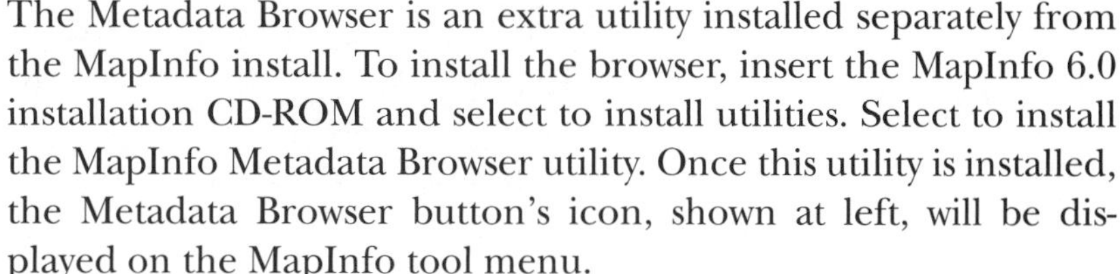

The Metadata Browser is an extra utility installed separately from the MapInfo install. To install the browser, insert the MapInfo 6.0 installation CD-ROM and select to install utilities. Select to install the MapInfo Metadata Browser utility. Once this utility is installed, the Metadata Browser button's icon, shown at left, will be displayed on the MapInfo tool menu.

MapInfo's MDB will help you formulate queries and search for metadata by way of over a hundred clearinghouses throughout the world. Each piece of metadata will describe the geospatial data you are looking for in a particular way, using federal metadata standards.

To start the MetaData Browser, click on the Metadata tool button. The startup splash screen, shown in the first of the following illustrations, will be displayed, followed by the initial query building screen, shown in the second of the following illustrations.

Metadata Browser splash screen.

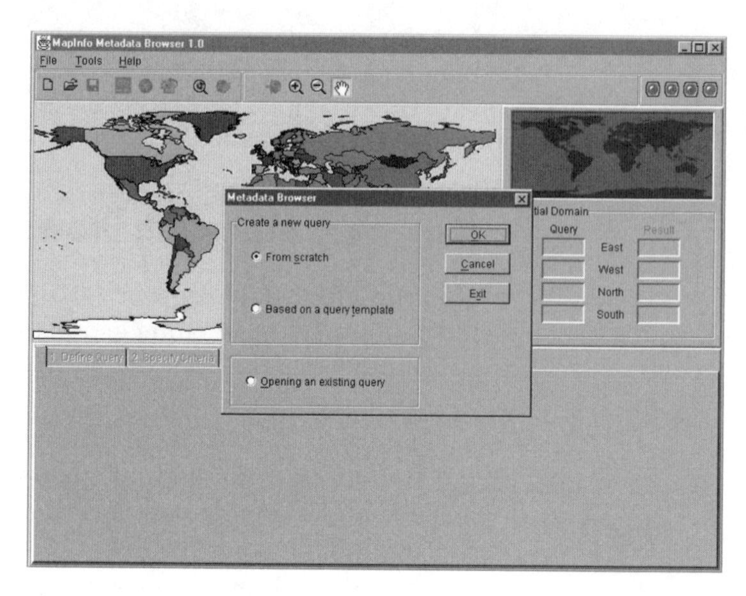

Metadata Browser initial query building screen.

Creating a metadata query is driven by a wizard-type approach. The first screen of the wizard, shown in the following illustration, asks you to name and describe the query.

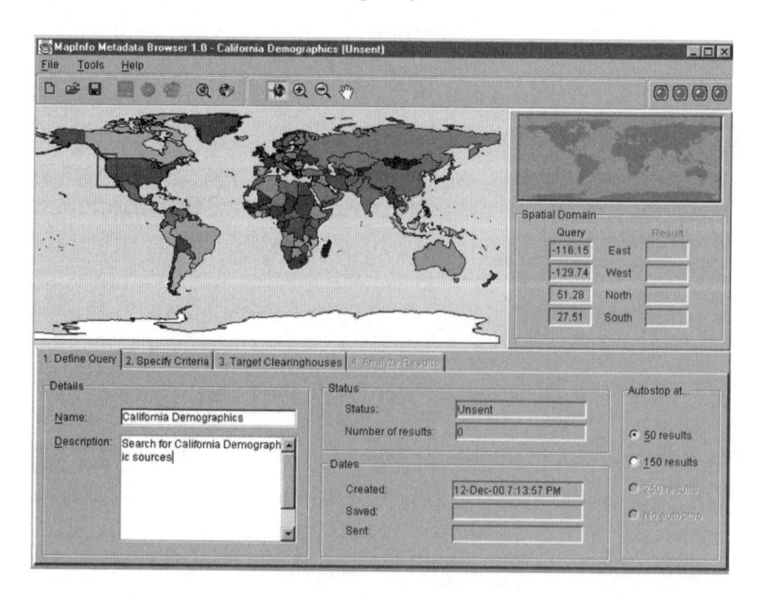

Metadata wizard screen for defining the query.

The second screen of the Metadata Query building application requires the user to insert the special keywords to look for in the various sections of metadata descriptions. The geographical region of your data search is also defined on this screen, shown in the following illustration.

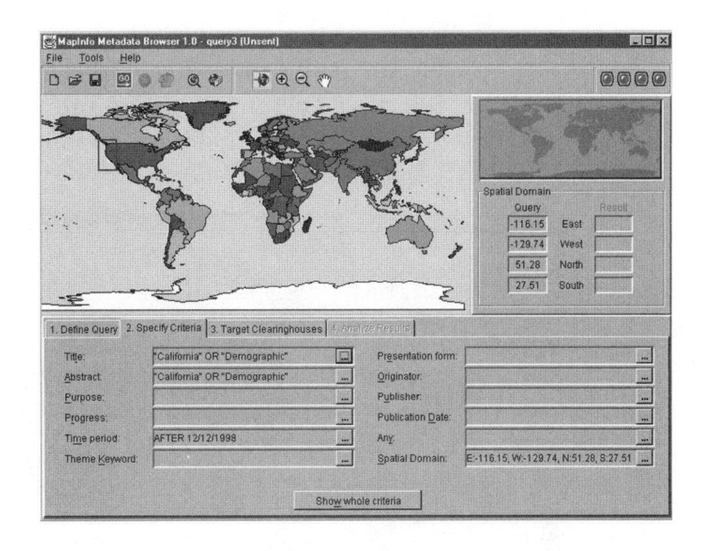

Metadata wizard screen for specifying criteria.

The third screen of the Metadata Query building application requires the user to select the data clearinghouses to search for the desired data. There is a list of available sources. The user will select which of the sources to add to the list of those that will perform the search.

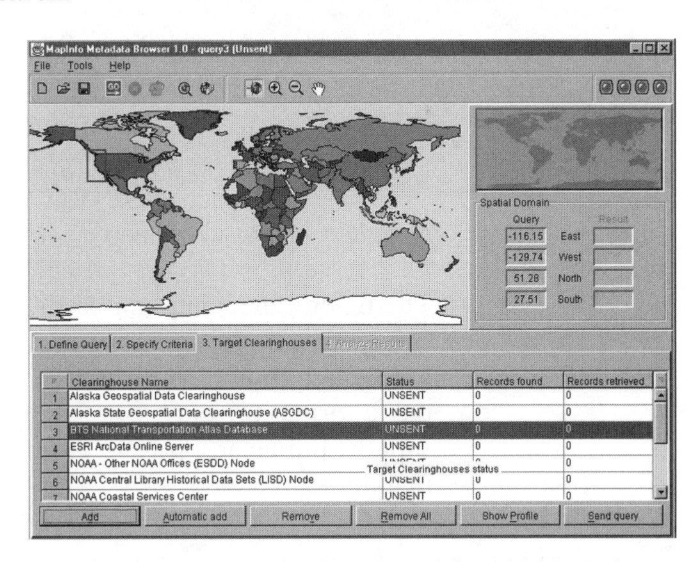

Metadata wizard screen for selecting clearinghouses.

Filling in this third screen of the metadata query completes the building portion of the application. You then click on the Send Query button in the lower right of the window. This submits the query over your Internet connection to perform the search within each of the specified clearinghouses.

Once all the queries have been attempted, the Analyze screen, shown in the following illustration, will show any found results. Selecting any one of the found items will provide you with additional information about the found data set and how to acquire it.

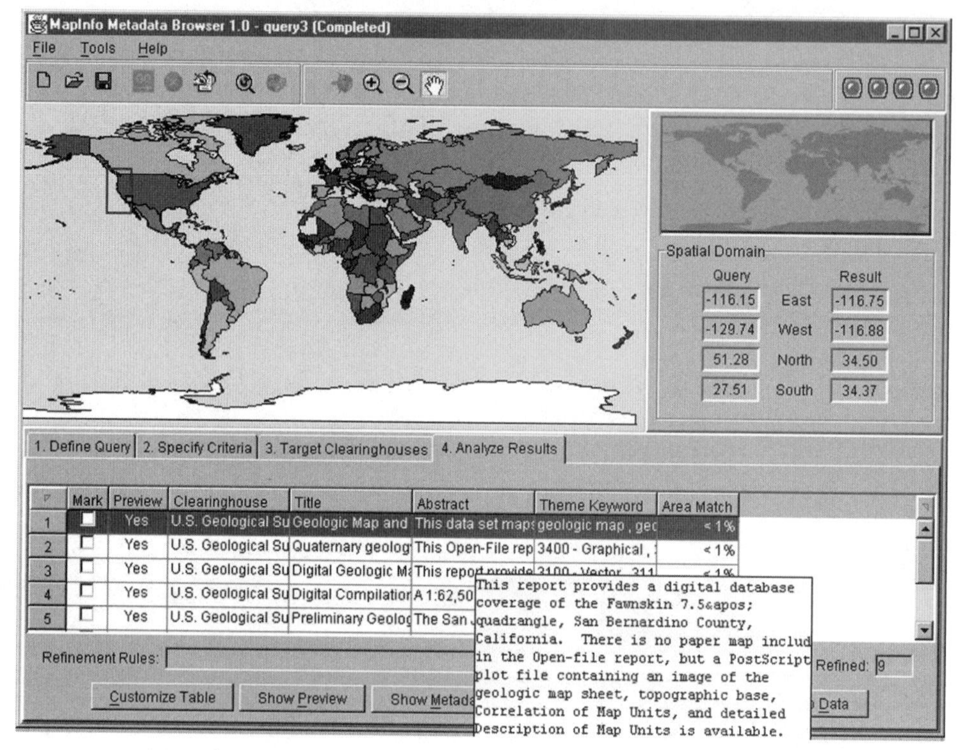

Metadata wizard screen for analyzing and displaying results.

The federal metadata standards and NSDI (National Spatial Data Infrastructure) clearinghouses, such as the NYS GIS Clearinghouse, help potential users effectively search for and find geospatial information and its related data. The NSDI mission is to support public and private sector applications of geospatial information.

Digitizing in MapInfo Professional

Although detailed step-by-step instructions on digitizing are found in the MapInfo Professional user documentation, the process is briefly summarized here. The first step in digitizing is to secure the paper map to the digitizing tablet. The paper map cannot move while digitizing, or the accuracy of your map will suffer dramatically. To begin digitizing preparation, select Map | Digitizer Setup from the Main menu bar. Select Projection from the Digitizer Setup dialog box to set the projection and coordinate system in which the paper map was created. Accuracy is paramount at this stage because the projection and coordinate system settings cannot be changed once you begin digitizing.

The next step is to register the map by selecting control points. The control points provide MapInfo with reference points for the tracing efforts to build the computerized map with the proper accuracy. Select Add from the Digitizer Setup dialog box, and then click on the paper map on the digitizing tablet to set the point. A dialog box will pop up, prompting you to enter a name and coordinates for the control point. The name is optional; the coordinates are not. Continue to Add until you have chosen several control points. (A minimum of four is recommended by MapInfo.)

If you wish to edit or remove a control point, select the point from the list and press the appropriate button. If you are digitizing an aerial photograph or something else that lacks a projection, you will need to select many more control points to ensure accuracy. (MapInfo recommends 30.) MapInfo will process the control point information by determining where you clicked on the map versus the coordinates you entered to provide an accuracy assessment for your map. The more control points you enter, the more likely you are to create an accurate map.

To begin digitizing points, select a drawing tool from the Drawing toolbar. Verify that a Map window is the active window and that the layer you want information digitized into is editable. From the Main menu, select Map | Change View and set the zoom to be larger than the area of the map you are digitizing. This action will ensure that you can see the entire map throughout the process. Now, enter the digitizing mode by pressing the <D> key on your keyboard while in the Map window. You will know you are in the digitizing mode if *DIG* appears on the status bar at the bottom of the MapInfo screen, and the screen cursor shape changes. You are now in position to begin tracing the map.

NOTE: *For digitizing options to be available, a WinTab or VTI digitizing driver must be loaded on your computer.*

Digitizing Example

The following tutorials involve digitizing a map of the state of Kansas. If you want to follow along, make a copy of the map shown in the following illustration, and attach it to your digitizing table.

Map of Kansas for digitizing tutorials.

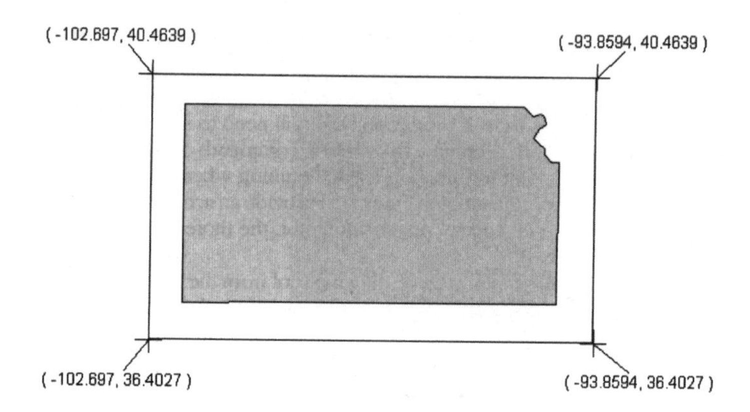

Create a new file from the File | New Table menu option. If you are using a MapInfo version prior to 4.0, remove the puck from the digitizing table when using the mouse, to avoid mouse/puck interference.

The object (Kansas map) will be digitized into the table, and the table must be the editable layer in the Map window. Make the layer editable in the Layer Control dialog box (accessed through the Map | Layer Control command). You are now ready to start with tutorial 11-1, which follows.

▼ TUTORIAL 11-1: ESTABLISHING AN EDITABLE LAYER

1. Select Map | Digitizer Setup from the Main menu bar.
2. Align the puck cross hair with the lower left corner of the Kansas map and press the first puck button.
3. Label this point as the lower left point with an X coordinate of *-102.697* and a Y coordinate of *36.4027.*

4 Continue locating and naming the remaining three corner points around the map. The Digitizer Setup window should resemble that shown in the following illustration.

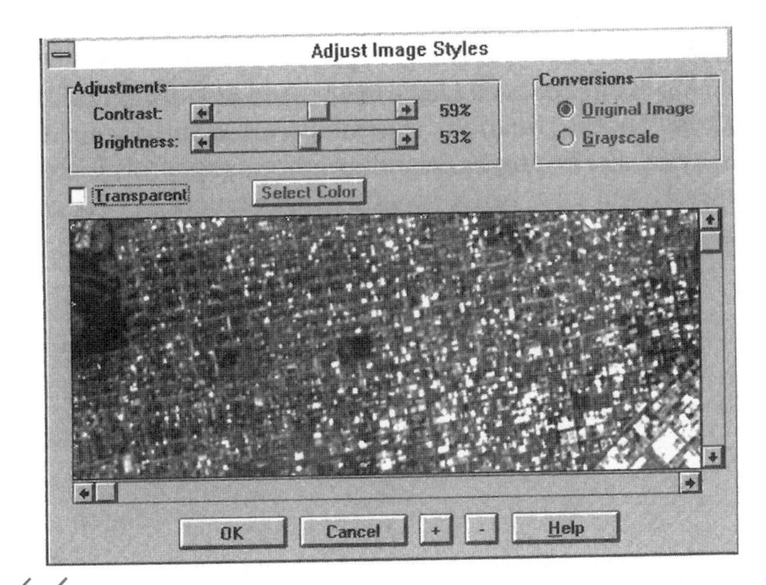

Digitizer Setup for Kansas map.

You are now ready to digitize the state's shape. Perform tutorial 11-2, which follows.

▼ Tutorial 11-2: Digitizing the Map Figure

1 Verify that the Map window is the active window.

2 Select the Polyline tool from the Drawing toolbar.

3 Press the <D> key on your keyboard. This key will toggle to and from digitizing mode. Note that *DIG* appears in the status bar at the bottom of the screen. If you are using a version of MapInfo prior to 4.0, you will also see that the cursor in the Map window has changed to a small circle with a cross hair. If you are using MapInfo Professional, the current location of the puck in relation to the Map window is shown with a horizontal and vertical line across the Map window, as shown in the following illustration.

4 Press the <S> key on the keyboard to turn the Snap mode on. This will ensure that nodes line up and connect.

5 Place the puck cross hair on the corner of the state of Kansas and click the first puck button. Proceed around the shape of Kansas while clicking the first puck button for each point you are able to identify.

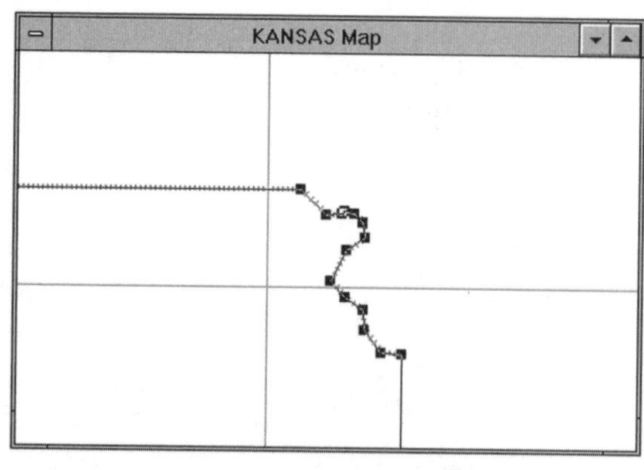

Puck location indicator when digitizing with MapInfo Professional.

6 When you return to the starting point, click the second puck button to indicate that you are finished drawing the polyline, as shown in the following illustration.

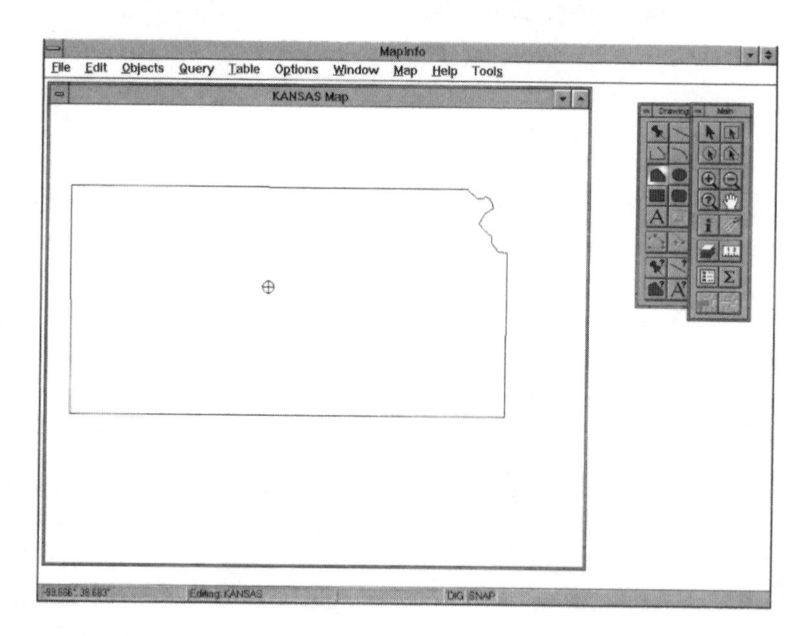

Digitizing the state of Kansas with the puck location indicator seen in versions prior to MapInfo Professional.

To check your digitizing output, open the *States* table. The newly digitized Kansas shape should lie directly over the *States* table representation of Kansas.

Alternatives to Digitizing for Creating Boundaries

For users who do not have access to digitizing equipment, there are alternatives for creating custom geographies if absolute accuracy of graphics is not required. These alternatives use MapInfo Professional's drawing, buffering, and aggregating capabilities to create new line and region geographies.

Drawing is a rather crude method, but will suffice for creating custom boundaries, such as trade areas. The Polygon drawing tool can be used to create custom boundaries, and the Line drawing tool to create line objects (such as roads). (See Chapter 7 for details on using these tools for editing purposes.)

Buffering and aggregation are suitable for creating custom boundaries using existing geographies. Detailed instructions for creating buffers appear in Chapter 10. Aggregation as described in Chapter 7 can be used to create new geographies from existing geographies such as block groups, census tracts, or zip code areas.

TIP: *The easiest way to create a new table containing aggregated geographies is to first select the geographies you wish to aggregate, and then save the selection as a new table, using the File | Save Copy As command. Open the table, and then modify its structure to contain the information you wish. Select the geographies in the new table, and then select the Objects | Combine command to create a single, new geography.*

Data Translators

Numerous tools are available in the Tools menu. This section expands on data translation capabilities found in the toolbar.

Using the Tool Manager

The Tool Manager allows you to easily add and access MapBasic programs (e.g., labeler, universal translator, and so on). For example, in tutorial 11-3, which follows, you will add the universal translator.

▼ TUTORIAL 11-3: USING TOOL MANAGER TO ADD A UNIVERSAL TRANSLATOR

1 Select Tools | Tool Manager. The Tool Manager dialog box, shown in the illustration at right, will display.

2 Click on the Add Tool button. The Add Tool dialog box, shown in the illustration below right, will display.

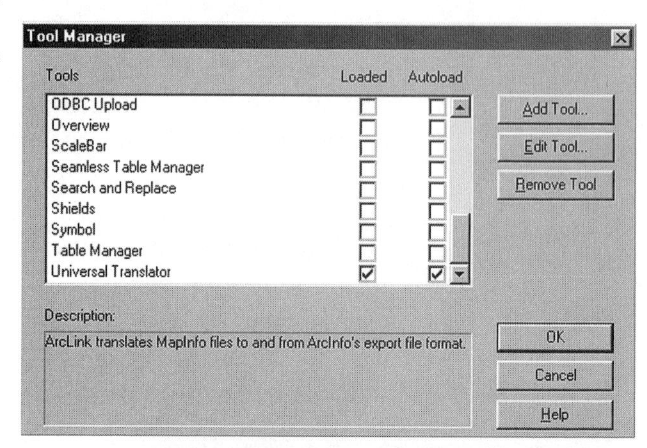

Tool Manager dialog box.

3 Enter *Universal Translator* as the Description.

4 Click on the location button and select the path for the MapBasic *mut.mbx* program. The program should reside in the *ut* directory, which is directly below the MapInfo directory.

5 Click on OK and the universal translator is added to the Tool Manager window.

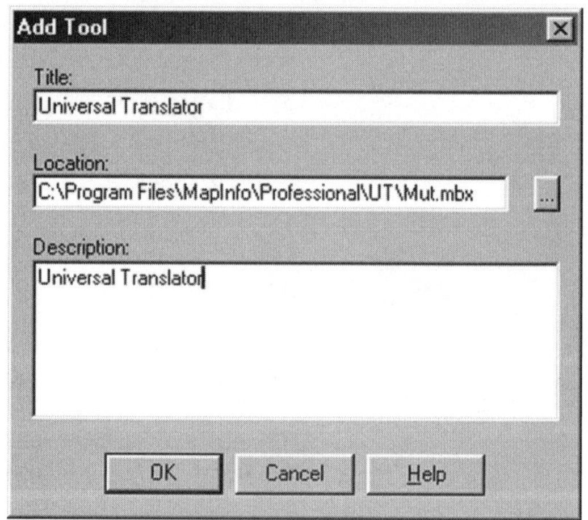

Add Tool dialog.

Universal Translator

The Universal Translator allows you to import and export MapInfo data using the following popular mapping file formats.

- Spatial Data Transfer Standard (SDTS)
- Vector Product Format (VPF)
- AutoCAD DWG/DXF
- ESRI Shape
- Intergraph/MicroStation design

- MapInfo MID/MIF
- MapInfo TAB

The universal translator is powerful and easy to use, and moves data quickly between common data formats. Check the Tools main menu bar to locate the Universal Translator menu item, as shown in the following illustration.

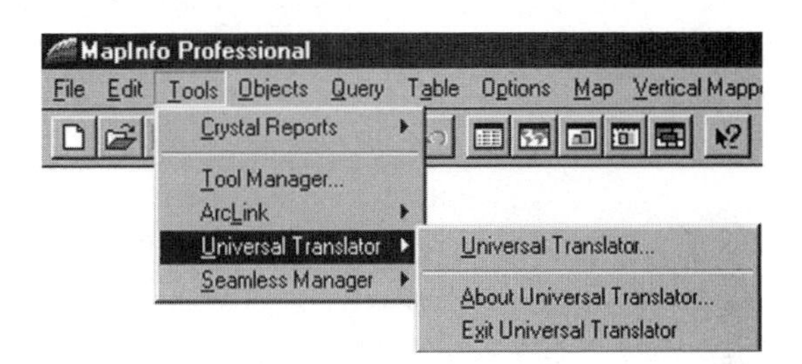

Universal Translator menu.

When you select Universal Translator, you will be provided with the Universal Translator dialog, shown in the following illustration. In the Universal Translator dialog, three types of information are specified. First is the description of your source data file. That is, identify the location and type of file you want the translator to work with. Next is defining the destination information about your request, or identifying the format of the resulting translated file, as well as its location. Finally, the log file is specified, which, after the translator finishes, can be viewed to check information and identify errors associated with the translation process.

Universal Translator dialog.

MapInfo Corporation provides the necessary links and documentation to describe how to create your own translators. With this information available, translators written by third-party developers will likely be created.

ArcLink

ArcLink is MapInfo Professional's bidirectional conversion utility program that translates ArcInfo export files to and from MapInfo TAB files. You can translate ArcInfo export files into either MapInfo format (TAB) or MIF/MID format (ASCII). ArcInfo points, arcs, nodes, annotations, tics, and polygons convert to MapInfo points, lines, polylines, text, and regions, respectively. Attribute values and relational data will also be translated. Upon completion of the conversion, the MapInfo TAB file can be displayed in MapInfo. The ArcLink menu is shown in the following illustration.

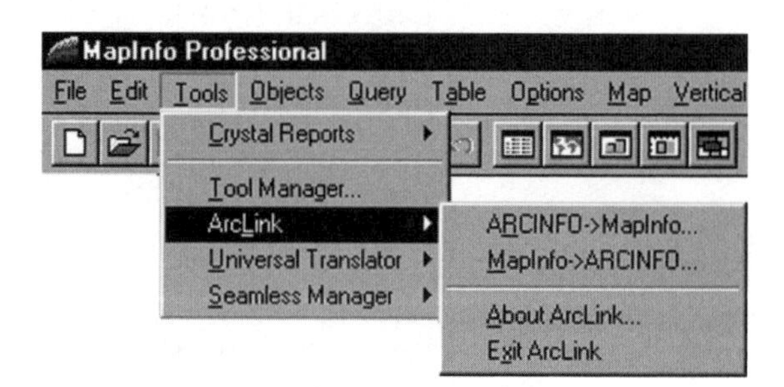

ArcLink menu.

You can choose to combine multiple TAB files into a single ArcInfo coverage. Upon completion of the conversion, the ArcInfo export file is ready to be directly imported into ArcInfo. Like the universal translator, ArcLink is found in the Tools menu. Upon selecting the Tools | ArcLink | ArcInfo>MapInfo menu item, you will be presented with the ArcLink dialog box (shown in the following illustration), which leads you through the translation process.

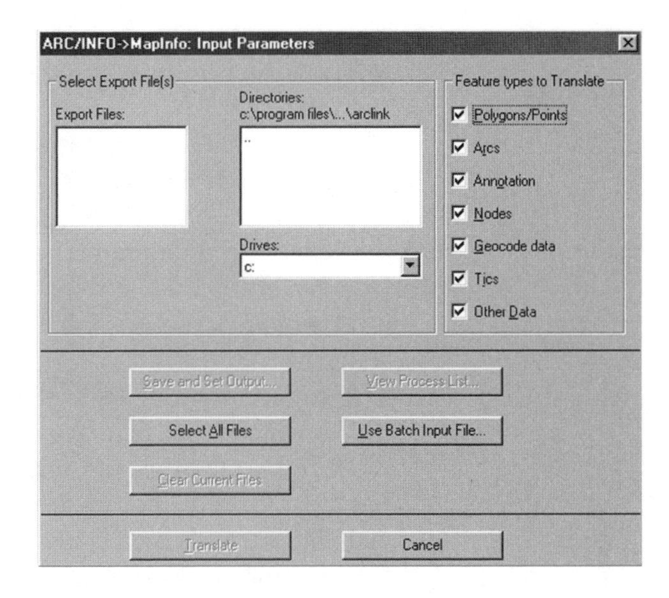

ArcLink *dialog box.*

Table 11-1, which follows, correlates features as they are translated from ArcInfo to MapInfo Professional.

Table 11-1: Translation from ArcInfo to MapInfo Professional

ArcInfo	MapInfo Professional
Points, nodes, tics	Points
Arcs	Lines or polylines
Polygons	Regions
Annotation	Text
Data	Data with no graphic objects

ODBC

ODBC (open database connectivity) allows you to access a remote database and update information in that database while working in MapInfo Professional. You can execute remote access of Oracle, Sybase, and Microsoft Access databases (to name a few) through linked tables. The ODBC functionality also allows you to assign coordinates to objects in a remote database so that the information can be mapped in MapInfo Professional.

Through ODBC functionality, the program creates a MapInfo table from data that has been downloaded directly from a remote ODBC database. You have the option of retaining the link between the new table and the remote database, which permits changes in MapInfo Professional to be reflected in the remote database. If the link is not retained, the new table is "standalone" in MapInfo.

The main limitation associated with linked tables is that if you do not download the primary key associated with the remote database, and you attempt to maintain the link with the remote database, you cannot modify the data in MapInfo Professional. Next, you cannot pack or modify the structure of a linked table. As a workaround, you can always download the database as a standalone table; however, changes made in MapInfo Professional will not be reflected in the remote database.

ODBC drivers must be installed in MapInfo Professional. If you cannot access the File | Open ODBC Table command, install the ODBC drivers from the MapInfo Professional installation CD-ROM or disks. When the ODBC drivers have been installed, you will see an additional ODBC toolbar, shown in the following illustration.

ODBC toolbar.

Opening an ODBC Table

To open an ODBC file, use the ODBC Open tool button or File | Open ODBC Table. Either of these actions will activate a set of wizard-type dialogs that lead you through the steps required to support downloading an ODBC table. Upon opening an ODBC table, and if you have not already connected to a data source, the SQL Data Sources dialog, shown in the following illustration, is the first dialog to display.

Select ODBC driver connection to use.

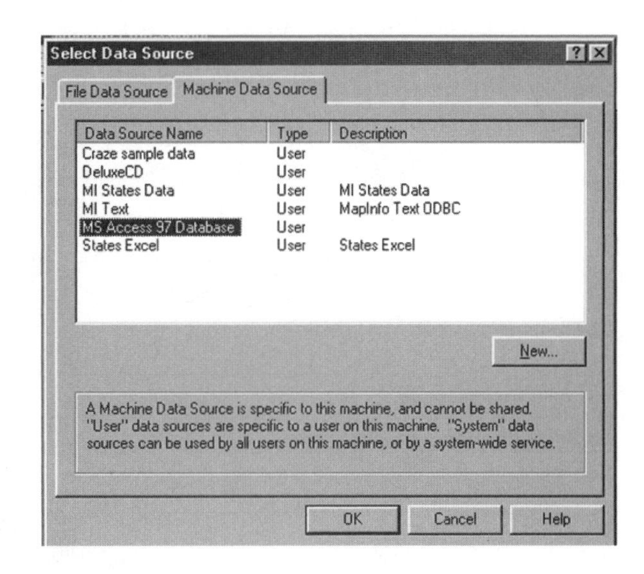

Choosing a Table

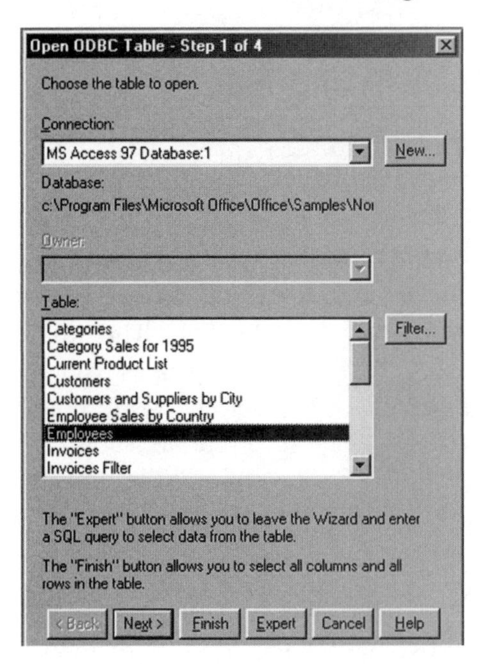

After a connection is established, the Open ODBC Table Wizard will be invoked. The wizard displays a series of four dialogs. The dialogs will assist you in opening an ODBC table. Each dialog will have an identical set of buttons along the bottom edge. The first dialog asks you to select a table from the connected data source, as shown in the illustration at left.

Select ODBC table.

Choosing Columns from a Table

After selecting a table, you must determine which data are to be extracted from the table. First, determine which columns to select. The second dialog asks you to select columns, as shown in the following illustration.

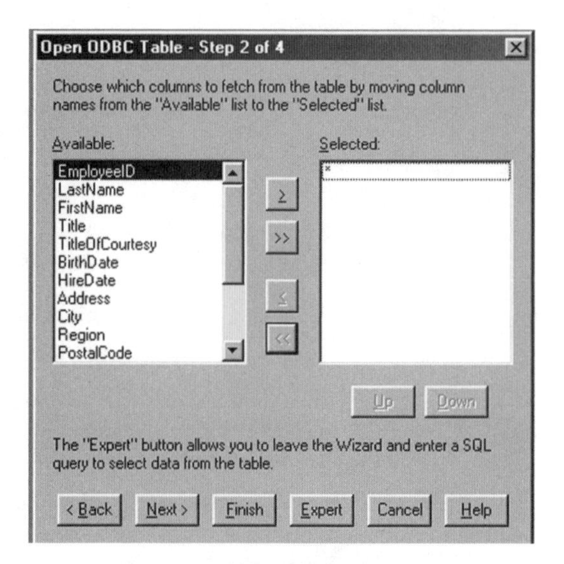

Selecting columns from an ODBC table.

Choosing Rows from Columns in a Table

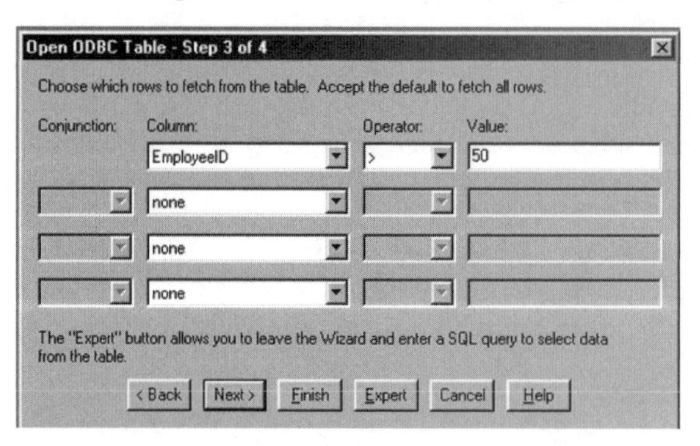

Selecting rows from ODBC table.

After selecting the columns to be downloaded from a table, you can filter the rows to be downloaded. The third dialog asks you to select rows to be filtered, as shown in the illustration at left.

Saving the Table Locally

Once the data to be downloaded have been identified, in the fourth dialog, you must determine the path name for the local table created, as shown in the following illustration.

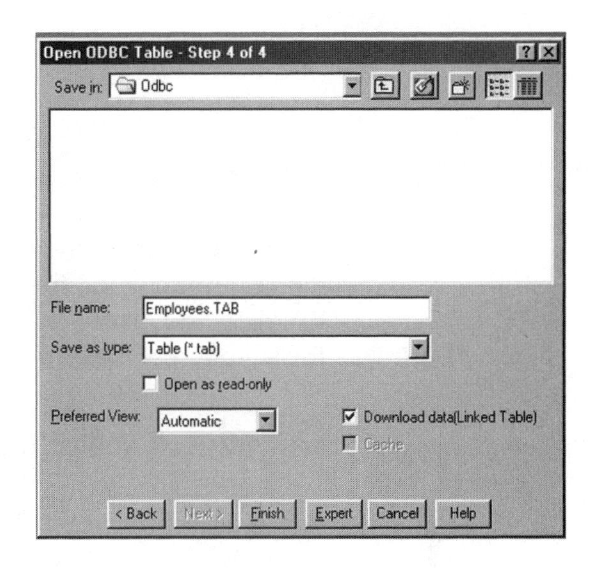

Saving ODBC table as local MapInfo table.

OLE Embedding

With the Drag and Drop button, a MapInfo map object can be embedded in MapInfo Professional files, as well as in other OLE container applications. MapInfo Professional's OLE embedding allows container applications to display maps and perform limited map manipulation functions. For example, you can embed a map in a Microsoft Word document. From within that Microsoft Word document, you can then click on the map and be able to use the following subset of MapInfo tools: Pan, Zoom In, Zoom Out, Info Tool, Change View, Drag Map Window, and Help. You will also have access to limited thematic and legend manipulations with this embedded map.

You can drag and drop a map object in the following OLE container applications: Microsoft Excel, PowerPoint, and Word. This drag-and-drop function uses the OLE protocol and permits you to modify the map within the other application. Because the OLE capability is available only when using 32-bit MapInfo Professional, you must be running Microsoft Windows NT, Windows 98, or Windows 95. If you are running 16-bit MapInfo (in Windows 3.11 or earlier versions), MapInfo maps can be embedded into other applications as bitmaps or metafiles, but they are not editable.

The map in the *container* application (destination application supporting OLE) can be modified because a limited version of MapInfo (MapInfo Map) will be activated in the container application. Either the menus of the container application will change to that of MapInfo Map, or the application will open a separate window to allow you to modify the map. MapInfo Map is equipped with most of the functionality of the Map menu in MapInfo; however, data cannot be modified or queried. After changing the map, you can drag and drop it back into MapInfo and save it as a workspace for later access.

Embedding a Map Object

Tutorial 11-4, which follows, tkes you through the process of using the drag-and-drop feature with Microsoft Word. The same procedure would be used to embed a map in Microsoft Excel or other OLE container applications.

▼ TUTORIAL 11-4: USING THE DRAG-AND-DROP FEATURE

1 Start a copy of Microsoft Word. Open a new Word document.

2 Start MapInfo Professional. Open the *States* table in a Map window.

3 Move the cursor over the *STATES* map. Once positioned over the map, the cursor will change to a hand and a handle.

4 Click on the Drag and Drop button, shown at right, on the Main tool bar.

5 Click and hold down the mouse button anywhere in the Map window. Drag the cursor to the lower task bar and position it over the Microsoft Word task. The cursor will change to a circle with a line through it. The Word task will become the active task for the window.

6 Release the mouse button; the map is dropped in the new location in the Word document, as shown in the following illustration.

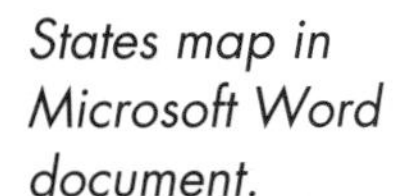
States map in Microsoft Word document.

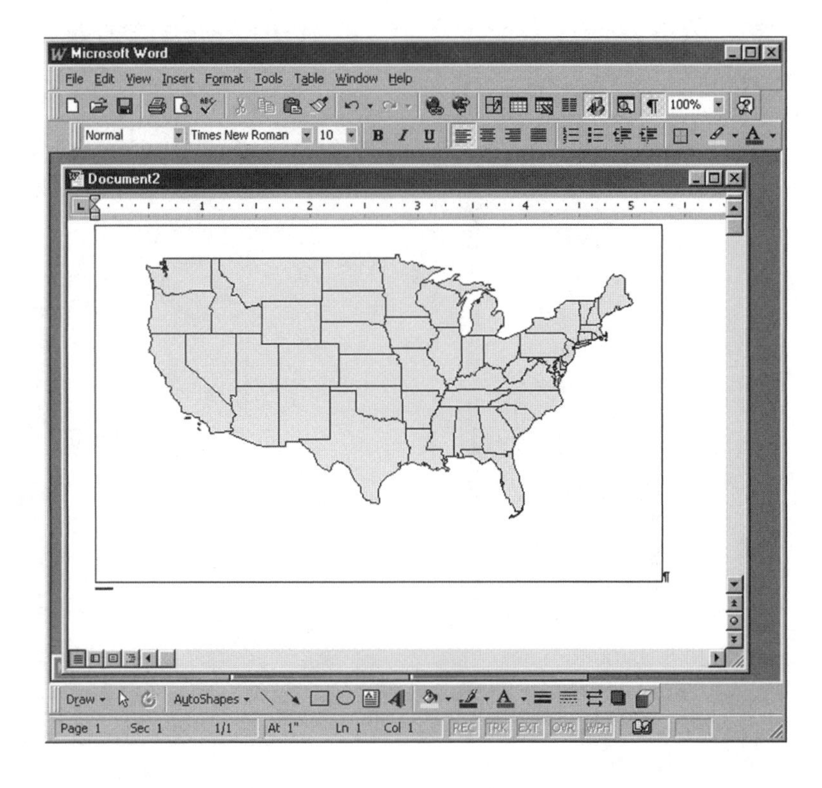

WARNING: *The Drag and Drop feature currently does not work in Microsoft PowerPoint. To achieve the same functionality, you must use the Copy Map Window option in the Edit menu to copy the map object. Then use <Ctl>+V or the paste function in the PowerPoint application to embed the map object.*

Manipulating Embedded Map Objects

Once a map object is embedded in a Microsoft Word document, you can activate the Map window by double clicking on the map object. This section explores selected embedded map capabilities. In tutorial 11-5, which follows, you will begin working with embedded map objects.

▼ TUTORIAL 11-5: WORKING WITH EMBEDDED MAP OBJECTS

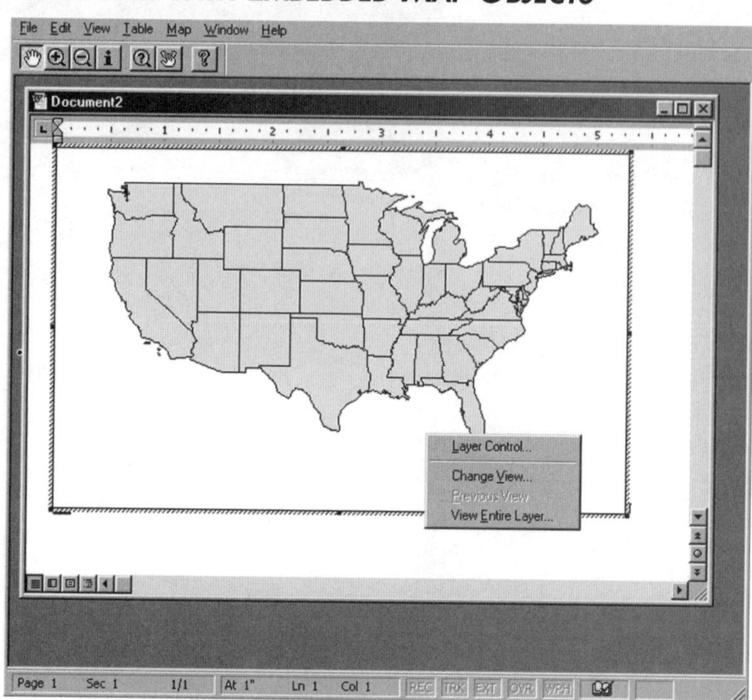

Embedded map object in Microsoft Word.

1 Double click on the map object contained within the Word document, which is shown in the illustration at right. This activates the subset of MapInfo tool buttons while in Microsoft Word.

2 Menus are also adjusted to reflect the active map object. Click the right mouse button while the cursor is positioned over the map. Layer control and view options are available via the mouse-button-activated menus. Note that the Microsoft Word Main menu bar has changed. Upon pulling down the Map menu bar, you will see additional functionality allowed for the map object, as shown in the following illustration.

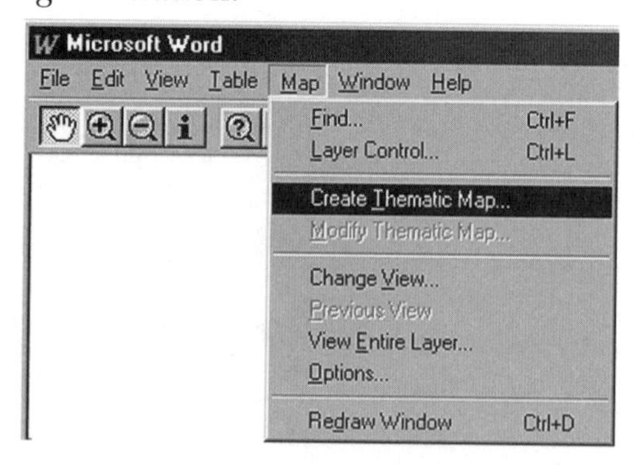

Map menu for embedded object.

3 Select the Map | Create Thematic Map menu bar item.

4 Select to create a ranged thematic map of the 1990 population. The thematic map creation works the same as for maps in the MapInfo environment.

5 Click and drag the left mouse button over the legend that was created. With these embedded maps, the legend "floats" within the Map window. You can position the legend where appropriate.

6 Select the Info tool from the tool bar, and click on the state of Missouri. Just as in the MapInfo application, the Info tool window displays the attribute information attached to the state of Missouri, as shown in the illustration at right.

7 Click outside the map object to reactivate the Microsoft Word tools.

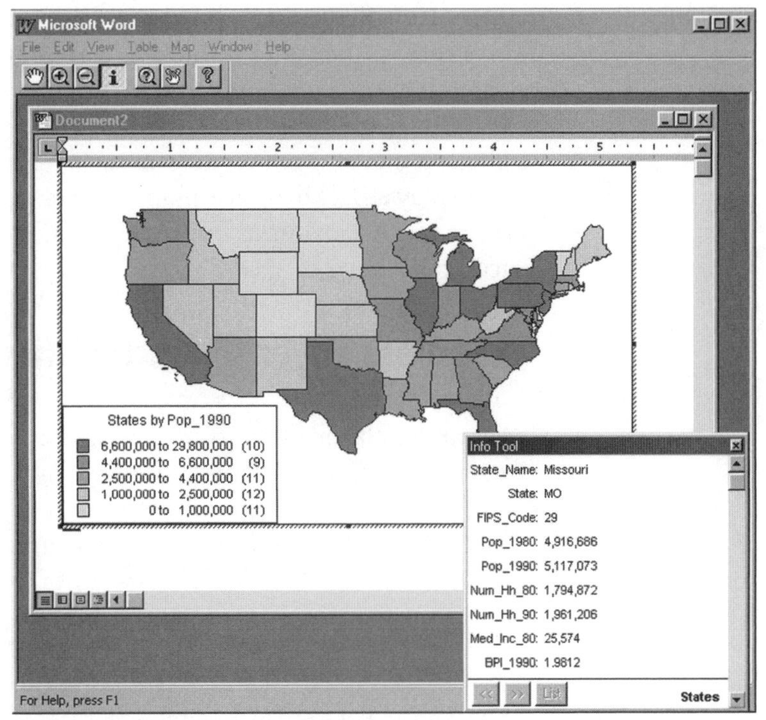

Thematic maps and attribute information in embedded maps.

NOTE 1: *Upon e-mailing a Word document containing a map object, a recipient using MapInfo Professional will have the same access to the map object as you. The map and data are attached to the document when it is saved.*

NOTE 2: *When using the MapInfo Info tool in Microsoft Word and other OLE container applications, the text in the Info box is sometimes very small.*

Seamless Map Layers

With the seamless map feature, a group of base tables is handled like a single table. You can change display attributes, apply labeling changes, or use the Layer Control dialog on several tables at once. Information can be retrieved from seamless tables with the Info tool in the same fashion as a regular MapInfo table. You can also select or browse any one of a seamless layer's base tables.

The seamless map layer feature is especially useful when you wish to display a backdrop for maps, such as in joining street or boundary maps. An example would be a seamless layer of county boundaries consisting of several county tables.

Seamless Layer Characteristics

A seamless layer includes the path name of each base table and a description that defaults to the table name. The following illustration shows the tables and associated description of each of the base tables contained in an *ok_city* seamless table.

Tables included in seamless table definition.

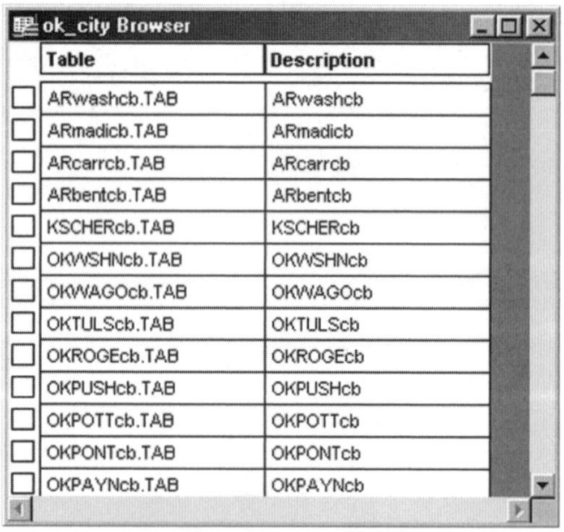

ok_city Browser

Table	Description
ARwashcb.TAB	ARwashcb
ARmadicb.TAB	ARmadicb
ARcarrcb.TAB	ARcarrcb
ARbentcb.TAB	ARbentcb
KSCHERcb.TAB	KSCHERcb
OKWSHNcb.TAB	OKWSHNcb
OKWAGOcb.TAB	OKWAGOcb
OKTULScb.TAB	OKTULScb
OKROGEcb.TAB	OKROGEcb
OKPUSHcb.TAB	OKPUSHcb
OKPOTTcb.TAB	OKPOTTcb
OKPONTcb.TAB	OKPONTcb
OKPAYNcb.TAB	OKPAYNcb

MapInfo functions that work with seamless map layers are described in the sections that follow.

Layer Control

Most of the functions in the Layer Control dialog work as usual on a seamless layer. You can Add, Remove, and Reorder layers, and set Display, Zoom Layering, and Label options for the seamless layer or for all base tables simultaneously. You cannot make a seamless layer editable or use the thematic mapping function.

Info Tool

The attribute information associated with base table objects can be retrieved with the Info tool.

Select Tools

Objects can be selected from the seamless layer. With selections you should understand that you will only be able to select groups of items residing in the same base table. When you use the Marquee or Radius select tools and the selected area spreads across two different base tables, MapInfo Professional selects the table in the center of either the circle or the polygon.

Browser Table

A Browser window can be displayed from a seamless table. When you select to browse a seamless layer, you will be prompted to select which base table in the seamless definition you wish to browse.

NOTE: *A base table can be any regular MapInfo table, but not a raster image.*

Creating or Modifying a Seamless Layer

If you find yourself opening numerous tables and map layers primarily for viewing purposes, consider creating a seamless map layer from such tables. To create a seamless layer, MapInfo Professional provides a tool called the Seamless Manager. In tutorial 11-6, which follows, you will create a seamless layer.

▼ TUTORIAL 11-6: CREATING A SEAMLESS LAYER

1 Open the tables you wish to combine into a seamless layer.

2 Select File | Run MapBasic Program.

3 Select *seammgr.mbx.*

4 From the MapInfo menu bar, select Tools | Seamless Manager | New Seamless Table.

5 The Seamless Manager, shown at right, will prompt you for information, and then create the seamless layer.

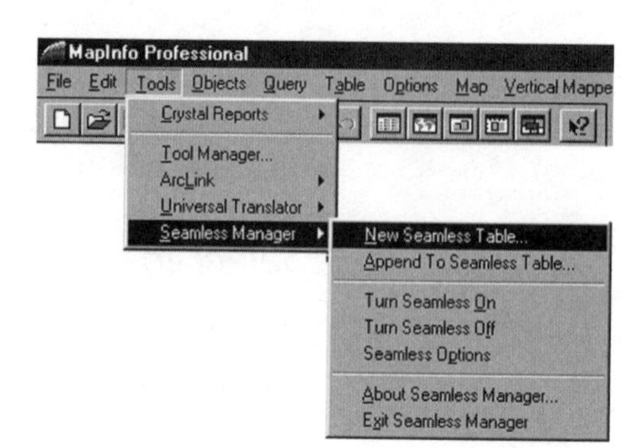

Seamless Manager menu items.

Select similar types of tables when creating a seamless table. In brief, use tables with the same projections and the same number of columns. For instance, you may wish for the seamless layer to contain several counties, each stored in a separate base table. Another example would be interstate highways that cross several state lines, where highways in states are stored in separate base tables.

NOTE: *Every time you edit the structure of a seamless table, you must recompile it using the Seamless Manager.*

Animation Layer

As mentioned in Chapter 5, the animation layer allows you to view changes in a data set in real time. MapInfo Professional positions the animation layer as the topmost layer in the Map window. Only this layer is redrawn when data in the table change, rather than the entire map. Consequently, redraws occur quickly. The animation layer is useful for real-time applications where features are repeatedly updated; GPS (global positioning system) technology is typically a part of using this layer.

A GPS allows you to pinpoint an exact location (coordinates) on Earth. Examples of applications that might require the animation layer and GPS technology are tracking and routing emergency vehicles as they are dispatched, tracking the location of police officers in the field, or helping rental car customers reach their destination by guiding them with the aid of an electronic map.

The animation layer is typically activated through MapBasic or the MapBasic window. It cannot be activated from within MapInfo

Professional. In addition, data related to this layer will not be saved in a workspace. You can use only one table at a time in a Map window as an animation layer. However, you can designate a different animation layer (one each) in multiple Map windows. In the following example, the MapBasic code lines open the *fire_trk* table and place it on the animation layer.

```
Open Table "fire_trk" Interactive
Add Map Layer fire_trk Animate
```

The table serving as the animation layer will not appear in the layer list in the Layer Control dialog box. In addition, you cannot access the animation layer via the Map window. This means that the Select tools and the Info tool are not available for use on the animation layer. If you attempt to add more than one table via a single *Add Map...Animate* command, the first table listed will be placed on the animation layer, and all others will be added as conventional layers. If you wish to add an animation layer to a Map window that already contains an animation layer, the new table will be placed on the animation layer, and the old table will be removed from the map. To remove the *fire_trk* table from the animation layer, you would use the following command:

```
Remove Map Layer fire_trk Animate
```

This statement will remove the animation layer from the map, but will not close the table. You can activate the animation layer even if you do not have access to MapBasic, and the layer can be activated within a workspace, even though information on the layer is not saved to a workspace. To activate the animation layer in a workspace, use a text editor (e.g., Microsoft Word or Notepad) and open the workspace file (*.wor*) in which you wish to include the animation layer. Remove the table you wish to animate from the Map *from* statement in the *.wor* file.

The Map *from* statement lists all files to be displayed on the map, and is in the *.wor* file located after all Open Table and Add column statements. If the table you wish to animate is not already opened within the workspace (i.e., included in the Open Table statement), add the following statement to the *.wor* file: *Open Table*

[table/layer name] Interactive. Use the same format as the Open Table statements already included in the workspace.

After you have either removed the table from the Map *from* statement or have opened the table, type in the following under the Map *from* statement: *Add Map Layer [table/layer name] Animate.* Save the workspace file in the text editor. When you open the workspace, the table will be placed on the animation layer. A word of caution: if you save the workspace file again within MapInfo, your changes will be lost and the table will not appear on the animation layer the next time you open the workspace.

The animation layer was designed to permit rapid updates to small areas of the Map window. To ensure the most rapid redraw speed possible, it is recommended that you do not display the Map in a Layout window. Next, verify that you have only one copy of the layer in the current Map window. If you animate a table already appearing in the Map window, screen updates will not occur as rapidly because the table also continues to appear as a conventional layer. In fact, the screen redraw rate will not improve at all. To ensure rapid redraws, turn off the visibility of the layer in the Map window (*set map layer fire_trk display off*) before adding it to the map as the animation layer.

Raster Images

Many users are intimidated by the terms *raster* and *vector.* Vector images resemble MapInfo tables in that they specify a line drawn between points. As you zoom in on vector files, you retain a clear image because the software continues "knowing" that it needs to draw a line between points.

A raster image is a type of picture consisting of row after row of tiny dots (pixels). Raster images are sometimes referred to as *bitmaps.* As you zoom in on a raster image, you will see square blocks of color (representing the individual pixels) grow larger.

In MapInfo Professional, raster images can refer to two types of images. The first is a digital representation of a feature of interest, such as a digital photo or a scanned document. The second is a specialized digital image that represents features of Earth (typically from an overhead view), such as those evident in satellite imagery and scanned aerial photography. MapInfo Professional can read monochrome, grayscale, and color raster images in the following formats.

- GIF (Graphics Interchange Format)
- JPG (or JPEG)
- TIF (or TIFF; Tagged Image File Format)
- PCX (PC Paintbrush)
- BMP (Windows bitmap)
- TGA (Targa)
- CUT (Halo CUT)

Aerial imagery can serve as a powerful background raster image for displaying data. The Map window in the following illustration shows the San Francisco street file, employed in previous chapters, with an aerial image as background. The amount of detail seen in an aerial or raster image depends on its resolution. Keep in mind that the higher the resolution of a raster image, the more disk storage space it will require. Raster images may be slow to display and update in a Map window.

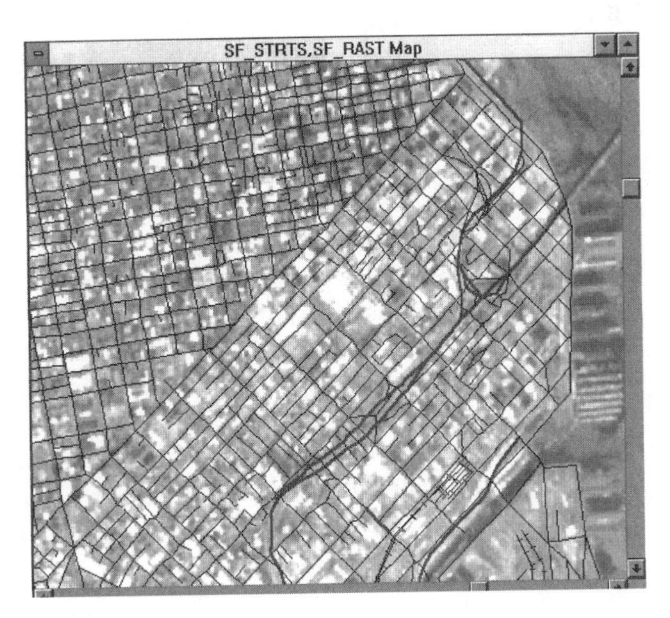

Raster image as a background for San Francisco.

Other types of raster images (such as scanned logos, photos, financial statements, and legal documents) may also be useful. For example, scanning photographs of a site location or place of interest provides a visual representation of what you would see at that geographical location, an example of which is shown in the following illustration.

Scanned photograph displayed in a Map window.

Changing the Appearance of a Raster Image

Adjusting a raster image's appearance can clarify a map. When you overlay additional map vector layers of data on top of a raster image, it may be difficult to discern lines that are part of the raster image from those that are part of the top layers. Adjusting the display style of the image can make it easier to differentiate the separate layers.

To adjust the appearance of the *Sf_rast* raster image (*samples* directory on the companion CD-ROM), select Table | Raster | Adjust Image Styles. The Adjust Image Styles dialog box, shown in the following illustration, will appear.

Changing contrast and brightness will change the appearance of the *Sf_rast* raster image. The settings in the Adjust Image Styles dialog box change image display, not the raster image itself. The new display parameters set with the dialog will be retained in the *Sf_rast.tab* file and displayed the same way the next time the table is opened.

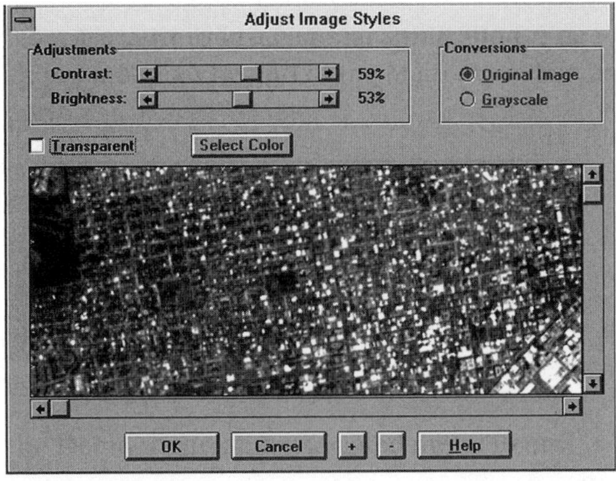

Adjust Image Styles dialog box.

Note the Transparent check box option in the Adjust Raster Image Styles dialog box. Use this option to make one of the colors in the raster image transparent.

Registering a Raster Image

The first time you open a raster image file in MapInfo Professional, you will be required to register the image so that MapInfo can position it properly in a Map window. During the registration process, MapInfo creates a *.tab* file for the raster image, in which it stores the coordinate information. Through the Image Registration dialog box, you inform MapInfo of the map coordinates that correspond to various points on the raster image. This coordinate information allows MapInfo to determine the position, scale, and rotation of the image.

You must have at least three coordinate control points for registering an image. The three points cannot be on a straight line. Accurate coordinates spread across the image allow MapInfo Professional to display the raster without distorting or improperly rotating the data. When MapInfo overlays a raster image with a vector image, it will distort and rotate the vector image so that both images line up properly.

The coordinates required to register a raster image can be determined from a point's coordinates on a paper map. With this method, the coordinates must be individually entered into the Image Registration dialog box, along with the point location on the image.

Alternatively, the points can be registered using an already existing MapInfo vector data file. Open the existing vector data file in a Map window, and the raster image in the Image Registration dialog box. Selecting Table | Raster | Select Control Points From Map from the Main menu bar allows selected coordinates from the Map window to be transferred to the Image Registration dialog box, shown in the following illustration. At this juncture, you reposition the new control point to its correct location on the raster. Selecting Table | Raster | Modify Image Registration when displaying a raster image invokes the Image Registration dialog box so that you can create new control points, or edit existing points.

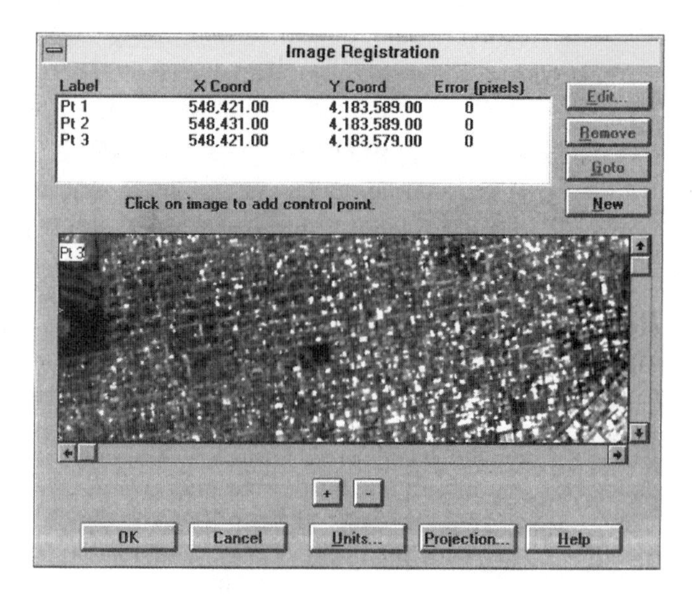

Image Registration dialog box.

When you open a raster image file, you have the option of displaying or registering the image. The option to display an unregistered image allows you to display a non-Earth raster image that will be used for presentation purposes only, such as the photograph shown previously. In 3.x versions of MapInfo, you must register these types of non-Earth raster images with at least three "dummy" control points. For example, a photograph may be registered with the upper left corner coordinate (0,0), upper right corner coordinate (1,0), and lower right corner coordinate (1,1).

TIP: *To register a raster image, map coordinates are entered in decimal degrees as opposed to degree/minute/second coordinates. MapInfo Profes-*

sional ships with a sample MapBasic program, longlats.mbx, *to convert degree/minute/second coordinates to decimal degrees.*

Global Positioning Systems

A global positioning system (GPS) allows you to calculate a precise position on the earth's surface (i.e., latitude and longitude). This technology, developed by the U.S. Department of Defense (DoD) to simplify accurate navigation, consists of a system of satellites (Navstar) at high-altitude orbits and handheld receivers on Earth. The exact coordinates of the GPS receiver are calculated by measuring the distance from Navstar satellites to the receiver. Although the federal government has invested billions of dollars in GPS, data from the Navstar system are available free of charge.

Common uses of GPS technology include vehicle tracking (e.g., truck fleets and emergency vehicles), and navigation (e.g., rental cars equipped with electronic maps to guide the driver to a specific destination, as well as aircraft and maritime navigation). Potential future uses of GPS include standard equipment in automobiles and other forms of transportation for preventing collisions, and accurate zero visibility landing systems for airplanes. For these types of applications, a transmitter must work in conjunction with the GPS receiver to relay data back to a central point.

Weather and tall buildings cause accuracy degradation of GPS readings. The DoD itself, however, is the major cause of degradation in its attempts to prevent hostile forces from obtaining accurate readings. Recently, the DoD was ordered to develop a plan for eliminating such intentional degradation, known as *selective ability.*

A *differential correction* process is applied to correct inaccuracies caused by degradation. This process involves using GPS coordinate readings over a known accurate point; the differences in actual versus GPS read points are employed to correct retrieved points.

The following grades of GPS receivers are currently available: navigation, mapping, and survey. As the accuracy requirement for exact positioning increases, so does the cost of the GPS receiver. Navigation GPS receivers are the least expensive, and are accurate to the length of a football field (approximately 100 yards). Navigation grade receivers do not interface well with PCs and are not equipped with differential correction capabilities. Mapping grade receivers are guaranteed accurate to the fraction of a meter. This grade of receiver is equipped to use differential correction, and receives signals from four satellites to calculate position. The most expensive receivers are survey grade, accurate to one centimeter. Survey receivers use more than four satellites to calculate position, and acquire sample readings over a long period of time.

The Geographic Tracker by Blue Marble Geographics Inc. (ships with MapInfo Professional) includes a MapBasic application (*MIGeoTrack.mbx*) that allows you to perform GPS tracking by showing a real-time derived position on top of a map background in MapInfo. The program also permits you to collect and incorporate field information (in real time) directly into MapInfo tables. Blue Marble Geographics calls this concept "GPS geocoding." You can practice making GPS simulation files for your locale with the *MakeGPS.mbx* utility in Geographic Tracker. In exercise 11-1, which follows, you have the opportunity to practice manipulating a raster image.

■ EXERCISE 11-1: RASTER IMAGE MANIPULATION

A common use for raster images is to draw map detail based on a raster image. For a quick look at this process, in this exercise you will open a raster image of the London area and trace over some of the streets on the raster image.

1 Open the *london.tab* file (*samples* directory on the companion CD-ROM).

2 Enlarge the Map window and zoom in (approximately) to a 3-mile zoom factor.

3 You may wish to select the Table | Raster | Adjust Image Styles dialog to lighten the color of the raster image.

4 Select File | New to create a new table for the streets you will be tracing.

5 Add a character column to the table for the street name, and name the new table *strts*.

6 The raster image is now set, as shown in the illustration at right, and you are ready to trace streets.

London raster image.

7 Select Options | Line Style and choose a line style that contrasts well with the raster image.

8 Press the <S> key to set the Map window to snap nodes.

9 Select the Polyline tool from the Drawing toolbar.

10 Use the Drawing tool to trace one of the streets you can distinguish on the raster image.

11 Once the street is drawn, move to the Browser window and type in a street name. The result is shown in the following illustration.

You could repeat this process to create a street network of the London area. Clear, high-quality raster images are necessary to produce useful street networks. You could also use this raster image technique to locate points of interest, trace flood areas, or outline fire damaged areas, among many other applications.

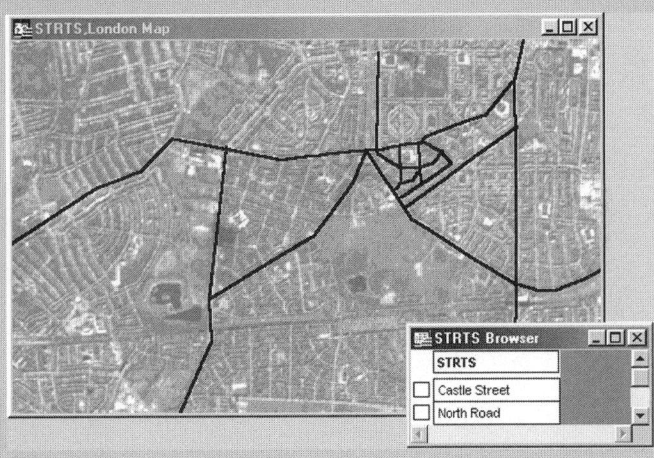

Drawing streets on top of a raster image.

Summary

This chapter reviewed several data manipulation options provided by MapInfo Professional: raster imagery, digitizing, real-time map updating (animation layer), data translators, and generating seamless map layers.

CHAPTER 12

CUSTOMIZING MAPINFO PROFESSIONAL

MAPINFO PROFESSIONAL CAN BE EXTENDED and controlled by developing programs in MapBasic, a development environment containing a text editor, a compiler, a linker, and on-line help. MapBasic is sold separately from MapInfo Professional.

The MapBasic language resembles most of the popular, structured programming languages. If you have worked with Pascal or C, the program structures for subroutines, loops, and *If, Then,* and *Else* functions will be familiar. MapBasic also retains Basic's English-like flavor and ease of use. If you have ever used Visual Basic, Qbasic, or any of the Basic derivative languages, you will find programming with MapBasic very similar.

MapBasic allows you to customize the client server environment in a manner much more elaborate than MapInfo workspaces. MapBasic was MapInfo's first true development tool, and although the company introduced additional products to help you automate specific functions and platforms, MapBasic remains the predominant tool for automating repetitive, systematic mapping tasks.

This chapter describes the development tools available to MapInfo users and focuses particularly on what you can do with MapBasic. Included in this chapter are sample programs and some very useful algorithms. All programs featured in the chapter, as well as other interesting utilities, are provided in the *samples\programs* directory on the companion CD-ROM.

Advantages of Software Development

Consider developing custom applications if you repeatedly use the same queries, table manipulations, or thematic maps. For example, assume that a city manager reviews crime locations throughout the city on a monthly basis. For this review the manager wants the same statistics and map hardcopy output from MIS every month. The MIS staff could write a custom application, and thus be assured that the monthly report is always created with the same information.

Types of Applications

The following application types can be developed with MapBasic.

- *Turnkey applications* are designed as a complete solution for a specific job.
- *Utilities* automate a tedious task.
- *Extensions* add new operations to the basic functionality of MapInfo Professional.

You may wish to develop a turnkey application to include new menu options or dialog boxes, or to eliminate existing menus. Many applications require ease of use for a very focused set of operations. To accomplish this, you need to be able to create options and queries specifically designed around your application. For example, MapBasic will allow you to provide users with a very specific crime analysis or facilities management application, rather than a generic mapping system.

Utility applications are aimed at automating simple and repetitive tasks. Suppose, for example, you want to include a grid of vertical and horizontal lines representing major latitudes and longitudes on most of your maps. You could draw such a grid by hand, but it would be a tedious and error-prone process. One of the sample MapBasic programs that ships with MapInfo Professional quickly and easily generates an accurate grid of latitude and longitude lines. If you regularly need to create grids, you can add this utility to the Main menu bar system.

Extension programs can be written in MapBasic to add entirely new functionality to MapInfo Professional. For example, MapBasic provides functions that allow you to locate the nearest point of interest from any point of reference.

Advantages of Custom Applications

By creating a custom application, you can achieve the following objectives.

- Ensure that a function is performed the same way every time without the risk of processing mistakes. The application can lead the user through the process.
- Modify a process by changing it in only one place (the program) rather than every place the process is used.
- Perform complex logic operations, such as *If/Then* statements and looping, that otherwise are impossible to handle.
- Handle errors in consistent ways defined by you.
- Include comments that allow you to document a complicated process.
- Control the user interface. In addition to setting your own system defaults via MapBasic, you can create your own menu items and dialog boxes.

TIP: *MapInfo Professional ships with a catalog containing applications written by MapInfo value-added resellers (VARs). Before spending a lot of time and effort on MapBasic program development, consider reviewing this catalog to determine whether the application you need is available.*

Introduction to MapBasic

MapBasic functions and programs range from utilities consisting of a few simple lines of code to complex multiple-function applications. The basic mechanisms for becoming familiar with MapBasic (the MapBasic window and workspaces) are discussed in the sections that follow.

MapBasic Window

The MapBasic window is another type of window that can be opened while in MapInfo Professional. Just as MapInfo executes tasks via menu bar selections, toolbar manipulations, and dialog boxes, the MapBasic commands used to execute these tasks are echoed in the MapBasic window.

To open the MapBasic window, select Options | Show MapBasic Window from the Main menu bar. When the window appears, a MapBasic menu option will also be added to the Main menu bar, as shown in the following illustration. The option allows you to clear the MapBasic window or save its content to a user-specified file.

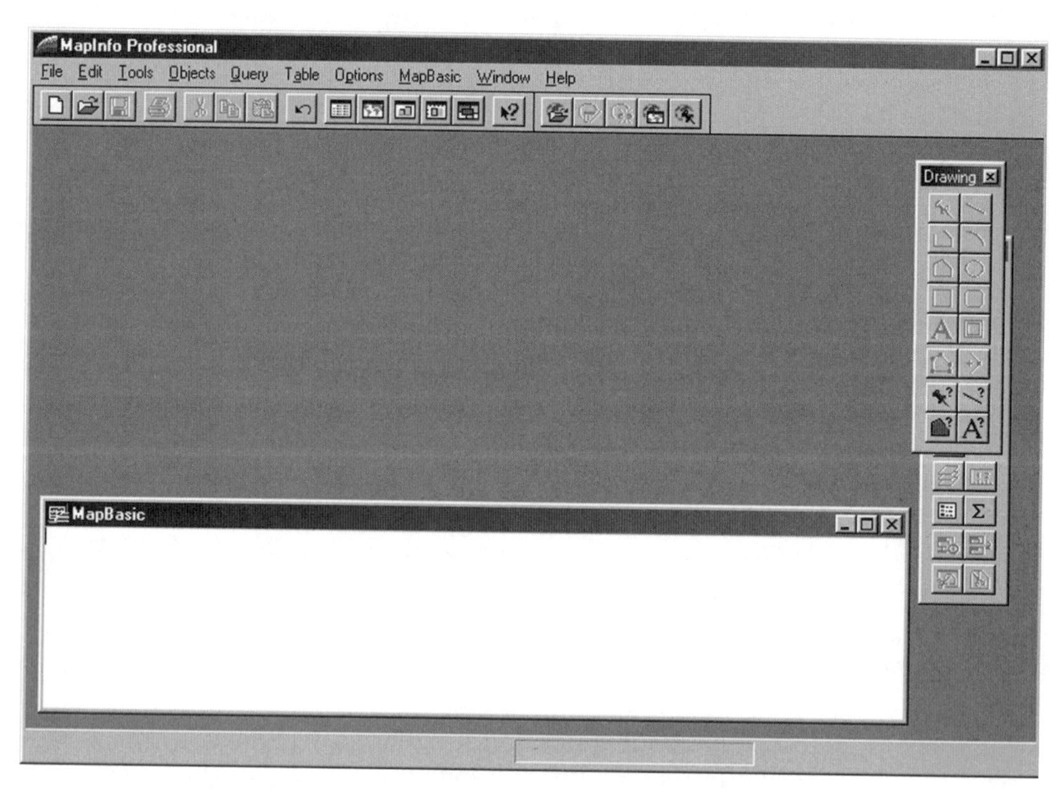

MapBasic window opened within MapInfo Professional.

In tutorial 12-1, which follows, you will execute a few MapInfo command sequences in order to view the MapBasic commands used.

▼ Tutorial 12-1: Viewing Commands Used

1 Select File | Open from the Main menu bar, and open the *States* table (*samples* directory on the companion CD-ROM).

2 When you click on the OK button, an Open command and a Map command display in the MapBasic window.

3 Select Window | New Browser Window from the Main menu bar. A Browse statement is added to the MapBasic window, as shown in the illustration at right.

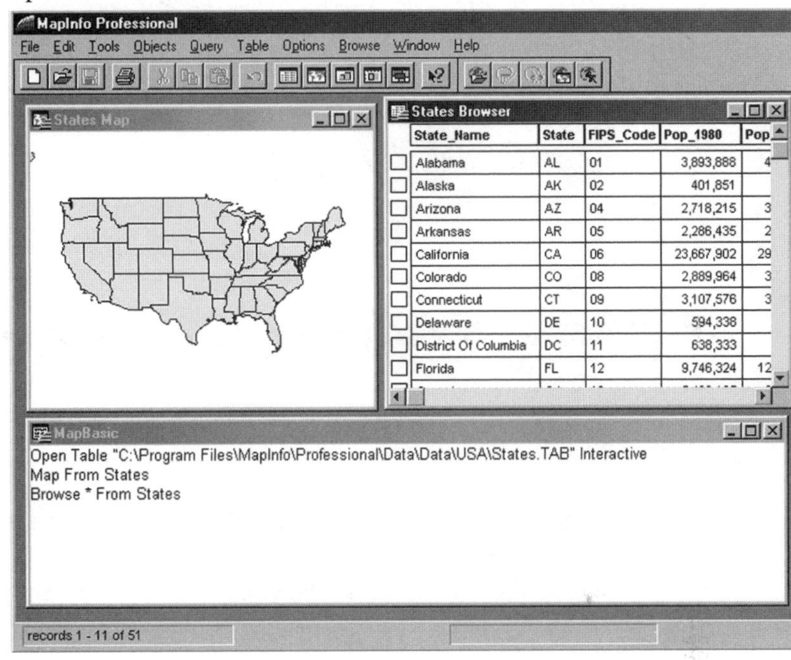

MapBasic window showing commands for opening the States *table in Map and Browser windows.*

In tutorial 12-2, which follows, you will continue the previous tutorial, here creating a thematic map based on the 1990 population attribute in the *States* table.

▼ Tutorial 12-2: Creating a Population Attribute Thematic Map

1 Make the Map window the active window.

2 Select Map | Create Thematic Map from the Main menu bar.

3 Create a thematic map using the Ranges option.

4 Set the dialog box for the *States* table and the *1990_pop* column of data.

5 After you click on the OK button, you will see a "Shade map" statement and a "Set legend" statement in the MapBasic window, as shown in the following illustration.

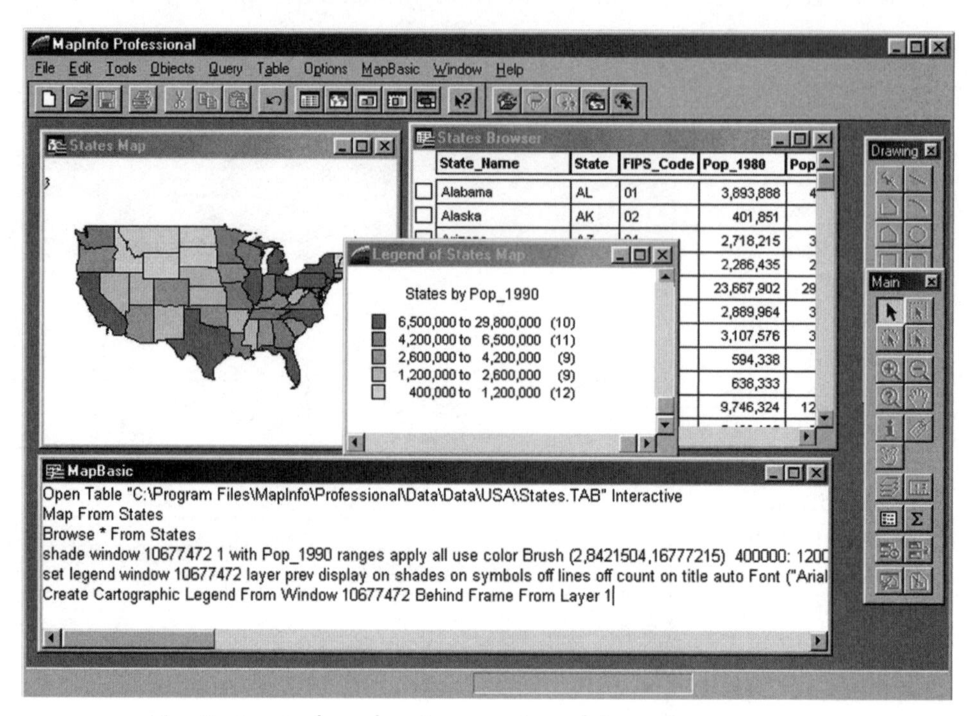

MapBasic window showing creation of thematic map.

The MapBasic window can be of great assistance when you are beginning to formulate statements to create a MapBasic program. Simply perform the actions with MapInfo you wish to include in the program, and then copy the commands displayed in the MapBasic window to the MapBasic editor. After a few minor changes, you can create an application ready for use whenever you want to repeat a particular function.

TIP: *Once you become familiar with MapBasic commands, you can enter any MapBasic statement in the MapBasic window. For example, assuming the states.tab file is located in the MapInfo data directory on the* c: *drive, entering the statement* Open Table C:\Program Files\MapInfo\Professional\ Data\Data\USA\States.TAB *will open the* States *table.*

Workspaces

In the simplest sense, a workspace is a collection of MapBasic statements. Open any workspace in a text editor and you will find a series of MapBasic commands used to recreate the environment that was saved as a workspace. With minor modifications, a workspace can become a program.

The workspace saved after the preceding session, in which the MapBasic window was demonstrated, is reproduced in the following. (See *states.wor* in the *samples\programs* directory on the companion CD-ROM.)

```
!Workspace
!Version 500
!Charset WindowsLatin1
Open Table "C:\Program Files\MapInfo\Professional\Data\Data\USA\
  States" As States Interactive
Set Window MapBasic
  Position (0.09375,2.80208) Units "in"
  Width 6.73958 Units "in" Height 1.375 Units "in"
Open Window MapBasic
Map From States
  Position (0.104167,0.09375) Units "in"
  Width 3.04167 Units "in" Height 2.38542 Units "in"
Set Window FrontWindow() ScrollBars Off Autoscroll On
Set Map
  CoordSys Earth Projection 1, 0
  Center (-95.844784,37.561978)
  Zoom 3815.369063 Units "mi"
  Preserve Zoom Display Zoom
  XY Units "degree" Distance Units "mi" Area Units "sq mi"
shade 1 with Pop_1990 ranges apply all use color Brush
(2,8421504,16777215)
  400000: 1200000 Brush (2,14737632,16777215) Pen (1,2,0) ,
  1200000: 2600000 Brush (2,13684944,16777215) Pen (1,2,0) ,
  2600000: 4200000 Brush (2,11579568,16777215) Pen (1,2,0) ,
  4200000: 6500000 Brush (2,10526880,16777215) Pen (1,2,0) ,
  6500000: 29800000 Brush (2,8421504,16777215) Pen (1,2,0)
  default Brush (2,16777215,16777215) Pen (1,2,0)
  # use 1 round 100000 inflect off Brush (2,16776960,16777215) at 3
  by 0 color 1 #
Set Map
  Layer 1
```

```
      Display Value
      Selectable Off
    Layer 2
      Display Graphic
      Label Line None Position Center Font ("Arial",0,9,0) Pen (1,2,0)
        With State_Name
        Parallel On Auto Off Overlap Off Duplicates On Offset 2
        Visibility On
set legend
    layer 1
      display on
      shades on
      symbols off
      lines off
      count on
      title auto Font ("Arial",0,9,0)
      subtitle auto Font ("Arial",0,8,0)
      ascending off
      ranges Font ("Arial",0,8,0)
        auto display off ,
        auto display on ,
        auto display on ,
        auto display on ,
        auto display on ,
        auto display on
Create Cartographic Legend
    Position (2.72917,0.989583) Units "in"
    Width 2.79167 Units "in" Height 1.54167 Units "in"
    Window Title "Legend of States Map"
    Portrait
    Frame From Layer 1
    Border Pen (0,1,0)
Browse * From States
    Position (3.28125,0.0208333) Units "in"
    Width 3.59375 Units "in" Height 2.44792 Units "in"
```

The workspace shows the use of the Open Table, Set Window, and Set Map commands, as well as commands for creating Map and Browser windows. Note that the commands in the workspace are lengthy and explicit, with information about the size of the window, the fonts used, and the X and Y positions of the windows. These details are required for precise reproduction of the workspace session, but are often not needed when building a MapBasic program.

In the workspace code block, the creation of the Browser window is accomplished through the following statement: *Browse * From States Position (3.28125,0.0208333) Units "in" Width 3.59375 Units "in" Height 2.44792 Units "in"*. In a MapBasic program that you design, you would most often use only the following fragment of the statement: *Browse * From States*. When you use the simpler statement, you are relying on MapInfo defaults to size and position the Browser window.

The preceding workspace can be transformed into a MapBasic program by adding subroutine syntax and compiling it. These topics are covered later in this chapter. First, let's take a look at the MapBasic development environment.

Development Environment

The MapBasic development environment contains a built-in text editor you can use to create and edit MapBasic programs. If you are familiar with the windowing environment, you will find the MapBasic user interface easy to use. To run MapBasic, double click on the MapBasic button, shown at left.

MapBasic button.

The MapBasic pull-down menus (File, Edit, Search, Program, Font, Style, and Window) provide you with everything you need to create, edit, and compile programs, as well as resolve syntax errors detected by the MapBasic compiler. The MapBasic development environment is shown in the following illustration.

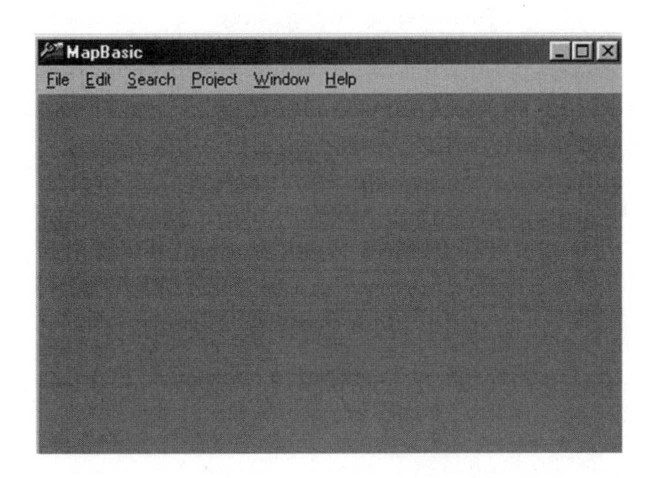

MapBasic development environment.

Sample MapBasic Program

In the tutorials that follow, you will build, compile, and run the following application. The application will loop five times, and the loop count is provided in the Message window. When the application is complete, a message box displays. The source code for this program follows. (See *first.mb* in the *samples\programs* directory on the companion CD-ROM.)

```
'*************************************************************
'*
'* Program: first.mb
'* Purpose: First example program.
'*
'*************************************************************
'*********************************
'* Declares
'*********************************
Declare Sub main
'*************************************************************
'* Function: main
'* Purpose: Count to five. Show users how to use the Print,
'*   Note, and Loop statements.
'*************************************************************
Sub main
 Dim x As Smallint
 '** Loop counting to 5
 For x = 1 to 5
  Print "The count is " +str$(x)
 Next
 '** Tell user processing is complete
 Note "Done Counting!"
End Sub
```

Tutorial 12-3, which follows, takes you through the process of building the foregoing application.

▼ TUTORIAL 12-3: BUILDING A MAPBASIC APPLICATION

1. Start the MapBasic development environment.
2. Select File | New from the MapBasic Main menu bar.

3 Type the code into the window.

4 Select File | Save As from the MapBasic Main menu bar.

5 Save the MapBasic source file as *first.mb.*

6 Select Project | Compile Current File from the MapBasic Main menu bar.

7 When the compile is complete, you will receive a successful compile message from MapBasic, as shown in the illustration above right.

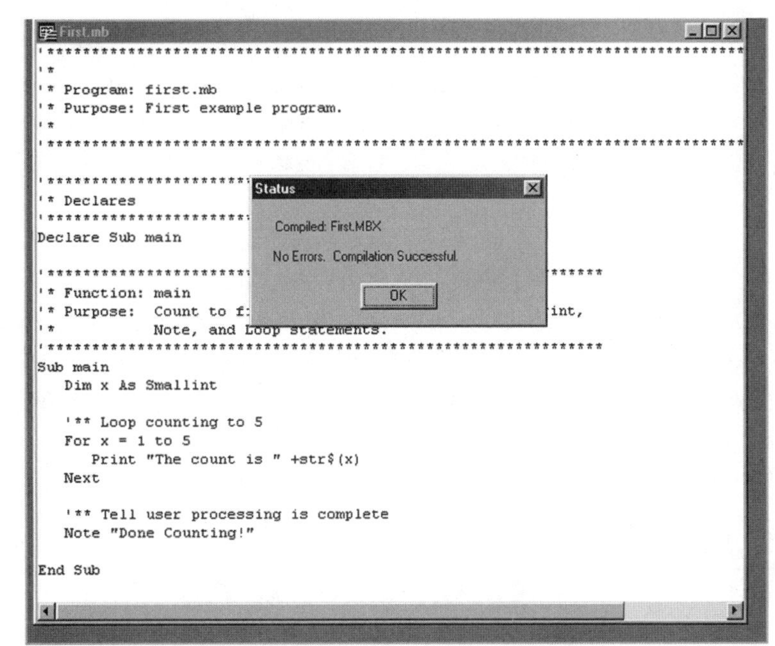

Successful compile of first.mb *program.*

Now you are ready to run the newly created application, *first.mbx.* You can run the application from the MapBasic Project | Run menu, or the MapInfo File | Run menu. For this example, the MapBasic menu selection is used. Tutorial 12-4, which follows, takes you through this process.

▼ TUTORIAL 12-4: RUNNING A MAPBASIC APPLICATION

1 Select Project | Run from the MapBasic Main menu bar.

2 If MapInfo Professional is already up and running, MapBasic will switch to MapInfo as the active application and run the *first.mbx* application. If MapInfo is not running, MapBasic will start the MapInfo program and run the *first.mbx* application.

3 When the *first.mbx* application runs, the information shown in the illustration at right appears in the MapInfo Professional window.

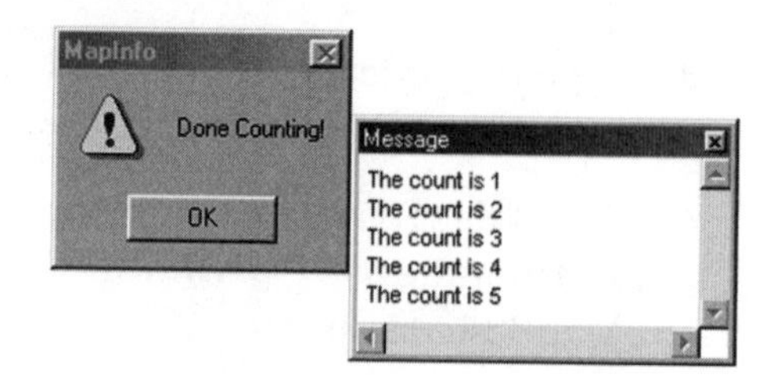

Results of running the first program.

TIP: *Unlike other development environments, the MapBasic environment does not include a debugger. Instead, MapBasic users rely heavily on the Print and Note statements for debugging MapBasic applications. In addition, a Stop statement can be used to halt processing of a program in order to check variables.*

Syntax Errors on Compiles

Typographical errors and incorrect statement construction are common occurrences. When you try to compile MapBasic code containing such errors, a notice of syntax errors appears, and the compiler informs you of the error type by line number. Assume that the looping statement in the *first.mb* source file contained *do* instead of *for*. The following illustration shows how the compiler would have flagged *do*.

In the MapBasic development environment, you can double click on the syntax error and the editing cursor will be positioned at the beginning of the line containing the error. You can also use the Search | Go To Line option from the MapBasic Main menu bar to position the cursor at the line number containing the error.

As seen in the preceding example, one small error can create many error messages. With experience, you will be able to identify errors and what needs to be done to fix them.

Example of syntax errors.

```
MapBasic
File Edit Search Project Window Help
First.mb
'*
'******************************************************************************

'***********************************
'* Declares
'***********************************
Declare Sub main

'*************************************************************
'* Function: main
'* Purpose:  Count to five.  Show users how to use the Print,
'*           Note, and Loop statements.
'*************************************************************
Sub main
   Dim x As Smallint

   '** Loop counting to 5
   Do x = 1 to 5
      Print "The count is " +str$(x)
   Next

   '** Tell user processing is complete
   Note "Done Counting!"

End Sub

(first.mb:22) Unrecognized command: to.
(first.mb:27) Unrecognized command: Next.
(first.mb:29) [End sub] found without corresponding sub statement.
(first.mb:29) Found [] while searching for [loop].
(first.mb:29) Sub/Function without End Sub/Function.
```

More MapBasic Programming

The purpose of this chapter is to familiarize you with MapBasic rather than present all elements found in a programming reference book, such as details on data types, operators, and language syntax. The best way to become familiar with the MapBasic language is to review sample programs. This section reviews the overall structure that MapBasic programs should have, and examines several sample applications.

TIP: *Numerous MapBasic sample applications become available when you install MapInfo Professional. Reviewing these applications can help familiarize you with MapBasic programming techniques.*

Program Organization

A typical MapBasic program contains several procedures that are sometimes referred to as *subprocedures.* To define a procedure, use the *Sub...End Sub* statement or the *Function...End Function* statement. Every program must contain a main procedure declared explicitly through a *Sub Main...End Sub* statement. When you run a MapBasic program, MapInfo Professional automatically executes the main procedure.

From within the main procedure, the *Call* statement is used to call another procedure. Although most statements in a program will be located within a procedure, the following statements must be located outside any procedure definition: *Type, Declare Sub, Declare Function, Define,* and *Global.* In addition, *Include* statements are usually located at the top of a program file, outside a procedure definition.

The following list summarizes the major program elements and shows the order in which these statements usually appear. Global-level statements appear at the top of the program.

- Include (e.g., *Include "mapbasic.def"*)
- Type...End Type
- Declare Sub
- Declare Function
- Define
- Global

Global-level statements are followed by the main procedure.

- Sub Main
- Dim statements
- Other statement types
- End Sub

The main procedure is followed by other procedure definitions.

- Sub . . .
- Dim statements
- Other statement types
- End Sub

In addition to other procedure definitions, the main procedure is followed by custom function definitions.

- Function . . .
- Dim statements
- Other statement types
- End Function

Menu Item Changes

When creating complete applications or utilities, you will almost always want to alter menus. In most cases you will simply add to the existing menu bar. However, some complete applications may involve removal of the menu bar and creation of a new one.

The following sample menu program creates and adds a simple menu at the end of the Main menu bar. The *menu.mb* program shows how to create subroutines invoked when the user selects one of the menu options. (See *menu.mb* in the *samples\programs* directory on the companion CD-ROM.) The *exit_menu* routine also shows how to return the menu bar to its original state and ensure a clean exit of the program. You can avoid confusing the user by cleaning up after the program has run.

```
'*********************************************************
'* Program: menu
'* Purpose: Experiment with program menu control.
'*********************************************************
'* Declares
Declare Sub main
Declare Sub open_initial_map
Declare Sub demo_report
Declare Sub demo_map
Declare Sub exit_menu
'**********************************************************
'* Function: main
'* Purpose: Main control.
'**********************************************************
Sub main
 '** Create another menu.
 Create Menu "&MenuFunctions" as
   "&Report" calling demo_report,
   "&Map" calling demo_map,
   "(-",
   "E&xit MenuFunctions" calling exit_menu
 Alter Menu Bar Add "&MenuFunctions"
 '** Open inital maps
 Call open_initial_map
End Sub
'**********************************************************
'* Function: open_initial_map
'* Purpose: Open the maps to be initially displayed.
'**********************************************************
Sub open_initial_map
 '** Open the states table
 Open Table "C:\MAPINFO\DATA\STATES.TAB" interactive
```

```
 Map From STATES
End Sub
'*********************************************************
'* Function: demo_report
'* Purpose: Open a browser report of all STATES table
'*    entries with a population larger than
'*    5,000,000.
'*********************************************************
Sub demo_report
 '** Select the records to be displayed into temp table
 Select * from STATES where pop_1990 > 5000000
   Into TEMP_STATES
 '** Display selected records in a browser window
 Browse * from TEMP_STATES
End Sub
'*********************************************************
'* Function: demo_map
'* Purpose: Create a thematic map of states table with
'*    pop_1990.
'*********************************************************
Sub demo_map
 '** Create thematic map layer.
 Shade STATES with Pop_1990 ranges apply all use color
  Brush (2,255,16777215)
  400000: 1500000 Brush (2,13697023,16777215) ,
  1500000: 3500000 Brush (2,9482495,16777215) ,
  3500000: 5500000 Brush (2,4215039,16777215) ,
  5500000: 29800000 Brush (2,255,16777215)
  default Brush (2,16777215,16777215)
End Sub
'*********************************************************
'* Function: exit_menu
'* Purpose: Exit this menu system, and restore the default
'*    MapInfo menu system.
'*********************************************************
Sub exit_menu
 '** Restore default menu bar.
 Create Menu Bar as Default
 '** Exit this application.
 Terminate Application "menu.mbx"
End Sub
```

After you create and compile the *menu.mb* program, you are ready to run it. The program adds a MenuFunctions option at the end of the Main menu bar, with Report, Map, and Exit MenuFunctions submenu options. The following illustration shows the appearance of these menus, as well as the Map and Report windows created from each of the menu options.

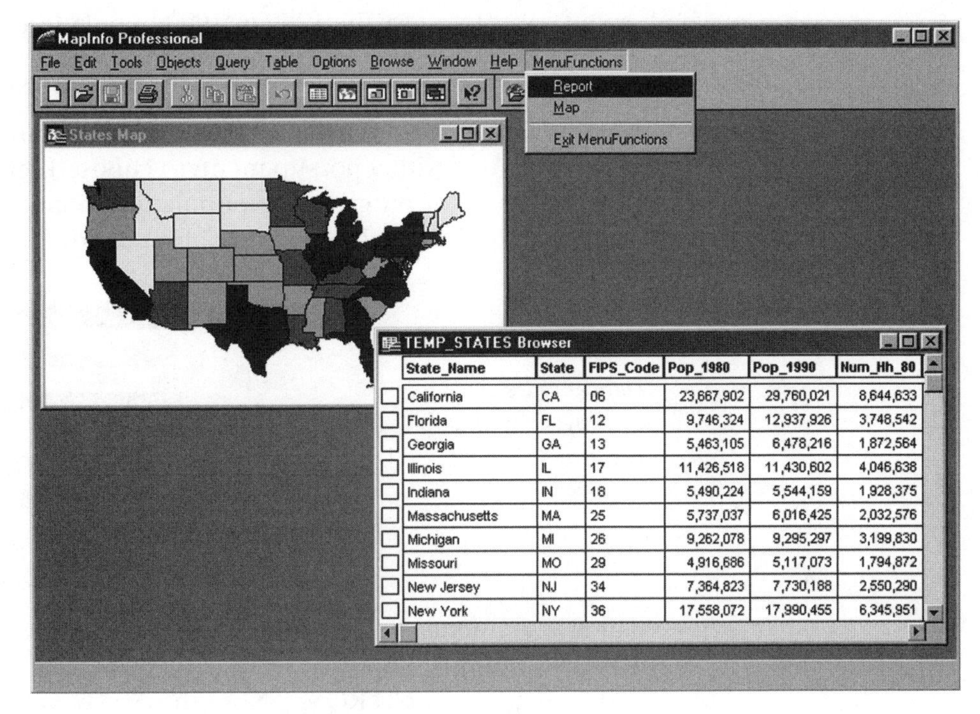

State_Name	State	FIPS_Code	Pop_1980	Pop_1990	Num_Hh_80
California	CA	06	23,667,902	29,760,021	8,644,633
Florida	FL	12	9,746,324	12,937,926	3,748,542
Georgia	GA	13	5,463,105	6,478,216	1,872,564
Illinois	IL	17	11,426,518	11,430,602	4,046,638
Indiana	IN	18	5,490,224	5,544,159	1,928,375
Massachusetts	MA	25	5,737,037	6,016,425	2,032,576
Michigan	MI	26	9,262,078	9,295,297	3,199,830
Missouri	MO	29	4,916,686	5,117,073	1,794,872
New Jersey	NJ	34	7,364,823	7,730,188	2,550,290
New York	NY	36	17,558,072	17,990,455	6,345,951

Menu program results.

If you have a similar need for developing additional functional menu items, you can use the *menu.mb* program as the framework for your program. Most programs are developed by copying and altering various pieces of code from existing programs.

TIP: *To run a MapBasic application automatically every time you start MapInfo Professional, add the MapBasic* .mbx *file name. For example,* mapinfow.exe menu.mbx *will automatically run* menu.mbx *when you start MapInfo Professional. MapBasic programs can also be added to workspaces with the Run Application command.*

Dialog Boxes

As you build applications or utilities you often need to ask users questions about exactly how they want to proceed with a particular process. Dialog boxes are the mechanism for requesting user input. In the following sample application, the *States* table is again queried for population data. However, this time the user is asked whether she wants to query the 1980 or 1990 population data, and which population value to use in the query.

The dialog box to be created is shown in the following illustration. The user is presented with a pop-up menu to choose 1980 or 1990 data, and a text box for entering the population value to be used in the query. OK and Cancel buttons are also provided.

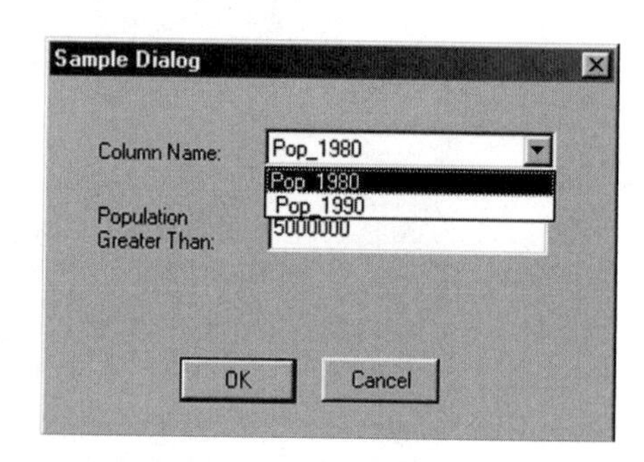

Dialog to be displayed by dialog program.

Several new statements are introduced in the *dialog.mb* sample program that follows. (See *dialog.mb* in the *samples\programs* directory on the companion CD-ROM.) Note the *System Includes* statement. The Include files contain definitions for the meaning of TRUE and FALSE used in the *demo_dialog* function. Next, *demo_dialog* is a function rather than a subroutine, which means that it returns a logical value to the calling routine. The *demo_dialog* function also shows how to pass parameters between routines. The final new item seen in the sample program is the use of a dialog statement. The dialog statement builds the dialog described previously, and the *demo_dialog* function also shows how to interpret user input for further program use.

```
'*****************************************************
'* Program: Dialog
'* Purpose: Experiment with dialog statement.
'*****************************************************
'* System Includes
```

```
Include "c:\Program Files\mapinfo\mapbasic\menu.def"
Include "c:\Program Files\mapinfo\mapbasic\mapbasic.def"
'* Declares
Declare Sub main
Declare Function demo_dialog(column_number As SmallInt,
        pop_value As Integer) As Logical
'**********************************************************
'* Function: Main
'* Purpose: Main control.
'**********************************************************
Sub main
 Dim ret_value As Logical
 Dim col_number As SmallInt
 Dim pop_cutoff As Integer
 '** Open the states table
 Open Table "C:\MAPINFO\DATA\STATES.TAB" interactive
 Map From STATES
 '** Display the dialog for user input
 ret_value = demo_dialog(col_number, pop_cutoff)
 '** If the user provided valid input in the dialog:
 If (ret_value = TRUE) Then
  '** If the user wants 1980 data:
  If (col_number = 1) Then
   Select * from STATES where pop_1980 > pop_cutoff
      Into STATES_1980
   Browse * from STATES_1980
  '** Else the user wants 1990 data:
  Else
   Select * from STATES where pop_1990 > pop_cutoff
      Into STATES_1990
   Browse * from STATES_1990
  End If
 Else
  Note "Operation cancelled by the user!"
 End If
 Terminate Application "dialog.mbx"
End Sub
'***************************************************************
'* Function: open_initial_map
'* Purpose: Open the maps to be initially displayed.
'***************************************************************
Function demo_dialog(column_number As SmallInt,
      pop_cutoff As Integer) As Logical
 Dim pop_text As String
 '** Build the dialog box
```

```
  Dialog
   Title "Sample Dialog"
   Width 209 Height 118
   Control StaticText
    Position 18, 20  Width 50  Height 13
    Title "Column Name:"
   Control PopupMenu
    Position 80, 16  Width 108  Height 12
    Title "Pop_1980; Pop_1990"
    Value 1
    Into column_number
   Control StaticText
    Position 18, 42  Width 56  Height 19
    Title "Population Greater Than:"
   Control EditText
    Position 81, 43  Width 104  Height 14
    Value "5000000"
    Into pop_text

   Control OKButton
    Position 49, 93  Width 43  Height 15
    Title "OK"
   Control CancelButton
    Position 102, 92  Width 43  Height 15
    Title "Cancel"
 '** Which button was selected by the user?
 If CommandInfo(CMD_INFO_DLG_OK) Then
  pop_cutoff = Val(pop_text)
  demo_dialog = TRUE
 Else
  demo_dialog = FALSE
 End If
End Function
```

After you create and compile the *dialog.mb* program, you are ready to run it. The program opens the *States* table and presents the user with a dialog for input. When the user clicks on the OK button, the appropriate query will be made. The window shown in the following illustration indicates the selection made from both the 1980 and 1990 population fields. By 1990, Missouri joined the ranks of states with populations larger than 5,000,000.

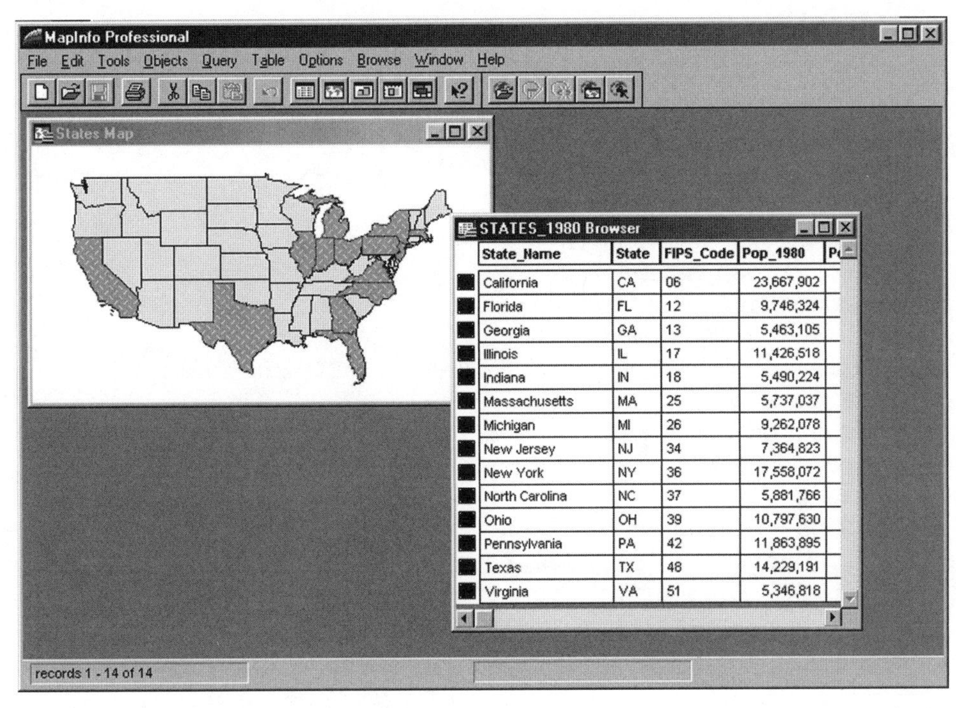

State_Name	State	FIPS_Code	Pop_1980
California	CA	06	23,667,902
Florida	FL	12	9,746,324
Georgia	GA	13	5,463,105
Illinois	IL	17	11,426,518
Indiana	IN	18	5,490,224
Massachusetts	MA	25	5,737,037
Michigan	MI	26	9,262,078
New Jersey	NJ	34	7,364,823
New York	NY	36	17,558,072
North Carolina	NC	37	5,881,766
Ohio	OH	39	10,797,630
Pennsylvania	PA	42	11,863,895
Texas	TX	48	14,229,191
Virginia	VA	51	5,346,818

Dialog program results.

Working with Tables

For most MapBasic applications you need to access certain rows and columns of data in a MapInfo table. The following *table.mb* example program simply loops through all records in the *States* table and totals the population. (See *table.mb* in the *samples\programs* directory on the companion CD-ROM.) Because this operation is easy to understand, you can concentrate on the statements needed to access data values. Writing a program to perform calculations on data values can save users a lot of time normally spent on performing a long series of select statements.

```
'********************************************************
'* Program: Table
'* Purpose: Experiment with table row and column values.
'********************************************************
'* System Includes
Include "c:\Program Files\mapinfo\mapbasic\menu.def"
Include "c:\Program Files\mapinfo\mapbasic\mapbasic.def"
```

```
'* Declares
Declare Sub main
'*************************************************************
'* Function: Main
'* Purpose: Main control.
'*************************************************************
Sub main
 Dim num_rows As Integer
 Dim total_pop As Integer
 Dim loop_count As Integer
 '** Open the dca table
 Open Table "C:\MAPINFO\DATA\STATES.TAB" interactive
 Map From STATES
 '** Get the number of rows in the table
 num_rows = TableInfo("STATES", TAB_INFO_NROWS)
 total_pop = 0
 '** Loop through rows of the table
 For loop_count = 1 to num_rows
  Fetch Rec loop_count from STATES
  total_pop = total_pop + STATES.Pop_1980
 Next
 Print "The Total US Population in 1980 is " +
     Format$(total_pop, ",#")
 '** Another way to loop through the rows of a table
 Fetch First from STATES
 Do While Not EOT(STATES)
  total_pop = total_pop + STATES.Pop_1990
  Fetch Next from STATES
 Loop
 Print "The Total US Population in 1990 is " +
   Format$(total_pop, ",#")
 Terminate Application "table.mbx"
End Sub
```

After you create and compile the *table.mb* program, you are ready to run it. The program opens the *States* table and loops through the table twice; the 1980 population is totaled first, and then the 1990 population. The totals are printed for the user in the Message window, as shown in the following illustration. The *table.mb* program shows two different methods of looping through the table. You must fetch the record or row you are ready to work with, and then the attributes for the record/row to make them available for access by the program.

Table program results.

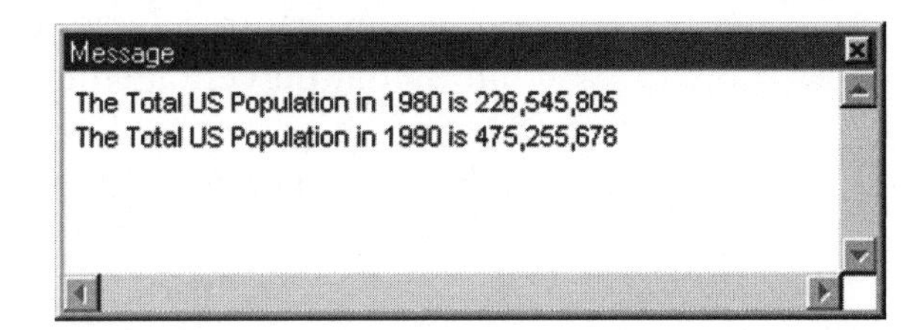

Message
The Total US Population in 1980 is 226,545,805
The Total US Population in 1990 is 475,255,678

TIP: *The MapBasic development environment does not contain extremely robust editing capabilities. If you plan on doing a lot of MapBasic development you may want to consider using a programmer's editor. When using another editor, the MapBasic compiler can be accessed using a command-line interface.*

Useful Geographic Operations and Algorithms

As you learn about MapInfo's MapBasic environment, we encourage you to particularly take note and leverage *spatial* operations of the language. Several features are particularly unusual (relative to other data management systems) and worth investigation. Generally, the most distinctive and powerful features are used in MapBasic SQL queries.

For example, experiment with "Distance" as a function in a query. The *Distance* function helps you determine the distance between features systematically without interactively invoking the distance icon.

Equally potent is the concept of using MapBasic to systematically employ *spatial* comparators in SQL *Where* clauses. Spatial queries return records based on geographic properties, in a manner that cannot be accomplished by comparing tabular columns. You can conduct "point in polygon" queries, for example, in order to determine points (such as customers) falling within polygons (such as a sales territory).

Other especially distinctive *spatial* features involve the concept of geographic *area*. *Area* is a spatial function used to return the physical area within a polygon, and can be particularly valuable for normalizing sales or demographic data according to the volume occurring per square mile or square kilometer. *AreaOverlap* is a function used to identify the net area that overlaps between two polygonal features. It can be especially useful if you need to determine what percentage of a feature falls within another. For example, you may want to determine what percentage of a block group falls within a circle around a business location.

A useful algorithm in MapInfo is one that systematically determines the distance from records in one file that are logically affiliated with records from a second file. For example, one might want to systematically determine the distance between a company's customers and their stores. Assuming that a store identifier can be found on the customer file, one can accomplish this need with the following combination of commands.

```
'**********************************************************
'* Sample Algorithm for computing distances
'* from Customers to a Store
'**********************************************************
Dim Storex, Storey as Float
 StoreX=Centroidx(Store.obj)
 StoreY=Centroidy(Store.obj)
 Update Customers Set StoreDist=distance(StoreX,StoreY,
Centroidx(obj),Centroidy(obj),"mi")
'**********************************************************
```

Another helpful algorithm in MapInfo is one that sums all the demographics or other geographic data from one file that is located within a certain distance from a specified location. For example, one might want to calculate the aggregate population within a distance from a potential store location. This can be accomplished with an algorithm similar to the following.

```
'**********************************************************
'* Sample Algorithm for Determining Aggregate Demographics
'* around a Store
'**********************************************************
Dim Storex, Storey as Float
 StoreX=Centroidx(Store.obj)
 StoreY=Centroidy(Store.obj)
 Update Demographics Set OverlapPct=
AreaOverlap(CreateCircle(StoreX,StoreY,Distance))/
Area(obj,"sq mi")
 Select sum(population*OverlapPct) from Demographics
Into AggregateResult
```

MapBasic Language Cross-reference

If you are familiar with other programming languages, the following MapBasic language summary tables can be invaluable for finding the MapBasic statements that correspond to the statements you are accustomed to using. These tables are also helpful in identifying keywords for use in the MapBasic Help system.

MapBasic Fundamentals

Tables 21-1 through 12-5, which follow, summarize MapBasic statements associated with fundamentals, including variables, looping and branching, procedures, output and printing, and error handling.

Table 12-1: MapBasic Variables Statements

Desired Statement	MapBasic Statement
Declare local variables	Dim
Declare global variables	Global
Resize array variables	ReDim
Determine array size	UBound( )
Declare custom data type	Type
Undefine a variable	UnDim

Table 12-2: MapBasic Looping and Branching Statements

Desired Statement	MapBasic Statement
Looping	For...Next, Exit For, Do...Loop, Exit Do, While...Wend
Branching	If...Then, Do Case, GoTo
Other flow control	End Program, Terminate Application, End MapInfo

Table 12-3: MapBasic Procedure (Main and Subs) Statements

Desired Statement	MapBasic Statement
Define a procedure	Declare Sub, Sub...End Sub
Call a procedure	Call
Exit a procedure	Exit Sub
Main procedure	Main
Special procedure names	System event handlers
React to selection	SelChangedHandler
React to window closing	WinClosedHandler
React to map changes	WinChangedHandler
React to window focus	WinFocusChangedHandler
React to DDE event	RemoteMsgHandler, RemoteQueryHandler
React to OLE Automation	RemoteMapGenHandler
Provide custom tool	ToolHandler
React to app termination	EndHandler
React if obtained or lost focus	ForegroundTaskSwitchHandler
Disable event handlers	Set Handler

Table 12-4: MapBasic Output and Printing Statements

Desired Statement	MapBasic Statement
Print window contents	PrintWin
Print text to Message window	Print
Set up Layout window	Layout, Create Frame, Set Window
Export window to file	Save Window

Table 12-5: MapBasic Error-handling Statements

Desired Statement	MapBasic Statement
Set up error handler	OnError
Return current error code	Err( )
Return current error string	Error$( )
Return from error handler	Resume
Simulate error	Error

Files (Non-table)

Tables 12-6 and 12-7, which follow, summarize MapBasic statements associated with file input and output, and file and directory names.

Table 12-6: MapBasic File Input/Output Statements

Desired Statement	MapBasic Statement
Open or create file	Open File
Close file	Close File
Delete file	Kill
Rename file	Rename File
Copy file	Save File
Read from file	Get, Seek, Input #, Line Input #
Write to file	Put, Print #, Write #
Determine file status	EOF( ), LOF( ), Seek( ), FileAttr( ), FileExists( )
Turn file into table	Register Table
Retry on sharing error	Set File Timeout

Table 12-7: MapBasic File and Directory Name Statements

Desired Statement	MapBasic Statement
Return system directories	ProgramDirectory$( ), HomeDirectory$( ), ApplicationDirectory$( )
Extract part of file name	PathToTableName$( ), PathToDirectory$( ), PathToFileName$( )
Return full file name	TrueFileName$( )
Let user choose file	FileOpenDlg( ), FileSaveAsDlg( )
Return temporary file name	TempFileName$( )

Functions

Tables 12-8 through 12-12, which follow, summarize MapBasic statements associated with functions.

Table 12-8: MapBasic Function/Procedure (Main and Subs) Statements

Desired Statement	MapBasic Statement
Define custom function	Declare Function, Function...End Function
Exit function	Exit Function

Table 12-9: MapBasic Data-conversion Function Statements

Desired Statement	MapBasic Statement
Convert strings to codes	Asc( )
Convert codes to strings	Chr$( )
Convert strings to numbers	Val( )
Convert numbers to strings	Str$( ), Format$( )
Convert strings to dates	StringToDate( )
Convert numbers to dates	NumberToDate( )
Convert object types	ConvertToRegion( ), ConvertToPline( )

Table 12-10: MapBasic Date and Time Function Statements

Desired Statement	MapBasic Statement
Obtain system time	Time( )
Obtain current date	CurDate( )
Extract parts of date	Day( ), Month( ), Weekday( ), Year( )
Read system timer	Timer( )
Convert string to date	StringToDate( )
Convert number to date	NumberToDate( )

Table 12-11: MapBasic Math Function Statements

Desired Statement	MapBasic Statement
Trigonometric functions	Cos( ), Sin( ), Tan( ), Acos( ), Asin( ), Atn( )
Geographic functions	Area( ), Perimeter( ), Distance( ), ObjectLen( )
Random numbers	Randomize, Rnd( )
Sign-related functions	Abs( ), Sgn( )
Truncating fractions	Fix( ), Int( ), Round( )
Other math functions	Exp( ), Log( ), Minimum( ), Maximum ( ), Round( ), Sqr( )

Table 12-12: MapBasic String Function Statements

Desired Statement	MapBasic Statement
Upper- and lower-case	UCase$( ), LCase$( ), Proper$( )
Find a substring	InStr( )
Extract part of string	Left$( ), Right$( ), Mid$( ), MidByte$( )
Trim blanks from string	LTrim$( ), RTrim$( )
Format dates as strings	FormatDate$( )
Format numeric value	Format$( ), Str$( ), Set Format, FormatNumber$( ), DeformatNumber$( )
Determine string length	Len( )

Table 12-12: MapBasic String Function Statements

Desired Statement	MapBasic Statement
Convert character codes	Chr$( ), Asc( )
Compare strings	Like( ), StringCompare( ), StringCompareIntl( )
Repeat string sequence	Space$( ), String$( )
Return unit name	UnitAbbr$( ), UnitName$( )

Map Objects and Layout Objects

Tables 12-13 through 12-17, which follow, summarize MapBasic statements associated with map objects and layout objects.

Table 12-13: MapBasic Map Layer Statements

Desired Statement	MapBasic Statement
Open map window	Map
Add layer to map	Add Map Layer
Remove map layer	Remove Map Layer
Label objects in layer	AutoLabel
Query map settings	MapperInfo( ), LayerInfo( )
Change map settings	Set Map
Create thematic layer	Shade, Set Shade, Create Ranges, Create Styles, Create Grid
Query map layer labels	LabelFindByID( ), LabelFindFirst( ), LabelFindNext( ), Labelinfo( )

Table 12-14: MapBasic Object Creation Statements

Desired Statement	MapBasic Statement
Creation statements	Create Arc, Create Ellipse, Create Frame, Create Line, Create PLine, Create Point, Create Rect, Create Region, Create RoundRect, Create Text, AutoLabel
Creation functions	CreateCircle( ), CreateLine( ), CreatePoint( ), CreateText( )

Table 12-14: MapBasic Object Creation Statements

Desired Statement	MapBasic Statement
Advanced operations	Create Object, Buffer( )
Store object in table	Insert, Update

Table 12-15: MapBasic Object Modification Statements

Desired Statement	MapBasic Statement
Modify object attribute	Alter Object
Change object type	ConvertToRegion( ), ConvertToPLine( )
Set editing target	Set Target
Erase part of object	Objects Erase, Erase( ), Objects Intersect, Overlap( )
Merge objects	Objects Combine, Combine( ), Create Object
Split objects	Objects Split
Add nodes at intersections	Objects Overlay, OverlayNodes( )
Control object resolution	Set Resolution
Store object in table	Insert, Update

Table 12-16: MapBasic Object Query Statements

Desired Statement	MapBasic Statement
Return calculated values	Area( ), Perimeter( ), Distance( ), ObjectLen( ), Overlap( ), AreaOverlap( ), ProportionOverlap( )
Return coordinate values	ObjectGeography( ), MBR( ), ObjectNodeX( ), ObjectNodeY( ), Centroid( ), CentroidX( ), CentroidY( ), ExtractNodes( ), Intersect-Nodes( )
Configure units of measure	Set Area Units, Set Distance Units, Set Paper Units, UnitAbbr$( ), UnitName$( )
Configure coordinate system	Set CoordSys( ), ChooseProjection$( )
Query map layer labels	LabelFindByID( ), LabelFindFirst( ), LabelFindNext( ), Labelinfo( )

Table 12-17: MapBasic Style Statements

Desired Statement	MapBasic Statement
Return current styles	CurrentPen( ), CurrentBrush( ), CurrentSymbol( ), CurrentFont( )
Return part of style	StyleAttr( )
Create style values	MakePen( ), MakeBrush( ), MakeSymbol( ), MakeFont( ), MakeCustomSymbol( ), MakeFontSymbol( ), Set Style, RGB( )
Query object style	ObjectInfo( )
Modify object style	Alter Object
Reload symbol styles	Reload Symbols
Store object in table	Insert, Update
Style clauses	Pen clause, Brush clause, Symbol clause, Font clause

Tables

Tables 12-18 through 12-20, which follow, summarize MapBasic statements associated with tables.

Table 12-18: MapBasic Table Creation/Modification Statements

Desired Statement	MapBasic Statement
Open existing table	Open Table
Close one or more tables	Close Table, Close All
Create new, empty table	Create Table
Turn file into table	Register Table
Import/export tables/files	Import, Export
Modify table structure	Alter Table, Add Column, Create Index, Drop Index, Create Map, Drop Map
Create report from table	Create Report From Table
Add, edit, delete rows	Insert, Update, Delete
Pack table	Pack Table

Table 12-18: MapBasic Table Creation/Modification Statements

Desired Statement	MapBasic Statement
Control table settings	Set Table
Save recent edits	Commit Table
Discard recent edits	Rollback
Rename table	Rename Table
Delete table	Drop Table
Retry on sharing error	Set File Timeout
Manage seamless tables	Set Table, TableInfo( ), GetSeamlessSheet( )
Create or modify metadata	Metadata

Table 12-19: MapBasic Table Query Statements

Desired Statement	MapBasic Statement
Position row cursor	Fetch, EOT( )
Select data from table	Select
Query selection table	SelectionInfo( )
Find map address	Find, Find Using, CommandInfo( )
Find map objects at location	SearchPoint( ), SearchRect( ), SearchInfo( )
Obtain table information	NumTables( ), TableInfo( )
Obtain column information	NumCols( ), ColumnInfo( )
Query table metadata	GetMetadata( ), Metadata

Table 12-20: MapBasic Remote Data Manipulation Statements

Desired Statement	MapBasic Statement
Communicate with data server	Server_Connect( )
Begin work with remote server	Server Begin Transaction
Assign local storage	Server Bind Column
Obtain column information	Server_columnInfo( ), Server_NumCols( )
Send an SQL statement	Server_Execute( )
Position row cursor	Server Fetch, Server_EOT( )
Save changes	Server Commit
Discard changes	Server Rollback
Free remote resources	Server Close
Make remote data mappable	Server Create Map
Change object styles	Server Set Map
Synchronize linked table	Server Refresh
Create linked table	Server Link Table
Unlink linked table	Unlink
Disconnect from server	Server Disconnect
Retrieve driver information	Server_DriverInfo( ), Server_NumDrivers( )
Get QELib connection handle	Server GetodbcHConn()
Get QELib statement handle	Server GetodbcHStmt()

User Interface

Tables 12-21 through 12-28, which follow, summarize MapBasic statements associated with manipulating the user interface.

Table 12-21: MapBasic Button Pad/Toolbar Manipulation Statements

Desired Statement	MapBasic Statement
Create new button pad	Create ButtonPad
Change button pad	Alter ButtonPad
Change button	Alter Button
Query status of pad	ButtonPadInfo( )
Respond to button use	CommandInfo( )
Restore standard pads	Create ButtonPads As Default
Make button active	Run Menu Command

Table 12-22: MapBasic Dialog Box Manipulation Statements

Desired Statement	MapBasic Statement
Standard dialog boxes	Ask( ), Note, ProgressBar, FileOpenDlg( ), FileSaveAsDlg( )
Custom dialog boxes	Dialog
Dialog box handler operations	Alter Control, TriggerControl( ), ReadControlValue( ), Dialog Preserve, Dialog Remove
Tell if user clicked OK	CommandInfo(CMD_INFO_DLG_OK)
Disable progress bars	Set ProgressBars
Modify MapInfo dialog box	Alter MapInfoDialog

Table 12-23: MapBasic Menu Manipulation Statements

Desired Statement	MapBasic Statement
Define new menu	Create Menu
Redefine menu bar	Create Menu Bar
Modify menu	Alter Menu, Alter Menu Item
Modify menu bar	Alter Menu Bar, Menu Bar
Invoke menu command	Run Menu Command
Query menu item status	MenuitemInfoByHandler( ), MenuitemInfoByID( )

Table 12-24: MapBasic Status Bar Manipulation Statements

Desired Statement	MapBasic Statement
Display or hide status bar	Status Bar

Table 12-25: MapBasic Window Manipulation Statements

Desired Statement	MapBasic Statement
Show or hide window	Open Window, Close Window, Set Window
Open new window	Map, Browse, Graph, Layout, Create Redistricter, Create Legend, Create Cartographic Legend
Determine window ID	FrontWindow( ), WindowID( )
Modify existing window	Set Map, Shade, Add Map, Remove Map, Set Browse, Set Graph, Set Layout, Create Frame, Set Legend, Set Cartographic Legend, Alter Cartographic Frame, Add Cartographic Frame, Remove Cartographic Frame, Set Redistricter
Return window settings	WindowInfo( ), MapperInfo( ), LayerInfo( )
Print window	PrintWin
Control window redrawing	Set Event Processing, Update Window
Count number of windows	NumWindows( ), NumAllWindows( )
Hide columns from Browser	Reproject
Get information about legends	LegendInfo(), LegendFrameInfo(), LegendStyleInfo()

Table 12-26: MapBasic System Event Handler Statements

Desired Statement	MapBasic Statement
React to selection	SelChangedHandler
React to window closing	WinClosedHandler
React to map changes	WinChangedHandler
React to window focus	WinFocusChangedHandler
React to DDE event	RemoteMsgHandler, RemoteQueryHandler( )
React to OLE Automation	RemoteMapGenHandler

Table 12-26: MapBasic System Event Handler Statements

Desired Statement	MapBasic Statement
Provide custom tool	ToolHandler
React to application termination	EndHandler
React if obtained or lost focus	ForegroundTaskSwitchHandler
Disable event handlers	Set Handler

Table 12-27: MapBasic Integrated Mapping and DDE Applications Statements

Integrated Mapping Applications	
Desired Statement	**MapBasic Statement**
Set MapInfo parent window	Set Application Window
Set Map window parent	Set Next Document
Running other Windows or UNIX applications	Run Program
Running other MapBasic applications	Run Application
Dynamic Data Exchange (DDE)	
Desired Statement	**MapBasic Statement**
Start DDE conversation	DDEInitiate( )
Send DDE command	DDEExecute
Send value via DDE	DDEPoke
Retrieve value via DDE	DDERequest$( )
Close DDE conversation	DDETerminate, DDETerminateAll
Respond to execute	RemoteMsgHandler, RemoteQueryHandler( ), CommandInfo(CMD_INFO_MSG)
Calling external Dynamic Link Libraries (DLLs)	Declare Function

Table 12-28: MapBasic Special-function Statements

Desired Statement	MapBasic Statement
Make beeping sound	Beep
Run string as interpreted command	Run Command
Save workspace	Save Workspace
Load workspace file	Run Application
Configure digitizing tablet	Set Digitizer
Return information about operating environment	SystemInfo( )
Set data value to be read by CommandInfo( )	Set Command Info
Set duration of drag object delay	Set Drag Threshold

TIP: *Developers may wish to use other development languages with MapInfo Professional and MapBasic, such as Visual Basic, Powerbuilder, Delphi, C, or C++. MapBasic supports both DDE and DLL interfaces used to connect with other applications. MapInfo Professional also supports OLE.*

MapX and Other MapInfo Development Tools

For some time MapBasic was the only option developers had for creating mapping applications with MapInfo Professional. Then MapInfo Corp. released MapX, a product that provides application developers with an easy, cost-effective option for embedding maps in new and existing applications for client/server environments.

MapX is an OCX component that can be quickly integrated into client-side applications. These client-side applications may be implemented with Visual Basic, PowerBuilder, Delphi, Visual C++, or other object-oriented languages, and in Lotus Notes (v4.5) using Lotus Script. Developers can work in the environments they are familiar with, and end users can access mapping through their familiar business applications.

More recently, MapInfo has also enhanced its product suite to include tools for the web and wireless environments. These products include MapXtreme, Routing J server, MapXsite, and MapXtend.

The product that has drawn the most attention among this group is MapXtreme, a mapping server that allows developers to distribute mapping capabilities over the Internet or corporate intranets. MapXtreme, available on both NT and Java platforms, positions developers to release customized interactive mapping applications over the Web.

The other products, although impressive in their own right, are positioned more for niche applications or platforms. Routing J is a Java-based turn-by-turn directions server for web-based applications. MapXsite is an interactive web-based application development toolkit for creating "Nearest Dealer" applications on a web site (site visitors can type in an address and receive a map portraying the area around the nearest site). MapXtend enables developers to create location applications on wireless personal digital assistants (PDAs). As you and your organization grow in your use of MapInfo, be aware that there are several tools beyond MapBasic that can help you distribute mapping around your organization. In exercise 12-1, which follows, you have the opportunity to practice writing a simple menu program.

■ *Exercise 12-1: Writing a Simple Menu Program*

The following exercise demonstrates how to write a simple menu program that displays workspaces when a menu item is selected by the user. Workspaces from the menu program to be created and displayed include a thematic map of the 1990 U.S. population, and a pie chart thematic map comparing urban and rural populations. Next, the MapBasic menu program is built and two different methods for incorporating workspaces into the menu program are presented. This type of program can be used for demonstrating proposed solutions to clients.

Creating the First Workspace

First, you need to create two workspaces to be displayed from the menu selections. The first workspace shows a simple thematic map of the 1990 U.S. population.

1 Select File | Open Table from the Main menu bar and open the *States* table.

2 With the *States* table displayed in the active Map window, select Map | Create Thematic Map.

3 Select the Ranges type of thematic map.

4 Set the next dialog box for the *States* table and the *Pop_1990* data column.

5 Set any colors and styles you like for the thematic map. The resultant map is shown in the illustration at right.

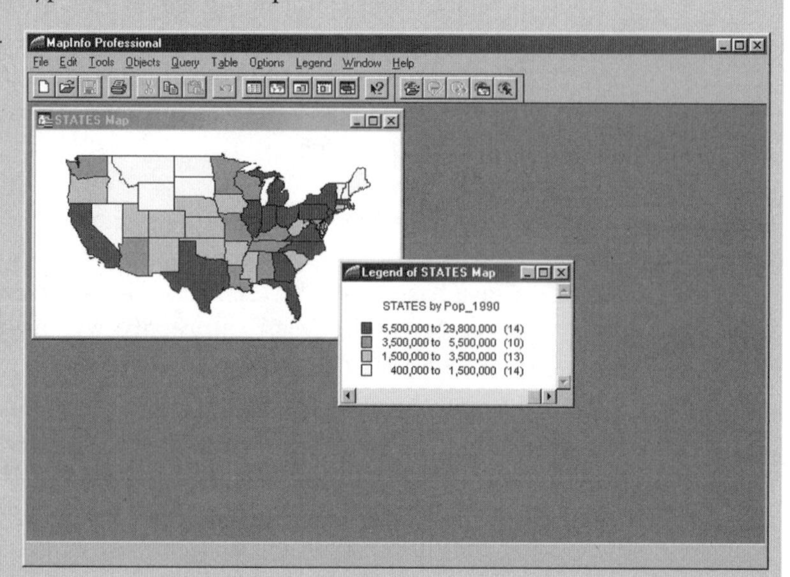

Population thematic workspace map.

6 Select File | Save Workspace from the Main menu bar.

7 Save the workspace as *work1.wor.* (This workspace is reproduced in a file of the same name in the *samples\programs* directory on the companion CD-ROM.)

Creating the Second Workspace

You are now ready to create the second workspace. For this workspace, you will use the same *States* table, and create a pie chart thematic map for comparing each state's urban and rural populations.

1 Select Map | Layer Control from the Main menu bar and remove the *Ranges by Pop_1990* layer from the map.

2 With the *States* table displayed in the active Map window, select Map | Create Thematic Map.

3 Select the pie chart Type of thematic map.

4 Set the next dialog box for the *States* table and the *pop_urban* and *pop_rural* data columns.

5 Set any colors and styles you like for the thematic map.

6 The default type of pie chart will be displayed as graduated symbols. This type of display does not clearly show the comparison of urban and rural populations. Select Map | Modify Thematic Map to change the pie chart style so that graduated symbols are not used. The resultant map is shown in the above illustration.

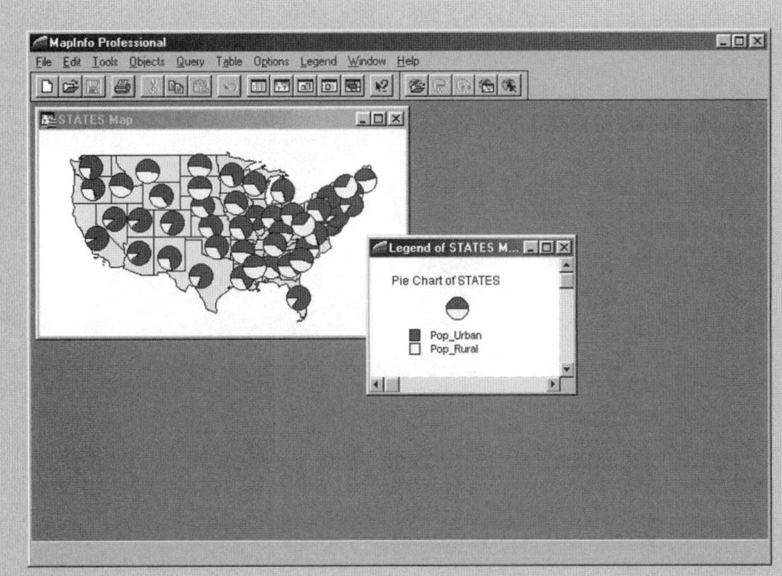

Urban and rural population thematic workspace map.

7 Select File | Save Workspace from the Main menu bar.

8 Save the workspace as *work2.wor.* (This workspace is reproduced in a file of the same name in the *samples\programs* directory on the companion CD-ROM.)

Preparing the MapBasic Program

Now you are ready to begin preparation of the MapBasic program. Continue with the following steps.

1 Start the MapBasic development environment.

2 Select File | New from the MapBasic Main menu bar.

3 Key in the following MapBasic source code.

```
'*****************************************************
'* Program: work
'* Purpose: Build a simple menu program to display
'*   workspaces for demonstration purposes.
'*****************************************************
'* Includes
Include "c:\Program Files\mapinfo\mapbasic\menu.def"
```

```
'* Declares
Declare Sub main
Declare Sub pop_thematic
Declare Sub pop_rural_urban
Declare Sub exit_menu
'**********************************************************
'* Function: main
'* Purpose: Main control.
'**********************************************************
Sub main
 '** Create another menu
 Create Menu "&Demo" as
   "&1990 Population Thematic" calling pop_thematic,
   "&Rural vs Urban Population" calling pop_rural_urban,
   "(-",
   "E&xit Demo" calling exit_menu
 '** Add DEMO to main menubar
 Alter Menu Bar Add "&Demo"
End Sub
'***************************************************************
'* Function: exit_menu
'* Purpose: Exit this menu system, and restore the default
'*   MapInfo menu system.
'***************************************************************
Sub exit_menu
 '** Remove the demo menubar
 Alter Menu Bar Remove "&Demo"
 '** Exit this application
 Terminate Application "work.mbx"
End Sub Choose File | Save As from the MapBasic main menu bar.
```

Adding the Demo Menu

Now that the source code has been entered, you can add the Demo menu to the MapInfo application. Continue with the following steps.

1 Select File | Save As from the MapBasic Main menu bar.

2 Save the file as *work.mb.*

3 Select Project | Compile Current File from the MapBasic Main menu bar.

4 If you do not receive a message reporting a successful compile, you have a syntax or typing error. If the compiler reports errors, you must correct them before recompiling the program.

5 Select Project | Run and switch to the MapInfo Professional application. The *work.mbx* program will add the Demo menu at the end of the Main menu bar, as shown in the following illustration.

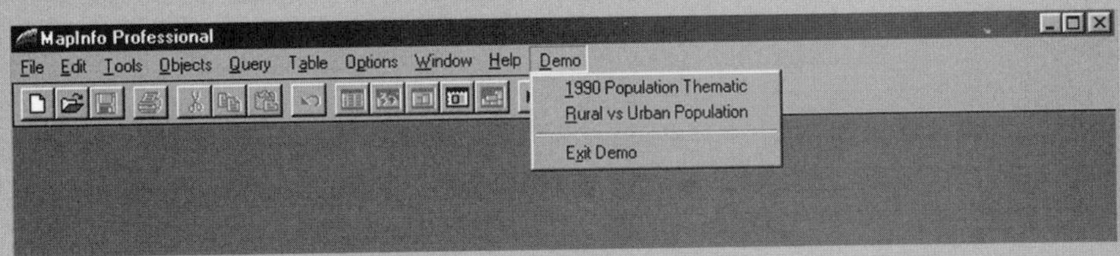

Demo menu added to MapInfo Professional application.

Thus far the program displays the Demo menu and associated menu items. The Exit Demo menu item works, but now you need to connect the workspaces to the 1990 Population Thematic and the Rural vs Urban Population menu items.

Adding Workspace Functionality

The following source code block adds the workspace functionality to the *work.mb* MapBasic application. (See *work.mb* in the *samples\programs* directory on the companion CD-ROM.) Note the addition of two subroutines for handling the two workspace calling menu items. At the beginning of each of these subroutines there is a call to *Run Menu Command* that requires inclusion of the additional *menu.def* file. This call works the same as the File | Close All option from the Main menu bar, and is used at the beginning of the workspace subroutines to remove any previous work from the MapInfo environment.

The first workspace routine, *pop_thematic*, contains the workspace commands. They have been copied from the workspace into the routine and modified slightly to compile correctly. The second workspace routine, *pop_rural_urban*, shows another approach using the *Run Application* command to run the workspace application.

```
'*********************************************************
'* Program: work
'* Purpose: Build a simple menu program to display
'*   workspaces for demonstration purposes.
'*********************************************************
'* Includes
Include "c:\Program Files\mapinfo\mapbasic\menu.def"
'* Declares
Declare Sub main
```

```
Declare Sub pop_thematic
Declare Sub pop_rural_urban
Declare Sub exit_menu
'**********************************************************
'* Function: main
'* Purpose: Main control.
'**********************************************************
Sub main
 '** Create another menu
 Create Menu "&Demo" as
   "&1990 Population Thematic" calling pop_thematic,
   "&Rural vs Urban Population" calling pop_rural_urban,
   "(-",
   "E&xit Demo" calling exit_menu
 '** Add DEMO to main menubar
 Alter Menu Bar Add "&Demo"
End Sub
'***************************************************************
'* Function: exit_menu
'* Purpose: Exit this menu system, and restore the default
'*   MapInfo menu system.
'***************************************************************
Sub exit_menu
 '** Remove the demo menubar
 Alter Menu Bar Remove "&Demo"
 '** Exit this application
 Terminate Application "work.mbx"
End Sub
'**********************************************************
'* Function: pop_thematic
'* Purpose: Display 1990 Population Thematic.
'**********************************************************
Sub pop_thematic
 '** First, close everything else in MapInfo
 Run Menu Command M_FILE_CLOSE_ALL
 '** Copy work1.wor into routine
 Open Table "STATES" As STATES Interactive
 Map From STATES
  Position (0.0520833,0.0520833) Units "in"
  Width 3.91667 Units "in" Height 2.14583 Units "in"
 Set Window FrontWindow() ScrollBars Off Autoscroll On
 Set Map
  CoordSys Earth Projection 1, 0
  Center (-95.844784,37.561978)
```

```
 Zoom 3815.369063 Units "mi"
 Preserve Zoom Display Zoom
 XY Units "degree" Distance Units "mi" Area Units "sq mi"
shade 1 with Pop_1990 ranges apply all use all Brush
(2,16777168,16777215)
 400000: 1500000 Brush (2,16777168,16777215) Pen (1,2,0) ,
 1500000: 3500000 Brush (2,16756880,16777215) Pen (1,2,0) ,
 3500000: 5500000 Brush (2,16744448,16777215) Pen (1,2,0) ,
 5500000: 29800000 Brush (2,16711680,16777215) Pen (1,2,0)
 default Brush (2,16777215,16777215) Pen (1,2,0)
Set Map
 Layer 1
 Display Value
 Selectable Off
 Layer 2
 Display Graphic
 Label Line None Position Center Font ("Arial",0,9,0) Pen (1,2,0)
  With State_Name
  Parallel On Auto Off Overlap Off Duplicates On Offset 2
  Visibility On
set legend
 layer 1
 display on
 shades on
 symbols off
 lines off
 count on
 title auto Font ("Arial",0,9,0)
 subtitle auto Font ("Arial",0,8,0)
 ascending off
 ranges Font ("Arial",0,8,0)
  auto display off ,
  auto display on ,
  auto display on ,
  auto display on ,
  auto display on
Create Cartographic Legend
 Position (3.34375,1.625) Units "in"
 Width 2.44792 Units "in" Height 1.26042 Units "in"
 Window Title "Legend of STATES Map"
 Portrait
 Frame From Layer 1
 Border Pen (0,1,0)
End Sub
```

```
'***************************************************************
'* Function: pop_rural_urban
'* Purpose: Display Rural vs Urban Thematic.
'***************************************************************
Sub pop_rural_urban
 '** First, close everything else in MapInfo
 Run Menu Command M_FILE_CLOSE_ALL
 '** Run the workspace application
 Run Application ApplicationDirectory$() + "work2.wor"
End Sub
```

1. Edit the *work.mb* source code, as in the previous block of code.
2. Select Project | Compile Current File from the MapBasic Main menu bar.
3. If you do not see a message informing you of a successful compile, correct the errors and recompile.
4. Select Project | Run and switch to the MapInfo application. The *work.mbx* program will add the Demo menu at the end of the Main menu bar. The 1990 Population Thematic and the Rural vs Urban Population menu items now display the workspaces you previously created.

Extras: Useful Utilities with Source on the Companion CD-ROM

As a bonus to *INSIDE MapInfo Professional* readers, we have included two very useful utilities on the companion CD-ROM. Each is fully documented and presented with source so that you can study how various functionality was implemented.

The "Distance Magic" program (*dmagic.mb*) provides a simple interface and back end functionality to quickly report the distance between features in two tables. Examine the code to better understand how distance operations can be implemented. Use the routine to systematize the process of reporting distances between features.

The "Join Magic" program (*jmagic.mb*) provides a multi-step "wizard"-like interface with supporting functionality to simplify the operation of joining different tables according to either tabular or spatial indices. This code demonstrates how MapInfo dialogs can be organized like "wizards," and indicates how Join operations can be automated. Use it for yourself or others if SQL query seems cumbersome.

Good luck with creating your own customized MapInfo world!

Summary

The MapBasic programming language allows you to control and extend MapInfo Professional by developing extensions, utilities, and complete applications. MapBasic provides you with a development environment that can be used to create MapBasic programs using the text editor, compiler, linker, and on-line help. Because the MapBasic language resembles most popular structured programming languages, it is relatively simple and easy to learn for users with programming experience. MapInfo also provides the MapBasic window and workspace mechanisms to help you become more familiar with MapBasic statements.

Several sample MapBasic applications reviewed in this chapter focused on the functional areas of menus, dialogs, and table manipulations. These sample applications are good starting points for experimenting with these statements and adding statements to create your own MapBasic applications. Should you ever seek alternative methods of distributing and/or automating MapInfo technology, you can evaluate several recent MapInfo products that help developers with web-based and wireless communications platforms.

PART III
CASE STUDIES

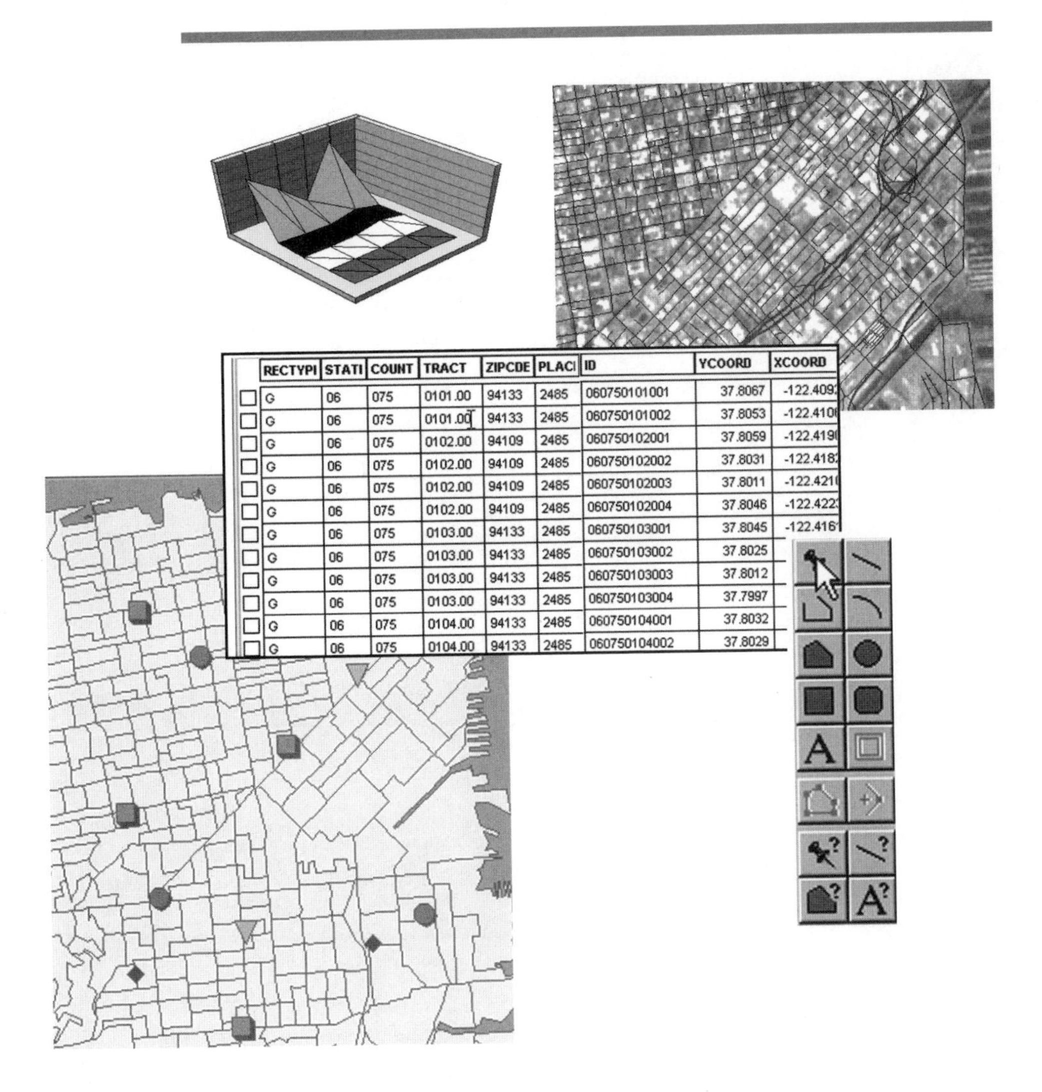

Case Study 1

First American Flood Data Services, Inc.

Product Diversification

As core products mature, businesses must attempt to diversify their portfolios in order to maintain and expand respective market offerings. The challenge lies in taking a well-justified business plan and developing it into consistent revenue streams.

The development environment for information products offers many obstacles to getting product to market effectively and efficiently. Creation and delivery of product are often what stands between success and failure in product line diversification. The tools used to implement ideas are key components to a successful business unit.

Established in 1996, First American Flood Data's New Products Department set out to create several new business units that would diversify their maturing product lines. After viable business units were created, the challenge became defining and developing products and then delivering them to market. Two departments at First American took advantage of MapInfo technology to create and deliver "new" information products: PropertyView and the California Property Disclosure data center.

PropertyView

PropertyView is a nationwide property and lifestyle information system conceived by First American Flood Data Services to deliver quality neighborhood information to families across the United States. The original concept had consumers accessing property information via toll-free numbers and the Web, whereas the product itself would take the form of several reports. Product information included school, crime, environmental, community (demographics), flood, and senior care data.

The Project

Early in the development of the product, the main challenge was how to illustrate neighborhood characteristics for relocating families. Ultimately, the solution was to generate maps that used census tract boundaries to delineate neighborhoods and illustrate neighborhood characteristics via thematic shading. For example, The Crime Report includes the propensity for crime at the census tract level. To effectively show how one side of town differs from another, MapInfo Professional was used to create color-coded images at the census level to reflect the crime rate, as depicted in the following illustration.

Different shades represent varying levels of crime in the neighborhood.

Conceptually, one might regard the development of this product as a relatively simple undertaking until one considers the number of images required to support its full manifestation. First American defined product requirements as follows: three map images for crime data, two map images for environmental information, seven images for general community information, four images for senior citizen data, and one image for floodplain considerations for each census tract in the country. In addition, more than seven images of "school" data were defined for each school district in the country. Altogether, with approximately 62,000 census tract and 14,000 school districts in the United States, the PropertyView concept required First American to support over 1 million "neighborhood" images.

Because PropertyView's initial concept predated relatively recent advancements in MapInfo technology related to the Internet, and First American was intent on delivering data on the Internet, the product team addressed its requirements by creating utilities that automatically generated images with MapInfo. A MapBasic batch application was written to create the maps and compose thematic reports according to predefined formats prescribed by product specifications.

Batch Image Generation

Batch image generation basically involved three steps: preprocessing data, applying geographic layers, and creation of the image. Each step is described in the sections that follow.

Preprocessing Data

Because the entire application is written in MapBasic, the data must be precoded to allow the application to work as efficiently as possible. For general feature representations, precoding involved defining acceptable "default" colors and line widths for each data type in MapInfo.

For thematic "neighborhood" properties, assigning a value to each respective level of shading was necessary. For example, in the previous image the different levels of crime are illustrated by assigning a corresponding shade of color. The actual levels of crime varied from 0.0 to 3.5, but to allow for efficient processing the data are precoded and assigned a 1 to 5 value representing the five different levels of crime in the area.

Applying Geographic Layers

The application begins by reading a census tract number from a corresponding data file. The census tract layer is then turned on in the application. Once the census tract number is found in the attribute table of the layer, the census tract centroid is centered and 2-1/2 times the area is then drawn into focus. Drawing the centroid into focus allows for a representative area of comparison to be shown. The point and line layers are then applied (automatically) and the streets and roads come into view.

Creation of the Image

Once all relevant layers are applied and the corresponding shading is complete, the application exports the view as a *.jpeg* image (stands for Joint Photographic Experts Group, a low-resolution, compressed graphic file format commonly used on the Web). In First American's process, the images are named according to census tract number, followed by a product designation. For example, the previous image is named *484530012cv2.jpg*. The census tract number followed product designation and image number.

PropertyView has been an exciting introduction to the First American product family. MapInfo has provided the vital software tools to enable the display of spatial information. Without MapInfo, it is difficult to envision how the product could effectively convey neighborhood "themes."

California Property Disclosure Data Center

A separate division of First American uses MapInfo Professional in analyzing properties in California. For every mortgage initiated in the state of California, state law mandates that the property be evaluated with respect to its earthquake, seismic, fire, flood, and tax facility district status. First American created the California Property Disclosure Data Center (CPDDC) to fulfill disclosure-related orders. The CPDDC has responded to its responsibilities with a combination of MapInfo and geocoding technologies (Centrus) from Sagent Corporation.

CPDDC's application essentially registers longitude/latitudes on the different hazard layers and outputs a determination as to property status. Hazard layers are enabled, and if a property is deemed to be too close to the hazard area it is output to a "sister" MapInfo Professional system that allows the user to overlay plat and parcel maps to make an accurate determination. An example of this is shown in the following illustration.

Floodplain areas in blue, and liquefaction in yellow.

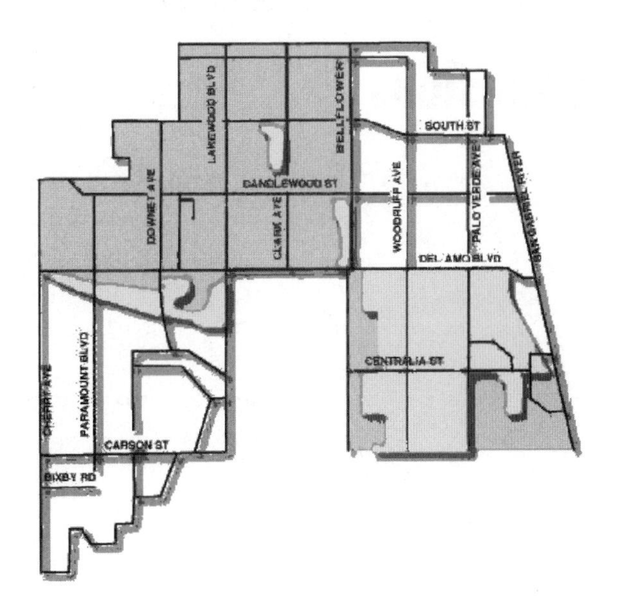

Where necessary, the longitude/latitudes of locations for which only addresses are known can be determined by the GDT/Centrus based geocoder products. Orders are batched through the Centrus geocoder, which compares street address information to GDT's digital street data, and longitude/latitudes are appended to location records.

Geocoded records are then portrayed in MapInfo. Based on the accuracy of the geocode as reported by Centrus, First American's routines reflect the geographic level of confidence in any particular point's position. The process of integrating the geocoder took less than four hours. Once again, Centrus provided the quick solution and a geocoding engine with horsepower.

First American's overall application was initially created to handle 200 to 500 orders a day. The system ultimately processed over 5,000 per day. Hundreds of records requiring geocoding are routinely processed in only a few hours. The MapInfo/Centrus combination has produced quite satisfactory performance.

By employing MapInfo technology, CPDDC successfully automated a spatial process that proved to be immensely efficient and effective. As requests for analysis are growing, other mapping workstations are being daisy-chained to the process to expand the organization's capacity for handling more manual determinations.

Conclusions

The development and delivery of product are crucial to the success of a new business unit. By taking advantage of the efficiencies MapInfo and related geographic technologies provide, First American Flood Data successfully created new product lines to help diversify its portfolio.

Contributions by Chris Roussel, formerly of First American Flood Data Services, Austin, Texas.

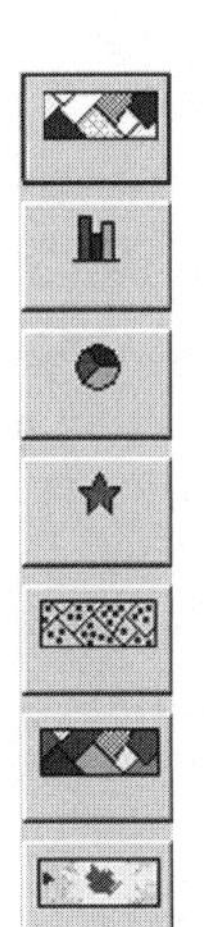

CASE STUDY 2

ELLER MEDIA COMPANY

MapInfo in Advertising

THE FOSTER AND KLEISER OUTDOOR ADVERTISING COMPANY was founded in 1901 by two native Californians, Walter Foster and George Kleiser. Outdoor advertising posters at that time were simply placed on any available surface: walls, fences, and poles. In time, Foster and Kleiser sought a more standardized method of operation, and developed uniform structures to upgrade the medium with attractive displays and landscaping. Their innovations and improvements brought immediate and long-lasting success. Over the years, Foster and Kleiser became established in major metropolitan markets throughout the United States, and were recognized as one of leaders of the outdoor advertising industry.

In August of 1995, Eller Media Company of San Antonio, Texas, acquired all facilities of Foster and Kleiser. Today, Eller Media has over half a million display points around the world, and access to over half the U.S. population. Eller Media continues leadership in the advancement of the outdoor advertising medium through research and innovations. The company manages data on the markets it serves and is making innovations in digital photo imaging and mapping. Eller Media continues to find more effective ways to market mobility that will help advertisers achieve their goals in San Antonio and other key metropolitan markets.

Eller Media Adopts MapInfo Professional

Along with the increased sophistication of the advertising clientele, Eller Media has sought to adopt technologies demonstrating that they can target locations, and demographic segments and lifestyles, via outdoor locations. Eller Media's first efforts at graphic solutions involved various projects with Market Information Services of America, Inc. (MISA). Together, Eller Media and MISA developed a software package to track the many billboard structures and advertisers on whom they rely.

This system provides the plant operator with a means of organizing and maintaining necessary data in a convenient form, and then easily produces the required charting plant status reports. MISA Charting Pro software, established in 1985 in close association with a number of outdoor and advertising plant operators, has been valuable in assisting outdoor advertising operators to schedule the placement of the advertising medium. It has been continually maintained and upgraded to assist advertising chartists with their most routine charting tasks.

For many years, Eller Media augmented the data on the Charting Pro system by locating sites on a manually generated map with text audit information describing the demographics for the client's products. Today, Eller Media uses Charting Pro in conjunction with MapInfo Professional to automatically display various data on maps. The Charting Pro database has been made accessible to MapInfo, and MapInfo accelerates and adds value to data display.

For example, MapInfo enables Eller Media to depict various actual and hypothetical "showings," or packages of poster locations sold for periods of one to twelve months. Using MapInfo, Eller Media can generate maps that depict how eight posters distributed throughout the city might display an advertisement for a period of two months, and then might be moved to eight different locations for a second two-month period.

Since adopting MapInfo, Eller Media has been able to depict the value of its media much more effectively than previously. They have embellished their "toolkit" with a variety of demographic, lifestyle, and traffic data and can present clients with persuasive images that portray how a relationship with Eller Media can help them attain revenue targets.

Current Use of MapInfo Research Tools

Eller Media Company analyzes markets by examining demographic profiles, buying power, socioeconomic information, and associated geographic distribution. Using Claritas's Compass for Windows, Eller Media also acquires detailed demographic data within zip codes and estimates the propensity of different market areas to purchase their clients' goods and services.

Claritas can locate and display concentrations of people who use various consumer products and identify specific demographic categories. This data can then be used to estimate potential exposure to an outdoor showing for selected audience segments. For example, by incorporating consumer information from an outside source, such as The Media Audit, Claritas can be used to determine the potential number of adults who drink soda in each zip code of the target market area, as indicated in the following illustration.

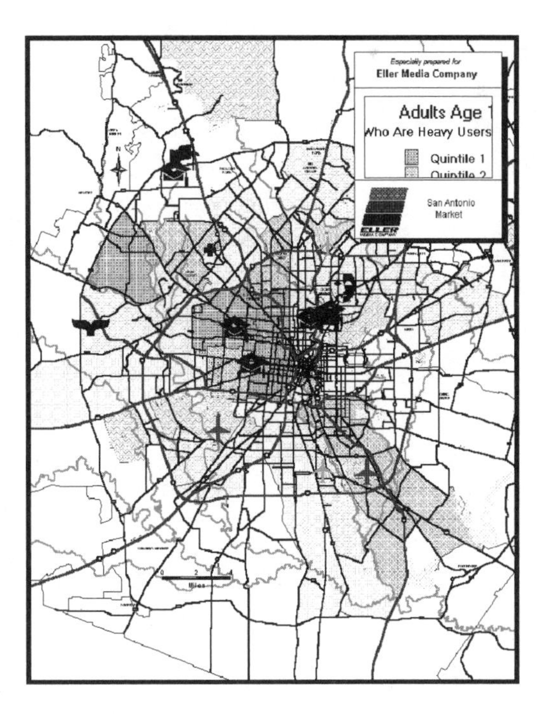

Map depicting soda consumers by zip code areas. (Image courtesy of Eller Media Company.)

Displaying the concentrations of a client's target audience in a market area is especially useful to an outdoor advertiser because it can facilitate preparation of an effective outdoor program and allows for a more efficient use of the advertising dollar. The most requested geodemographic map in San Antonio focuses on the Hispanic population. Claritas information is often matched with zip code boundaries and grouped in 20% quintiles. Each quintile has been coded with a color ranging from red ("hot") to white ("cold"). Although each quintile represents approximately 20% of the target audience in the marketplace, quintile 1 is usually considered the "hottest" quintile because it contains the heaviest concentration of target audience persons.

Such capabilities have made an obvious impact on Eller Media's clients. As Lauren Carlson, media director for Cavette & Company in Minneapolis, Minnesota, remarks: "We've had the greatest success with outdoor as an awareness builder. With the latest mapping capabilities available from outdoor companies, we can now track our customers with pinpoint accuracy, thereby focusing on the geographic areas that match our audience best."

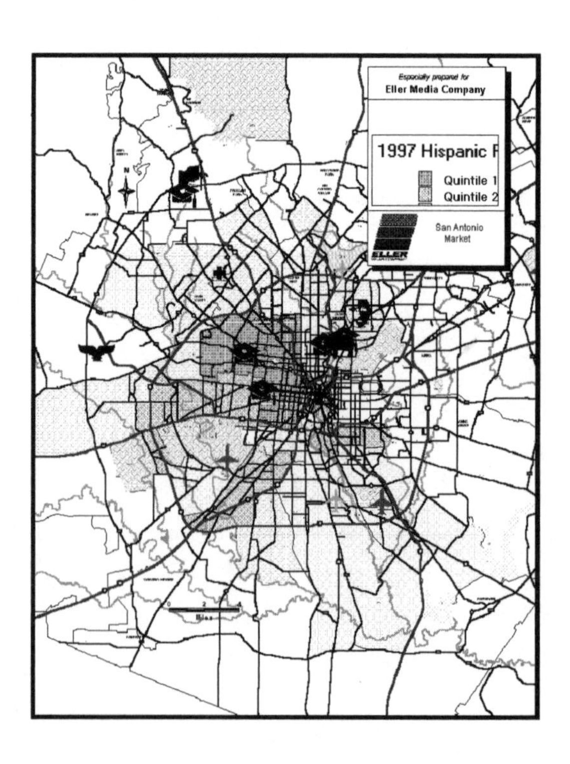

Map depicting concentrations of Hispanic population in San Antonio, Texas. (Image courtesy of Eller Media Company.)

Another noteworthy MapInfo application in use at Eller Media combines Demotracker and ChartLink to schedule outdoor advertising and thematically map demographics to traffic flow passing billboards. Customized mapping software exclusive to Eller Media Company incorporates the data derived from Claritas, Media Audit, and Simmons Survey to examine counties at the street level. The software allows for the mapping of demographic analysis, as well as specific locations, such as retail chains, hospitals, convenience stores, and so on.

Eller Media's Design and Draw Process

Eller Media typically prepares a "base map" by opening a workspace for geodemographics, and opening a dBase *DBF* file selected for the appropriate target audience. Using MapInfo's Create Thematic Map option, for example, a zip code table is selected as *TXZip96* and joined with the demographic file previously opened. Ranges are set to five and the style is custom. The styles are set to red, orange, yellow, green, and white, and a customized legend is entered to show the quintile levels, as depicted in the following illustration.

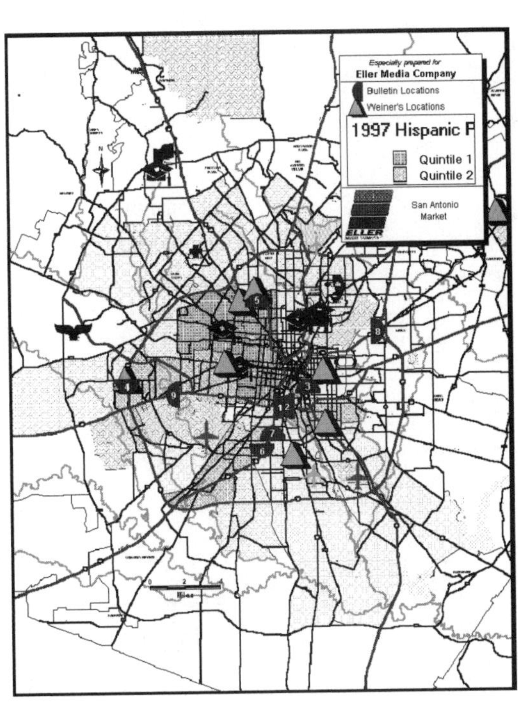

Map depicting symbols representing pinpoint locations. (Image courtesy of Eller Media.)

MapInfo Professional produces all maps of Custom media showings, Claritas thematic maps, retail/point of interest locations, targeted showings, and charted media showings. Corporate standards of symbols, fonts, and colors have been established for use with all markets across the United States.

Generally, the basic map layers contain zip codes, street files, highways, bodies of water, landmarks, and respective labels. Once information is added, such as billboard locations or retail outlets, the base map becomes a customized map. Symbols are used to graphically represent a point object, such as customer locations, competitor locations, or billboard locations. Labels are added to symbols and correspond to a list of locations. The map, along with a location list of the showings and/or competitor locations, constitutes an effective visual marketing tool.

Other Uses for MapInfo at Eller Media

Eller Media also uses MapInfo to create "photoliths," or pictures of bulletins, along with a very small map used to display a single bulletin location. Information about the approach, description of the bulletin, and description of the location is very helpful when individuals cannot visit bulletin sites nor view the bulletin itself, as indicated in the following illustration.

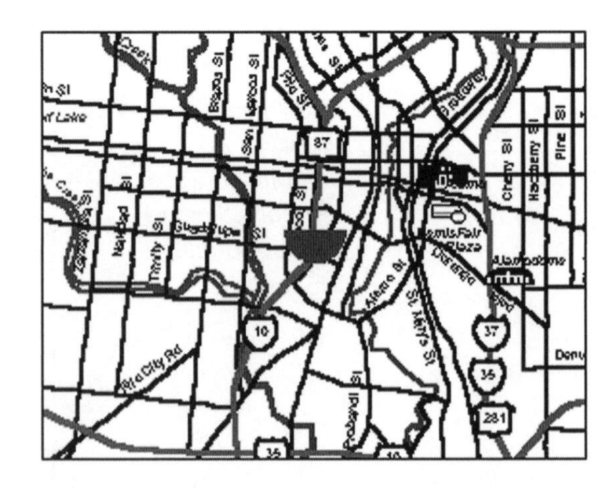

Part of photolith showing precise location and surroundings of bulletin. (Image courtesy of Eller Media.)

To create photoliths, Eller Media MapInfo users open an inventory of "bulletins." A specific location is identified with the Find command. The cosmetic layer is set to editable and a symbol is placed over the tagged location. The zoom level is set to 6, and the layout map size is set to 2" x 3". The small map is printed, cut, and attached to the picture with text describing the billboard location.

Recognizing that outdoor advertising is based on consumer mobility, Eller Media also creates "Mobility Maps" based on Claritas data showing highway traffic flow. Daytime commuters are subtracted from the nighttime totals to display drivers on the streets and highways. Heavy traffic flow on expressways, primary arteries, and major intersections are displayed as "red" areas. Locating bulletins at red areas permits the advertiser to reach the target audience more efficiently and with lower advertising cost.

Mobility Maps have proven very effective. In reference to an Altoids Mint campaign, Leo Burnett's senior vice president Gary Singer said, "Through geodemographic targeting, we were able to identify not only the markets with the most potential, but also the zip codes within those markets where our WIMPs (white-collar, independent mint users with purchasing power) worked and played" (July 1996 issue of TAB newsletter, *Inside Out of Home*).

To develop a program around client locations, buffers are useful. The client selects the best target option, and that option is then charted within the radius, including client locations, so that the client can view the total program. An example of the use of buffers is shown in the following illustration.

The flexibility of MapInfo Professional allows Eller Media to draw maps of diverse sizes, ranging from 2" x 3" to 24" or 36". The geodemographic base layer allows text data to be displayed in a colorful graphic representation. Adding specific medium locations to the map ensures that advertisers receive a well-distributed showing reaching the targeted audience.

MapInfo Professional is a strategic tool for Eller Media in that it shows clients how using the company's services can accomplish their goals. MapInfo enables advertisers to get a clearer picture of Eller Media's value. Whether targeted at advertisers' locations,

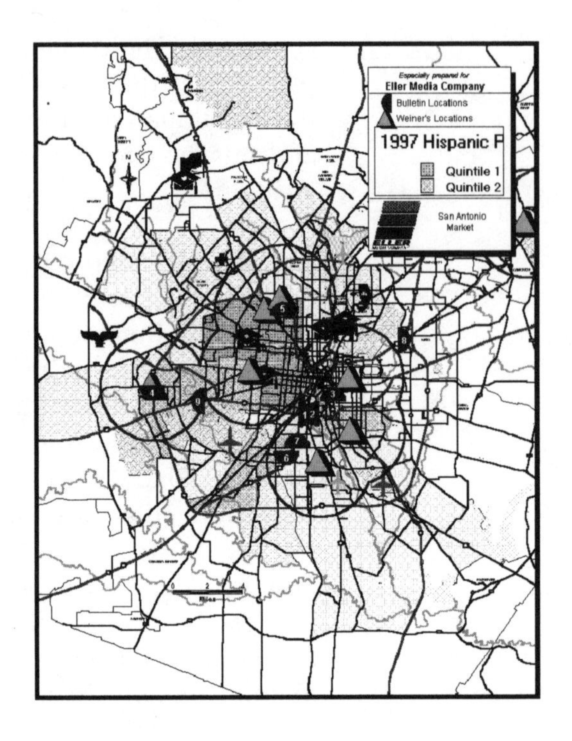

Map depicting buffers, or areas of surrounding client locations. (Image courtesy of Eller Media.)

demographic targets, or lifestyle destinations (malls, schools, and so on), MapInfo creates a detailed picture that makes it easy for clients to understand how Eller Media can help them meet advertising challenges.

Contributions by Cindy Gabel, Eller Media Company, San Antonio, Texas.

CASE STUDY 3

GRUBB & ELLIS

MapInfo Professional as a Strategic Tool in Providing Real Estate Services

GRUBB & ELLIS IS ONE OF THE NATION'S LARGEST publicly traded commercial real estate services firms. With approximately 8,000 professionals and staff, Grubb & Ellis provides transactional, management, financial, and strategic services to clients in more than 25 countries worldwide. Through its Management Services subsidiary and affiliates, Grubb & Ellis manages a total portfolio of over 125 million square feet.

In 1995, Grubb & Ellis wanted to enhance and expand on the existing copies of MapInfo Professional technology they acquired earlier that year. The company needed to find easy-to-use technology that would be capable of providing maps and reports commonly required for real estate analysis on behalf of their clients and prospective customers. The main goal was to show customers a tremendous amount of data quickly and effectively.

Grubb & Ellis sought a mapping system integrator, one that could assist them in the acquisition of an easy-to-use software tool, management of all previously purchased and proprietary Grubb & Ellis databases, and deployment and support of a large, multi-office installation. Within a month, Grubb & Ellis identified Integration Technologies (IT), a Newport Beach, California, based integration firm specializing in multi-user business mapping solutions. With experience serving hundreds of customers in real estate, restauranteuring, and retail, IT was able to provide the right solution quickly and efficiently. IT integrated existing MapInfo Professional software into an easier-to-use software solution, combined all acquired and created data tables, and solved the multi-office deployment issues necessary to meet the requirements of the Grubb & Ellis offices.

To solve Grubb & Ellis's problem, IT supplied its product AnySite Professional, a real estate business mapping software solution built on MapInfo Professional. AnySite Pro enables brokers and analysts to take Grubb & Ellis's data, purchased data (from sources such as National Decisions Systems and National Research Bureau), and over 200,000 other commercially available layers of information (such as traffic volume, enhanced streets, and so on) and transform *all* of it into a *single*, turnkey solution that is easy to use.

According to Lisa Ackerman, vice president of mapping solutions at Integration Technologies: "Grubb & Ellis had a common problem. The ability to get mapping technology into the right hands (brokers and analysts) so they could demonstrate to customers why they should work with Grubb & Ellis and which real estate business decisions to make. With our real estate experience, mapping knowledge, experienced staff, and deployment strategies, we knew we could make this happen for Grubb & Ellis, and happen fast."

By mid-1998, Grubb & Ellis had deployed AnySite Pro in over 40 of its 87 offices nationwide. Data analysts can now quickly create maps and reports for transaction professionals to share with customers, facilitating the decision-making process and making choices transparent.

"Showing customers we understand their business is the most important service we can provide," says Robert Bach, vice president of marketing at Grubb & Ellis. "Once we get in front of them with the mapping technology, our clients are ready and excited to see more."

Content

The maps and reports produced in AnySite Pro on MapInfo Professional provide customers and brokers with instant access to the wide variety of data available at Grubb & Ellis. The single most effective way to help customers make the right decision regarding their real estate needs is to take a broker's knowledge and combine it with the available data to create a map that shows customer-focused information, such as the following.

- Available locations
- Comparable sales
- Market saturation
- Competitors
- Demographics
- Shopping centers
- Employment data
- Traffic volume
- Dollar demand for a variety of products and services

"Without the mapping technology, we used to spend a great deal more time sorting through huge amounts of data. Now, using AnySite Pro, we can analyze all of the data available at Grubb & Ellis, including data sources such as National Decision Systems, to develop answers that are much easier to determine and to present to each customer," says Bach. "Maps and demographic reports have become an absolute requirement for *every* presentation package that is sent to a customer."

Examples

Several examples demonstrate how Grubb & Ellis customers benefited from using AnySite Pro and MapInfo Professional's mapping technology. For Diedrich Coffee, Grubb & Ellis utilized mapping technology to display multiple variables simultaneously. A standard package for Diedrich Coffee included data reports, as well as maps displaying education, income, residential and daytime population, and drive-time polygons, an example of which is shown in the following illustration.

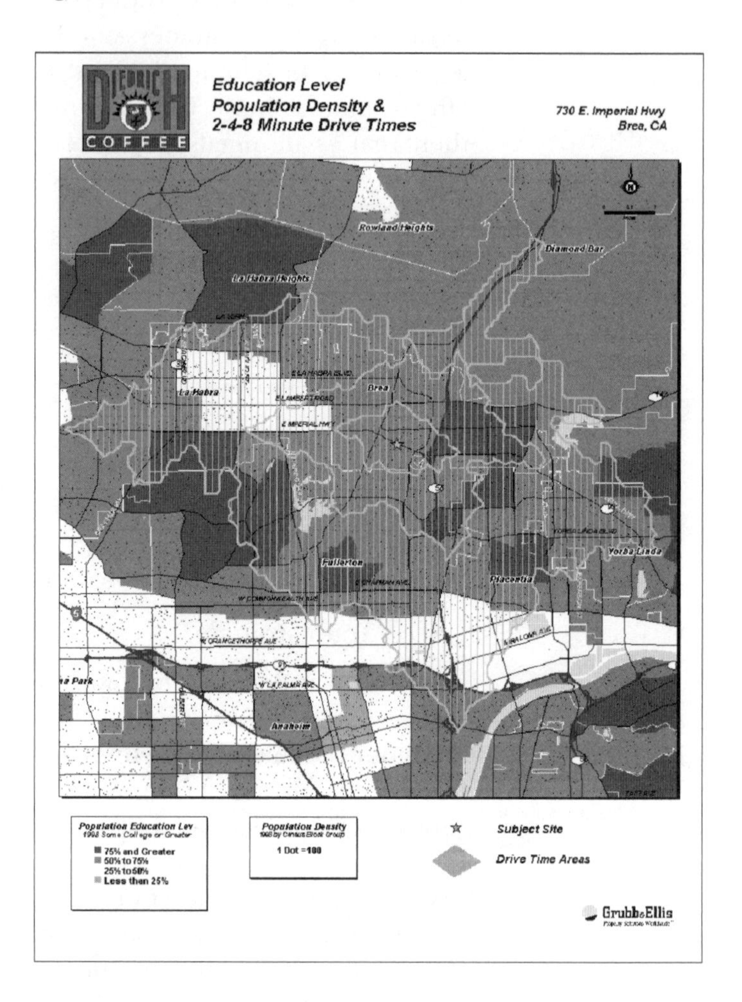

Sample map depicting education and income levels, residential and daytime population, and drive-time polygons. (Image courtesy of Grubb & Ellis.)

This mapping technology enabled Grubb & Ellis to display the population density, education level, *and* drive-time polygons *simultaneously* to demonstrate to Diedrich why certain sites were better locations than others. For XYZ Theaters, Grubb & Ellis used mapping technology to show drive-time polygons around an existing movie theater and its competition, an example of which is shown in the following illustration.

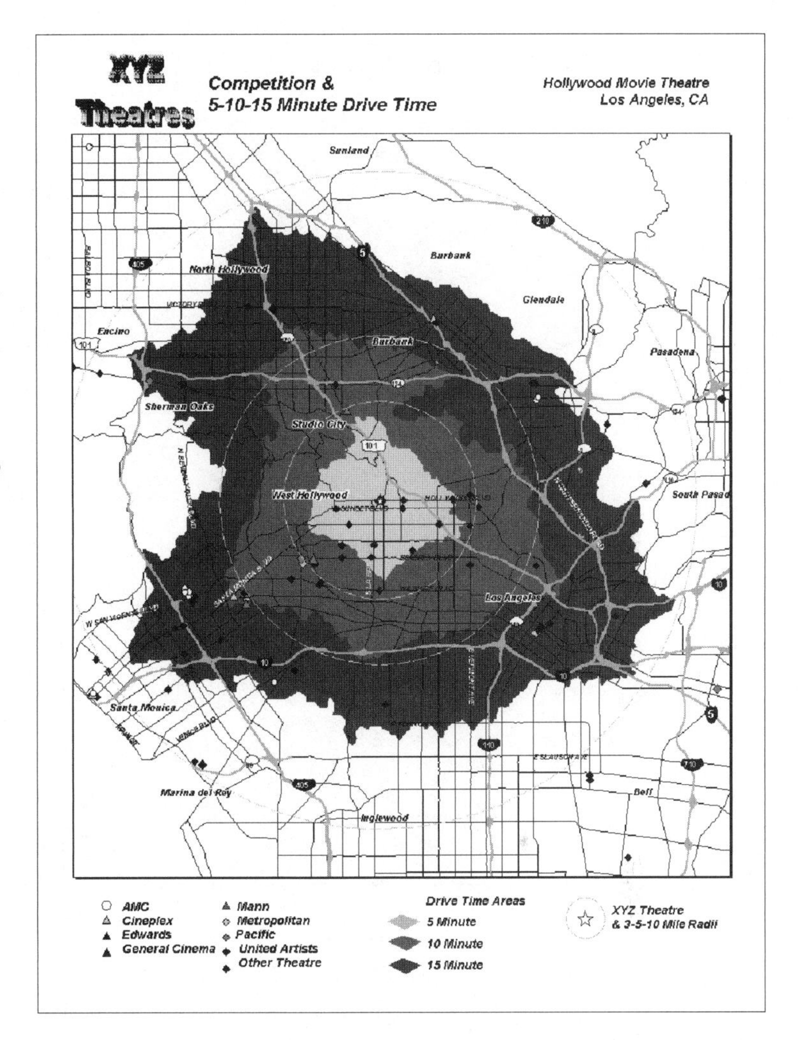

Sample map depicting drive-time polygons surrounding a vendor location, along with competitors. (Image courtesy of Grubb & Ellis.)

In this example, the flexibility of the mapping technology enabled Grubb & Ellis to address the number of competitors in the area by displaying the points thematically with the use of different shapes and colors, to allow the reader to distinguish between the theaters. The map clearly identifies the right real estate solution for this customer. For several clients, Grubb & Ellis has utilized MapInfo to display traffic volume, an example of which is shown in the following illustration.

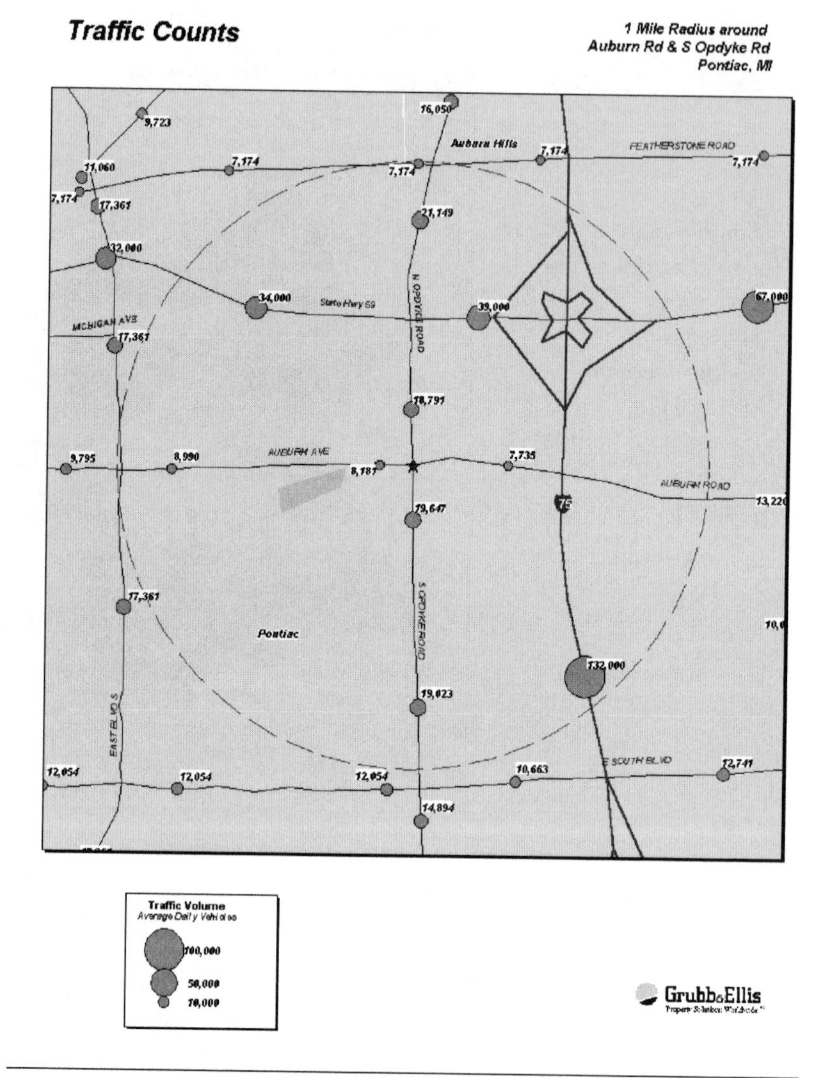

Map depicting drive times. (Image courtesy of Grubb & Ellis.)

Grubb & Ellis commonly uses mapping technology to create a graduated symbol thematic map of traffic counts. Such maps allow readers to intuitively see the traffic flow in relation to a specific geographic region, a significant factor in the decision-making process. For ABC Company, Grubb & Ellis used mapping technology to geographically display the location of the company's employees, as shown in the final illustration.

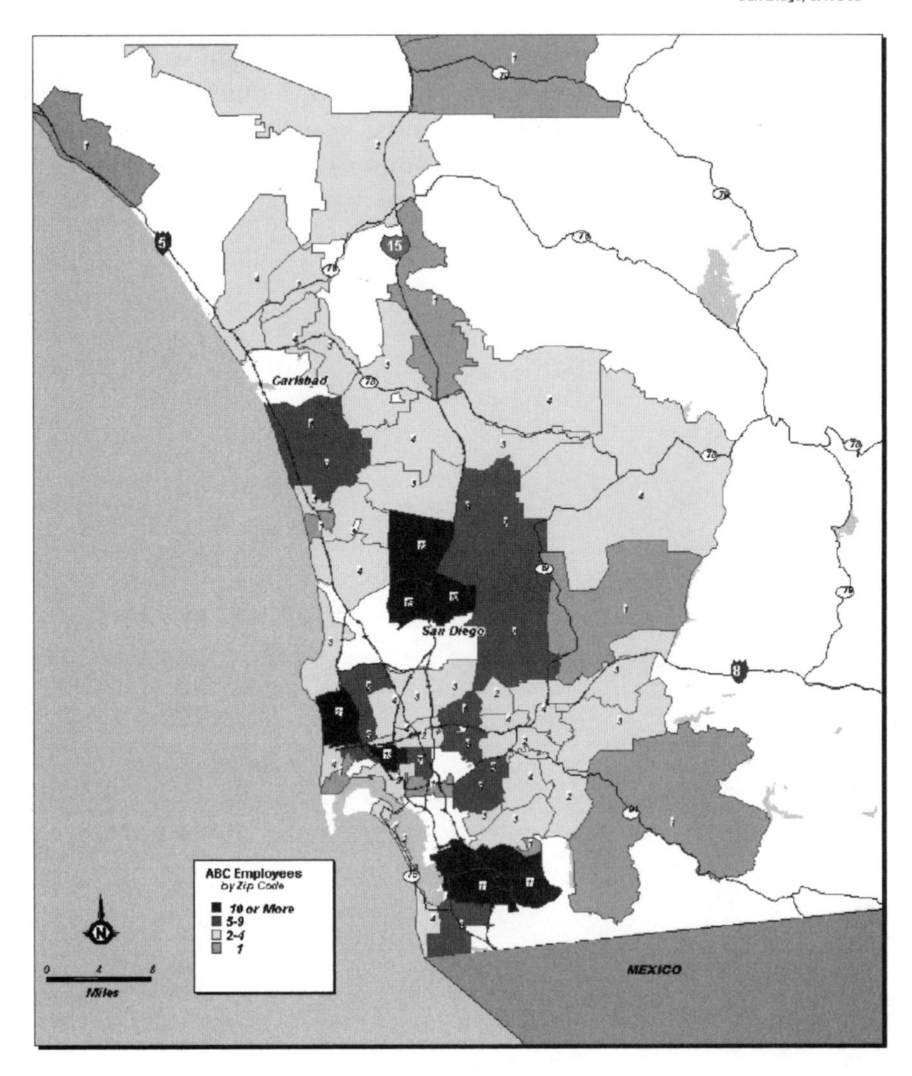

Map depicting residences of company employees. (Image courtesy of Grubb & Ellis.)

Viewing the data as a map helps the customer make a clear choice when selecting new facility locations.

Conclusions

"Solving customers' real estate business problems is what we do. Using mapping technology is the most effective solution for all of our client presentations," concludes Bach. At Grubb & Ellis, MapInfo has become an important tool for providing clients a highly effective, high-quality service.

CASE STUDY 4

MARKETMAP

A Geodemographic Real Estate Valuation System

REAL ESTATE AGENTS SUMMARIZE THE FACTORS that determine a property's value in three words: location, location, location. Clearly, there are few applications more linked to the geodemographic environment than the valuation of real estate. Of course, these three words are a qualified concept. In practice, it is good for a house to be near a highway, but not right next to it. It is great to have a supermarket close by, but not in your backyard. Regional malls, parks, fire stations, health care facilities, and other geographic area attributes can all contribute value, as can natural attributes, such as lakes, rivers, or mountain views.

Once again, proximity to certain natural attributes can also be a negative factor if they are associated with floods or unstable land. The lifestyle and socioeconomic context of your neighbors play a large part in determining the value of a home, such that one home's value could be increased or reduced by proximity to more valuable homes or less valuable homes, respectively. Furthermore, despite the imperative of location, the intrinsic characteristics of a home are likewise critical to its value: lot size, size of the home, number of bedrooms and baths, age, and condition can add or subtract value.

When real estate agents give clients a suggested list price and expected sale price based on memory, they are in fact employing a fairly sophisticated mathematical model. They are trading off the characteristics of the home versus others on the market, evaluating the locational benefits, and factoring in the socioeconomic context and trends in order to derive current market value. Typically, experience and intuition drive this exercise. With the use of GIS tools, however, this calculation can be made explicit, with all terms (variables) laid out in a linear regression model, which is accessed and run in MapInfo.

This case study highlights an application developed in MapBasic that performs a current real estate valuation for a residential property in order to deliver a real-time valuation to a loan officer at a financial institution. MarketMap, an international geodemographic systems development firm, developed this application for a building society (the equivalent of a U.S. savings and loan institution) in South Africa. The application was developed for the bank's home equity lending department to facilitate processing initial property valuations over the phone, and delivery of preapprovals to potential borrowers within minutes.

The driving force behind the development was to make the building society more responsive to consumers by removing a critical time-consuming component in the lending process: the time required to calculate current market value. With a faster loan approval determination (albeit preliminary, and pending a visual inspection to confirm the facts), a larger proportion of loan inquiries should lead to closed deals.

Deciding on Data and Level of Geographic Resolution

The factors determining the value of residential real estate are well known. Prior to the widespread development of GIS-compatible data sets, however, few of these factors existed in a format that enabled the real estate valuation process to be automated. As previously noted, many factors are defined in terms of distance or geographic proximity: access to a highway, distance to a primary school or a supermarket, and presence of a park within a neighborhood. Many other factors that exist in available data sets—such as geophysical attributes (terrain and flood plains, for instance), crime statistics, and demographic trends—must be referenced to a common, local-level geography in order to be related systematically to property value.

Statistical models, moreover, require a commonly referenced database in which causal factors (independent variables) can be directly linked to the sale price outcome (dependent variable). In South Africa, as in many countries, sale prices—along with the attributes of the residence (bedrooms, baths, square meterage, and so on)—are recorded in a Multiple Listing Service (MLS) database used by real estate agents. The MLS data for properties that have sold are summarized each month in a subscription database: the South African Property Transfer Guide. All of the country's major banks subscribe to this database, which also contains street addresses and cadastral property references that can be readily geocoded.

Once the property transfer records are geocoded, they can be matched to demographic and socioeconomic data on the area, as well as point data on the locations of specific attributes (schools, supermarkets, and so on). The MLS data encompass the vast majority of formal real estate transfers, including Western-style suburban homes, sectional title (condominium) units, and homes in the former Black townships, many of which were transferred from city council ownership to freehold title as part of the unraveling of the country's apartheid structure.

One major issue in developing the valuation model was the need for locational precision. Many real estate GIS systems are built around precise mapping of the property location and boundaries, accomplished via cadastral boundary reference maps. However, although this may be critical for applications such as mapping utility line easements or determining flood planes, it is not necessary for most valuation tasks. What is more important is the *relative* location: where is a particular property in relation to other facilities that make up its geodemographic context? In most instances, such relative measures can be accomplished at a moderate level of geographic resolution, such as a neighborhood indicator.

Because South Africa does not have street number referencing files similar to the U.S. Census Bureau's TIGER files, point-level geocoding of a street address is not easily carried out. The smallest census unit geography is the Enumeration Area (EA), defined for the 1996 Census to consist of areas of 100 to 200 households. However, the digital boundaries of EAs have only recently become available, and very few other data sets are referenced to this level of geography. Furthermore, in the absence of a TIGER-like system, it would be very difficult to place an individual property into an EA during a live, real-time interaction over the phone.

The next largest unit is known as a *suburb*, roughly similar to the census tract in the United States. This proved to be the appropriate level of geography to use. Municipalities within South Africa's ten major metropolitan areas (as well as large, nonmetropolitan towns) are subdivided into suburbs. Unlike census tracts, however, the names and boundaries of suburbs are well known to South Africans. Their boundaries are routinely sign posted, and many people refer to them as part of their home address. These neighborhood names and respective boundaries are explicitly defined in South Africa, unlike U.S. urban neighborhoods, which often have amorphous boundaries. These boundaries (with major and minor roads, parks, rivers, and other attribute layers) are produced in MapInfo Professional-ready format by Map Studio, a national map book publisher based in Johannesburg.

Much more importantly, a wide range of geodemographic data is commercially available at the suburb level. Included are current population estimates from the Demographic Information Bureau, a current income model developed for MarketMap by WEFA Southern Africa, and the South African version of United Kingdom based Experían Corporation's MOSAIC neighborhood cluster classification system. These could readily be overlaid with supermarket, shopping center, school, hospital, and other point data, which MarketMap has compiled.

The initial system focused on the eastern suburbs of Pretoria, the capital of the country, and a city of about 800,000 people. The system was then extended to the Western Metropolitan SubStructure of Johannesburg, an area of some 2 million people that includes the former Black township of Soweto, as well as some of the country's most affluent suburbs. The design of the system enabled it to be implemented in all ten metropolitan areas and about 40 large secondary towns, which account for the vast majority of South Africa's urbanized, "first world" population. The system, however, could not be used to calculate value for farmland, game reserves, tribal communal land, or other rural village property.

Data Integration and Statistical Modeling

Continuing with the South Africa project, although all requisite data for a statistical model were readily available, they were not all geographically referenced. The South Africa Public Transfer Guide (SAPTG) data contained street addresses, but they were not "fielded." This process is not as straightforward as it would be in the United States, due to the lack of TIGER-like files or a standard for addressing such as the Coding Accuracy Support System (CASS). MarketMap, however, has developed several software tools used to clean, format, and geocode address data to the suburb level. Assignment to a suburb is constrained to possible choices within the four-digit postal code, because suburb names are often duplicated across municipalities. Spelling variants between English and Afrikaans, and alternative names for a suburb, are also handled.

This process automatically geocodes about 80% of SAPTG records. Remaining addresses are subjected to an increasingly manual process, which seeks to match the cadastral description to a suburb (a master matching database would be highly useful, but as yet does not exist). Of course, because the property transfer data are being used in a statistical model, 100% geocoding accuracy is not essential, provided that the geocoded properties yield a large, representative sample. In other words, it is assumed that records that cannot be geocoded are randomly distributed in terms of important characteristics. For purposes of model development, a 12-month historical file of property transfers was obtained from SAPTG. The following are among the data available from this file.

- List price
- Sale price
- Age of the building
- Condition of the building
- Square meterage of the building
- Square meterage of the property
- Number of bedrooms
- Number of baths
- Number of common rooms
- Presence of a pool
- Presence of a cottage (e.g., in-law quarters) on the property
- Presence of domestic servant quarters
- Type of water (city or well/borehole)
- Electrical hookup
- Rates and taxes
- Name of mortgage bond holder

For each area, a suburb-level attribute file was then developed, which was match/merged to the property transfer database by linking to the geocode. The following are the suburb level attributes computed in MapInfo Professional.

- Number of neighborhood shopping centers in the suburb
- Number of regional shopping malls within 10 kilometers of the suburb centroid
- Distance from suburb centroid to a highway interchange
- Distance from suburb centroid to a hospital
- Presence of a primary school within suburb
- Current demographic profile (number of households, age distribution, race distribution, and so on)
- Percent distribution of suburb in each of eight "Living Standards Measure" categories, a socioeconomic status scale widely used in South Africa for marketing studies

- Current median income and percent in each of seven income ranges
- Percent change in population and median income from the 1991 South African Census
- Lifestyle cluster code

The flat file was written out from MapInfo to a *.dbf* format file, which was read in to SPSS (Statistical Package for the Social Sciences, by SPSS Inc. of Chicago, Illinois). SPSS enables several options for development of linear regression models. Several models were tested to develop a function that yields the prediction of sale price as a function of residence and area attributes. In this process, some of the variables (such as distance to a highway interchange) were transformed into quadratic forms, so that they would optimize at an "in between" point, and yield a lower number if the subject property was too close or too far away. Several other variables were collapsed into fewer response categories in order to capture more of the variation within a measure having a normal distribution, as required for regression analysis.

Finally, in the initial suburban Pretoria model development, 516 property transfers in the 12-month file were found to be sales of plots only; these were excluded from the analysis. Also excluded were two sales of properties valued in excess of ZAR1 million, which were outliers representing custom estate homes.

The remaining cases were processed through a series of stepwise linear regression models, first using forward entry, and then backward removal, with varying thresholds for inclusion of variables in the final equations. Forward entry reveals variables that add the most to the model, whereas backward removal reveals those that most detract from predictive validity. A final model was run using the subset of variables that proved significant through either of the two methods at the 0.10 level, after adjustment of response coding to correct for any heteroskedasticity of error terms.

Collinearity diagnostics were then performed to ensure that the independent variables being used were not in themselves determinants of each other. From this exercise it was discovered that removing the "land area of the property" variable led to the number of bathrooms and bedrooms becoming significant. In other words, the latter two are linearly related to building area, which leads to collinearity problems. It was decided that it would be far easier for a person inquiring about a home equity loan over the phone to state the number of bedrooms and baths than to recall square meterage.

Model Results

The statistical model proved highly effective at yielding an accurate calculation of the list and sale price. The exercise demonstrated that real estate agents and parties to a property transaction by and large do act rationally and consistently in valuing a property when it comes up for sale.

The ultimate model derived for the overall area has an R^2 value of almost 75%, and an F statistic of 145 significant at the 99% confidence level. The R^2 of 0.75 reveals a highly linear pattern, particularly given the natural variance inherent in a number such as a home sale price, which has a wide range of substantial variance from one property to the next. The low to high range of the predicted sale price was R99,650 to R789,904 compared to actual sale price figures of R95,000 to R940,000, respectively. Every variable included in the model is statistically significant at the 90% level of confidence. The overall model is significant at the 99% level of confidence. To provide an example of variable weightings generated by the model, the following are coefficients for one set of suburbs to the east of Pretoria.

- Each additional year of age of the home leads to a decrease of R417.
- Each bathroom adds R9,891.
- Each square building meter adds R846.
- Each square plot meter adds R35.
- Each increment in building condition code adds R12,799.
- A pool adds R15,443.

- If Blacks account for more than 15% of the suburb population, the value is increased by R11,005. (In context, this is an indicator of the presence of domestic servants, because this area was predominantly White and many homes are equipped with domestic servant quarters.)
- Each buffer level of distance to a highway decreases price by R21,016 (i.e., those closer to the highway have a higher value).
- Curiously, the number of schools was associated with a decrease in home values by a reduction of R14,965 (which may indicate a temporal effect whereby the suburbs with schools were previously high market-value areas but have lost ground to newer developments).

NOTE: *In October of 1998, the South African rand traded at 6.5 to the U.S. dollar. At the time this model was developed, the rand traded at approximately 4.0 to the U.S. dollar. Real estate prices in South Africa tended to be about one-third the price of comparable properties in the United States.*

Implementing the Model in MapInfo

Once this model was developed in SPSS, the coefficients of the model could be applied to tables of data maintained at the suburb level in MapInfo Professional, in order to yield a calculation for a residence of any given characteristics in any suburb within the study area. This functionality was built in to a MapBasic program that could be accessed from within a master real estate loan inquiry system running in the bank's telephone loan support center.

The MapBasic program called MapInfo and automatically loaded several layers of data, including regional boundaries, suburb boundaries, rivers and lakes, major road networks, and selected landmark point attributes. It also reconfigured the MapInfo menu system to offer a Valuation menu pull-down, as well as several other pull-downs for summary data analysis (e.g., home mortgage bond value by suburb, mortgage bonds by issuing bank, and three-year mortgage lending volume trends). These latter pull-downs were intended primarily for use by corporate marketing and strategic planning managers. The following illustration shows the "look and feel" of the MapInfo screen appearing before the telephone loan counselor.

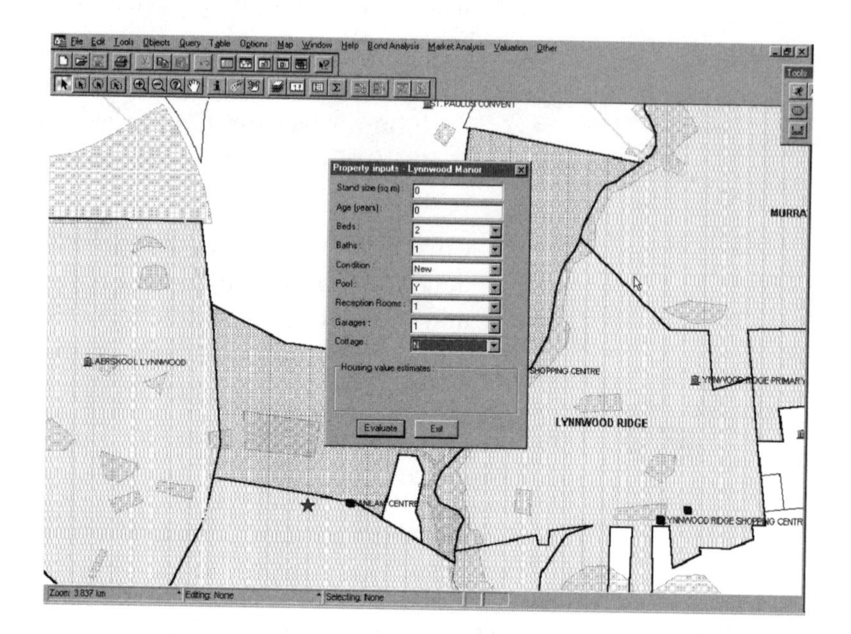

MapInfo workspace upon initiation of valuation system.

The task of the loan counselor is simply to ask the caller to state which suburb he or she lives in. A pull-down menu provides an alphabetical list of choices for the user to select from. This leads the MapBasic program to automatically zoom and reconfigure the screen to display that suburb at the center of the screen, with suburb-level attributes such as schools and shopping centers. The zoom distance is automatically adjusted for scale and viewability.

The next step on the menu system opens a window wherein the user enters the property's specific details, obtained live from the caller. Upon clicking on the Evaluate button, the system multiplies the values entered by their respective coefficients, obtains the geographic context variables for that suburb from the MapInfo table and multiplies them by their coefficients, adds in the model's constant term, and returns the current market value. This process is accomplished within one second. An example of the resultant screen is shown in the following illustration.

Because the name of the mortgage bond holder is also contained in the SAPTG file, the Bond Analysis menu enabled the bank's

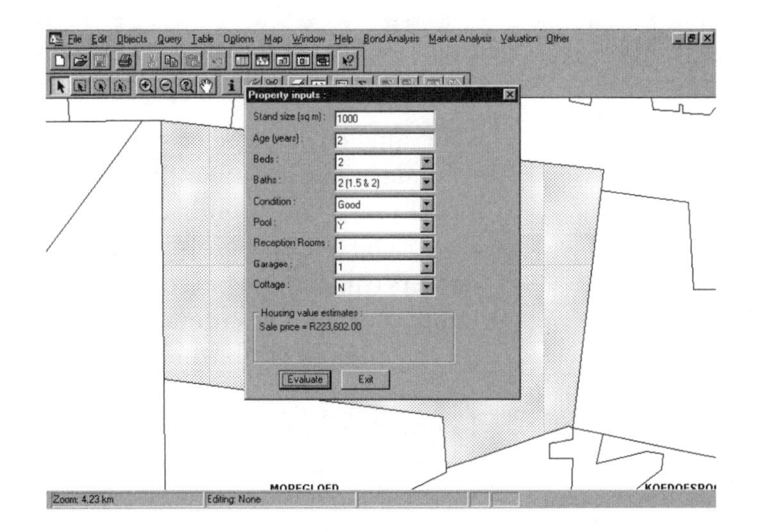

Valuation program result screen.

marketing department to access data from monthly property transfer files and view the relative volume and market share of mortgages outstanding across each market area. The map displays for this analysis were prepackaged in the MapBasic program to facilitate ease of use. This type of analysis, shown in the following illustration, is helpful to a bank in determining areas where it is succeeding or falling short of achieving market share goals.

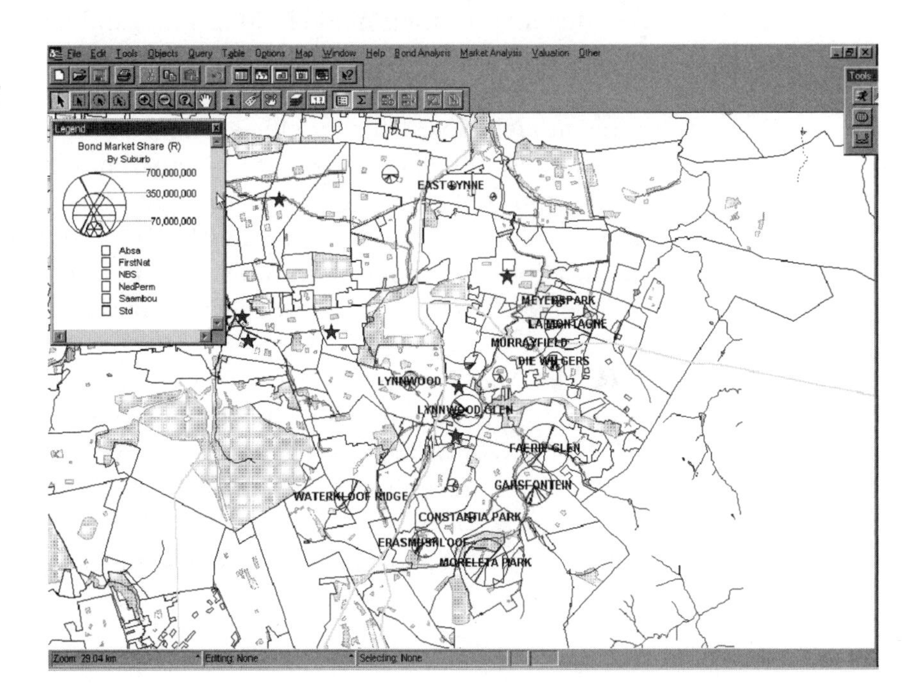

Mortgage bond market share analysis.

Updating and Maintaining the System

The value of this type of system is largely dependent on the ability to update and maintain the statistical models on which it is based. Local real estate trends can shift over relatively short periods of time. The construction of new highways, schools, or shopping centers can change perceptions of value, and in the South African context, squatter settlements could sometimes appear overnight on the edge of established neighborhoods.

In order to implement this system, MarketMap positioned itself to serve as a third-party processor of monthly real estate transaction records. Each month's SAPTG records would be received by MarketMap, geocoded, appended to the rolling 12-month history file, scanned for land-only sales and other exceptions, and then fed into the statistical software for recalibration of the model's coefficients. On an annual basis, the fundamental terms of the model would be revalidated to determine if consumers' tradeoff structures were changing (a longer term process that can be caused by the aging of a community, for instance, which will tend to downgrade the importance of nearby schools). Adjustments to model coefficients are then made in MapBasic, and a new version of the valuation model is delivered to the client site.

This model proved highly efficient and effective, and delivered a tool that would enable a bank to be much more competitive in the evaluation and granting of home equity lines of credit. Further development of the concept may also be undertaken to produce applications for property insurance value calculations, real estate market profiles for prospective sellers or buyers, and retail planning.

NOTE: MarketMap, Inc., was recently merged with the Daniel Consulting Group of Austin, Texas. Questions or comments about MarketMap or its techniques should now be directed to Frederick Barber, Executive Vice President, Conclusive Strategies, 141 Loop 64, Suite B, Dripping Springs, TX 78620. Phone 512/894-4880, fax 512/894.4881, or make contact via the Web at *www.conclusivestrategies.com.*

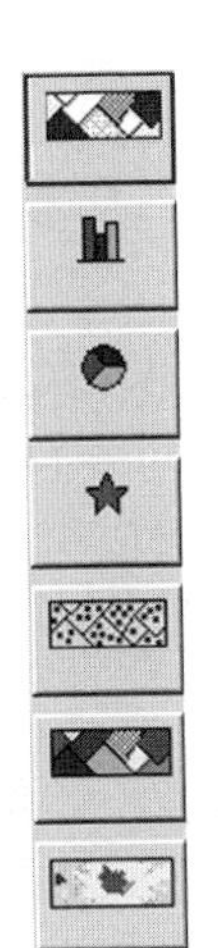

CASE STUDY 5

DESKMAP SYSTEMS, INC.

Supplying the Railroad Industry with Mapping Technology

THE U.S. RAIL INDUSTRY HAS CHANGED DRAMATICALLY. Large and small companies have merged or have acquired other interests. Track is bought and sold or placed out of service. Current rail mapping has become critical for the industry.

DeskMap utilizes MapInfo Professional and extensive mapping databases of the rail network to provide current mapping to the industry. The company has produced and maintains four rail databases, designed for use with MapInfo Professional or other GIS software packages.

The databases (U.S. Railroad Map Database, U.S. Railroad Major Systems, Canadian Railroad Map Database, and the Mexican Railroad Map Database) consist of two major components. One component is a graphical layer of data that represents each company's railroad lines. Second, each railroad line is linked to a database that contains information identifying the railroad company and the area in which the line exists. Each database also includes station names and a file of reporting marks with corresponding railroad company names. All four rail databases contain critical information on every rail line.

HazMat/Emergency Response Project

A major railroad company contracted with DeskMap Systems for assistance in complying with a Pipeline Safety Advisory Bulletin (Railroad-Pipeline Emergency Plans Coordination). The railroad company needed a database and related maps to display where their track crossed all pipelines.

To comply with the advisory bulletin, the company required information on all pipelines that shared a common right-of-way, ran parallel to the company's right-of-way, or crossed the company's right-of-way. DeskMap was requested to provide research, database development, and mapping services to create a digital map database of pipeline crossings. The project required creation of a database that would identify each pipeline company and emergency contact, pipeline commodity (e.g., natural gas, crude oil), and pipeline diameter. Key railroad data would also be provided, including milepost and operating division information.

Prototype

The entire HazMat project would involve 23 states representing the operating area in which the railroad company owned track. It was decided that one state would be used as a prototype. Using the U.S. Railroad Map Database, a query was performed to select all of the railroad company's main line track in the state of Indiana.

Development Process

Database development to identify where pipelines crossed the railroad tracks involved the creation of several map layers. First, the railroad lines were constructed with associated attribute information about the lines. Second, railroad stations and mileposts were added. Third, the pipelines that crossed or were near the railroads were identified. Fourth, the pipeline intersection points with required information were created. Together, these four map layers provided the information required to accurately identify pipeline crossings.

Developing the Rail Map Layer

To verify the accuracy of DeskMap's database against the railroad company's records of its rail lines, the company forwarded data in a *DXF* format from its CAD system. DeskMap converted the data to a MapInfo format. DeskMap's railroad database was found to be extremely accurate compared to the company's data. In some cases modifications were made to detailed areas. A newly acquired rail line was added to the database.

Additional data from the railroad company was then added to the database. The company's divisions, districts, and other information were added.

Rail Station Points

Railroad stations with respective mileposts were required to determine an operational reference location on the track. In some cases additional rail stations were added at intersections and division boundaries. Track charts and timetables were researched to identify the required stations.

Developing Pipeline Map Layer

A pipeline database was imported into MapInfo Professional and added to the project. Using MapInfo's buffering capabilities, DeskMap created a buffer around each rail segment to focus only on the pipelines that would affect the company's track. All pipeline segments within 500 feet on both sides of the track were identified, as shown in the following illustration.

Creating Pipeline Intersection Points

DeskMap then developed a program using MapBasic to locate pipeline and rail intersections. The program also added a point at the intersection to identify the crossing.

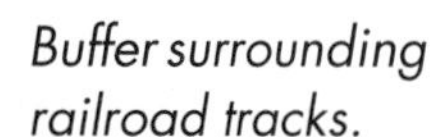

Buffer surrounding railroad tracks.

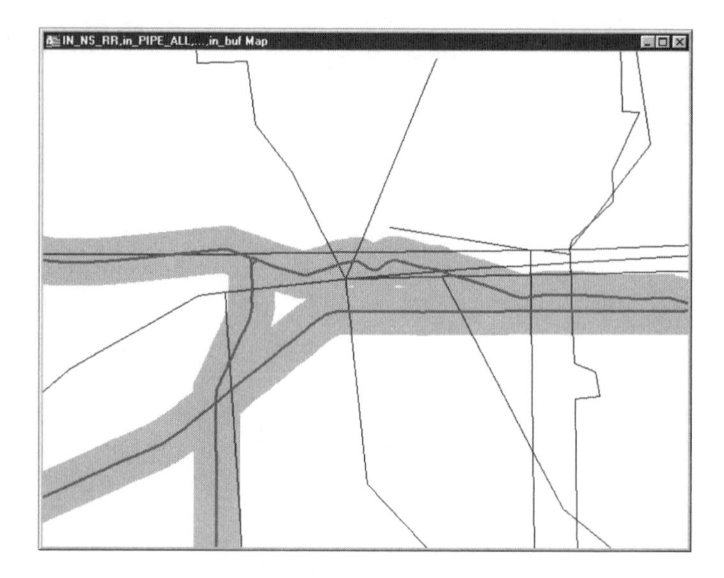

A unique number was added to the pipeline intersection points so that each could be independently identified. The numbering system had a state code. A numeric suffix was added after the number if a pipeline had multiple rail crossings.

Calculating the milepost of each crossing for every point on each line required the creation of a second MapBasic program. The milepost of each railroad station point was used as a control point for a known milepost. Two control points were selected and the program calculated the milepost for each pipeline crossing between the two points. For pipelines that ran parallel to the track, the program was used to calculate "begin" and "end" points.

Identifying Pipeline Crossings

At this stage, the database had four separate map layers: railroad stations, pipeline intersection points, pipe, and rail, as shown in the first of the following illustrations. The station layer contained name and milepost fields. The rail file contained all lines operated by the company. The pipe file was a graphic line file consisting of attributes pertaining to every pipeline, such as company name, contact information, commodity, and pipeline diameter. The pipeline intersection point file, an example of which is shown

in the second of the following illustrations, contained the calculated milepost and the name of the nearest station.

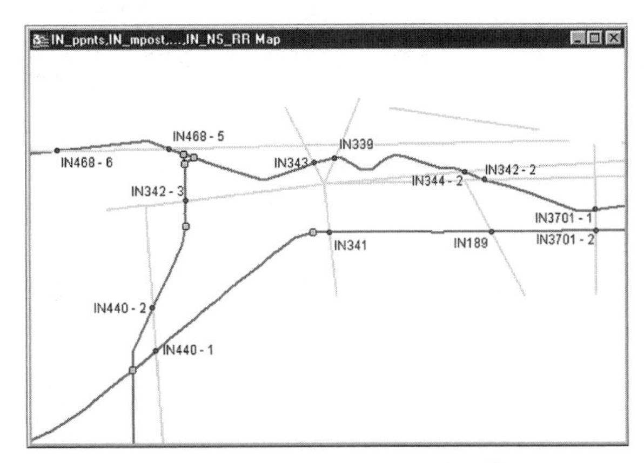

Sample pipeline crossing point.

Info Tool

unique_id:	IN341
multi_id:	0
Pipeline_Co:	NIPSCO
Emergency_Phone_AM:	(800) 464-7726
Emergency_Phone_PM:	
Cmdty_Group:	Natural Gas
Spec_Cmdty:	
Line_Type:	Delivery Line
Diameter:	
Division:	LAKE
District:	Huntington
MP_Prefix:	D
Milepost:	146
MP_Suffix:	
State:	IN

<< >> List IN_ppnts

Info Tool window showing archived information about pipeline crossing.

Once the pipeline point layer was updated, a map of Indiana was produced with each point labeled by a unique ID. The map included all of the company's rail in Indiana, milepost control points, pipelines, pipe points, and a state boundary. A report was then generated of all pipe points, as well as a listing of the pipeline companies in Indiana that impact on the railroad company.

Final Phase

After the prototype was completed and approved, the remaining databases and maps were created for the remaining 22 states, so that all pipelines would be identified in and around the railroad right-of-way for the entire route system. The same steps to produce the prototype were conducted for the remaining states. After the databases were constructed, maps and reports were produced.

The composition of each map was dependent on the density of intersecting pipe and rail points. Each intersection was labeled, and in highly congested areas, insets were required to display detail. In some cases, because several states required many inset maps, small state location maps were placed in the corner of each inset map for further clarification.

Once all maps were finalized and printed, a final report for each state was completed and delivered to the rail company. Each state report included all pipeline intersection points and a listing of the pipeline companies in that state. The customer was provided with a complete database, system maps, and state maps. A 36" x 60" map was created to display the company's total rail system. MapInfo Professional software and training were provided for the continued maintenance and viewing of the database.

Using the Database

The company has developed an ongoing ability to run queries in MapInfo to select areas of interest. In the following illustration, for example, a user has queried for all pipeline crossing points in the state of Indiana that belong to the NIGAS pipeline company. The user can also run queries selecting all pipeline crossings for a specified district, division, and commodity, among many other attributes.

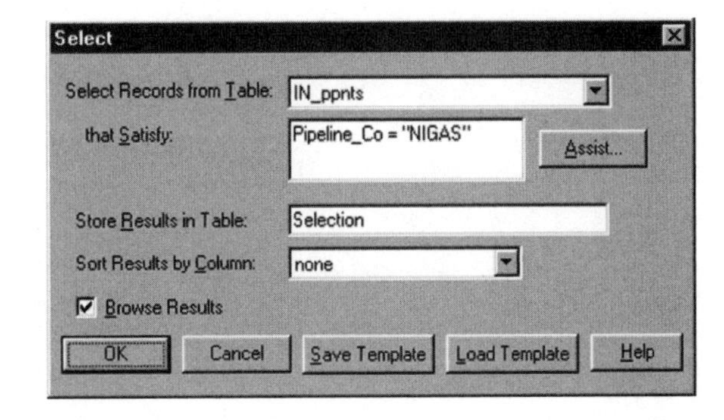

Sample query of pipeline crossing points.

Conclusions

In the event of a rail incident, the company can now reference the database and maps and, if a pipeline is present, coordinate emergency response planning with the pipeline company. Years ago, the development of this type of system would have required a mind-boggling effort to coordinate paper maps with tabular records pertaining to locations of track, rail incidents, pipelines, and so on. Moreover, the result of such a task would likely have been a cumbersome product that would frustrate users who sought to perform rapid, ongoing data analysis.

Today MapInfo Professional makes the integration tasks manageable and yields intuitive, easily interpreted, highly effective output. As the rail industry changes, rail data in MapInfo applications are essential for companies that wish to manage change in a sensible manner.

About DeskMap

DeskMap Systems, Inc., of Austin, Texas, develops mapping solutions for a wide variety of business customers. One of the company's primary areas of concentration is the railroad industry, and it produces complete digital map databases of rail systems, performs special projects and application development, and supplies stock and custom maps.

Contributions by Steve Stewart, DeskMap Systems, Austin, Texas.

APPENDIX

DATA FUNDAMENTALS AND SOURCES

Data – The Fuel for Analysis

A BIT OF CONVENTIONAL WISDOM heard frequently among geographic information systems (GIS) professionals is that applications are only as powerful as the data inside the software. Data is the fuel necessary to run any analysis engine, and MapInfo is no exception. This means that an analysis using MapInfo is only as good (i.e., reliable) as the information fed into the analysis. If you attempt an analysis with incomplete or unreliable data, MapInfo will provide solutions based on this data. It will *not* make up for shortcomings in the data. The old adage "Garbage in, garbage out" aptly applies to GIS analysis.

With so much data readily available, many businesses now face information overload. MapInfo is a solution to information overload, providing a method of integrating, analyzing, and distilling massive amounts of data.

What Works in MapInfo? What Is Quality Data?

Assuming you can convert information to common PC formats, nearly any data set can be depicted in MapInfo. The issue in selecting information for MapInfo is not so much "Can the data work?" as much as "Can I perform quality analysis with the data?"

Quality, by definition, is a subjective trait. When we apply the term *quality* toward artwork, "quality" is in the eye of the beholder and it is entirely possible that different individuals will view the same artwork very differently in terms of quality. In many ways, this is true of data. What passes as "perfect" for one's use might be profoundly flawed for another.

Think of data as differentiated by its dating, accuracy, and resolution. The first term, *dating*, is perhaps most intuitive: the older a data source, the more likely the data will be unsuitable for use. The newer the data source, the more likely it will reflect today's circumstances and situation. For example, you would not want to choose 1980 road files to represent fast-growing southern U.S. cities accurately in 2001. Also for example, in evaluating a new retail business site, one would want only the latest demographic data, with coming-year projections. Needs for timely data are best addressed through subscription licensing arrangements, whereby provisions exist for updating the user's data as the data provider updates its own files.

Be aware, though, that "currency" issues do not apply everywhere. Be careful not to waste time and energy securing the latest releases of inherently static features. Most businesspeople would not, for example, be concerned about securing the latest topographic or elevation databases, nor should they concern themselves with securing subscriptions on U.S. Census geography, in that it only changes once per decade. When securing data, be sure that you understand which of your requirements are time sensitive and which are not.

The accuracy issue is perhaps more specific to "geographic data" than other forms of data, and may be less intuitive. Accuracy, as applied to MapInfo data, can be split into two categories: relative and absolute. Relative accuracy refers to the placement of objects (such as customer locations, boundary lines, and streets) in comparison to one another.

Data is said to possess relative accuracy if the positions of features relative to one another are correct. Absolute accuracy, however, implies that objects are positioned on a map or data set in such a manner that one could pinpoint features on Earth with the help of survey equipment. To assist readers with directions, most road maps need only be relatively accurate. To assist a utility with work orders and underground pipeline work, one would aspire for data possessing absolute accuracy.

Related to accuracy is the issue of resolution. By using different source data, providers can deliver information with varying levels of detail, or resolution, to different geographic features' representation.

Consider being asked to draw a map of a state you are not entirely familiar with from an 8.5-inch x 11-inch satellite image of the entire North American continent. It would be very difficult to provide more than rough approximations of the state. Coastlines and borders would probably be depicted by simple line forms, and if you dared to indicate any geographic features within the state, they would be only crudely depicted. The resultant map would provide only coarse resolution for the features you were intending to represent.

However, if you received an 8.5-inch x 11-inch aerial photograph for the same state, your map would likely delineate the geographic borders and features around your state with much greater detail. You would have sufficient detail in your "base" materials to generate a stronger approximation of the state's features.

Data vendors encounter similar issues when preparing data for commercial distribution. Sources for the same geographic data can be located with different resolutions. You can purchase a county map, for example, in which individual counties are depicted as combinations of 8 to 12 straight lines, or you can purchase this data with resolution indicating that it consisted of hundreds of intricate connections.

Currency, accuracy, and resolution profoundly affect how data will work in your application and can contribute mightily to the success or failure of your use with MapInfo. They are also pivotal determinants of price, discussed in material to follow.

Securing Data

When advancing with MapInfo, most organizations will inventory their data needs and make a decision about how much internal data they want to connect to MapInfo and how much external data they wish to purchase for use with MapInfo. There is a proliferation of company data: POS systems, customer databases, internal facilities information, as well as external data sets, including demographics, psychographics, graphic information, business lists, and so on. If you are responsible for decisions of this nature, you should pause to consider your data and purchasing options.

Internal Data

Internal data refers to proprietary data belonging to an individual or organization. In the past, internal data was typically stored on paper, on a street map, or in someone's mind. It was not easily accessible or transferable throughout an organization. However, over the past decade or two, the advent of POS systems, more in-depth marketing activity, and "due diligence" tasks have resulted in a proliferation of computerized and accessible internal data. Most businesses have a variety of sales information, customer databases, and facility data well suited for analysis in MapInfo.

If you have data that is suited to MapInfo but is not yet computerized, know that there are organizations and technologies to help you computerize your information. Universities, temporary agencies, and scanners can help make this process more manageable. Temps can be hired to handle labor-intensive tasks, such as inputting basic and repetitive information (e.g., survey results).

Scanning technology can also be employed for this task, if the survey forms are set up correctly. University students can help with digitizing information or creating unique geographies (through aggregating or disaggregating available information) with the aid of a GIS. With the raster imaging capabilities of MapInfo, scanners also allow aerial photographs and store layouts to be incorporated into your electronic data.

Preparing Internal Data for Use in MapInfo: Geocoding

Of all the different ways users might convert existing data into MapInfo graphic formats, the most common is undoubtedly through MapInfo geocoders. The geocoding process—adding correct longitude and latitude or appropriate national projection coordinates to database records—is the first step in any mapping, data visualization, or spatial analysis application.

MapInfo offers several geocoding products with excellent accuracy, and their products (listed at the end of this section) enable geocoding to be deployed on the desktop, on the server, as a standalone geocoding application, or as an OCX/Active X for integration into any business application (such as point-of-sale transactions or customer service applications).

MapInfo Geocoding Products

The following are descriptions of some of MapInfo's primary geocoding products.

MapMarker PLU.S.: MapMarker PLU.S. is MapInfo's premier geocoding solution. In a single pass, MapMarker PLU.S. intelligently corrects and standardizes address data while assigning latitude and longitude coordinates. The MapMarker PLU.S. address-matching dictionary incorporates address data from the U.S. Postal Service (with enhanced street geometry) and data from GDT, Inc., to ensure the highest possible hit rates.

MapMarker: MapMarker utilizes the same powerful geocoding engine as MapMarker PLU.S., but with somewhat fewer processing options and street data not quite as current as MapMarker PLU.S. MapMarker's address-matching dictionary is based on address data from the U.S. Postal Service, street geometry from the U.S. Census Bureau TIGER 98 files, and ZIP+4 centroids from GDT, Inc. Products similar to MapMarker are available in several other countries, including Germany, The Netherlands, and the United Kingdom.

Products can be found all over the globe, including the following.

GeoLoc Australia: GeoLoc Australia is a nationwide geocoder product that matches addresses with latitude/longitude points and tags the geocoded records with the 1996 Census Collection District boundary for Australia.

GeoLoc New Zealand: GeoLoc New Zealand is a nationwide geocoder for New Zealand that matches addresses with latitude/longitude points to the exact address location instead of matching to the approximate position in the address range. Source data to the house number address allows this level of precision for geocoding.

Digitization and Other Graphic Operations

MapInfo also has a capability to "digitize" or convert hardcopy maps through detailed tracing procedures into graphic data. Digitizing, in MapInfo, is the process of taking a paper map, tracing its features, and turning the information into a MapInfo layer. Today, it is conducted most often in government and university settings and rarely performed in commercial settings.

This digitization process creates a vector file that will allow the information to be changed, edited, and queried, unlike a scanned raster image. Digitizing is very useful in creating new boundaries or roads that are unique to your application or that are not commercially available. Census boundaries, street files, and tax parcels are examples of files that have been digitized. Scanning information is much more rapid than digitizing data, but you pay a price: you cannot manipulate or query raster images.

Digitizing in MapInfo requires several items: a digitizing mouse (also called a puck), a digitizing tablet, and a device driver that MapInfo can use. Early versions of MapInfo only supported the Virtual Tablet Interface (VTI) for digitizing. Since version 4.0, MapInfo has supported both the VTI and the Wintab interfaces. The graphics tablet, also known as a digitizer, has long been in use in professional CAD circles. The digitizer consists of a table and a puck. Electronic circuitry in the digitizing table senses the location of this puck as it is moved over the surface. MapInfo supports several types of hardware setups, and it is best before purchasing one to consider calling MapInfo support to discuss your options.

Apart from digitizing, you might also find it useful to create data by using MapInfo's drawing, buffering, and aggregating capabilities to create new line and region geographies. Many of these operations can produce graphic features in formats suitable for saving and future use.

External Data

External data refers to information available from commercial data providers. Particularly for private organizations, external data may be the most prominent type of information you use.

Many commonly used data sets are already in a computerized format that can be imported or used directly by MapInfo. Commercially available data includes boundaries such as census tracts, block groups, states, or zip codes. It also includes the corresponding information relating to boundaries, such as census data, lifestyle segmentation systems, market research data, business lists, and electronic street files.

External Data Sources

There are many providers of external data. As of this writing, the three biggest data providers in the industry are MapInfo, Claritas, and National Decision Systems (NDS). Many others provide more specific data sets; for example, RL Polk (automotive registration information), MRI and Simmons (marketing research and expenditure estimates), and InfoUSA (business/competitive lists). MapInfo has amassed strategic relations with many of these organizations, and when you order data from MapInfo or from other providers MapInfo format, the information has typically already been formatted for MapInfo.

Purchasing Geographic Features from External Data Providers

Few organizations find that they can manually convert or digitize all of the data they want or need to be successful. As a result, they look to the expanding range of products provided by MapInfo and other data houses as a means of addressing their need for data.

Bear in mind always that the type of data required will depend on the type of analysis you wish to perform. Budget and time constraints will also play a part in data acquisition. For example, different data providers offer vastly different pricing structures for the same data. However, there are also varying levels of reliability and maintenance. Cost may also be affected by the number of users that require access to the data. The more users that require access, typically, the greater the cost. Network licenses for data can also be acquired. Shop the different vendors, asking about the charges and service levels.

Be aware, too, that the type of hardware and software you are running can also affect the price and availability of data. MapInfo data providers are plentiful, but another GIS program not as widely distributed may not have as many data options. For example, data is typically not as readily available for UNIX environments as for Windows platforms.

Streets Data

MapInfo offers very comprehensive, accurate, and high-quality street data products for use in routing, drive-time area studies, and other MapInfo applications. Streets include information such as street maps, highways, water features, railroad tracks, bridges, points of interest (e.g., hospitals, schools, prisons), area landmarks (e.g., parks, golf courses), city boundaries, and more. A few MapInfo street products are described in the following.

StreetPro: Created using GDT Dynamap 2000, StreetPro is MapInfo's most comprehensive and up-to-date street file for the United States. It is updated and released quarterly with an average of 400,000 new address segments, and contains over 50% more addressed street segments than TIGER. Exclusive to StreetPro, MapInfo has added MapBasic tools such as the Autoloader, Shields Manager, Street Append, and SIFTER, which allow users to quickly analyze and change the look of their data. StreetPro includes seamless state maps, workspaces, and geosets. These are available in both display and with addresses. Products of scope similar to the U.S. Streetpro are available in many other countries, including the following:

- Austria
- Belgium
- Germany
- Spain
- Europe in general
- France
- Italy
- Netherlands
- Portugal
- Switzerland

The following describe other types of Street products.

StreetInfo: StreetInfo is MapInfo's TIGER-based street file with addresses for the United States. It is updated and released as often as TIGER is released. StreetInfo comes with MapBasic tools that allow the user to quickly analyze and change the look of data.

CompuStreets: MapInfo is the distributor of CompuStreets, a Canadian street file developed by CompuSearch. It is updated semiannually and contains address segments.

StreetWorks Australia: Created using Public Sector Mapping Agency data, Australia StreetWorks provides users with 13 layers of data, including addressed segmented streets.

Australia StreetWorks Display: A fast, compact, nationwide street reference map with 13 layers of data for Australia. MapInfo has processed the data to combine multiple segments that represent the same street and has removed the address information to create a fast-display map. Used in conjunction with GeoLoc, Australia StreetWorks Display delivers a quick, efficient mapping solution.

New Zealand StreetWorks Display: A fast, compact, nationwide street reference map with several layers of data for New Zealand. As for Australia StreetWorks, MapInfo has processed the data to combine multiple segments that represent the same street and has removed the address information to create a fast-display map. Used in conjunction with GeoLoc New Zealand, this package also provides a quick, efficient mapping solution.

Marketing, Real Estate, and Telecommunications Data Considerations

Many different types of GIS applications can help advertising and marketing departments. Marketers often blend demographic, customer, and sales information to make product, service, and advertising decisions. By overlaying customer and store/product information, for example, advertisers can determine improved media buy strategies. They can determine optimized merchandise mix strategies by adjusting the mix based on the demographics of stores' trade area. Using GIS analysis, it is possible to develop highly customized promotional campaigns that provide special offers to carefully targeted neighborhoods. Manufacturers can demonstrate to suppliers where they can find the consumers that will purchase the manufacturers' products.

GIS has become perhaps even more prominent for real estate professionals. Many real estate professionals rely on GIS in performing their jobs. Corporate real estate professionals typically use GIS for site selection purposes. In Chapter 2, a site selection case was used to introduce topics discussed in the remainder of the book. This analysis (though very basic) is typical of the site selection decision process. Most site selection professionals use traffic count, current store location, and customer and competitor information as inputs into proprietary models used to help them determine sales potential of new and remodeled locations.

Some models are very sophisticated, using complex algorithms and statistics to project potential sales, whereas others are not much more than a professional's instincts. In addition, commercial data is available from such sources as Woods & Poole, Claritas, and Equifax on current and modeled sales of certain products by several geographic breakdowns. Real estate brokers' use of GIS is typically limited to providing others with data. Customers of commercial brokers have come to expect their broker to provide GIS-based information as a value-added service. Such data includes traffic counts, competitor locations (for those selecting retail/service sites), demographic reports, and lifestyle data.

Fortunately, there is an abundance of sources for commercial data and typically an abundance of internal data available for large corporate installations. Such data includes the following categories: customer, sales/transaction, location, demographic, lifestyles, business location, tax parcel, traffic count, sales potential, and geographic.

Key MapInfo Marketing and Real Estate Products

Combining mapping applications with demographic data—information relating to population, income, expenditure, retail activity, employment, and lifestyle segmentation data—creates a clear picture of any business situation, and allows organizations to make informed and accurate decisions. MapInfo has a wide selection of worldwide demographic data products and packaged applications to enhance analysis. Examples of packaged demographic applications are described in the following.

TargetPro: TargetPro is an advanced data access and presentation tool for the business user who wants to perform quick and accurate demographic analysis. Custom demographic reports can be easily created in a variety of media, thereby enabling users to access and distribute information throughout their enterprise. TargetPro is an easy-to-use package that provides to a wide range of personnel, from staff to management, the capability of performing on-the-fly market analyses.

TargetPro Cluster Analyzer: TargetPro with Cluster Analyzer is designed to allow organizations of any size to profile a customer database by demographic cluster. By determining the profile of a database and comparing it to the profile of other database, including geographic areas, the Cluster Analyzer is able to determine the market potential of a user's own products. Market Potential can analyze a variety of databases, including geographic areas, mailing lists, syndicated surveys, and other customer databases, based on similarities in the cluster profile of each database. The end result of an analysis is a score in the form of an index, or a count of households and population, that represents the size of the user's potential market in a set of geographic areas.

MapInfo and its partner community have links to many types and sources of data that can be useful in marketing and real estate applications. For example, many customers will find value in traffic count data from such sources as MapInfo and GDT. Many customers will find phone list products useful. Sources such as ABI provide annually verified data by SIC code, whereas off-the-shelf (unverified), nationwide yellow page databases (such as Pro CDs Select Phone) are available for under $100. Marketing and advertising applications have grown increasingly more common. Speak with your MapInfo representative to learn about the newest items on the shelves.

Telecommunications Products

Using MapInfo software technology and data, telecommunications professionals can gain a complete view of their marketscape and position themselves to respond quickly and efficiently to changing market conditions. With everything from wireless boundaries to tower sites to exchange boundaries, MapInfo supplies users with tools and data sets *updated on a monthly basis* in order to visualize the marketscape, analyze competitive strategies, perform site analysis, and implement a solid growth plan. MapInfo's telecommunications data products include raster data (aerial photography, satellite imagery, and U.S. Geological Survey products), elevation and terrain data (from DEM and LULC), as well as trademarked products, such as the following.

- ExchangeInfoPlus
- MSA/RSAInfo
- AreaCodeInfo
- Obstacle
- PCSInfo
- WirelessInfo

Finding Data

New data sets are constantly being released. There are many different sources of information and many places from which it might be retrieved. The savvy user, however, will recognize that MapInfo's business interests dictate that they provide a variety of competitive data options for the user. Contact your local MapInfo reseller or MapInfo representative to accelerate your learning curve about data sources and assemble options that might otherwise be difficult to locate.

The more you use MapInfo, the more it will likely become evident that you cannot learn enough about data sources. Provided in the sections that follow are additional sources for a wide variety of data. With the fast pace of the industry, many links or sources will undoubtedly change before this book goes to press. Keep searching, and we wish you the best of luck with your program and search for data!

Satellite and Aerial Data/Imagery Sources

The following are among the many satellite and aerial data and imagery sources found on the Internet and elsewhere.

American Society for Photogrammetry and Remote Sensing
Washington, D.C.
301/493-0290
http://www.asprs.org/asprs

Autometric, Inc.
Alexandria, VA
703/658-4000
http://www.autometric.com

Earth Satellite Corporation
Rockville, MD
301/231-0660
http://www.earthsat.com

EarthWatch, Inc.
Longmont, CO
303/682-3800
http://www.digitalglobe.com

EDRAS, Inc.
Atlanta, GA
444/248-9000
http://www.edras.com

EROS Data Center, U.S. Geological Survey
Sioux Falls, SD
605/594-6151
http://www.edcwww.cr.usgs.gov/eros-home.html

OrbImage
Dulles, VA
703/406-5436
http://www.orbimage.com

Remote Sensing Resources on the Web
University of Manchester
Manchester, United Kingdom
http://www.man.ac.uk/Arts/geography/rs/rs.html

Resource 21
Denver, CO
303/768-0015
http://www.tec.army.mil/CCIO/RESOURCE21.htm

Space Imaging EOSAT
Thornton, CO
303/254-2000
http://www.spaceimage.com

SPOT Image Corporation
Reston, VA
703/715-3100
http://www.spot.com

Data Over the Web: Internet Sites

The tables that follow provide a host of Internet data sources by category. A special thanks is due Dr. Dennis Fitzsimons, who help create our original list of GIS-related interesting Internet sites toward building these tables. Contact information for Dr. Fitzsimons follows.

Dr. Dennis Fitzsimons
Department of Geography & Planning
Southwest Texas State University
San Marcos, TX 78666-4616
e-mail: df02@swt.edu

Agencies

1990 U.S. Census LOOKUP	http://cedr.lbl.gov/cdrom/doc/lookup_doc.html
ACSM Home Page	http://www.landsurveyor.com/acsm/
Bureau of the Census	http://www.census.gov/
Central Intelligence Agency	http://www.odci.gov/cia/
ERIN	http://kaos.erin.gov.au/erin.html
Federal Geographic Data Committee (FGDC)	http://www.fgdc.gov/
FEMA	http://www.fema.gov/homepage.shtml
Govt. Publications on the Web	http://www.library.nwu.edu/gpo/
NASA Homepage	http://www.gsfc.nasa.gov/ NASA_homepage.html
National Geodetic Survey	http://www.ngs.noaa.gov/index.html
Natural Resources Canada (NRCan)	http://www.nrcan.gc.ca/home/nrcanhpe.htm
NIMA homepage	http://164.214.2.59/nimahome.html
NOAA Coastal & Estuarine	http://www-ceob.nos.noaa.gov/
State and Local Government	http://www.piperinfo.com/piper/state/states.html
U.S. Federal Govt. Agencies	http://www.lib.lsu.edu/gov/fedgov.html
U.S. Fish and Wildlife	http://www.fws.gov/
U.S. Geological Survey	http://www.usgs.gov/
USGS - Water Resources of the United States	http://h2o.usgs.gov/
USGS Mapping Information	http://www-nmd.usgs.gov/
USGS Publications Index Guide	http://andriot.com/USGS.htm
World Health Organization	http://www.who.ch/

Cartography

1992 Natl. Resources Inventory Atlas	http://www.nhq.nrcs.usda.gov/nriatlas.html
Abbreviations & Acronyms	http://www.lib.berkeley.edu/EART/abbrev.html
ACMLA-Assoc. of Canadian Map Libraries	http://www.sscl.uwo.ca/assoc/acml/acmla.html
Alexandria Digital Library	http://alexandria.sdc.ucsb.edu/
Baltic GIS Database	http://www.grida.no/baltic/
Barcelona Map Exhibit	http://www-nais.ccm.emr.ca/ barcelona_map_exhibit/estart.htm
Berkeley Map Collection	http://www.lib.berkeley.edu/EART/digital/ tour.html
Boston Transport Services	http://www.mbta.com/~imagemap/ GIFBAR?139,5

Bowen US maps	http://130.166.124.2/USpage1.html
California Atlas (Bowen)	http://130.166.124.2/CApage1.html
Cart/Geog Web Sources - U of Georgia	http://www.libs.uga.edu/maproom/ahtml/mchpcr1.html
Cartographic Communication	http://www.utexas.edu/depts/grg/gcraft/notes/cartocom/toc.html#3.3
Cartographic Glossary	http://www.lib.utexas.edu/Libs/PCL/Map_collection/glossary.html
Cartographic Images - Siebold	http://www.iag.net/~jsiebold/carto.html
Cartographic Materials–Waterloo	http://www.lib.uwaterloo.ca/discipline/Cartography/cart.html
Cartographic Records (Digital) - Pilot	http://www.bcars.gs.gov.bc.ca/cartogr/general/maps.html
Cartographic Reference Books - Philadelphia Print Shop	http://www.philaprintshop.com/cartrftx.html
Cartographic references - PCL	http://www.lib.utexas.edu/Libs/PCL/refserv/geography/Cartographic_reference.html
Cartography - Calendar of Events	http://www.cyberia.com/pages/jdocktor/
Cartography - Indiana State	http://www.indstate.edu/gga/gga_cart/index.html
Cartography Info Center–LSU	http://www.cadgis.lsu.edu:80/cic/
Cartography Resources - GMU	http://geog.gmu.edu/gess/jwc/cartogrefs.html
Census Index of /mapGallery/images/	http://www.census.gov/ftp/pub/geo/www/mapGallery/images/
Centennia	http://www.clockwk.com/
CGRER NetSurfing: Maps and References	http://www.cgrer.uiowa.edu/servers/servers_references.html
Chicago 1990 Census Maps	http://www.lib.uchicago.edu:80/LibInfo/Libraries/Maps/chimaps.html
Color Landform Atlas - US	http://fermi.jhuapl.edu/states/states.html
Color Use Guidelines	http://www.gis.psu.edu/Brewer/CBColorHTML/CBColorTop.html
Country maps from W3 servers in Europe	http://www.tue.nl/europe/
Digital Chart of the World	http://ilm425.nlh.no/gis/dcw/dcw.html
Digital Wisdom	http://www.digiwis.com/
DMSP City Lights	http://web.ngdc.noaa.gov/dmsp/IMAGERY/newols-app-city.html
Election results (Clinton)	http://www.geog.ucsb.edu/~lawson/election.html

Encyberpedia's MAPS and geography	http://www.montecristo.com:80/map1.htm
GeoSystems: Map Skills	http://www.geosys.com/cgi-bin/genobject/mapskills/tigdd27
GMU Geo 310 Maps	http://geog.gmu.edu/gess/jwc/student_projects.html
How far is it?	http://gs213.sp.cs.cmu.edu/prog/dist
IASBS Digital Map Files	http://www.usm.maine.edu/~maps/iasbs/digmaps.html
ICA: U.S. Nat'l Comm	http://www.gis.psu.edu/ica/ICAusnc.html
Imaginary Maps	http://www.gsi-mc.go.jp/tizu/animap.html
Java Atlas Home page	http://www.ggr.ulaval.ca/JAVA/Java.html
Links to Maps	http://www.cco.caltech.edu/~salmon/maps.html
London Maps	http://multimap.com/london/
Map Images On the Web	http://www.cadgis.lsu.edu:80/cic/mapsnet.html
Map Links	http://www.lbl.gov/Web/Maps.html
Map Room - Oxford	http://www.bodley.ox.ac.uk/users/nnj/
MapArt Gallery	http://www.map-art.com/map-art/software/gallery.html
MapFinder	http://fieber-john.campusview.indiana.edu/mapfinder/
MapLink: Online Directory	http://www.maplink.com/Mldir1.htm
Mapmaker	http://loki.ur.utk.edu/ut2kids/maps/map.html
Maps & GIS INFOMINE Search Screen	http://logic17.ucr.edu//mapsinfo.html
Maps - Nottingham	http://acorn.educ.nottingham.ac.uk/ShellCent/maps/welcome.html
Mapville Home Page	http://www.mapville.com/
Multi-Scale Maps	http://www.c3.lanl.gov/~cjhamil/browse/main.html
National Atlas Info-Canada	http://www-nais.ccm.emr.ca/naisgis.html
New York City Maps	http://www.soc.qc.edu/Maps/
New York Map Portfolio	http://www.sunysb.edu/libmap/nymaps.htm
No. Am. Breeding Birds - Ranges	http://www.npsc.nbs.gov:80/resource/distr/birds/breedrng/breedrng.htm
NSW SoE 1995 - Maps	http://www.epa.nsw.gov.au/soe/95/listmaps.htm
NYC Subway Map Picker	http://www.mediabridge.com/nyc/transportation/subways/picker.html

Oddens's Bookmarks	http://kartoserver.frw.ruu.nl/html/staff/oddens/oddens.htm
Oversized Color Maps	http://www.cc.columbia.edu/imaging/html/largemaps/oversized.html
PCL Map Collection	http://www.lib.utexas.edu/Libs/PCL/Map_collection/Map_collection.html
Pilot - Cartographic Records	http://www.bcars.gs.gov.bc.ca/cartogr/general/maps.html
Presidential Elections in Maps	http://www.lib.virginia.edu/gic/elections/index.html
Project Argus - Visualization	http://severn.geog.le.ac.uk/argus/
Road Map Collectors of America	http://falcon.cc.ukans.edu/~dschul/rmca/rmca.html
Shand - Map-Related Web Sites	http://www.geog.gla.ac.uk/sites/mapsites.htm
Southern California Area Maps	http://artscenecal.com/Maps.html
UC San Diego - Map Room	http://gort.ucsd.edu/mw/maps.html
United States Thematic Maps	http://oseda.missouri.edu:80/graphics/us/pop/
University Campus Maps	http://www.lib.uwaterloo.ca/discipline/Cartography/campus.html
Useful Sites about maps	http://research.umbc.edu/~roswell/mipage.html
Vinland Map	http://www.mcri.org/vm_image.html
Vinland Map and Shroud Updates	http://www.mcri.org/vm_shroud_update.html#anchor544696
Vinland Map – largest photo	http://www.digalog.com/viking/vinland/m/l/vmap.htm
Yahoo: Maps	http://www.yahoo.com/Science/Geography/Maps

Commercial

Adobe Systems	http://www.adobe.com/
adobe.mag	http://www.adobemag.com/
Adventurous Traveler Bookstore - Map List	http://www.gorp.com/atb/maps.htm
Agfa Home	http://www.agfa.com/
Apple Computer Technical Support	http://www.support.apple.com/
Argus Technologies Homepage	http://www.argusmap.com/
Avenza Software Marketing Inc. MAP-MAC Page	http://www.avenza.com/map-mac.html
Buydirect.Com	http://www.buydirect.com/
Cartesia	http://www.map-art.com/

Commercial Geography Resources	http://lorax.geog.scarolina.edu/geogdocs/otherdocs/comm.html
Computer Companies Info Center	http://library.microsoft.com/compcos.htm
DeLorme	http://www.delorme.com/home.htm
ESRI	http://www.esri.com/
Etak Incorporated	http://www.etak.com/
Fodor's	http://www.fodors.com/
Forefront-eduspecial	http://www.ffg.com/edu/eduspecial.html
Four One Company Ld	http://www.icis.on.ca/fourone/
GIS World, Inc.	http://www.gisworld.com/index.html
Government Technology	http://www.govtech.net/
Intergraph	http://www.intergraph.com/
Mac Zone Home	http://www.maczone.com/maczone?mzstart@55884jhkm
MacAddict	http://www.macaddict.com/
MacWarehouse	http://www.warehouse.com/MacWarehouse/
Macworld Online Buyers Guide	http://www.macworld.com/buyers/hot.deals/index.html
MAGELLAN Geographix	http://www.magellangeo.com/
MapInfo	http://www.mapinfo.com/homepage.html
Mercator's World Home Page	http://www.mercatormag.com/hihome.htm
Mountain High Maps	http://www.digiwis.com/
OMNI Resources (Topo sheets)	http://www.omnimap.com/catalog/index.htm
Power Computing Corporation	http://www.powercc.com/
Rand McNally	http://www.randmcnally.com/home/
Raven Maps & Images	http://www.ravenmaps.com/
Silicon Graphics' Silicon Surf	http://www.sgi.com/
Simm.Net	http://194.72.252.2:80/simmnet/
Sirs, Inc.	http://www.sirs.com/
Steven Gordon Cartography	http://www.clearwater.com/gordonmaps/
Sure!MAPS Digital Mapping	http://www.horizons.com/suremaps
Terra Data/Geocart	http://hudson.idt.net/~terrainc/
The Gold Bug	http://www.goldbug.com/
ThinkSpace/Map Factory	http://www.thinkspace.com/
United Computer Exchange Corporation	http://www.uce.com/
World of Maps	http://www.WorldofMaps.com/

Data Sources

1990 U.S. Census LOOKUP	http://cedr.lbl.gov/cdrom/doc/lookup_doc.html
Assoc. of Research Librarians (ARL) Statistics and Information	http://www.lib.virginia.edu/socsci/arl/test-arl/index.html
Bay Area Regional Database	http://bard.wr.usgs.gov/
Bureau of the Census	http://www.census.gov/
Bureau of Transportation Statistics	http://www.bts.gov/btsprod/order.html
CGRER NetSurfing: Maps and References	http://www.cgrer.uiowa.edu/servers/servers_references.html#interact-generalCIE
SIN Home Page	http://www.ciesin.org/
CIESIN: Directory /pub/census	ftp://ftp.ciesin.org/pub/census/
Demography & Population	http://coombs.anu.edu.au/ResFacilities/DemographyPage.html
ENRM prototype server	http://enrm.ceo.org/home.pl
EROS Data Center	http://edcwww.cr.usgs.gov/doc/edchome/datasets/edcdata.html
FGDC Subcommittee on Cultural and Demographic Data	file://www.census.gov/pub/geo/www/standards/scdd/index.html
Global Land Information System	http://edcwww.cr.usgs.gov/webglis/
Government Information Sharing Project	http://govinfo.kerr.orst.edu/index.html
GPO-Federal Locator Services	http://www.access.gpo.gov/su_docs/dpos/adpos400.html
Mable/Geocorr Home Page	http://ts2.ciesin.org/plue/geocorr/
Mable/Geocorr 2.01 Home Page	http://www.oseda.missouri.edu/plue/geocorr/
National Geospatial Data	http://nsdi.usgs.gov/nsdi/
Northern Prairie Science Center	http://www.npsc.nbs.gov/
Office of Social & Economic Data Analysis - OSEDA	http://www.oseda.missouri.edu/
OSU-Spatial Data Sources	http://ncl.sbs.ohio-state.edu/5_sdata.html
Population Reference Bureau	http://www.prb.org/prb/index.html
PSU-Earth and Mineral Sciences Library	http://vector.gis.psu.edu/emsltop.html
Regional Economic Information System, 1969-1993	http://www.lib.virginia.edu/socsci/reis/reis1.html
SEDAC-Socioeconomic Data	http://sedac.ciesin.org/
U.S. Census Data-Lawrence Berkeley Natl Lab	http://cedr.lbl.gov/mdocs/LBL_census.html
UAr-Japan GIS/Mapping Sciences Resource Guide	http://www.cast.uark.edu/jpgis/

UConn Map Library-MAGIC	http://magic.lib.uconn.edu/
UN gopher	gopher://gopher.undp.org:70/11/ungophers/popin
USA Counties 1996	http://govinfo.kerr.orst.edu/usaco-stateis.html
USGS NSDI - DEMs	http://nsdi.usgs.gov/nsdi/products/dem.html
World Factbook Master Home Page - CIA	http://www.odci.gov/cia/publications/nsolo/wfb-all.htm

Earth Sciences

ABAG Earthquake Maps and Information	http://www.abag.ca.gov/bayarea/eqmaps/eqmaps.html
Caltech: Frequently-Used Resources	http://www.caltech.edu/caltech/Frequent.html
Central MI Univ. - Geography and Earth Science	http://www.cmich.edu/~3nrwbhg/homepage.htm
CIESIN Gateway WWW Interface	http://wwwgateway.ciesin.org/
CSC Earth Science Topics	http://www.csc.fi/earth_science/earth_science.html
Earth Pages	http://starsky.hitc.com/earth/earth.html
Earth Viewer	http://fourmilab.ch/earthview/vplanet.html
Earth2 Project	http://www.ems.psu.edu/Earth2/E2Top.html
Earthmap Home Page	http://www.gnet.org/earthmap/
Earthquake Info - USGS	http://quake.wr.usgs.gov/
EarthView	http://www.ldeo.columbia.edu/EV/EarthViewHome.html
EdWeb Home Page	http://edweb.cnidr.org:90/
Eisenhower National Clearinghouse	http://enc.org/
GAIA Alert	http://www.newciv.org/millennium_matters/gaia
Gap Analysis Home Page	http://www.gap.uidaho.edu/gap/index.html
Geosciences Resources	http://www.cc.columbia.edu/cu/libraries/indiv/geosci/offsite.html
GeoWeb Home Page	http://wings.buffalo.edu/geoweb/
Gisnet - Online Resources for Earth Scientists	http://www.gisnet.com/gis/ores/gis/hyper.html
McKnight Test Pages	http://www.prenhall.com/divisions/esm/mcknight/public_html/
Meteorology Obs: References	http://www.ems.psu.edu/~fraser/Meteo471/Meteo471wRefMeteo.html
NASA - EOS Project Science Office	http://eospso.gsfc.nasa.gov/

Northern Prairie Science Center	http://164.159.215.66/
NOSC - Planet Earth home page	http://www.nosc.mil/planet_earth/everything.html
Physical Geography Resources	http://feature.geography.wisc.edu/phys.htm
PSU - Earth and Mineral Sciences Library	http://vector.gis.psu.edu/emsltop.html
Sirs, Inc.	http://www.sirs.com/
Tasa Exchange Earth Science Links	http://www.swcp.com/~tasa/links.html
Tasa Graphics Earth Exchange	http://www.swcp.com/~tasa/
The GeoSphere Project Report	http://www.infolane.com/geosphere/
The NASA/JPL Imaging Radar Home Page	http://southport.jpl.nasa.gov/
The World Lecture Hall	http://www.utexas.edu/world/lecture/
UC Berkeley - Internet Resources in the Earth Sciences	http://www.lib.berkeley.edu/EART/EarthLinks.html
UCI Science Education Programs Office	http://www-sci.lib.uci.edu/SEP/SEP.html
UIUC - Online Guide To Meteorology	http://covis.atmos.uiuc.edu/guide/guide.html
USGS - Earth and Environmental Science	http://www.usgs.gov/network/science/earth/earth.html
USGS/Pasadena Home Page	http://www-socal.wr.usgs.gov/
USRA - Earth System Science Education	http://www.usra.edu/esse/ESSE.html
Views of the Solar System	http://bang.lanl.gov/solarsys/
Virtual Library - Earth Sciences Resources	http://www-vl-es.geo.ucalgary.ca/VL/html/es-resources.html
W.M. Keck Foundation Seismological Observatory Earthquake Information	http://www.baylor.edu/~Geology/keck_eq.html
WebEarth	http://www.hyperreal.com/~mpesce/we/

Environmental

Environmental Organization WebDirectory	http://www.webdirectory.com/
Global Information Locator	http://www.g7.fed.us/gils.html
Greenpeace International Home Page	http://www.greenpeace.org/
IGC: EcoNet	http://www.igc.org/igc/econet/index.html
National Audubon Society	http://www.audubon.org/
National Wildlife Federation	http://www.nwf.org/
Nature Conservancy	http://www.tnc.org/
Rainforest Action Network Home Page	http://www.igc.apc.org/ran/
Sierra Club Home Page	http://www.sierraclub.org/

Geography

Geographic & related data resources	http://www.thomas.com/othergeo.html
Geography and GIS Resources	http://www.clark.net/pub/lschank/web/geo.html
MiSU - Geography-Related Servers	http://www.ssc.msu.edu/~geo/geoglinks.html
NGS - National Geographic Online	http://www.nationalgeographic.com/
The Microstate Network	http://microstate.com/cgi-win/mstatead.exe/listregions
UB GIAL - Internet Geography Information	http://www.geog.buffalo.edu/GIAL/netgeog.html
UCR - INFOMINE	http://lib-www.ucr.edu/
UT-Austin: Geography Resource Center	http://www.utexas.edu/depts/grg/virtdept/resources/contents.htm
Virtual Library: Geography	http://www.icomos.org/WWW_VL_Geography.html

GIS/Remote Sensing

AGI GIS Dictionary - Free Edition	http://www.geo.ed.ac.uk/agidict/welcome.html
Anaglyphs	ftp://geog.ucsb.edu/pub/tmp/
Arizona Geographic Information Council	http://www.state.az.us/gis3/agic/agichome.html
Autometric, Inc. Home Page	http://www.autometric.com/
Baltic GIS Database	http://www.grida.no/baltic/
CALMIT Nebraska-Lincoln	http://www.calmit.unl.edu/calmit.html
Declassified Satellite Photos	http://edcwww.cr.usgs.gov/Webglis/glisbin/search.pl?DISP
Digital Land Systems Research	http://www.mira.net.au/dlsr/
Earth Observation Magazine	http://www.eomonline.com/
EarthRISE	http://earthrise.sdsc.edu/earthrise/maps/
Erdas	http://www.erdas.com/
EROS home page	http://edcwww.cr.usgs.gov/eros-home.html
EUROGI Homepage	http://www.frw.ruu.nl/eurogi/eurogi.html
GeoWeb For GIS/GPS/RS	http://www.ggrweb.com/
GIS and GIS-Related Net Sites	http://www.hdm.com/gis3.htm
GIS Ftp Resource List	http://www.geo.ed.ac.uk/home/gisftp.html
GIS Gopher Resource List	http://www.geo.ed.ac.uk/home/gisgopher.html
GIS/Cart. (Norway)	http://www.iko.unit.no/gis/gisen.html

GIS/Mapping Sciences Resource Guide	http://www.cast.uark.edu/jpgis/jpgsittmf.html
GISnet BBS' MapInfo Support Page	http://www.csn.net/gis/mapinfo/
GPS - Peter Dana (UT-Austin)	http://www.utexas.edu/depts/grg/gcraft/notes/gps/gps.html
GPS Navigation	http://www.inmet.com/~pwt/gps_gen.htm
GPS World Home Page	http://www.gpsworld.com/
Guide to DEMs	http://www.truflite.com/text/demguide.htm
ImageNet	http://www.coresw.com/
MIT + MassGIS Digital Orthophoto Project	http://ortho.mit.edu/
NASA - JSC Digital Image Collection	http://images.jsc.nasa.gov/html/home.htm
NCGIA	http://www.ncgia.ucsb.edu/
REGIS - Envir Planning/GIS at Berkeley	http://www.regis.berkeley.edu/
Remote Sensing and GIS Information	http://www.gis.umn.edu/rsgisinfo/rsgis.html
RSL World Wide Web (WWW)...	http://www.gis.umn.edu/
Space Imaging... 1-Meter Satellite Imagery	http://www.spaceimage.com/
The DEM Reader Page	http://www.electriciti.com/~brianw/DEM_Reader.html
The Kingston Centre for GIS	http://giswww.kingston.ac.uk/menu.html
Thoen's Web	http://www.gisnet.com/gis/index.html
U of Edinburgh: GIS WWW Resourse List	http://www.geo.ed.ac.uk/home/giswww.html
Virtual Library: Remote Sensing	http://www.vtt.fi/aut/ava/rs/virtual/other.html

Historical Cartography

A G S Collection	http://leardo.lib.uwm.edu/
Appalachian Arts Maps	http://www.athens.net/~aarts/
Behaim Globus	http://www.ipf.tuwien.ac.at/veroeffentlichungen/ld_p_ch96.html
Brock University Map Library Home Page	http://www.brocku.ca/maplibrary/
Carta Historica	http://www.jyu.fi/tdk/hum/historia/carta/index.html
Cartographic Arts	http://www.dogstar.com/carto
Exploring the West from Monticello: Home	http://www.lib.virginia.edu/exhibits/lewis_clark/home.html
FINFO: Finland 500 years on the Map of Europe	http://www.vn.fi/vn/um/mapseng.html
Harry Ransom Humanities Research Center	http://www.utexas.edu/depts/grg/classes/grg374/resource/hrcmaps/maps/hrcmaps.html

Harvard Map Collection's Home Page	http://icg.harvard.edu/~maps/
Heritage Map Museum	http://www.carto.com/
Historic maps of the Netherlands	http://grid.let.rug.nl/~welling/maps/maps.html
Historica	http://www.zynet.co.uk:8001/beacon/html/livhis.html
HIstorical maps (amateur)	http://maps.linex.com/map.html
History of Cartography	http://ihr.sas.ac.uk/maps/mapsmnu.html
Int'l Map Trade Assoc.	http://www.maptrade.org/
Jim Seibold Home Page	http://www.iag.net/~jsiebold/carto.html
Map Libraries - Buffalo	http://wings.buffalo.edu/libraries/units/sel/collections/maproom.html
Map sellers	http://192.160.127.232/cgi-bin/category.pl?=Cartography
MapHist Discussion Group	http://kartoserver.frw.ruu.nl/HTML/STAFF/krogt/maphist.htm
MapHist Indexed Hard Copies	http://kartoserver.frw.ruu.nl/HTML/STAFF/krogt/maph_hc.htm
Maps at Duke University	http://www.lib.duke.edu/pdmt/maps.html
Matthew H. Edney's Links	http://www.usm.maine.edu/~maps/edney/links.html
Menotomy Maps Video	http://users.aol.com/videomap/disp/video.htm
NYC Library Map Division	http://www.nypl.org/reearch/chss/map/map.html
Osher Map Library	http://www.usm.maine.edu/~maps/oml/
Paulus Swaen	http://www.swaen.com/
Perseus Atlas Project	http://perseus.holycross.edu/PAP/Atlas_project.html
Rare Map Collection - U of Georgia	http://www.libs.uga.edu/darchive/hargrett/maps/maps.html
Robert Ross & Co.	http://www.abaa-booknet.com/usa/ross/
Rohrbach Library Map Collection	http://www.kutztown.edu/library/maps/map.html
RYHINER-Project at the University Library of Berne	http://www.stub.unibe.ch/stub/ryhiner/ryhiner.html
Steve Bartrick Antique Prints & Maps - Early British maps	http://www.antiquarian.com/bartrick/prints/rare_uk_maps.html
The Cartographic Creation of New England	http://www.usm.maine.edu/~maps/exhibit2/

The Map Case - Oxford, UK	http://www.bodley.ox.ac.uk/guides/maps/mapcase.htm
UMn - Borchart - Map Libraries	http://www-map.lib.umn.edu/map_libraries.html
University of Georgia Map Collection	http://www.libs.uga.edu/maproom/ahtml/mchpi1.html
WebMuseum: Map	http://www.oir.ucf.edu/wm/map/
Yale Map Collection	http://www.library.yale.edu/MapColl/front.htm

Interactive Cartography

BWCAW Internet Mapserver	http://www.gis.umn.edu/bwcaw/mapping/mapping.html
CDF Map Making Facility	http://spp-www.cdf.ca.gov/mapmaker/
Census Data Access Tools	http://www.census.gov/ftp/pub/main/www/access.html
Census Map Stats	http://www.census.gov/datamap/www/index.html
Census Thematic Mapping System	http://www.census.gov/themapit/www/
CIESIN DDViewer	http://sedac.ciesin.org/plue/ddviewer/htmls/whtst.html
CIESIN-SEDAC's DDCartogram	http://plue.sedac.ciesin.org/plue/ddcarto/
CIESIN: Access to U.S. Demographic Data	http://www.ciesin.org:2222/nii.html
CIESIN: Index of /	http://www.ciesin.org:2222/
Clickable State Maps	http://govinfo.kerr.orst.edu/pub/map.html
Coastline Extractor	http://crusty.er.usgs.gov/coast/getcoast.html
Design Map - Outlines	http://life.csu.edu.au/cgi-bin/gis/Map
DOOGIS: A Dynamic GIS Interface	http://doogis.dis.anl.gov/
ETOPO-5 Map Generator	http://www.evl.uic.edu/pape/vrml/etopo/
Flagstaff USGS Shaded Relief Maps	http://wwwflag.wr.usgs.gov/USGSFlag/Data/shadedRel.html
Great Lakes Map Server	http://epawww.ciesin.org/arc/map-home.html
ICE MAPS : Interactive Calif Envir Mgt	http://ice.ucdavis.edu/ice_maps/
Interactive Mapper for Arkansas	http://www.cast.uark.edu/products/MAPPER/
Interactive Spatial Data Browser (DLG Data) – UVA Library	http://www.lib.virginia.edu/gic/spatial/dlg.browse.html
Interactive species mapping (ERIN)	http://www.erin.gov.au/database/WWW-Fall94/species_paper.html
Map-It - A GMT3 Map Generator	http://crusty.er.usgs.gov:80/mapit/

Menotomy MVideo Page	http://users.aol.com/videomap/disp/video.htm
NAIS Map Home Page	http://ellesmere.ccm.emr.ca/naismap/naismap.html
Oakland Map Room	http://199.35.5.101/index1.htm
Online Map Creation	http://www.aquarius.geomar.de/omc/
Pennsylvania Statistics by County	http://www.maproom.psu.edu/cbp/
Rapid Imaging Software's Home Page	http://www.landform.com/
The Profile Maker v1.0	http://www.geo.cornell.edu/geology/me_na/profile_maker/profile_maker.htmlTiger
Map Service	http://tiger.census.gov/cgi-bin/mapbrowse
TMS Experimental Browser	http://tiger.census.gov/cgi-bin/mapbrowse
TOPO! Interactive Maps	http://www.topo.com/
TruFlite's 3D World	http://www.truflite.com/
Up-to-the-minute Southern California Earthquake Map	http://www.crustal.ucsb.edu/scec/webquakes/
USGS Elevation Database	http://wxp.atms.purdue.edu/usgs.html
Virginia County Interactive Mapper Map Creation Form	http://www.lib.virginia.edu/gic/mapper/tigermap.html
Zip2 Interactive Map	http://www.zip2.com/Scripts/map.dll?java=yes&type=htm&usamap.x=1&searching=yes

Interesting

DEOS Altimetry Atlas	http://dutlru8.lr.tudelft.nl/altim/atlas/
Earth and Moon Viewer	http://www.fourmilab.ch/earthview/vplanet.html
Interactive Connecticut Map	http://www.cs.yale.edu/HTML/YALE/MAPS/connecticut.html
Metro Denver Temporal GIS Project	http://www.sni.net/~castagne/index.html
UVa Library Geographic Information Center Homepage	http://www.lib.virginia.edu/gic/

Locators

BigBook: Map Search	http://www.bigbook.com/showpage.cgi/1b2dda27-1-0-0?page=navigator_map
City.Net Maps	http://city.net/indexes/top_maps.html
CitySearch: U.S. Map	http://www.citysearch.com/
DeLorme: CyberRouter	http://route.delorme.com/

EtakGuide	http://www.etakguide.com/
GeoCities	http://www.proximus.com/geocities/
Infoseek	http://www.infoseek.com/ Facts?pg=maps.html&sv=N3
MapBlast	http://www.mapblast.com/
MapQuest	http://www.mapquest.com/
Maps On Us	http://www.mapsonus.com/
Power Search (ATMs)	http://visa.infonow.net/powersearch.html
U.S. Gazetteer	http://www.census.gov/cgi-bin/gazetteer
US Gazetteer SUNY-Buffalo	http://wings2.buffalo.edu/cgi-bin/gazetteer
Vicinity Services	http://www.vicinity.com/vicinity/services.html
Virtual Map - Dynamic Street Maps	http://www.virtualmap.com/
Xearth HTML Front End	http://www-bprc.mps.ohio-state.edu/xearth/ xearth.html
Xerox PARC Map Viewer	http://pubweb.parc.xerox.com/map
Yahoo! Maps	http://maps.yahoo.com/yahoo/

Map Projections

Great Globe Gallery	http://hum.amu.edu.pl/~zbzw/glob/ glob1.htm
John Snyder's Map Projection	http://orc.dev.oclc.org:9000/ LOGIN:entityClearLimits=1:entityChooseDb=1:se ssionid=0:dbname=mapbib:next=html/ mapbib_simple_search.html
Map Projection Home Page	http://everest.hunter.cuny.edu/mp/index.html
Map Projection Overview - Peter Dana (UT-Austin)	http://www.utexas.edu/depts/grg/gcraft/ notes/mapproj/mapproj.html
Peters Projection	http://www.webcom.com/~bright/ petermap.html

Search

All4One Search Engines	http://www.all4one.com/
All-in-One Search Page	http://www.albany.net/allinone/
AltaVista Technology	http://www.altavista.com/
Argus Clearinghouse	http://www.clearinghouse.net/
AT&T Internet Toll Free 800 Directory	http://www.tollfree.att.net/dir800/
Bartlett Familiar Quotations	http://www.columbia.edu/~svl2/bartlett/

BigBook Directory Search	http://www.bigbook.com/
Bigfoot Home Page	http://bigfoot.com/
British Library	http://www.bl.uk/
CIESIN - Information Resources	http://www.ciesin.org/home-page/library.html
CIESIN Gateway	http://wwwgateway.ciesin.org/
City.Net	http://www.city.net/
Cosmix Mother Load - Insane Search	http://www.cosmix.com/motherload/insane/
Crayon	http://sun.bucknell.edu/~boulter/crayon/
E-Zines Database Query	http://www.dominis.com/Zines/query.shtml
Educational Hotlists	http://sln.fi.edu/tfi/hotlists/hotlists.html
EINet Galaxy	http://galaxy.einet.net/galaxy.html
Excite Netsearch	http://www.excite.com/
Four11 White Page Directory	http://www.four11.com/
Genealogy Resource	http://godzilla.westworld.com:80/~mgunn/
GEONAME	http://www.gdesystems.com/IIS/SlipSheets/GEONAME.html
GNN Home Page	http://www.gnn.com/
HotBot	http://www5.hotbot.com:5555/
Hotlist: Geography	http://sln.fi.edu/tfi/hotlists/geography.html
InfoSeek	http://www.infoseek.com/
Infoseek Ultra	http://ultra.infoseek.com/
Jumbo	http://www.jumbo.com/
Libraries on the Web	http://chehalis.lib.washington.edu/libweb/usa.html
LookUP!	http://www.lookup.com/
Lycos, Inc. Home Page	http://lycos.cs.cmu.edu/
MetaCrawler Searching	http://metacrawler.cs.washington.edu:8080/index.html
Movie Database	http://www.msstate.edu/Movies/search.html
NAISMap Home Page	http://ellesmere.ccm.emr.ca/naismap/naismap.html
Nice Geography/GIS Servers	http://www.frw.ruu.nl/nicegeo.html
OKRA	http://okra.ucr.edu/okra/
Open Text	http://www.opentext.com/
ProFusion	http://www.designlab.ukans.edu/ProFusion.html

SavvySearch	http://guaraldi.cs.colostate.edu:2000/./
Search InfoMac Archives	http://www.netam.net/~baron/infomac/
Search Univ. of Michigan Software Archives	http://www.netam.net/~baron/umich/
search.com	http://www.search.com/
Starting Point	http://www.stpt.com/
SUNY Buffalo Libraries	http://wings.buffalo.edu/libraries/units/sel/electronic.html
Switchboard	http://www.switchboard.com/
Virtual Library: Subject Catalogue	http://www.w3.org/pub/DataSources/bySubject/Overview.html
WebCrawler Searching	http://webcrawler.com/
Welcome to Pathfinder	http://pathfinder.com/@@Y4YH6aFmlgMAQPBa/pathfinder/welcome.html
WhoWhere? PeopleSearch	http://www.whowhere.com/
WWW File Library	http://www.axes.co.jp/~hisashin/lib/index.html
WWW Virtual Library: Geography	http://hpb1.hwc.ca:10002/WWW_VL_Geography.html
Yahoo	http://www.yahoo.com/
Zip2 Main Search	http://www.zip2.com/

Societies

AAG	http://www.aag.org/index.html
AAG GIS Specialty Group	http://www.gis.sc.edu/gis/aaggis.html
American Congress on Surveying and Mapping	http://www.landsurveyor.com/acsm/
ASPRS	http://www.us.net/asprs/
Canadian Map Libraries & Archives	http://www.sscl.uwo.ca/assoc/acml/acmla.html
Cartographica	http://utl1.library.utoronto.ca/www/utpress/journal/jour5/car_lev5.htm
Cartography Specialty Group	http://www.csun.edu/~hfgeg003/csg/
Friends of the Earth	http://www.foe.co.uk/index.html
ICA: U.S. National Committee	http://www.gis.psu.edu/ica/ICAusnc.html
Map Societies Around the World	http://www.csuohio.edu/CUT/MapSoc/Index.htm
MapHist Discussion Group	http://kartoserver.frw.ruu.nl/HTML/STAFF/krogt/maphist.htm

NACIS Sixteenth Annual Meeting	http://maps.unomaha.edu/NACIS/Conference.html
NCGE Home Page	http://multimedia2.freac.fsu.edu/ncge/
Society of Cartographers	http://www.shef.ac.uk/uni/projects/sc/
Transactions of the Institute of British Geographers	http://ppt.geog.qmw.ac.uk/
URISA	http://www.urisa.org/

Weather

Blue Skies Java	http://cirrus.sprl.umich.edu/javaweather/
Current Weather Map	http://www.mit.edu:8001/usa.html
Current Weather Maps/Movies	http://wxweb.msu.edu/weather/
EarthWatch Weather on Demand	http://www.earthwatch.com/
Intellicast	http://www.intellicast.com/
Interactive Weather Browser	http://wxweb.msu.edu/weather/interactive.html
Interactive Weather Information Network	http://iwin.nws.noaa.gov/iwin/graphicsversion/main.html
Interactive weather maps	http://wxp.atms.purdue.edu/interact.html
Live Access to Climate Data	http://ferret.wrc.noaa.gov/fbin/climate_server
Real-Time Weather Data: Surface Page	http://http.rap.ucar.edu/weather/surface.html
Surface Data Details	http://wxp.atms.purdue.edu/surface_det.html#depict
Texas A&M Meteorology Weather Center	http://www.met.tamu.edu/weather.shtml
The Weather Visualizer	http://covis.atmos.uiuc.edu/covis/visualizer/
Tornadoes	http://cc.usu.edu/~kforsyth/Tornado.html
TV Weather Dot Com- The Weather Supersite	http://www.tvweather.com/
Weather Channel	http://www.weather.com/
Weather Channel - Teachers' Resources	http://www.weather.com/weather_whys/teachers_resources/
WeatherNet	http://cirrus.sprl.umich.edu/wxnet/
World Meteorological Organization	http://www.wmo.ch/

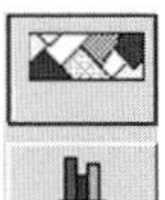

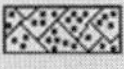

INDEX

H

I

L

U

V

W

X

Z

Companion CD-ROM Credits and Copyrights